Biofuels: Green Energy and Technology

Biofuels: Green Energy and Technology

Editor: Alice Wheeler

R CALLISTO REFERENCE

www.callistoreference.com

Callisto Reference,
118-35 Queens Blvd., Suite 400,
Forest Hills, NY 11375, USA

Visit us on the World Wide Web at:
www.callistoreference.com

ISBN: 978-1-64116-249-4 (Hardback)

Trademark Notice: Registered trademark of products or corporate names are used only for explanation and identification without intent to infringe.

Cataloging-in-Publication Data

Biofuels : green energy and technology / edited by Alice Wheeler.
    p. cm.
Includes bibliographical references and index.
ISBN 978-1-64116-249-4
1. Biomass energy. 2. Clean energy. 3. Biomass energy industries--Technological innovations.
4. Clean energy industries--Technological innovations. I. Wheeler, Alice.
TP339 .B56 2020
333.953 9--dc23

Table of Contents

Preface

The purpose of the book is to provide a glimpse into the dynamics and to present opinions and studies of some of the scientists engaged in the development of new ideas in the field from very different standpoints. This book will prove useful to students and researchers owing to its high content quality.

A fuel that is produced from biomass is referred to as biofuel. It primarily includes the liquid or gaseous fuels that are used for transportation. It is considered as a form of renewable or green energy if the biomass used in the production of biofuel regrows quickly. Biodiesel and bioethanol are the most common types of biofuels. Biodiesel is manufactured from oils or fats through transesterification. It is often used in vehicles as a fuel as well as a diesel additive for the reduction of carbon-monoxide and hydrocarbons from diesel-powered vehicles. Bioethanol is produced by fermentation. It is made up of carbohydrates that are produced in sugar or starch crops like sugarcane and corn. This book unfolds the innovative aspects of biofuels which will be crucial for the holistic understanding of the subject matter. It elucidates new techniques and their applications in a multidisciplinary approach. This book will serve as a valuable source of reference for those interested in this field.

At the end, I would like to appreciate all the efforts made by the authors in completing their chapters professionally. I express my deepest gratitude to all of them for contributing to this book by sharing their valuable works. A special thanks to my family and friends for their constant support in this journey.

Editor

Productivity and resistance to weed invasion in four prairie biomass feedstocks with different diversity

JESSICA E. ABERNATHY[1]*, DUSTIN R. J. GRAHAM[1]*, MARK E. SHERRARD[1] and DARYL D. SMITH[2]

[1]*Department of Biology, University of Northern Iowa, 144 McCollum Science Hall, Cedar Falls, IA 50614, USA*, [2]*Tallgrass Prairie Center, 2412 West 27th Street, Cedar Falls, IA 50614-0294, USA*

Abstract

High-diversity mixtures of native tallgrass prairie vegetation should be effective biomass feedstocks because of their high productivity and low input requirements. These diverse mixtures should also enhance several of the ecosystem services provided by the traditional monoculture feedstocks used for bioenergy. In this study, we compared biomass production, year-to-year variation in biomass production, and resistance to weed invasion in four prairie biomass feedstocks with different diversity: one species – a switchgrass monoculture; five species – a mix of C_4 grasses; 16 species – a mix of grasses, forbs, and legumes; and 32 species – a mix of grasses, forbs, legumes, and sedges. Each diversity treatment was replicated four times on three soil types for a total of 48 research plots (0.33–0.56 ha each). We measured biomass production by harvesting all plant material to ground level in ten randomly selected quadrats per plot. Weed biomass was measured as a subset of total biomass. We replicated this design over a five-year period (2010–2014). Across soil types, the one-, 16-, and 32-species treatments produced the same amount of biomass, but the one-species treatment produced significantly more biomass than the five-species treatment. The rank order of our four diversity treatments differed between soil types suggesting that soil type influences treatment productivity. Year-to-year variation in biomass production did not differ between diversity treatments. Weed biomass was higher in the one-species treatment than the five-, 16-, and 32-species treatments. The high productivity and low susceptibility to weed invasion of our 16- and 32-species treatments supports the hypothesis that high-diversity prairie mixtures would be effective biomass feedstocks in the Midwestern United States. The influence of soil type on relative feedstock performance suggests that seed mixes used for biomass should be specifically tailored to site characteristics for maximum productivity and stand success.

Keywords: agroenergy, bioenergy, biomass, productivity, switchgrass, tallgrass prairie, weed resistance

Introduction

Rising global energy use and decreasing fossil fuel reserves have increased the need for renewable sources of energy. Many of the current bioenergy crops (e.g., corn, soybeans, oilseed rape, sugarcane, and willow) require fertilizer and pesticide inputs and compete with food crops for land. These shortcomings have increased interest in alternative bioenergy crops, such as switchgrass (*Panicum virgatum* L.) and *Miscanthus* (*Miscanthus x giganteus*), which are highly productive and can grow on marginal farmland (Lewandowski *et al.*, 2003; Heaton *et al.*, 2004; Khanna *et al.*, 2008). Another viable bioenergy crop, particularly in the Midwestern United

States, is a mixture of native perennial tallgrass prairie vegetation (Hector *et al.*, 1999; Balvanera *et al.*, 2006; Tilman *et al.*, 2006; Cardinale *et al.*, 2007). Experiments focusing on the diversity–productivity relationship suggest that high-diversity prairie mixtures produce more bioenergy than corn on marginal land (Tilman *et al.*, 2006), produce more biomass than perennial monocultures (Tilman *et al.*, 2006; Cardinale *et al.*, 2007; Fornara & Tilman, 2009), and sustain high yields for decades without fertilizer (Glover *et al.*, 2010). While the economic and ecological benefits of high-diversity prairie mixtures for bioenergy seem attractive, more research is needed to determine the feasibility of growing these crops on a production-level scale.

Diversity–productivity experiments suggest that unfertilized high-diversity biomass crops will be more productive than unfertilized low-diversity biomass crops because of greater niche differentiation and/or better facilitation (i.e., the 'complementarity effects';

*First authorship is shared.

Correspondence: Mark E. Sherrard
e-mail: mark.sherrard@uni.edu

Loreau & Hector, 2001; Cardinale *et al.*, 2007; but see Hooper *et al.*, 2005 for alternative mechanisms). High-diversity mixtures are more morphologically and phenologically variable than low-diversity mixtures, which should increase total resource acquisition (Wilsey, 2010). For example, high-diversity prairie mixtures have greater variation in root depth and root architecture than low-diversity mixtures, which should increase water and nutrient uptake in these communities (Fornara &Tilman, 2009; Postma & Lynch, 2012). Also, high-diversity mixtures typically have higher functional diversity (i.e., more functional groups: cool-season C_3 grasses, warm-season C_4 grasses, and forbs) than low-diversity mixtures, expanding the time frame in which resources are acquired during the growing season (Diaz & Cabido, 2001; Fargione & Tilman, 2005). One example of enhanced facilitation in high-diversity mixtures is the inclusion of legumes. Legumes form symbiotic associations with nitrogen-fixing rhizobial bacteria. These associations increase nitrogen availability within the community.

High-diversity biomass crops should be more resistant to weed invasion than low-diversity biomass crops because they provide fewer resources for potential invaders (Knops *et al.*, 1999; Levine, 2000; Hooper *et al.*, 2005; Balvanera *et al.*, 2006). For example, Fargione & Tilman (2005) compared five treatments with different diversity and found that the high-diversity mixtures were less susceptible to weed invasion because they captured a greater proportion of available soil nitrates. High-diversity mixtures also tend to have greater absolute cover than low-diversity mixtures, which reduces light availability (Levine, 2000) and helps minimize weed invasion (Davis *et al.*, 2000). Minimizing weed invasion is important for maximizing yield in biomass feedstocks. Although weed invasion increases diversity, the addition of exotic species does not have the same positive influence on productivity as the addition of native species in tallgrass prairie systems (Isbell & Wilsey, 2011). These exotic species may not be adapted to local conditions and occupy space that would otherwise contain prairie species with higher productivity. From a management perspective, the invasion of woody species would be particularly costly if targeted removal is required.

High-diversity prairie mixtures should also enhance several of the ecosystem services provided by the traditional monoculture feedstocks used for bioenergy. Two concurrent studies at our research site have shown that high-diversity biomass mixtures provide better nesting habitat for birds (Myers *et al.*, 2015) and more resources for butterflies (Myers *et al.*, 2012) than switchgrass monocultures. High-diversity mixtures are also less susceptible to yield loss via specialized pests than monocultures (Knops *et al.*, 1999). For example, the gall midge pest *Chilophaga virgati* specializes on switchgrass and decreases productivity and fitness in infected monocultures (Boe & Gagne, 2011). Further, high-diversity mixtures should display lower year-to-year variation in any particular ecosystem service than monocultures because they have species with differing levels of stress tolerance (i.e., the insurance effect, Yachi & Loreau, 1999; Hooper *et al.*, 2005). This interspecific variability will ensure a certain level of ecosystem service in extreme climatic years and could help maintain consistent rates of belowground carbon sequestration over the timeframe necessary to mitigate climate change (Hooper *et al.*, 2005).

The potential value of high-diversity prairie mixtures as biomass feedstocks has encouraged some to examine the feasibility of growing these crops on a production-level scale. In particular, three recent studies examined whether Conservation Reserve Program (CRP) lands in Iowa, unfertilized polycultures, and reconstructed prairies might be useful biomass feedstocks. These experiments all supported the potential utility of diverse prairie for bioenergy, finding that CRP land and switchgrass monocultures have similar theoretical ethanol yields (Jungers *et al.*, 2013) and that unfertilized polycultures (31 species, Jarchow & Liebman, 2013) and restored prairies (Zilverberg *et al.*, 2014) are sufficiently productive (9.1 and 7.3 Mg ha^{-1} respectively). However, to the best of our knowledge, no one has compared the productivity and ecosystem services of high-diversity vs. low-diversity prairie mixtures specifically designed for biomass on a production-level scale.

In this study, we compare biomass production, year-to-year variation in biomass production, and resistance to weed invasion in four prairie biomass feedstocks with different diversity (one, five, 16, and 32 species). We predict that the high-diversity treatments (16 and 32 species) will produce more biomass, display lower year-to-year variation in biomass production, and be more resistant to weed invasion than the low-diversity treatments (one and five species).

Materials and methods

Research site

This study was conducted at the Cedar River Ecological Research Site in Blackhawk County, Iowa (42°23N, 92°13W). The 40 ha site is on marginal farmland with a flat slope (0–2%) and a corn suitability rating (CSR) of 50–79 (Natural Resource Conservation Service, 2014). CSR is an index (0–100) that ranks all soils in the state of Iowa based on their potential row crop productivity. There are three soil types at the site: (i) an excessively drained Flagler sandy loam (CSR = 50); (ii) a well-drained

Waukee loam (CSR = 79); and (iii) a somewhat poorly drained Spillville–Coland alluvial complex (CSR = 60; Natural Resource Conservation Service, 2014). The relative amounts of sand, silt, and clay vary between soils: Flagler sandy loam – 73.8% sand, 17.0% silt, and 9.2% clay; Waukee loam – 66.2% sand, 20.9% silt, and 12.8% clay; Spillville–Coland alluvial complex – 42.1% sand, 35.9% silt, and 22.0% clay (Natural Resource Conservation Service, 2014). These soils will henceforth be referred to as the 'sand', 'loam', and 'clay' soils, respectively. The sand soil has the lowest nutrient availability and water holding capacity (Myers et al., 2015; Sherrard et al., 2015). The loam and clay soils have similar nutrient availability but the clay soil has higher water holding capacity (Myers et al., 2015; Sherrard et al., 2015).

In spring 2009, four diversity treatments were seeded at the site: (i) one species – a switchgrass (Panicum virgatum L.) monoculture; (ii) five species – a mixture of C_4 grasses; (iii) 16 species – a mixture of C_3 and C_4 grasses, forbs, and legumes; and (iv) 32 species – a mixture of C_3 and C_4 grasses, sedges, forbs, and legumes (see Table S1 for species list). Each diversity treatment contains all species from treatments of lesser diversity plus additional species. Four replicate plots (0.33–0.56 ha each) of each diversity treatment were randomly established on each soil type for a total of 48 research plots (four replicates × four diversity treatments × three soil types; see Sherrard et al., 2015 or Myers et al., 2015 for site map). The size of our plots provides a realistic representation of a production-level biomass crop and should generate reliable estimates of productivity (with minimal edge effects), wildlife use (e.g., Myers et al., 2012, 2015), and susceptibility to weed invasion in the different treatments. To minimize the likelihood of contaminating diversity treatments during establishment, the plots were seeded from least to most diverse using a Truax native seed drill. Prior to seeding, all plots were seeded with Roundup ready soybeans in July 2008 and glyphosate was applied in July/August 2008. Other site management during the study period included: establishment mowing (June 2009) to reduce competition with annual weeds, burning (April 2011), haying (March 2012), and burning (April 2014). A small patch of crown vetch and reed canary grass was treated with glyphosate in 2014 to prevent spread; otherwise, no fertilizers, herbicides, pesticides, weeding, or irrigation have been applied to the treatment plots.

The species composition of each diversity treatment was selected based on its potential utility as a biomass feedstock. Switchgrass was chosen as the monoculture because it has been recommended as a bioenergy crop by the U.S. Department of Energy (McLaughlin et al., 1999). We used source identified class yellow label seed for the switchgrass monoculture to ensure that the genotype of all seeds originated from remnant prairies in Iowa. The 'yellow tag' designation indicates that the Iowa Crop Improvement Association has verified the seed source in accordance with standards set by the Association of Official Seed Certifying Agencies (AOSCA). In pilot research, we found that switchgrass plots grown from Iowa 'yellow tag' seed produced more biomass than plots grown from cultivar seed (D. Smith, pers. obs.). The five C_4 grass treatment was selected because all five species are highly productive in tallgrass prairies. We used Iowa 'yellow tag' seed for the five species in this treatment as well. The 16-species treatment was

chosen based on nine a priori criteria: (i) a statewide distribution; (ii) high aboveground biomass production; (iii) availability of Iowa 'yellow tag' seed; (iv) ease of establishment from seed; (v) ability to maintain standing vegetation through winter; (vi) ability to grow in a variety of soil moisture conditions; (vii) variable phenologies and life histories – species that produce biomass at different times; (viii) long life span; and (ix) ability to coexist with other species. Many of the species in the 32-species treatment were selected based on the above criteria; however, some were selected because they are commonly seeded species in native tallgrass prairie restorations. The seeding rate of the one- and five-species treatments was 561 pure live seeds m^{-2} (Table S1), which was based on recommendations for establishing switchgrass as a bioenergy crop (Natural Resource Conservation Service, 2009). The 16- and 32-species treatments contained the same number of graminoid seeds as the one- and five-species treatments plus seeds of other functional groups for a total of 829 and 869 pure live seeds m^{-2}, respectively. These seeding rates are consistent with recommendations for prairie restorations in Iowa. Because our diversity treatments are perennial, they do not need to be reseeded after establishment.

Climate data

During our five-year study (2010–2014), the average growing season (April–October) temperature for the region was 16.9 °C and the average growing season precipitation was 698 mm (data collected from nearest weather station: Waterloo Airport, 15.5 km, Fig. S1). The site experienced a drought in 2012 (growing season precipitation = 443.2 mm). The clay and loam soil experienced severe flooding in spring 2013 (clay: submerged for ~two weeks, max height = 1.8 m; loam: submerged for two days, max height = 50 cm) and spring 2014 (clay: submerged for one week, max height = 1.3 m; loam: submerged for two days, max height = 30 cm). The sand soil did not experience flooding during the study.

Experimental design

To compare biomass production between treatment combinations, we harvested biomass in each year of the study (2010–2014) between August 25 and September 27 (dates within this range differ between years based on the timing of plant senescence). This is the timing of maximum yield in switchgrass biomass crops (Heaton et al., 2004). In 2010–2012, ten 0.1-m^2 quadrats were randomly selected in each plot and all standing biomass was cut to ground level. The duff layer (senesced vegetation from the previous year) was omitted from harvest. In 2013 and 2014, we increased the quadrat size to 0.3 m^2 to obtain more plant tissue. After harvest, the biomass was divided into functional groups: C_4 grasses, C_3 graminoids, forbs, legumes, and weeds dried to a constant mass (min. 65 °C for 72 h) and weighed. Harvested biomass was used to estimate plot-level productivity in Mg ha^{-1}. We used the portion of weeds from the harvested biomass to estimate % weed biomass in each plot. Any species that was not included in the original seed mix of that plot was classified as a weed. Consequently, a weed could either be a species from another diversity treatment ('treatment' weeds) or a species that was not seeded at the site ('nontreatment' weeds). We acknowledge that our low-diversity treatments have a higher probability of containing 'treatment' weeds than our high-diversity treatments with this approach. For example, the 32-species treatment, by definition, can not contain any 'treatment' weeds. To account for this bias, we performed an additional statistical analysis that compared % weed biomass between treatments using 'nontreatment' weeds only. 'Nontreatment' weed biomass was estimated from the basal area coverage of each weed group in 2014 (see below).

To examine changes in species composition over the five-year study, basal area coverage of every species was measured each year in July. Two 10 m transects were established in random positions in each plot (one transect oriented North–South, one transect oriented East–West). A 0.1-m^2 quadrat (20 cm × 50 cm) was placed at one meter intervals along each transect and basal area coverage of each seeded species was estimated one inch above the ground by comparing the total area of live material to 0.006 cm^2 standardized squares. From 2010 to 2013, the presence of weeds was noted during this analysis but not quantified. We modified this design in 2014 and quantified the basal area coverage of every weed species to characterize the relative % of 'treatment' vs. 'nontreatment' weeds. The % of bare ground was measured in 2012–2014 during this sampling period to assess vulnerability to weed invasion. Ground covered in plant litter was not classified as bare ground. % bare ground was higher in 2014 because of the spring burn that year.

Statistical analysis

Aboveground biomass, % weed biomass, and % bare ground were analyzed with repeated measures ANOVAS with diversity treatment and soil type as fixed factors and year as the repeated measure. Aboveground biomass met the assumption of normality, but % weed biomass and % bare ground were log (1+x)- and square-root–transformed, respectively, to meet this assumption. All three measures violated the homogeneity of variance assumption. Aboveground biomass and weed biomass were corrected using the Greenhouse–Geisser epsilon ($\varepsilon = 0.630$ and 0.708 respectively). % bare ground data was corrected using the Huynh Feldt correction because the Greenhouse–Geisser correction was too conservative for these data ($\varepsilon > 0.75$; Girden, 1992). All post hoc analyses were performed according to Loftus & Masson (1994) using confidence intervals calculated according to Hollands & Jarmasz (2010).

To correct for bias associated with differences in the amount of 'treatment' weeds between diversity treatments, we compared the % of 'nontreatment' weeds between treatment combinations using a 2-way ANOVA with diversity treatment and soil type as fixed factors. This analysis was performed on 2014 data only as this was the only year in which the basal area coverage of 'treatment' vs. 'nontreatment' weeds was quantified.

To examine year-to-year variation in biomass production, we calculated coefficients of variation for each treatment combina-

tion and compared these coefficients using ANOVA with diversity treatment as a fixed factor and soil type as a random factor.

We used nonmetric multidimensional scaling (NMDS) to examine changes in species composition in the five-, 16-, and 32-species treatments on each soil type over the five-year study. We used the Manhattan dissimilarity index after comparing it to other dissimilarity indices with the rank index function in R. A 2-dimensional solution was used after comparing stress and goodness of fit. Permutational multivariate analysis of variance (PERMANOVA; Anderson, 2001) was used to test for significant differences between diversity treatments, years, and soil types.

All statistics were performed using the 'VEGAN' package (v. 2.0-10; Oksanen et al., 2013), the 'EZ' package (v.4.2-2; Lawrence, 2013), or the 'NLME' package (v. 3.1-117; Pinheiro et al., 2014) of R (v. 3.1.1; R Core Team, 2014).

Results

Biomass production

Aboveground biomass production differed between diversity treatments, soil types, and years (Figs 1 and S2, Table 1). On average, more biomass was produced in the one-species treatment (8.24 Mg ha^{-1} yr^{-1}) than the five-species treatment (7.17 Mg ha^{-1} yr^{-1}, Fig. 1). The 16- and 32-species treatments produced 8.03 Mg ha^{-1} yr^{-1} and 7.91 Mg ha^{-1} yr^{-1}, respectively, which did not differ significantly from the other two diversity

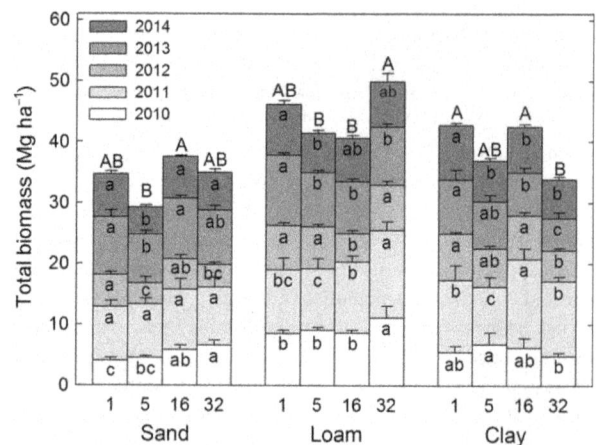

Fig. 1 Cumulative biomass production of each soil type × diversity treatment combination during the five-year study (2010–2014). The bars in each stack represent mean annual biomass production (+ 1SE). Post hoc analyses compare biomass production between diversity treatments within a soil type. Capital letters indicate significant differences in cumulative biomass production between diversity treatments over the five-year study. Lower case letters indicate significant differences in biomass production between diversity treatments in a given year.

Table 1 Repeated-measures ANOVA comparing aboveground biomass, % weed biomass, and % bare ground between treatment combinations. 'Plot' represents variation between factors (diversity treatment and soil type) and 'Within' represents variation within factors across the repeated measure (year)

	Biomass			% weed biomass†			% bare ground‡		
	df	MS	F	df	MS	F	df	MS	F
Plot									
Diversity treatment (T)	3	12.86	2.88*	3	0.0052	10.68***	3	15.97	12.13***
Soil type (S)	2	86.44	12.41***	2	0.0026	5.34**	2	0.11	0.09
T × S	6	13.60	3.05*	6	0.0001	0.28	6	2.38	1.81
Residuals	36	4.45		36	0.0005		36	1.32	
Within									
Year (Y)	4	196.20	52.24***	4	0.0028	9.22***	2	496.80	366.60***
T × Y	12	8.89	2.37***	12	0.0001	3.32***	6	7.80	5.77***
S × Y	8	21.37	5.69***	8	0.0003	1.13	4	4.40	3.24*
T × S × Y	24	2.36	0.63	24	0.0005	1.67*	12	0.40	0.327
Residuals	144	3.76		144	0.0003		72	1.40	

Reported values are: degrees of freedom (df), mean squares (MS), and F-statistics (F).

*$P < 0.05$; **$P < 0.01$; ***$P < 0.001$.

†Data log(1+x)-transformed.

‡Data square-root-transformed.

treatments. More biomass was produced on the loam soil (8.90 Mg ha^{-1} yr^{-1}) than on the sand soil (6.82 Mg ha^{-1} yr^{-1}, Fig. 1). Biomass production on the clay soil (7.79 Mg ha^{-1} yr^{-1}) did not differ significantly from the other two soil types. More biomass was produced in 2011 than in 2010, 2012, and 2014 (Figs 1 and S2). Biomass production in 2013 did not differ significantly from any other year.

The rank order of the four diversity treatments differed between soil types (treatment × soil term Table 1). On the sand soil, the 16-species treatment produced more biomass than the five-species treatment but not more than the one- or 32-species treatments (Fig. 1). On the loam soil, the 32-species treatment produced more biomass than the five- and 16-species treatments but not more than the one-species treatment (Fig. 1). On the clay soil, the one- and 16-species treatments produced more biomass than the 32-species treatment but not more than the five-species treatment (Fig. 1).

The diversity treatment that produced the most biomass varied between years (treatment × year term Table 1). In 2011, the 16- and 32-species treatments produced more biomass than the one- and five-species treatments (Fig. S2). In 2013, the one-species treatment produced more biomass than the five- and 32-species treatments but not more than the 16-species treatment. In 2014, the one-species treatment produced more biomass than the five-species treatment, but not more than the 16-and 32-species treatments. In 2010 and 2012, all diversity treatments produced the same amount of biomass.

Year-to-year variation in biomass production differed between soil types ($F = 7.007$; $P < 0.05$). The coefficient of variation for biomass production across years was 0.292 on the loam soil, 0.403 on the clay soil, and 0.381 on the sand soil. Year-to-year variation in biomass production did not differ between diversity treatments ($F = 1.609$; $P = 0.284$); however, there was a nonsignificant trend suggesting that variability increased with diversity. Specifically, the coefficient of variation for each diversity treatment was as follows: one-species: 0.332, five-species: 0.333, 16-species: 0.373, 32-species: 0.398.

Weed biomass

In the basal area coverage survey conducted at the end of the five-year study (2014), most weeds were 'nontreatment' weeds (species that were not seeded in any treatment at the site). Nontreatment weeds represented 82.8% (one-species), 74.2% (five-species), 83.3% (16-species), and 100% (32-species) of total weed coverage.

Percent weed biomass ('treatment' + 'nontreatment' weeds) differed significantly between diversity treatments, soil types, and years (Fig. 2, Table 1). Weed biomass was higher in the one-species treatment than in the five-, 16-, and 32-species treatments (7.33%, 3.10%, 2.46%, and 2.53% respectively, Table 1). 'Nontreatment' weed biomass was also higher in the one-species treatment than in the five-, 16-, and 32-species treatments ($F = 8.611$, $P < 0.001$, only 2014 data analyzed). Weed biomass was higher on the clay soil (5.47%) than on the sand soil (2.84%, Fig. 2, Table 1). Weed biomass was

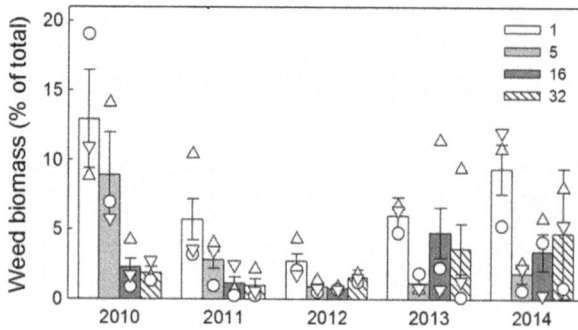

Fig. 2 Percent weed biomass in each diversity treatment. Bars represent mean % weed biomass (± 1SE) of each diversity treatment pooled across soil types and symbols represent mean % weed biomass within a soil type (circle = sand; down triangle = loam; up triangle = clay). Standard error bars omitted from within soil type means for clarity.

3.25% on the loam soil, which did not differ significantly from either other soil type. Weed biomass was higher in 2010 (during the early establishment of the site) than in 2011 and 2012. Weed biomass increased in 2013 and 2014 after flooding on the loam and clay soils (Fig. 2).

The significant treatment × soil type × year term for weed biomass (Table 1) was likely driven by severe flooding on the clay soil in 2013. On the clay soil in 2013, weed biomass was highest in the 16- and 32-species treatments (Fig. 2). In contrast, weed biomass was highest in the one-species treatment on the sand and loam soils in most years.

Bare ground

Percent bare ground differed between diversity treatments and years (Table 1). There was less bare ground in the 32-species treatment than in the one-, five-, and 16-species treatments (Fig. S3). There was significantly more bare ground in 2014 (85.8%) than in 2013 (18.7%) and significantly more bare ground in 2013 than in 2012 (13.1%). Percent bare ground was higher in 2014 because of the spring burn.

Differences in % bare ground between diversity treatments varied across years (treatment × year term, Table 1). In 2012 and 2014, % bare ground was lowest in the 16- and 32-species treatments but in 2013, % bare ground was lowest in the five-species treatment (Fig. S3). Percent bare ground was highest in the one-species treatment every year.

Species composition

The species composition of the five-, 16-, and 32-species treatments changed over the five-year study (Table 2, Fig. 3). The species composition of the 16- and 32-spe-

Table 2 Three factor nonparametric PERMANOVA reporting differences in species composition between treatments, soil types, and years in the 5-, 16-, and 32-species treatments

	df	MS	F
Year (Y)	4	0.843	9.45*
Soil type (S)	2	0.846	9.48*
Diversity treatment (T)	2	3.412	38.25*
Y × S	8	0.322	3.61*
Y × T	8	0.267	2.99*
S × T	4	0.232	2.60*
Y × S × T	16	0.102	1.14
Residuals	135	0.089	

*$P < 0.001$.

Reported values are as follows: degrees of freedom (df), mean squares (MS), and F-statistics (F).

cies treatments also differed between soil types (Fig. 3). The most dramatic change in species composition occurred in the 16- and 32-species treatments on the clay soil after the flooding in 2013 (Fig. 3).

In the 16- and 32-species treatments, years in which *Andropogon gerardii* (big bluestem) and *Sorghastrum nutans* (Indian grass) had high basal area coverages were years with high productivity and years in which *Schizachyrium scoparium* (little bluestem) had high basal area coverage were years with low productivity (Fig. 4). The basal area coverages of *Desmodium canadense* (showy tick-trefoil) and *Heliopsis helianthoides* (oxeye sunflower) decreased after 2011. The basal area coverage of *Panicum virgatum* (switchgrass) increased after the flooding in 2013 (Fig. 4).

Discussion

Diversity–productivity experiments have helped foster the hypothesis that high-diversity prairie mixtures would be effective bioenergy crops (e.g., Tilman *et al.*, 2006). To test this hypothesis, we compared biomass production, year-to-year variation in biomass production, and resistance to weed invasion in four treatments of tallgrass prairie vegetation with different diversity. Our results indicate that high-diversity prairie mixtures produce the same amount of biomass as a switchgrass monoculture and are more resistant to weed invasion on a range of soil types. Collectively, these results support the conclusion that high-diversity prairie mixtures would be effective biomass feedstocks in the Midwestern United States. In contrast with the insurance effect (Yachi & Loreau, 1999; Hooper *et al.*, 2005), we found that year-to-year variation in biomass production was equal in all diversity treatments.

In contrast to other diversity–productivity experiments, biomass production did not increase with species

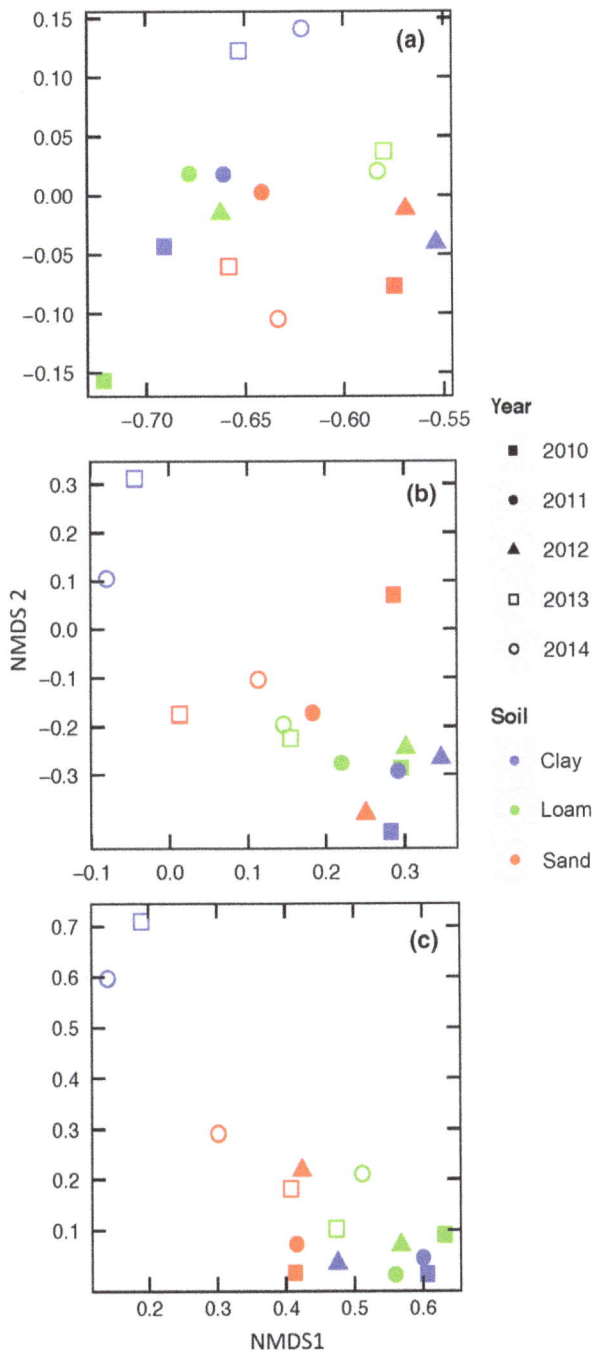

Fig. 3 Nonmetric multidimensional scaling (NMDS) plot depicting changes in community composition in the five-, 16-, and 32-species treatments during the five-year study. The NMDS plot was separated by treatment for clarity: 5-species (a), 16-species (b), and 32-species (c) treatment. Soil type is represented with different colors (blue = clay; green = loam; red = sand), and year is represented by symbol. 2D-stress: 0.16, linear R^2: 0.981, nonmetric R^2: 0.926.

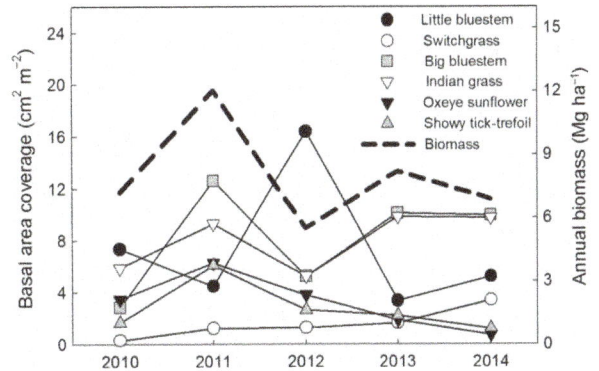

Fig. 4 Basal area of all species with coverages >5 cm^2 m^{-2}. Values represent the mean basal area coverage of species in the 16- and 32-species treatment plots on all three soil types. Mean annual biomass production of all treatment combinations is provided for reference.

whereas most diversity–productivity studies are based on random species assemblages (e.g., Hector *et al.*, 1999; Tilman *et al.*, 2006; Cardinale *et al.*, 2007). A synthesis of diversity–productivity experiments found that high-diversity mixtures often produce more biomass than monocultures on average (i.e., over-yielding is common), but rarely produce more biomass than the most productive monoculture (i.e., transgressive over-yielding is rare; Cardinale *et al.*, 2007). Because switchgrass is a highly productive monoculture (McLaughlin *et al.*, 1999), our experimental design was perhaps more consistent with a test of transgressive over-yielding. Based on this comparison, the equal productivities of the 16- and 32-species treatments and the switchgrass monoculture actually supports the value of these high-diversity mixtures for bioenergy (Sanderson *et al.*, 2004). The estimated yields of our high-diversity treatments (average = 7.8 Mg ha^{-1} yr^{-1}) were twice those reported for low input high-diversity prairies in the US Billion Ton Update (3.9 Mg ha^{-1} yr^{-1}) and are consistent with reported yields for unfertilized diverse prairies in Iowa (9.1 Mg ha^{-1} yr^{-1}; Jarchow & Liebman, 2013). The estimated yield of our switchgrass monocultures (8.24 Mg ha^{-1} yr^{-1}) was higher than those reported for unfertilized fields of the 'Cave in Rock' cultivar in southern Iowa (3.9 Mg ha^{-1}; Lemus *et al.*, 2008) and comparable to the average productivity of 20 fertilized switchgrass cultivars on fertile (CSR = 75) soils in southern Iowa (Lemus *et al.*, 2002).

Another factor that might have impacted our ability to detect a positive effect of diversity on productivity was the high nutrient content of our soils. Many diversity–productivity experiments are conducted on low nutrient soil (Lambers *et al.*, 2004; Tilman *et al.*, 2006, 2012; Fornara & Tilman, 2009; Isbell *et al.*, 2011; Jungers *et al.*, 2013), which increases the likelihood of detecting the benefits of niche differentiation and facilitation for

diversity in our study. The most likely reason for this distinction was that our seed mixes were specifically designed for their potential value as biomass feedstocks biomass production in high-diversity mixtures (e.g., Dybzinski *et al.*, 2008). For example, reported values of initial soil nitrogen (N) at Cedar Creek Ecosystem Science Reserve, home of the Biodiversity II experiment, range from 0.09 to 1.1 g kg^{-1} (Wedin & Tilman, 1993) and 0.378 – 0.701 g kg^{-1} (Tilman, 1987). At the beginning of our study, total N in the surface soil (0–15 cm) was 2.13 g kg^{-1}, 2.00 g kg^{-1}, and 1.28 g kg^{-1} in the clay, loam, and sand soils, respectively (Sherrard *et al.*, 2015). This indicates that our lowest N soils had ~16% higher N than the highest N soils at Cedar Creek Ecosystem Science Reserve. Because of our higher initial soil N content, it may take longer than five years for nutrient depletion to begin limiting productivity in the low-diversity treatments. Supporting this interpretation, long-term diversity–productivity studies have shown that the superior yields of high-diversity vs. low-diversity mixtures often become more pronounced with time (Cardinale *et al.*, 2007; Fornara & Tilman, 2009).

Our results indicate that soil type influences the relative productivity of our biomass feedstocks, as the rank order of the four diversity treatments differed between soil types (treatment × soil type term, Table 1). Other diversity–productivity studies have noted that soil fertility can influence the relationship between species richness and productivity (Hooper *et al.*, 2005; Balvanera *et al.*, 2006; Ma *et al.*, 2010), which might explain some of the variation observed in our study. In natural systems, low phosphorous/high potassium soils, such as our loam soil (Myers *et al.*, 2015; Sherrard *et al.*, 2015), tend to support communities of greater species richness (Janssens *et al.*, 1998). This could explain the strong performance of our 32-species treatment on the loam soil (Fig. 1). From a management perspective, the contrasting performance of our four diversity treatments on different soil types suggests that seed mixes designed for bioenergy must be specifically tailored to the soil characteristics of a site for maximum productivity and stand success.

The five-species treatment performed poorly on all three soil types suggesting that a C$_4$ grass mixture is not an ideal biomass feedstock on marginal farmland in the Midwestern United States. Our results are consistent with Wilsey (2010), who used the same five-species mixture and found that it produced less biomass than switchgrass and big bluestem monocultures (nonsignificant trend). The low productivity of the five-species treatment in our study may have been caused by higher rates of N depletion in this treatment. In a concurrent study examining plant tissue N content, switchgrass plants in the five-species treatment had lower leaf N, lower photosynthesis, lower chlorophyll content, and lower capacity for light capture (FvFm) than switchgrass plants in other diversity treatments (Sherrard

et al., unpub ms). Also supporting this interpretation, the poor performance of the five-species treatment was most evident on the sand soil (Fig. 1), which had the lowest initial N content and the highest probability of ultimately becoming N deficient. The 16- and 32-species treatments contain legumes, which have likely slowed the rate of N depletion in these treatments. The switchgrass monoculture does not contain legumes, but big bluestem, Indian grass, and little bluestem all have faster rates of N uptake than switchgrass (Fargione & Tilman, 2006), which might account for a slower rate of N depletion in the monoculture. The low productivity of the five C$_4$ grass mixture is disappointing because Conservation Reserve Program (CRP) land in Iowa often has a similar species composition and has the potential to be a large existing source of biomass for bioenergy (Adler *et al.*, 2009; Jungers *et al.*, 2013).

Although we had two floods (2013 and 2014) and a drought year (2012) during the study period, our results did not support the hypothesis that high-diversity mixtures have more consistent annual yields than low-diversity mixtures (i.e., the insurance effect, Yachi & Loreau, 1999; Hooper *et al.*, 2005). Instead, we detected a nonsignificant trend of higher year-to-year variation in biomass production with increasing diversity. Pfisterer & Schmid (2002) suggest that species-poor systems can be more resistant to disturbance than species-rich systems because they are statistically less likely to contain a species that will be greatly affected by disturbance and because the positive effects of niche differentiation may be minimized in disturbance years. Switchgrass is drought and flood tolerant, which could be why this treatment maintained the most consistent year-to-year biomass production in our study. Conversely, the 16- and 32-species treatments contained species that were less resistant to disturbance. The species composition of these treatments changed rapidly after the drought and floods at our site (Fig. 3), which likely influenced the productivity of these treatments. In terms of ecosystem services, our results suggest that high diversity does not necessarily ensure more consistent year-to-year production in biomass feedstocks. This is particularly true for feedstocks grown on marginal farmland in a floodplain.

Establishment time, annual precipitation, and changes in species composition may have contributed to year-to-year variation in biomass production during the five-year study. 2011 was the year in which biomass production was highest (Fig. S2) because there was high rainfall (Knapp & Smith, 2001), no flooding, and it was not during the early establishment of the site. Other years were less productive because they were either early in site establishment (2010), a drought year (2012), or a flood year (2013 and 2014). Changes in basal area coverage of big bluestem and Indian grass may have influenced aboveground biomass production in the 16- and 32-species treatments (Fig. 4). Flooding on the clay soil in 2013 and 2014 reduced the abundance of these two highly productive species and likely reduced biomass production in these years. Oxeye sunflower and showy tick-trefoil are both early establishment species (Camill *et al.*, 2004) and their decreasing abundance over the course of the study may be part of the reason that biomass production was higher in 2011 than in 2012–2014.

Our results suggest that weed biomass was influenced by variation in % bare ground, and to lesser extent, variation in soil N between treatment combinations. The one-species treatment had the highest % bare ground (Fig. S3), which likely contributed to higher weed biomass in this treatment (Fig. 2; Levine, 2000). In 2014, the five-species treatment had the same % bare ground as the one-species treatment but fewer weeds suggesting that bare ground was not the only factor influencing weed biomass at our site. Weed biomass may have been lower in the five-species treatment because there is less soil N to facilitate weed invasion in this treatment. This interpretation is consistent with our previous conclusion that efficient N uptake by other species in this diversity treatment (Fargione & Tilman, 2006) has accelerated the rate of N depletion. The presence of nitrogen-fixing legumes should make the 16- and 32-species treatments more vulnerable to weed invasion, but higher plant coverage in these treatments (Fig. S3) offsets this vulnerability. Weed invasion can reduce yield in bioenergy crops (Palmer & van der Maarel, 1995) because an increase in exotic species diversity does not have the same positive influence on productivity as an increase in native species diversity in tallgrass prairie systems (Isbell & Wilsey, 2011).

Management implications

For landowners interested solely in biomass production, our results suggest that a switchgrass monoculture is the best choice for a biomass feedstock. It has the lowest seed cost (one-species: $158 ha^{-1}; five-species: $282 ha^{-1}; 16-species: $1643 ha^{-1}; 32-species: $2354 ha^{-1}), it is productive on a variety of soils (Fig. 1), and it maintains consistent annual yields because of high resistance to disturbance. Two weaknesses of a switchgrass monoculture for bioenergy are that it is more susceptible to weed invasion (Fig. 2) and that it will likely require more fertilizer than high-diversity prairie bioenergy crops to maintain our reported yields. This study was conducted on relatively high N soil and not of sufficient length to showcase N depletion

in the one-species treatment but such an effect would likely occur with annual fall harvests.

For landowners interested in additional ecosystems services, the 16-species treatment would be the best choice. This mixture is highly productive and should maintain high yields with minimal fertilizer because of enhanced niche differentiation and facilitation (Loreau & Hector, 2001; Cardinale *et al.*, 2007). This mixture provides better habitat for birds and pollinator resources for butterflies than a switchgrass monoculture (Myers *et al.*, 2012, 2015) and annual post frost harvests should not affect the species and functional group composition (Jungers *et al.*, 2013). For landowners that are particularly interested in ecosystem services, perhaps at the expense of some productivity, the 32-species treatment would be the best choice. This treatment would be a good candidate for multifunctional on farm use (e.g., the STRIPS program in Iowa - which integrates prairie strips with row crops in watersheds to reduce nutrient runoff and erosion, or, the Buffer Initiative in Minnesota). The additional diversity of this treatment should increase nutrient retention and provide even better habitat for wildlife (Myers *et al.*, 2012, 2015). However, this mixture should not be planted at sites that flood frequently. Flooding alters the species composition of this treatment, which will reduce the diversity-based environmental benefits of the costly seed mix. For example, white wild indigo was the only legume that survived the 2013 and 2014 floods on the clay soil.

In our study, we used a site management strategy that maximized stand establishment and habitat value for wildlife. Establishment mowing and burning helps control weed abundance and fosters productivity in prairie restorations (Smith *et al.*, 2010). Harvesting biomass in spring maintains fall and winter habitat for birds (Fargione *et al.*, 2009) but reduces biomass yield relative to fall harvest. State, federal, and private landowners seeking to balance the provisioning of ecosystem services (e.g., wildlife habitat, soil and water conservation, and recreation) with economic returns would likely use a comparable management model. Consequently, our results might apply best to county-owned recreational land or CRP land (Adler *et al.*, 2009).

Landowners that prioritize biomass production would likely use a different management strategy (e.g., no burning/complete, annual fall harvests immediately after stand establishment) resulting in different productivity, weed resistance, and wildlife benefit values than those reported in our study and in Myers *et al.* (2012, 2015). Sites that are not burned early in establishment would have more weed biomass than our research plots, but the differences between diversity treatments reported in our study (Fig. 2) would likely still persist because switchgrass monocultures naturally provide more light to invading weeds than high-diversity prairie mixtures (Fig. S3). Although we hayed our site in spring, we estimated productivity from quadrats harvested in fall, and therefore, our data should provide a realistic estimate of fall biomass production values. However, ground-level hand clipping can overestimate harvestable biomass with field-scale baling, which leaves ~12 cm stubble (Zilverberg et al., 2014). Future research at the site will include baling to examine the % reduction in biomass production across treatments. Fall harvests also remove more tissue N than spring harvests (Dohleman et al., 2012), which would accelerate the rate of soil N depletion (particularly in biomass feedstocks that lack legumes) and ultimately reduce yield.

In conclusion, our results suggest that high-diversity mixtures of native prairie vegetation would be effective biomass feedstocks in the Midwestern United States. In comparison to one of the leading bioenergy crops in the United States (a switchgrass monoculture), these mixtures produce the same amount of aboveground biomass, display similar year-to-year consistency in their biomass production values, and are more resistant to weed invasion. Companion studies at our site suggest that high-diversity mixtures also provide better habitat and resources for wildlife (Myers et al., 2012, 2015). Future research at the site will examine rates of belowground carbon sequestration, which could represent another significant advantage of high-diversity vs. low-diversity biomass feedstocks.

Acknowledgements

We thank Kenneth Elgersma, Mark Myers, and Dave Williams for helpful discussions on the manuscript. We thank Molly Schlumbohm, Jessica Riebkes, Jordan Young, Andrew Ridgway, Heather Chamberlain, Jordan Koos, Hallie Kuchera, Haley Bloomquist, Zachary Kockler, Richard Knar, Dave Williams, and the Tallgrass Prairie Center for assistance in the field and laboratory. This work was supported by the Iowa Power Fund, Iowa EPSCoR under NSF Grant Number EPS-1101284, NRCS, and the University of Northern Iowa.

References

Adler PR, Sanderson MA, Weimer PJ, Vogel KP (2009) Plant species composition and biofuel yields of conservation grasslands. Ecological Applications, 19, 2202–2209.

Anderson MJ (2001) A new method for non-parametric multivariate analysis of variance. Austral Ecology, 26, 32–46.

Balvanera P, Pfisterer A, Buchmann N, He J, Nakashizuka T, Raffaelli D, Schmid B (2006) Quantifying the evidence for biodiversity effects on ecosystem functioning services. Ecology Letters, 9, 1146–1156.

Boe A, Gagne RJ (2011) A new species of gall midge (Diptera: Cecidomyiidae) infesting switchgrass in the Northern Great Plains. Bioenergy Research, 4, 77–84.

Camill P, McKone MJ, Sturgis S et al. (2004) Community- and ecosystem-level changes in a species-rich tallgrass prairie restoration. Ecological Applications, 14, 1680–1694.

Cardinale BJ, Wright JP, Cadotte MW et al. (2007) Impacts of plant diversity on biomass production increase through time because of species complementarity. Proceedings of the National Academy of Sciences of the United States of America, 104, 18123–18128.

Davis MA, Grime JP, Thompson K (2000) Fluctuating resources in plant communities: a general theory of invisibility. Journal of Ecology, 88, 528–534.

Diaz S, Cabido M (2001) Vive la difference: plant functional diversity matters to ecosystem processes. Trends in Ecology and Evolution, 16, 646–655.

Dohleman FG, Heaton EA, Arundale RA, Long SP (2012) Seasonal dynamics of above- and below-ground biomass and nitrogen partitioning in Miscanthus × giganteus and Panicum virgatum across three growing seasons. GCB Bioenergy, 4, 534–544.

Dybzinski R, Fargione JE, Zak DR, Fornara D, Tilman D (2008) Soil fertility increases with plant species diversity in a long-term biodiversity experiment. Oecologia, 158, 85–93.

Fargione JE, Tilman D (2005) Diversity decreases invasion via both sampling and complementarity effects. Ecology Letters, 8, 604–611.

Fargione JE, Tilman D (2006) Plant species traits and capacity for resource reduction predict yield and abundance under competition in nitrogen-limited grassland. Functional Ecology, 20, 533–540.

Fargione JE, Cooper TR, Flaspohler DJ et al. (2009) Bioenergy and wildlife: threats and opportunities for grassland conservation. BioScience, 59, 767–777.

Fornara DA, Tilman D (2009) Ecological mechanisms associated with the positive diversity-productivity relationship in an N-limited grassland. Ecology, 90, 408–418.

Girden ER (1992) ANOVA: Repeated Measures. Sage University Papers Series on Quantitative Applications in the Social Sciences. Sage Publications, Thousand Oaks, CA.

Glover JD, Culman SW, Dupont ST et al. (2010) Harvested perennial grasslands provide ecological benchmarks for agricultural sustainability. Agriculture, Ecosystems and Environment, 137, 3–12.

Heaton E, Voigt T, Long SP (2004) A quantitative review comparing the yields of two candidate C_4 perennial biomass crops in relation to nitrogen, temperature and water. Biomass and Bioenergy, 27, 21–30.

Hector A, Schmid B, Beierkuhnlein C et al. (1999) Plant diversity and productivity experiments in European grasslands. Science, 286, 1123–1127.

Hollands JG, Jarmasz J (2010) Revisiting confidence intervals for repeated measures designs. Psychonomic Bulletin and Review, 17, 135–138.

Hooper DU, Chapin FS III, Ewel JJ et al. (2005) Effects of biodiversity on ecosystem functioning: a consensus of current knowledge. Ecological Monographs, 75, 3–35.

Isbell F, Wilsey BJ (2011) Increasing native, but not exotic, biodiversity increases aboveground productivity in ungrazed and intensely grazed grasslands. Oecologia, 165, 771–781.

Isbell F, Calcagno V, Hector A et al. (2011) High plant diversity is needed to maintain ecosystem services. Nature, 477, 199–202.

Janssens F, Peeters A, Tallowin JRB, Bakker JP, Bekker RM, Fillat F, Oomes MJM (1998) Relationship between soil chemical factors and grassland diversity. Plant and Soil, 202, 69–78.

Jarchow ME, Liebman M (2013) Nitrogen fertilization increases diversity and productivity of prairie communities used for bioenergy. Global Change Biology: Bioenergy, 5, 281–289.

Jungers JM, Fargione JE, Sheaffer CC, Wyse DL, Lehman C (2013) Energy potential of biomass from conservation grasslands in Minnesota, USA. PLoS ONE, 8, e61209.

Khanna M, Dhungana B, Clifton-Brown J (2008) Costs of producing Miscanthus and switchgrass for bioenergy in Illinois. Biomass and Bioenergy, 32, 482–493.

Knapp AK, Smith MD (2001) Variation among biomes in temporal dynamics of aboveground primary production. Science, 291, 481–484.

Knops JMH, Tilman D, Haddad NM et al. (1999) Effects of plant species richness on invasion dynamics, disease outbreaks, insect abundance and diversity. Ecology Letters, 2, 286–293.

Lambers JHR, Harpole WS, Tilman D, Knops J, Reich PB (2004) Mechanisms responsible for the positive diversity-productivity relationship in Minnesota grasslands. Ecology Letters, 7, 661–668.

Lawrence MA (2013) ez: Easy analysis and visualization of factorial experiments. R package version 4.2-2. http://CRAN.R-project.org/package=ez

Lemus R, Brummer EC, Moore KJ, Molstad NE, Burras CL, Barker MF (2002) Biomass yield and quality of 20 switchgrass populations in southern Iowa, USA. Biomass and Bioenergy, 23, 433–442.

Lemus R, Brummer EC, Burras CL, Moore KJ, Barker MF, Molstad NE (2008) Effects of nitrogen fertilization on biomass yield and quality in large fields of established switchgrass in southern Iowa, USA. Biomass and Bioenergy, 32, 1187–1194.

Levine JM (2000) Species diversity and biological invasions: relating local process to community pattern. *Science*, **288**, 852–854.

Lewandowski I, Clifton-Brown JC, Andersson B *et al.* (2003) Environment and harvest time affects the combustion qualities of *Miscanthus* genotypes. *Agronomy Journal*, **95**, 1274–1280.

Loftus GR, Masson ME (1994) Using confidence intervals in within-subject designs. *Psychonomic Bulletin and Review*, **1**, 476–490.

Loreau M, Hector A (2001) Partitioning selection and complementarity in biodiversity experiments. *Nature*, **412**, 72–76.

Ma W, He J, Yang Y *et al.* (2010) Environmental factors covary with plant diversity–productivity relationships among Chinese grassland sites. *Global Ecology and Biogeography*, **19**, 233–243.

McLaughlin S, Bouton J, Bransby D *et al.* (1999) Developing switchgrass as a bioenergy crop. In: *Perspectives on New Crops and New Uses* (ed. Janick J), pp. 282–299. ASHS Press, Alexandria, VA, USA.

Myers MC, Hoksch BJ, Mason JT (2012) Butterfly response to floral resources during early establishment at a heterogeneous prairie biomass production site in Iowa, USA. *Journal of Insect Conservation*, **16**, 457–472.

Myers MC, Mason JT, Hoksch BJ, Cambardella CA, Pfrimmer JD (2015) Birds and butterflies respond to soil-induced habitat heterogeneity in experimental plantings of tallgrass prairie species managed as agroenergy crops in Iowa, USA. *Journal of Applied Ecology*, **52**, 1176–1187.

Natural Resource Conservation Service (2009) Planting and managing switchgrass as a biomass energy crop. Technical Note No. 3. Available at: http://www.nrcs.usda.gov/Internet/FSE_DOCUMENTS/stelprdb1042293.pdf (accessed 1 July 2015).

Natural Resource Conservation Service (2014) Web soil survey. Available at: http://websoilsurvey.sc.egov.usda.gov/App/WebSoilSurvey.aspx (accessed 13 August 2014)

Oksanen J, Blanchet FG, Kindt R *et al.* (2013) vegan: Community Ecology Package. R package version 2.0-10. http://CRAN.R-project.org/package=vegan

Palmer MW, van der Maarel E (1995) Variance in species richness, species association, and niche limitation. *Oikos*, **73**, 203–213.

Pfisterer AB, Schmid B (2002) Diversity-dependent production can decrease the stability of ecosystem functioning. *Nature*, **416**, 84–86.

Sherrard ME, Joers LC, Carr CM, Cambardella CA (2015) Soil type and species diversity influence selection on physiology in *Panicum virgatum*. *Evolutionary Ecology*, **29**, 679–702.

Smith D, Williams D, Houseal G, Henderson K (2010) *The Tallgrass Prairie Center Guide to Prairie Restoration in the Upper Midwest*. University of Iowa Press, Iowa City, IA.

Tilman D (1987) Secondary succession and the pattern of plant dominance along experimental nitrogen gradients. *Ecological Monographs*, **57**, 189–214.

Tilman D, Hill J, Lehman C (2006) Carbon-negative biofuels from low-input high-diversity grassland biomass. *Science*, **314**, 1598–1600.

Tilman D, Reich PB, Isbell F (2012) Biodiversity impacts ecosystem productivity as much as resources, disturbance, or herbivory. *Proceedings of the National Academy of Sciences of the United States of America*, **109**, 10394–10397.

Wedin D, Tilman D (1993) Competition among grasses along a nitrogen gradient: initial conditions and mechanisms of competition. *Ecological Monographs*, **63**, 199–229.

Wilsey B (2010) Productivity and subordinate species response to dominant grass species and seed source during restoration. *Restoration Ecology*, **18**, 628–637.

Yachi S, Loreau M (1999) Biodiversity and ecosystem productivity in a fluctuating environment: the insurance hypothesis. *Proceedings of the National Academy of Sciences of the United States of America*, **96**, 1463–1468.

Zilverberg CJ, Johnson WC, Owens V *et al.* (2014) Biomass yield from planted mixtures and monocultures of native prairie vegetation across a heterogeneous farm landscape. *Agriculture, Ecosystems and Environment*, **186**, 148–159.

Environmental costs and benefits of growing *Miscanthus* for bioenergy in the UK

JON P. MCCALMONT[1], ASTLEY HASTINGS[2], NIALL P. MCNAMARA[3], GOETZ M. RICHTER[4], PAUL ROBSON[1], IAIN S. DONNISON[1] and JOHN CLIFTON-BROWN[1]

[1]*Institute of Biological, Environmental and Rural Sciences (IBERS), Aberystwyth University, Gogerddan, Aberystwyth, Wales SY23 3EQ, UK,* [2]*Institute of Biological and Environmental Science, University of Aberdeen, 24 St Machar Drive, Aberdeen AB24 3UU, UK,* [3]*Centre for Ecology & Hydrology, Lancaster Environment Centre, Library Avenue, Bailrigg, Lancaster LA1 4AP, UK,* [4]*Rothamsted Research, West Common, Harpenden, Hertfordshire AL5 2JQ, UK*

Abstract

Planting the perennial biomass crop *Miscanthus* in the UK could offset 2–13 Mt oil eq. yr^{-1}, contributing up to 10% of current energy use. Policymakers need assurance that upscaling *Miscanthus* production can be performed sustainably without negatively impacting essential food production or the wider environment. This study reviews a large body of *Miscanthus* relevant literature into concise summary statements. Perennial *Miscanthus* has energy output/input ratios 10 times higher (47.3 ± 2.2) than annual crops used for energy (4.7 ± 0.2 to 5.5 ± 0.2), and the total carbon cost of energy production (1.12 g CO_2-C eq. MJ^{-1}) is 20–30 times lower than fossil fuels. Planting on former arable land generally increases soil organic carbon (SOC) with *Miscanthus* sequestering 0.7–2.2 Mg C4-C ha^{-1} yr^{-1}. Cultivation on grassland can cause a disturbance loss of SOC which is likely to be recovered during the lifetime of the crop and is potentially mitigated by fossil fuel offset. N_2O emissions can be five times lower under unfertilized *Miscanthus* than annual crops and up to 100 times lower than intensive pasture. Nitrogen fertilizer is generally unnecessary except in low fertility soils. Herbicide is essential during the establishment years after which natural weed suppression by shading is sufficient. Pesticides are unnecessary. Water-use efficiency is high (e.g. 5.5–9.2 g aerial DM (kg $H_2O)^{-1}$, but high biomass productivity means increased water demand compared to cereal crops. The perennial nature and belowground biomass improves soil structure, increases water-holding capacity (up by 100–150 mm), and reduces run-off and erosion. Overwinter ripening increases landscape structural resources for wildlife. Reduced management intensity promotes earthworm diversity and abundance although poor litter palatability may reduce individual biomass. Chemical leaching into field boundaries is lower than comparable agriculture, improving soil and water habitat quality.

Keywords: biodiversity, bioenergy, crop modelling, energy crops, GHG, land-use change, *Miscanthus*, perennial grasses, plant ecophysiology, renewable energy

Introduction

The IPCC 5th report (IPCC, 2014) makes clear that it is *extremely likely* that cumulative anthropogenic greenhouse gas emissions have led to unequivocal climate warming with temperature and precipitation extremes seen since the 1950s that are unprecedented over millennia. The report states with *high confidence* that without additional mitigation efforts, climate warming will more likely than not exceed 4 °C above pre-industrial levels by 2100; extremes of weather

McCalmont and Hastings are joint first authors.

Correspondence: John Clifton-Brown
e-mail: jhc@aber.ac.uk

resulting from this would lead to 'substantial species extinctions, global and regional food insecurity, and consequential constraints on human activities' with the highest relative price to be paid by those least responsible for the problem. The IPCC states categorically that limiting these impacts to more manageable levels requires net global CO_2 emission to decrease to zero in the next few decades through rapid decarbonization of energy production. Sustainable biomass offers the almost unique opportunity to provide storable, flexible use of fuel that can be readily converted to heat, electricity, or even liquid transport fuels and is the single option that might provide a future mechanism to remove atmospheric carbon by capture and storage (CCS) (ETI 2015). Over the past 20 years, fairly comprehensive field data have become available for the clone-based interspecies hybrid *M. giganteus*. This review examines the environmental benefits and trade-offs associated with large-scale planting of *Miscanthus* for UK bioenergy, now made possible by recent breeding of seed-based hybrids in the UK and USA. The review focusses on environmental impacts of field production; it does not cover wider economic analyses that are also important determinants of commercial uptake.

Renewable fuels

Use of renewable energy in the UK mix was up 30% from 2012 to 2013 (see Fig. 1); it supplied 14.9% of UK electricity which was 5.2% of total energy (DUKES 2014a), well short of the UK 2012 Bioenergy Strategy target of 15% of total energy by 2020 (DECC, 2012). Plant biomass supplied 21.6% of total renewables which offset 2079 kt of oil equivalent electricity and 339 kt oil eq. heat (DUKES 2014b). Despite the replacement of 2.4 million tonnes of oil use, domestic biomass production remains at a low level compared to total energy demand. Only 0.05 Mha, or 0.8%, of UK arable land was used for bioenergy production in 2013 with 0.03 Mha being used to produce maize for anaerobic digestion (DEFRA 2014a). Results from the OFGEM biomass sustainability data set (Ofgem 2014), excluding liquid feedstocks or anaerobic digestion, showed the UK burnt 3.9 Mt of plant biomass for electricity in 2013; of this, 1.8 Mt was produced in the UK with the majority being wood products. Home grown dedicated energy crops provided only 56 kt: 47 kt of *Miscanthus* and 9 kt of SRC willow.

Biomass crops

While the initial premise regarding bioenergy was that carbon recently captured from the atmosphere into plants would deliver an immediate reduction in GHG emission from fossil fuel use, the reality proved less straightforward. Studies suggested that GHG emission from energy crop production and land-use change might outweigh any CO_2 mitigation (Searchinger et al., 2008; Lange, 2011). Nitrous oxide (N_2O) production, with its powerful global warming potential (GWP), could be a significant factor in offsetting CO_2 gains (Crutzen et al., 2008) as well as possible acidification and eutrophication of the surrounding environment (Kim & Dale, 2005). However, not all biomass feedstocks are equal, and most studies critical of bioenergy production are concerned with biofuels produced from annual food crops at high fertilizer cost, sometimes using land cleared from natural ecosystems or in direct competition with food production (Naik et al., 2010). Dedicated perennial energy crops, produced on existing, lower grade, agricultural land, offer a sustainable alternative with significant savings in greenhouse gas emissions and soil carbon sequestration when produced with appropriate management (Crutzen et al., 2008; Hastings et al., 2008, 2012; Cherubini et al., 2009; Dondini et al., 2009a; Don et al., 2012; Zatta et al., 2014; Richter et al., 2015). Greenhouse gas mitigation is a primary concern in bioenergy production but is not the only consideration, particularly for large-scale land-use transitions. Bioenergy supply chains must be energetically favourable, maintain or increase soil carbon, and be cost effective and environmentally sustainable without

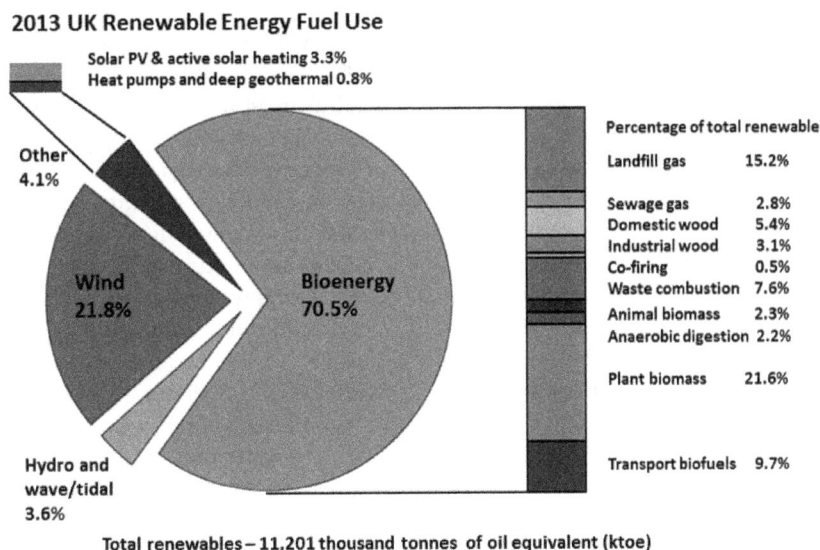

Fig. 1 Current UK renewable energy use as of end 2013, up 30% between 2012 and 2013, supplying 14.9% of UK electricity. Bioenergy contributes 70.5% of total renewable with plant biomass alone contributing 21.6% at 3.9 Mt which was primarily imported.

interfering with essential food production (Tilman *et al.*, 2009; Valentine *et al.*, 2012). *Miscanthus* × *giganteus* (hereafter *Miscanthus*), a low-input, fast-growing perennial energy grass, is seen to offer an attractive alternative biomass crop (Lewandowski *et al.*, 2003; Harvey, 2007; Heaton *et al.*, 2008, 2010; Zhuang *et al.*, 2013) with energy output/input ratios around ten times that of annual energy crops (Felten *et al.*, 2013) and significant potential to reduce fossil fuel CO_2 emission (Clifton-Brown *et al.*, 2004, 2007; Hillier *et al.*, 2009). Felten *et al.* (2013) compared energy balances for oil seed rape (OSR), maize, and *Miscanthus* and found output/input ratios of 4.7 ± 0.2, 5.5 ± 0.2, and 47.3 ± 2.2, respectively.

Potential UK land availability

The 2007 UK biomass strategy (DEFRA, 2007) set a target of 0.35 Mha of UK agricultural land growing perennial energy crops by 2020; this would be part of an overall one million hectares in biofuel and energy crop production. The 2012 UK Bioenergy Strategy (DECC, 2012) suggests that the potential land available specifically for *Miscanthus* that would not impinge on food production is in the range of 0.72–2.8 Mha which is well above the 2007 target (Fig. 2 puts the 0.35 Mha into context by showing current UK agricultural land use and 5 year trends). These strategy reports stress that while some energy crops may reduce soil erosion, improve biodiversity, and aid fuel security, production must take place '...in those parts of the UK where it makes sense...'

Lovett *et al.* (2009) reported that 0.35 Mha of *Miscanthus* could be easily accommodated in the UK. Initially considering just England, using a GIS approach, they produced a constraint map based on 11 preclusion factors covering biophysical, social, and environmental considerations, for example high soil carbon content, cultural or natural heritage, and urban centres, with a final constraint that only poorer quality land in agricultural land class (ALC) grades 3 or 4 would be considered, excluding higher grades 1 and 2 and the worst grade 5. Results suggested potential land availability for *Miscanthus*, of 3.12 Mha, around one-quarter of total English land area. The authors point out that the 0.35 Mha target represents only 11.6% of this and planting *Miscanthus* on this more marginal agricultural land would not impinge on essential food production. Lovett *et al.* (2014) expands the GIS constraint mapping to include Scotland and Wales (see Fig. 3a). Here, results suggested 8.5 Mha potentially suitable for growing *Miscanthus* or SRC willow/poplar, applying the extra restriction to ALC grade 3 or worse reduced this to 6.4 Mha. Grade 3 agricultural lands represent the majority of UK

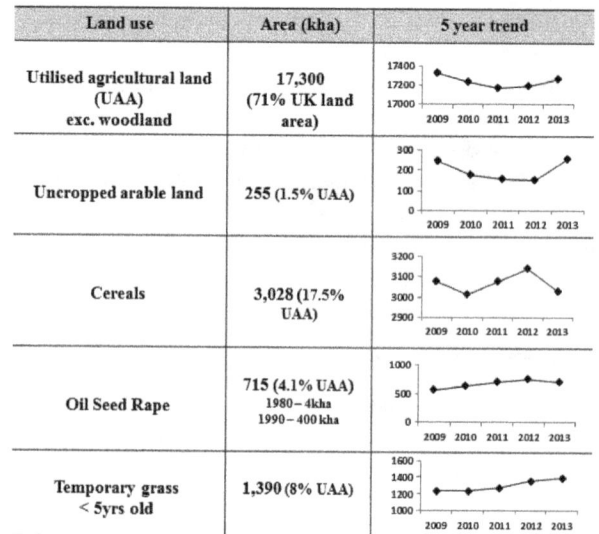

Land use	Area (kha)	5 year trend
Utilised agricultural land (UAA) exc. woodland	17,300 (71% UK land area)	
Uncropped arable land	255 (1.5% UAA)	
Cereals	3,028 (17.5% UAA)	
Oil Seed Rape	715 (4.1% UAA) 1980 – 4kha 1990 – 400 kha	
Temporary grass < 5yrs old	1,390 (8% UAA)	

Fig. 2 2013 extent and 5-year trends in major UK agricultural land areas, 0.26 Mha of arable were uncropped in 2013 due to poor weather in 2012 preventing annual cultivation.

farmland and covered around 59% of the total 8.5 Mha identified in this study, excluding this and restricting planting to the very worst agricultural land grades 4 and 5 left 1.4 Mha, four times the 0.35 Mha target.

Land-use change

Land-use change is central to UK agriculture, field crop species and farming practices are in constant flux, and land usage will typically follow an economic rationale within the constraints of the EU's Common Agricultural Policy. In 1993, to curb EU overproduction of food, farmers set aside a minimum of 15% of their cropped land; by 2000, this had dropped to 10% and was zero by 2008. In England alone, from 2000 to 2006, the average land area set aside was 0.57 Mha (DEFRA 2014b). The days of set aside are over, agricultural production must increase worldwide to meet growing demands for both food and energy; accommodating this while avoiding increased exploitation of natural lands requires a move towards more sustainable intensification (Tilman *et al.*, 2011; Godfray & Garnett, 2014). Concentrating agronomic effort and resources away from the least productive 10% of farmed area to more productive land while retaining a low-input, high-output perennial energy crop on these poorer areas could offer a mechanism for both intensification and diversification within farms. Farmers might identify areas of their farms where yields of more conventional crops are at their lowest, and bought at the expense of high effort and chemical input while detracting effort from their more productive land, and give this area over to at least one economic

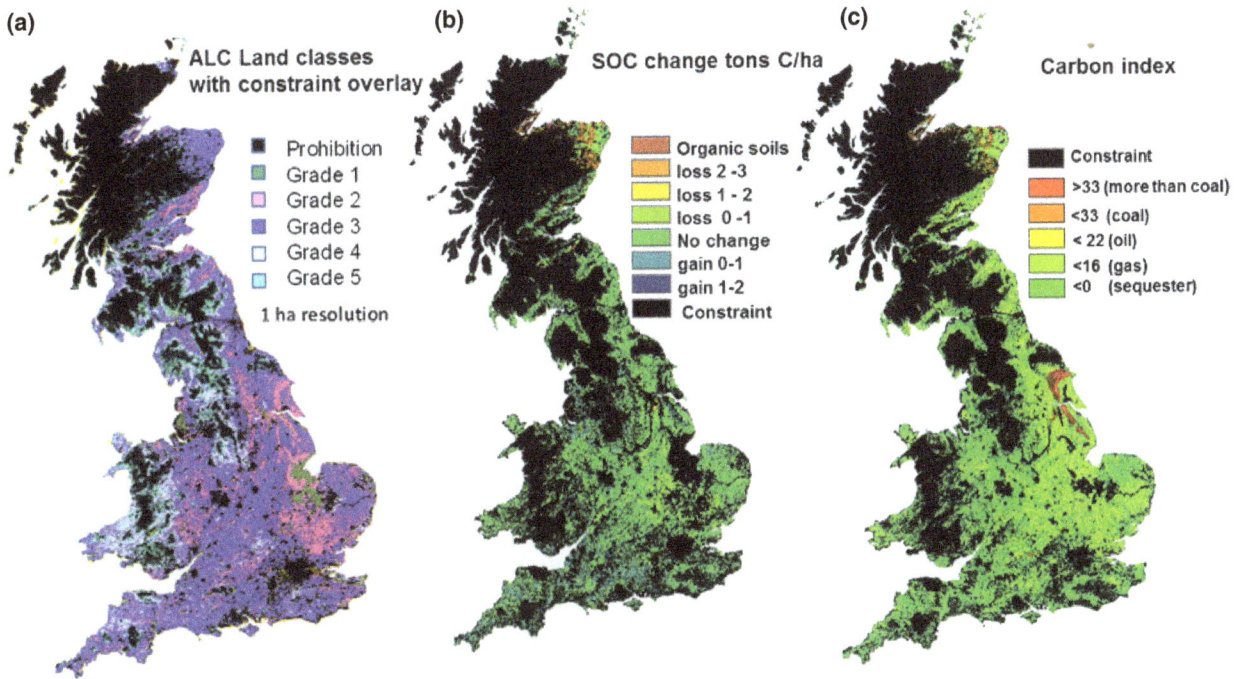

Fig. 3 (a) shows distribution of agricultural land classes (ALC) with excluded areas in black following Lovett *et al.* (2014). (b) shows the map of modelled annual change in soil carbon following land-use change from existing agriculture to *Miscanthus* outside these areas from Milner *et al.* (2015), and (c) shows the potential carbon intensity index, in g CO_2-C equivalent per MJ energy in the furnace, compared to coal (33), oil (22), and North Sea gas (16). The *Miscanthus* carbon intensity is calculated considering rhizome propagation, 2-year establishment, pelletized fuel, and 100 km of transportation.

cropping cycle of a low-input perennial. *Miscanthus* can improve overworked or difficult soils by acting as a long-term break crop, increasing soil carbon, organic matter, and earthworm diversity (Kahle *et al.*, 2001; Hansen *et al.*, 2004; Felten & Emmerling, 2011). Perhaps an ideal situation would see the cycling of this long-term break crop around the farm area with conventional crop rotations following to take advantage of the improved organic matter and soil structure. Figure 2 shows a summary of some of the main UK agricultural land uses on areas that might be suitable for *Miscanthus* production and the variability of their extent between 2009 and 2013 (DEFRA 2014c). Of the 17.3 Mha of utilized agricultural area (UAA) in the UK, around 3 Mha (17.5%) were in cereal production in 2013, and this area varied between 3 and 3.2 Mha over the last 5 years with a not unusual 0.26 Mha being completely uncropped in 2013 due to high rainfall, preventing planting in autumn 2012. The area of temporary grassland less than 5 years old, perhaps a prime candidate for *Miscanthus* production in western areas of the UK, has been steadily increasing over the last 5 years, currently 1.39 Mha (8% of UAA), up from 1.24 Mha in 2009. The current extent of oil seed rape (OSR) production, at 0.72 Mha, is more than double the *Miscanthus* target. Land cover for this conspicuous, high-input crop had risen from being

almost unknown in 1980 to 0.40 Mha by 1990 and covering more than 0.70 Mha by 2013 (DEFRA 2014d). Modelling studies (Lovett *et al.*, 2009; Hastings *et al.*, 2013) show that the mean yield of *Miscanthus* on grade 3b, 4, and 5 land outside the excluded areas is around 10 tons DM ha^{-1} yr^{-1}, and the 2007 Biomass Strategy target of 0.35 Mha could produce up to 70 PJ energy, equivalent to 1.67 Mt of oil or 1.17% of total UK energy. The 2012 Bioenergy Target range of 0.72–2.8 Mha would produce 3.44–13.38 Mt oil eq. (2.41–9.39% of total energy). Agricultural land is a finite resource in the UK, providing a range of ecosystem services: from food and energy to culture and leisure with impacts on water quality and natural habitats. Milner *et al.* (2015) offer a comprehensive 'threat matrix' quantifying the potential impact on a range of ecosystem services across the UK from land-use change to perennial biomass crops. They concluded that there was little difference between *Miscanthus* and SRC when planted on lower grade land with both offering positive improvements in service provision although climate-driven yield estimates and previous land-use were key factors. Optimization of land utilization at a national scale is essential although currently lacking, if land is to be used for producing energy, then policy should aim to produce the maximum amount of energy per ha within the context of wider ecosystem service provision. In terms of energy production intensity, *Miscanthus* biomass produces more net energy per hectare than other bioenergy crops at around 200 GJ ha^{-1} yr^{-1}, especially arable [maize for biogas 98, oil seed rape for biodiesel 25, wheat and sugar beet ethanol 7–15 (Hastings *et al.*, 2012)]. Felten *et al.* (2013) calculated similar figures, reporting 254 GJ ha^{-1} yr^{-1} for *Miscanthus*. Energy production intensity calculated for woody perennials can vary significantly by area (Bauen *et al.*, 2010) with yield predictions largely driven by future climate projections (Hastings *et al.*, 2013). Tallis *et al.* (2013) showed that in the right circumstances, even old varieties of SRC willow can exceed 150 GJ ha^{-1} yr^{-1}, suggesting that planting combinations of crops may be most efficient for overall energy production.

Soil carbon

More than twice as much carbon is stored in the world's soils compared to vegetation or atmosphere (Post *et al.*, 1990; Lal, 2004; Cox *et al.*, 2011; Scharlemann *et al.*, 2014). It is critically important to understand the impact of large-scale agricultural land-use change on these storage reservoirs. Although extending the capacity of European soils to sequester carbon may be limited relative to overall CO$_2$ emissions (Smith *et al.*, 1997; Mackey *et al.*, 2013), the potential for losing soil carbon through misplaced land-use change could be far more significant. Any soil disturbance, such as ploughing and cultivation, is likely to result in short-term respiration losses of soil organic carbon, decomposed by stimulated soil microbe populations (Cheng, 2009; Kuzyakov, 2010). Annual disturbance under arable cropping repeats this year after year resulting in reduced SOC levels. Perennial agricultural systems, such as grassland, have time to replace their infrequent disturbance losses which can result in higher steady-state soil carbon contents (Gelfand *et al.*, 2011; Zenone *et al.*, 2013). Upscale predictions of these carbon deltas rely on measurements of sample data informing and validating process models (Dondini *et al.*, 2009a; Zatta *et al.*, 2014; Agostini *et al.*, 2015). However, collecting enough samples to determine the significance level of any observed change can be challenging (Kravchenko & Robertson, 2011) and it is extremely rare to find baseline soil samples taken prior to land-use change. Adjacent reference sites of the previous land-use are generally taken to represent baseline soil conditions although these may not necessarily represent initial conditions accurately. Richter *et al.* (2015) had the rare opportunity to compare *Miscanthus* soils after 14 years to both archived baseline and an adjacent reference site, though only at 0–30 cm; they found that using the reference site would have suggested greater

declines in original SOC than were seen when compared to the actual baseline. With these limitations in mind, we report here results from empirical sample data and discuss whether some clear trends emerge.

Table 1 summarizes nine soil sampling studies of land-use transition from both arable and grassland with SOC stocks compared to adjacent land. Across these nine studies, 21 comparisons were made between *Miscanthus* plantations and grassland (7) or arable (14) with plantation ages ranging from 3 to 19 years. Direct comparisons of absolute numbers between studies are problematic. Some studies (Hansen *et al.*, 2004; Clifton-Brown *et al.*, 2007; Schneckenberger & Kuzyakov, 2007; Dondini *et al.*, 2009b; Poeplau & Don, 2014) sampled only single sites within each comparison or age class, while others (Felten & Emmerling, 2012; Zimmermann *et al.*, 2012) had multiple sites within each comparison. Only two studies (Zatta *et al.*, 2014; Richter *et al.*, 2015) had access to a baseline soil archive collected prior to the land-use change; both these studies investigated the impact on soil carbon of different *Miscanthus* genotypes, while others were limited to *M x giganteus*. Sample depths also varied widely as did management regimes with some sites fertilized and others not. Despite this variability, it seems likely that arable land converted to *Miscanthus* will sequester soil carbon; of the 14 comparisons, 11 showed overall increases in SOC over their total sample depths with suggested accumulation rates ranging from 0.42 to 3.8 Mg C ha^{-1} yr^{-1}. Only three arable comparisons showed lower SOC stocks under *Miscanthus*, and these suggested insignificant losses between 0.1 and 0.26 Mg ha^{-1} yr^{-1}.

The grassland to *Miscanthus* comparisons showed three increases, three decreases, and one no change in soil C stocks. no doubt complicated by the *Miscanthus* being planted on former arable, arable/fallow, or grassland despite being compared to long-term grassland, whereas all comparisons to arable were planted on former arable land. The range of gains and losses was relatively small, −1 to 0.94 Mg C ha^{-1} yr^{-1} with only the increase of 0.94 Mg ha^{-1} yr^{-1} shown to be significant (Hansen *et al.*, 2004). Another study, not included in Table 1 due to incompatibility of units (Kahle *et al.*, 2001), primarily compared plant derived organic matter between *Miscanthus* and grassland but also sampled for organic carbon at four sites in Germany over multiple years making a total of 12 comparisons. Of these, 8 showed higher SOC stocks under *Miscanthus* compared to the grassland with 5 of these being shown to be highly significant ($P < 0.01$), only one site showed lower concentrations of SOC across 2 years of sampling. One literature review (Anderson-Teixeira *et al.*, 2009) felt confident to suggest that conversion of temperate grassland to *Miscanthus* would see an eventual increase in

Table 1 Baselines and changes in soil organic carbon (SOC), soil core analysis under land-use change to Miscanthus from arable and grassland. Results compared to concurrent reference sites taken to represent baseline conditions before conversion in all cases except Zatta et al. (2014) who compared results to baseline soil archive. Richter et al. (2015) also offer baseline soil archive comparison but not to total sample depth, and results shown here are from their reference site (see text for details). Individual results shown are as reported from individual sites or means from multiple sites in the same transition class/age group; superscripts in location field indicate site numbers in each comparison

Location	Mean annual air temp. (°C)/ rainfall (mm)	Soil texture (0–30 cm)	Mean harvest yield (Mg DM ha^{-1} yr^{-1})	Total sample depth (cm)	Plantation age (years)	Land-use prior to Miscanthus plantation	Comparison land use	Total SOC under Miscanthus to sample depth (Mg C ha^{-1})	Total SOC under control to sample depth (Mg C ha^{-1})	Direction of suggested SOC change	Net SOC (C3 and C4) accumulation rate since planting (over total sample depth) (Mg C ha^{-1} yr^{-1})	C4-C accumulation over sample depth (Mg C ha^{-1} yr^{-1})	References
Ireland[1]	9.9/1004	Loam/sandy loam	13.4	30^{FD}	15	Grassland	Grassland	64	59.7	up	0.29^{ns}	0.6	Clifton-Brown et al. (2007)
Ireland[8]	9.3/822	Sandy loam	na	30^{FD}	3	Grassland	Grassland	79.6	80.625	down	−0.34^{ns}	0.9	Zimmermann et al. (2012)
UK[1]	9.9/1216	Sandy clay loam	14.5	30^{ESM}	6	Grassland	Grassland	71.6	78.8	down	−1.2^{ns}	1.25	Zatta et al. (2014)
Germany[1]	8.7/679	Loam	na	100^{FD}	9	Grassland	Grassland	112	121	down	−1^{ns}	Mg ha^{-1} na	Schneckenberger & Kuzyakov (2007)
Germany[1]	8.7/548	Sandy loam	na	100^{FD}	12	Arable/fallow	Grassland	70	64	up	0.5^{ns}	Mg ha^{-1} na	Schneckenberger & Kuzyakov (2007)
Denmark[1]	7.4/705	Coarse loamy sand	12.6	100^{FD}	9	Arable	Grassland	91	91	no change	0^{ns}	0.78	Hansen et al. (2004)
Denmark[1]	7.4/706	Coarse loamy sand	14.1	100^{FD}	16	Arable	Grassland	106	91	up	0.94*	1.13	Hansen et al. (2004)
Ireland[1]	9.3/830	Sandy loam	16	60^{FD}	14	Arable	Arable	131.3	105.8	up	1.82*	3.2	Dondini et al. (2009a,b)
Germany[4]	10.5/761	Loamy sand	15	150^{FD}	2	Arable	Arable	92.9	90.5	up	1.2^{ns}	1.4	Felten & Emmerling (2012)
Germany[2]	10.5/762	Loamy sand	15	150^{FD}	5	Arable	Arable	109.5	90.5	up	3.8^{ns}	0.68	Felten & Emmerling (2012)

(continued)

Table 1 (continued)

Location	Mean annual air temp. (°C)/ rainfall (mm)	Soil texture (0–30 cm)	Mean harvest yield (Mg DM ha⁻¹ yr⁻¹)	Total sample depth (cm)	Plantation age (years)	Land-use prior to Miscanthus plantation	Comparison land use	Total SOC under Miscanthus to sample depth (Mg C ha⁻¹)	Total SOC under control to sample depth (Mg C ha⁻¹)	Direction of suggested SOC change	Net SOC (C3 and C4) accumulation rate since planting (over total sample depth) (Mg C ha⁻¹ yr⁻¹)	C4-C accumulation over sample depth (Mg C ha⁻¹ yr⁻¹)	References
Germany[7]	10.5/763	Loamy sand	15	150^{FD}	16	Arable	Arable	112.5	90.5	up	1.375*	1.03	Felten & Emmerling (2012)
Denmark[1]	7.4/707	Coarse loamy sand	12.6	100^{FD}	9	Arable	Arable	91	92	down	−0.1^{ns}	0.78	Hansen et al. (2004)
Denmark[1]	7.4/708	Coarse loamy sand	14.1	100^{FD}	16	Arable	Arable	106	92	up	0.88*	1.13	Hansen et al. (2004)
Netherlands[1]	8/760	na	13	80^{FD}	11	Arable	Arable	93.9	77.6	up	1.48^{ns}	0.70	Poeplau & Don (2014)
Germany[1]	7.3/550	na	15	80^{FD}	15	Arable	Arable	147.8	150.3	down	−0.17^{ns}	0.37	Poeplau & Don (2014)
Denmark[1]	7.9/859	na	14	80^{FD}	18	Arable	Arable	96.5	89	up	0.42^{ns}	0.54	Poeplau & Don (2014)
Switzerland[1]	8.4/1185	na	14	80^{FD}	17	Arable	Arable	133	116.2	up	0.99*	0.80	Poeplau & Don (2014)
Germany[1]	9/707	na	15	80^{FD}	19	Arable	Arable	94.8	77.6	up	0.91*	0.93	Poeplau & Don (2014)
Switzerland[1]	9/860	na	14	80^{FD}	15	Arable	Arable	108.7	85.2	up	1.57*	0.88	Poeplau & Don (2014)
Ireland[8]	9.3/823	Sandy loam	na	30^{FD}	3	Arable	Arable	64.52	60.51	up	1.34^{ns}	0.62	Zimmermann et al. (2012)
UK[1]	na	Silty clay loam	8.75	100^{ESM}	14	Arable	Arable	106.9	110.6	down	−0.26^{na}	1.10	Richter et al. (2015)

FD, fixed depth sampling; ESM, equivalent soil mass; ns, not significant; na, unavailable.

Zatta and Richter are mean results from several genotypes.

Superscript number in location field indicates number of sites included in the result.

Superscript in SOC accumulations indicates significance of change where available. * = $P < 0.05$.

SOC, but the results here would suggest that there are enough uncertainties to prevent making such an outright assertion, and it is perhaps safer to suggest that over the lifetime of the crop SOC stocks in the soil profile would be at least maintained. In unpublished work, R. Rowe, A.M. Keith, D. Elias, M. Dondini, P. Smith, J. Oxely and N.P. McNamara (2015, in submission) investigated multiple paired comparisons between *Miscanthus* and grassland (nine sites, mean age 7 years) and *Miscanthus* and arable (11 sites, mean age 6.5 years) and reported the results of soil carbon modelling from soil cores taken at these sites. They report lower soil carbon stocks under *Miscanthus* compared to both arable and grassland control sites although these differences were only found to be significant in the 0–30 cm layer, where arable transition (mean plantation age 6.5 years) suggested a reduction in soil carbon of -0.93 Mg C ha^{-1} yr^{-1} and grassland transition (mean age 7 years) at -3.17 Mg C ha^{-1} yr^{-1}; for the 0–100 cm depth, these became 0.05 Mg C ha^{-1} yr^{-1} and -0.69 Mg C ha^{-1} yr^{-1} although differences were not found to be significant when considered over this depth. R. Rowe, A.M. Keith, D. Elias, M. Dondini, P. Smith, J. Oxely and N.P. McNamara (2015, in submission) suggest sampling limitations in many previous studies which typically sample to a fixed depth and do not account for changes in soil bulk density due to land-use change, and they employed an equivalent soil mass sampling (ESM) strategy in their study, as outlined by Gifford & Roderick (2003). In this technique, sample depth is adjusted to account for soil surface uplift due to belowground Miscanthus biomass. R. Rowe, A.M. Keith, D. Elias, M. Dondini, P. Smith, J. Oxely and N.P. McNamara (2015, in submission) found that apparently larger SOC stocks under *Miscanthus* compared to controls were not seen when using ESM sampling. However, while care was taken to account for soil bulk density changes under Miscanthus, no similar accommodation is made to account for possible erosion losses under annual arable cultivations used as surrogates for baseline.

Soil carbon turnover

Even where results suggest maintained or increased SOC, initial disturbance losses after planting *Miscanthus* will still occur. It is important to note, as comprehensively discussed by Agostini *et al.* (2015), that despite SOC changes being generally reported as an annual mean over the age of the plantation, these deltas are unlikely to be constant over time. SOC derived from crop inputs will be lower during the early years of establishment (Zimmermann *et al.*, 2012) with disturbance losses of resident C3 carbon outpacing C4 inputs

when planted into grassland. Litter drop and root exudates are a function of yield and biomass and will build and reverse this over time although long-term studies over the potential 15–20 year crop lifetime are notably lacking. Sources of SOC can be investigated using isotopic analysis (Balesdent *et al.*, 1987). C4 plants such as *Miscanthus* discriminate less against ^{13}C than native C3 plants, and therefore, samples of SOC sequestered under *Miscanthus* will show less depletion of this isotope when compared to an atmospheric standard. Zatta *et al.* (2014) showed that while SOC after conversion of grassland to *Miscanthus* showed no significant difference after 6 years, the isotopic signature showed a clear C4 source demonstrating a rapid turnover in soil carbon with mobilized C3 carbon being quickly replaced, results also shown in other such studies (e.g. Poeplau & Don 2014; Richter *et al.*, 2015). Litter input to the soil plays a vital role in sequestering carbon in a mature crop, and overwinter litter drop in *Miscanthus* is reasonably consistent at around 30–35% of aboveground biomass production (Lewandowski *et al.*, 2000; Clifton-Brown *et al.*, 2004). The dropped litter accounts for most of the reduction in yield during ripening but is a gain for soil carbon and organic matter and significantly improves overall combustion quality of the harvested biomass. Hansen *et al.* (2004) used stable isotope analysis of soils under two *Miscanthus* plantations to calculate a coefficient of retention for input of carbon from this litter at 26% of the total carbon input for their 9-year-old plantation, increasing to 29% for the longer, 16-year plantation. The data collated in Table 1 suggest a reasonable range of this C4-C sequestration rate in the top 30 cm to be between 0.5 and 1.5 Mg ha^{-1} yr^{-1} with one outlier at 3.2 Mg ha^{-1} yr^{-1} (Dondini *et al.*, 2009a). Combining the, albeit limited, published sample data suggests that stocks of SOC in the top 30 cm on converted grasslands is likely to be higher than converted arable land (Fig. 4), but that the accumulation of SOC occurs faster in conversions from arable soils (Fig. 5). The correlation between plantation age and SOC can be seen in Fig. 6, although the wide scatter ($R^2 = 0.2$) likely reflects limited data; the trendline suggests a net accumulation rate of 1.84 Mg C ha^{-1} yr^{-1} with similar levels to grassland at equilibrium.

Soil carbon spatial modelling across the UK

Soil carbon stocks are a balance between the soil organic matter decomposition rate and the organic material input each year by vegetation, animal manure, or any other organic input. In *Miscanthus* plots, the difference between peak yield and harvest offtake can be used to calculate the soil carbon input from leaf fall and stubble after harvest and estimates of root turnover (Hansen

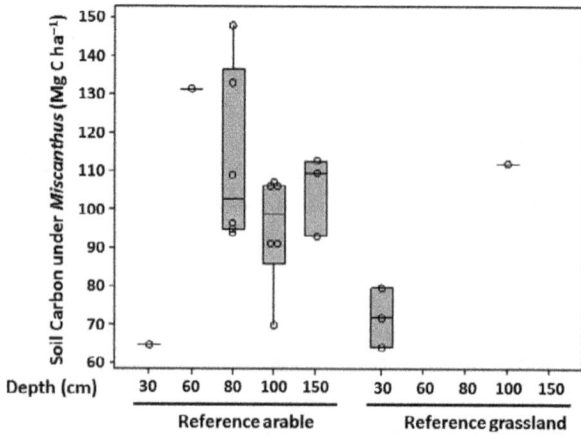

Fig. 4 Boxplot of soil organic carbon stocks found under *Miscanthus* results from Table 1. The categories are land use (arable or grassland) and depth of soil that is considered in the SOC content reported in the literature. Varying sample depths reported reflects limited data at greater depths from previous grassland.

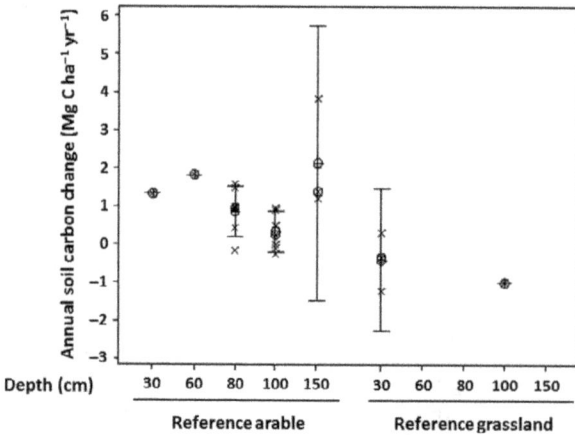

Fig. 6 Plantation age vs. SOC stocks under *Miscanthus* from Table 1; slope is 1.84 Mg ha^{-1} yr^{-1} ($R^2 = 0.2$).

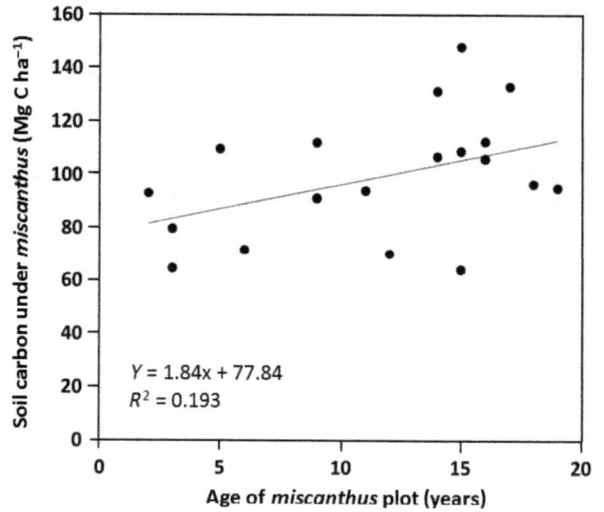

Fig. 5 Annual change in soil organic carbon (SOC) under *Miscanthus* from Table 1; as Fig. 3, limited data at greater depths for previous grassland.

et al., 2004). If the previous land use and soil organic carbon level is known, then the new soil carbon content of land converted to *Miscanthus* plantations can be estimated using models calibrated by field experiments, either in arable land (i.e. Dondini *et al.*, 2009a) or in pasture (i.e. Zatta *et al.*, 2014); this can be carried out spatially for the entire UK land area and verified by the other published data in Table 1. The cohort model (Bosatta & Ågren, 1985) is used in Fig. 3b to calculate the mean annual SOC change for each km^2 grid square and its spatial distribution, and the model uses initial soil carbon from the Harmonized World Soil Data Base (HWSD) and the predicted SOC input over 15 years of *Miscanthus* cropping. Figure 7 shows the histogram of

SOC change for the UK on land not excluded by Lovett *et al.* (2014) for the first three 15-year crop cycles of *Miscanthus*. Milner *et al.* (2015) give more details of this and suggest that 99.6% of land within these constraints planted with *Miscanthus* following economic scenarios of Alexander *et al.* (2014) would see net gains in SOC between 1.5 and 2.5 Mg C ha^{-1} yr^{-1}, and the slope of Fig. 5 at 1.84 Mg ha^{-1} yr^{-1} fits well within this range. Hillier *et al.* (2009) give a detailed comparison of GHG emissions and SOC changes based on yield predictions for SRC poplar, *Miscanthus*, OSR, and winter wheat with clear overall GHG benefits being seen with *Miscanthus* and SRC on both arable and grassland.

Chemical requirements

Fertilizer

Nutrient offtakes at spring harvest in *Miscanthus* are low, and it is therefore generally unfertilized in commercial production except possibly during establishment where initial soil nutrient status is low. Unnecessary use of nitrogen fertilizer reduces the sustainability of biomass production and GWP offset; therefore, understanding where application is necessary and at what specific rates is important. Cadoux *et al.* (2012) reviewed nutrient offtake in mature *Miscanthus* harvests in 27 studies over 10 countries and found a median content of 4.9 g N (kg DM)$^{-1}$ when harvested in the early spring. Given a typical UK offtake of 10–15 Mg DM ha^{-1} yr^{-1}, the annual export of organic nitrogen from a site in harvest material would range between 49 and 73.5 kg N ha^{-1}. Accounting for an atmospheric N deposition rate of 35–50 kg N ha^{-1} yr^{-1}

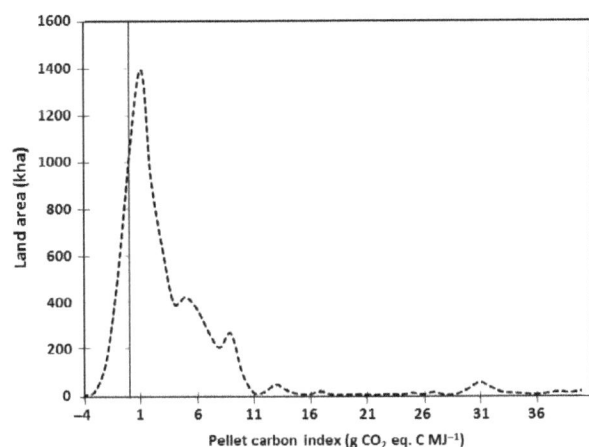

Fig. 7 Carbon intensity of *Miscanthus* pellets produced in the UK outside excluded areas described in Lovett *et al.* (2014) and mapped in Fig. 3. Units are g CO_2-C equivalent per MJ energy in the furnace. X-axis indicates potential area of land that could produce *Miscanthus* at the carbon intensity index indicated on the y-axis.

(Goulding *et al.*, 1998) suggests that *Miscanthus* is unlikely to benefit greatly from inputs of N unless it was being established in very low fertility soils; for example, an optimum application of 100 kg N ha^{-1} was seen to give significant yield benefits on a low-fertility sandy loam soil by Shield *et al.* (2014). Shield *et al.* (2014) emphasized that soil nutrient status prior to establishment was key to determining the need for fertilizer, as are ongoing circumstances. Lewandowski & Schmidt (2006) reported that compared to triticale or reed canary grass, *Miscanthus* showed far higher N-use efficiency and did not respond well to high concentrations of N fertilizer. Maximum yields were observed with no fertilizer but with existing plant available N in the soil (mineralized) at 50 kg N ha^{-1}; higher applications of N fertilization (above 114 kg N ha^{-1} yr^{-1}) were detrimental to crop performance, particularly where soil water was in short supply. Lewandowski *et al.* (2000) reviewed 19 *Miscanthus* field trials across Europe and reported that there was little response to N fertilizer after the second or third year, although there was some suggestion that early rhizome development may benefit from a low level of application where soils may be low in available N to begin with. This very low demand for added fertilizer was reported by Christian *et al.* (2006) who used 15N isotope-enriched nitrogen fertilizer applied at 60 kg N ha^{-1} to study uptake during the establishment phase following planting. Only 23 kg N ha^{-1} of the total 117 kg N ha^{-1} taken up by the developing crop was found to have come from the fertilizer, and 80% had come from mineralization of soil organic matter of the former grassland or atmospheric deposition. There is a growing body of evidence to sug-

gest some level of bacterial nitrogen fixation associated with *Miscanthus* (Davis *et al.*, 2010; Dohleman *et al.*, 2012). Nitrogenase activity has been found in both rhizomes and surrounding soil bacteria (Eckert *et al.*, 2001; Miyamoto *et al.*, 2004) with isotope analysis revealing high levels of biologically fixed nitrogen in *Miscanthus* biomass, particularly in the first year of establishment (Keymer & Kent, 2013). Christian *et al.* (2008) followed their *Miscanthus* crop for 14 years under three application regimes, zero, 60, and 120 kg N ha^{-1} yr^{-1}, and concluded that there was no yield response from the application of the N fertilizer although monitoring of soil fertility and offtake did suggest, in these soils at least, a benefit from additions of phosphate (7 kg P ha^{-1} yr^{-1}) and potassium (100 kg K ha^{-1} yr^{-1}). One trade-off to this low nitrogen requirement is that emissions and leaching may initially rise following planting into highly fertilized land or grassland killed in preparation for conversion (Christian & Riche, 1998; Behnke *et al.*, 2012) as *Miscanthus* is unlikely to utilize all the available nutrients in the establishing year. There may, as mentioned in Heaton *et al.* (2010), be a case made for trials of some suitable cover crop to be planted during the transition to take advantage of these resources.

Pesticide

Despite studies finding some incidence of agricultural disease in *Miscanthus* (Christian *et al.*, 1994; O'Neill & Farr, 1996; Ahonsi *et al.*, 2010), it does not appear to have become a significant problem after more than a decade of commercial growing in the UK, and pesticides are still not generally considered necessary. Lamptey *et al.* (2003) found that while *Miscanthus* (*M. sinensis* in this case) was susceptible to yield losses from infection with Cereal Yellow Dwarf Virus after being inoculated with them in laboratory experiments, it was more resistant than other energy grasses in their study. All were difficult to infect once the plants had got past the seedling stage but of 18 *Miscanthus* plants none were found to become infected when exposed to the virus after stem extension. Even during its susceptible seedling stage infection was only 33% despite deliberate inoculation, compared to almost 100% for *Phalaris arundinacea* and *Echinochloa crus-galli*. Lamptey *et al.* (2003) did warn, though, that conventional rhizome propagation and translocation of *Miscanthus* run the risk of disease transfer between sites; care must be taken that crops sourced for rhizomes are disease free as pesticide control on field crops would be uneconomic and undesirable for sustainable biomass production. The results of such infections were seen in a *Miscanthus* field trial in central Italy (Beccari *et al.*, 2010) where 90% of the transplanted rhizomes failed to establish due to fungal infection by *Fusarium spp.* and *Mucor hiemalis.* Field contamination and improper rhizome storage (high temperature and humidity) were cited as likely factors and *Miscanthus* litter buried in soil have been found previously to contain *Fusarium* spores (Gams *et al.,* 1999). Despite these possible challenges, disease incidence is extremely low with the 14-year production life with no pesticide application of Christian *et al.* (2008) being typical.

Herbicide

Once established, *Miscanthus* competes vigorously with weed species, litter build up below the canopy aids suppression, and the fast closing canopy reduces light available to competitors (Lewandowski *et al.,* 2000; Christian *et al.,* 2008). In the establishing years however, and particularly where grassland is converted, chemical weed control is essential (Jørgensen, 2011). Control is generally accomplished through conventional pre- and postemergent herbicides, sometimes combined with timely application of glyphosate immediately prior to new *Miscanthus* shoot emergence, allowing weed species some time to develop before application. Competition from grassland weeds in this type of land-use transition can be challenging in the early years (Clifton-Brown *et al.,* 2007), and land-use change follows the conventional practice of glyphosate spraying of the existing vegetation, sometimes in two rounds, before ploughing, soil preparation, and planting. Christian *et al.* (2008) offer rare documentation of their complete herbicide history over a 14-year *Miscanthus* study, demonstrating that herbicide weed control was not necessary every year with the bulk of their herbicide mixes applied in years one and four with spring application of glyphosate only in years 4 and 13, and they note the effectiveness of the *Miscanthus* canopy structure and litter layer in natural weed suppression.

Soil nitrous oxide (N₂O) emission

N$_2$O has a global warming potential 298 times greater over 100 years than CO$_2$ (IPCC, 2007), and agriculture is the largest producer of this gas (Williams *et al.,* 2010; Reay *et al.,* 2012). When comparing *Miscanthus* to more usual annual crop rotations, it generally presents lower N$_2$O emission although it is well known that N$_2$O can be particularly challenging to scale from highly variable individual measurements to landscape sums (Rochette & Eriksen-Hamel, 2008; Jones *et al.,* 2011). Drewer *et al.* (2012) compared both *Miscanthus* and SRC willow to arable rotations of wheat and oil seed rape (OSR), reporting that despite this high variability in the results,

the mean N$_2$O flux rates were around five times higher under the annual crops than under the unfertilized perennial bioenergy crops with differences being highly significant. They also investigated the effects of adding fertilizer to the *Miscanthus* and OSR control plots at 50 kg N ha^{-1}. Before application, *Miscanthus* flux rates were around zero compared to OSR at 300 μg N$_2$O-N m^{-2} h^{-1}; emissions began to rise within 24 h of the treatment with the highest flux rate measured after 36 h. *Miscanthus* N$_2$O emission rates rose to a maximum of 330 μg N$_2$O-N m^{-2} h^{-1} compared to 2350 μg N$_2$O-N m^{-2} h^{-1} from the OSR. Emissions declined steadily from there, and after 8 days, no significant difference could be found between the two sites. This transient increase in N$_2$O following fertilization has been reported in several studies, and Gauder *et al.* (2012) measured emissions rising by a factor of four between fertilized and unfertilized *Miscanthus*, although these were still only around 30% of a fertilized maize comparison. Jørgensen *et al.* (1997), however, measured flux rates from fertilized *Miscanthus* at twice that of winter rye during April and November, though still only around 6% of the gross fossil fuel CO$_2$ offset potential they did represent about 1.5% of the mass of N application, exceeding the IPCC tier 1 expectation of 1% (IPCC, 2007) this was corroborated by Behnke *et al.* (2012) who found N$_2$O emissions between 1.1 and 2.4% of applied N. Roth *et al.* (2015) calculated the yield response to fertilizer necessary to offset the GWP of this increased N$_2$O production. They carried out application trials at 63 and 125 kg N ha^{-1} on a 15-year-old crop and concluded that the increased biomass yields they observed did outweigh increased N$_2$O emissions; however, it must be considered that their harvest was from a single year and took place in November where biomass could be in the region of 30% greater than the more usual spring harvest. This might suggest that yield gains are found in increased leaf biomass rather than stem which would not translate into harvested biomass for energy or figure in fossil fuel offset as in commercial practice, leaves are ideally lost over winter.

Roth *et al.* (2013) compared newly established and long-term *Miscanthus* plantations to 18-year-old grassland in Ireland. They found that N$_2$O flux rates from unfertilized *Miscanthus* are similar to unfertilized, ungrazed *Lolium* grassland although they do suggest higher rates during the early establishment period when planted into previous grassland. They calculated cumulative fluxes under newly established *Miscanthus* at 614 g N$_2$O-N ha^{-1} yr^{-1}, less from the established long-term *Miscanthus* at 378 g N$_2$O-N ha^{-1} yr^{-1} and lowest from the grassland at 217 g N$_2$O-N ha^{-1} yr^{-1}. However, being both unfertilized and ungrazed, this grassland is an unrealistic comparison for commercial agriculture. Clayton *et al.* (1997) calculated a figure 10 times higher at 2.94 kg N_2O-N ha^{-1} yr^{-1} for fertilized, ungrazed *Lolium* grassland, while Oenema *et al.* (1997), studying intensively managed, grazed grassland, found emissions rising still further, ranging from 10 to 40 kg N_2O-N ha^{-1} yr^{-1} as the effects of urine, trampling and dung release around three times as much N_2O as mown grassland (Velthof *et al.*, 1996; Rafique *et al.*, 2011).

Carbon intensity in energy production – life cycle assessment

In theory, burning biomass for energy should be carbon neutral as carbon released to the atmosphere was previously fixed from it during photosynthesis. Greenhouse gas benefits lie in reduced fossil fuel use and associated CO_2 emission. In the case of crops or forest managed specifically for bioenergy, there are additional energy inputs and associated GHG costs required for the production process that must be considered. Any anthropological intervention in the process of growing vegetation, changes in land cover, and tillage disturbance, using agrochemicals or altered water balances, leads to changes in the soil's physical and chemical properties. Therefore, the cultivation of feedstock for bioenergy will create some GHG emissions which need to be compared to the land use they replace to estimate the net impact on the atmosphere. The embedded carbon in the machinery and plant manufacture, energy use in cultivation, agrochemicals, transport, and processing/conversion of feedstock into fuel also need to be added to the GHG cost of bioenergy in life cycle analyses (LCA).

Hastings *et al.* (2009) compared *Miscanthus* production to fossil fuels using a metric of g CO_2-C equivalent emissions per MJ of energy at the furnace. For fossil fuels, they included the cost of exploration, production, processing, and delivery to the furnace and for *Miscanthus* biomass; it was plant propagation to the furnace. Their LCA included the impact on soil carbon per ha of land and used crop yields as reported in Hastings *et al.* (2013) to calculate soil input and energy yield. Establishment costs were spread over a 15-year crop lifetime and followed the current practice of rhizome propagation with full tillage and two herbicide applications during establishment. It assumed annual cutting, drying in the field in a swath, high-density bailing and pelleting, and nitrogen fertilizer sufficient to balance the harvest offtake minus N deposition. The IPCC tier 1 N_2O emission factor of 1% of applied N fertilizer was used with the assumption that production emissions are from European producers. This results in an amortized GHG establishment cost of 124 kg C ha^{-1} y^{-1} and

a yield-related annual GHG cost of 57 kg C Mg^{-1} of crop. The results in Fig. 3c show most of the land in the UK could produce *Miscanthus* biomass with a carbon index that is substantially lower, at 1.12 g CO_2-C equivalent per MJ energy in the furnace, than coal (33), oil (22), LNG (21), Russian gas (20), and North Sea gas (16) (Bond *et al.*, 2014), thus offering large potential GHG savings over comparable fuels even after accounting for variations in their specific energy contents. Felten *et al.* (2013) found *Miscanthus* energy production (from propagation to final conversion) to offer far higher potential GHG savings per unit land area when compared to other bioenergy systems. They found *Miscanthus* (chips for domestic heating) saved 22.3 ± 0.13 Mg CO_2-eq ha^{-1} yr^{-1} compared to rapeseed (biodiesel) at 3.2 ± 0.38 and maize (biomass, electricity, and thermal) at 6.3 ± 0.56. Only the low-input *Miscanthus* was found to be effectively a CO_2 sink. Styles & Jones (2007) calculated GHG savings for Miscanthus in Ireland at 35 Mg CO_2-eq ha^{-1} yr^{-1}, while Brandao *et al.* (2011) gave a figure of 11.01 for the UK. Of course these savings are determined by the specific energy source they offset and comparisons do not always account for displaced production. Styles *et al.* (2015) investigated the effects of indirect land-use change, that is considering GHG emissions from the production of food displaced by bioenergy feedstock production. They found only *Miscanthus* and rotational maize offered GHG savings when these indirect land-use change (iLUC) impacts were considered and the percentage of displaced production that was directly replaced determined a threshold. Typically replacing 2–14% for food crops or grassland diverted into anaerobic digestion negated potential GHG savings, whereas it was around 85% for pelletized *Miscanthus*. The GHG benefits for rotational maize were, however, heavily offset by ecosystem service impacts due to intensive production, and of the six bioenergy crop systems investigated, *Miscanthus* was shown to offer the greatest benefits in ecosystem service provision. It was stressed, though, that these positive effects could be localized, consideration needed to be given where production might be displaced to and the impacts of any land-use changes incurred. The importance of understanding indirect land-use change was also highlighted by Tonini *et al.* (2012) who used sensitivity analysis to show that uncertainties around this were significant determinants in LCA results. They compared four conversion pathways (AD, gasification, small-scale CHP, and large-scale cofiring with coal) for ryegrass, willow, and *Miscanthus* and found that when considering their Danish systems, only large-scale cofiring of *Miscanthus* and willow offered real GHG savings compared to fossil fuel alternatives.

Water balance

Water-use efficiency

Miscanthus has higher water-use efficiency (WUE) compared to more conventional C3 crop species, and even some other C4 crops which typically produce more biomass per unit of water transpired (Long, 1983). Beale et al. (1999) investigated WUE in field trials of Miscanthus and another potential C4 biomass crop, Spartina cynosuroides, under both rain-fed and irrigated conditions; they estimated the ratio of aboveground biomass to water use for Miscanthus under rain-fed conditions at 9.2 g DM $(kg H_2O)^{-1}$ compared to 6.8 g DM $(kg H_2O)^{-1}$ for S. cynosuroides. Both crops appeared to become less efficient under irrigation, down by 15% for Miscanthus to 7.8 g DM $(kg H_2O)^{-1}$ and 25% for S. cynosuroides to 5.1 g DM $(kg H_2O)^{-1}$, possibly reflecting greater canopy evaporation under the irrigation regime. Beale et al. (1999) compared their results to the water-use efficiency of a C3 biomass crop, Salix viminalis, reported in Lindroth et al. (1994) and Lindroth & Cienciala (1996), and suggest that WUE for Miscanthus could be around twice that of this willow species. Clifton-Brown & Lewondowski (2000) reported figures from 11.5 to 14.2 g total (above- and belowground) DM $(kg H_2O)^{-1}$ for various Miscanthus genotypes in pot trials, and this compares to figures calculated by Ehdaie & Waines (1993) with seven wheat cultivars who found WUE between 2.67 and 3.95 g total DM $(kg H_2O)^{-1}.$ Converting these Miscanthus values to dry matter biomass per hectare of cropland would see ratios of biomass to water use in the range of to 78–92 kg DM ha^{-1} $(mm H_2O)^{-1}$. Richter et al. (2008) modelled harvestable yield potentials for Miscanthus from 14 UK field trials and found soil water available to plants was the most significant factor in yield prediction, and they calculated a DM yield to soil available water ratio at 55 kg DM ha^{-1} $(mm H_2O)^{-1}$, while just 13 kg DM ha^{-1} was produced for each 1 mm of incoming precipitation, likely related to the high level of canopy interception and evaporation. Even by C4 standards these efficiencies are high, as seen in comparisons to field measurements averaging 27.5 ± 0.4 kg aboveground DM ha^{-1} $(mm H_2O)^{-1}$ for maize (Tolk et al., 1998).

Soil water balance

However, despite impressive efficiency figures, accumulating biomass at the rapid rate that makes Miscanthus interesting as an energy crop will inevitably lead to increased demand for water and consideration needs to be given to water availability when locating plantations (Vanloocke et al., 2010). When Yaeger et al. (2013) compared Miscanthus to corn and soya bean grown in the American mid-West, they found that Miscanthus had effectively a 2-month longer season of transpiration which meant a reduction in soil water reserves during low rainfall. This reduction can be exacerbated by the dense canopy of Miscanthus which intercepts more rain allowing more evaporation at the leaf level and less throughfall to the soil compared to some other crops. Stephens et al. (2001) modelled reductions in hydrologically effective rainfall (HER), that is rainfall that becomes incorporated into the soil under Miscanthus and willow compared to permanent grass and winter wheat. Results showed reductions in HER under Miscanthus were lower than those for willow SRC but still large at between 100 and 120 mm yr^{-1}; using their estimate for 0.10 Mha of energy crops reducing HER by 150 mm (average across both crops was 140–180 mm yr^{-1}) meant 0.35 Mha would reduce rainfall reaching the soil by 0.7% of total UK rainfall. Blanco-Canqui (2010) point out that this water-use and nutrient efficiency can be a boon on compacted, poorly drained acid soils, highlighting their possible suitability for marginal agricultural land. The greater porosity and lower bulk density of soils under perennial energy grasses, resulting from more fibrous, extensive rooting systems, and reduced ground disturbance, improves soil hydraulic properties, infiltration, hydraulic conductivity, and water storage compared to annual row crops. There may be potentially large impacts on soil water where plantation size is mismatched to water catchment or irrigation availability but note that increased ET and improved ground water storage through increased porosity could be beneficial during high rainfall with storage capability potentially increased by 100 to 150 mm. There is also a benefit of reduced chemical inputs and nitrate leaching associated with Miscanthus, significantly improving water quality running off farmland (Christian & Riche, 1998; Curley et al., 2009). McIsaac et al. (2010) reported that inorganic N leaching was significantly lower under unfertilized Miscanthus (1.5–6.6 kg N ha^{-1} yr^{-1}) than a maize/soya bean rotation (34.2–45.9 kg N ha^{-1} yr^{-1}). They also reported that soils under Miscanthus were drier and calculated increased evapotranspiration from the Miscanthus at 104 mm yr^{-1}. Finch et al. (2004) studied UK energy grasses and willow SRC and compared them to existing grassland and arable. They found that in years of sufficient rainfall, Miscanthus is likely to use less or the same water as existing agricultural land. In drought years, although Miscanthus was likely to impact more on soil water deficits due to increased interception and rooting depth, this could lead to reduced groundwater recharge rates in drier years or reduced winter run-off in wetter conditions.

Biodiversity

As *Miscanthus* is an agricultural crop, its place in an agricultural landscape of fields, margins, and farm woodland should be considered in terms of its potential to increase or decrease resources for wildlife in land-use transitions. How does it compare to existing land use or other potential energy crops? Felten & Emmerling (2011) compared earthworm abundance under a 15-year-old *Miscanthus* plantation in Germany to cereals, maize, OSR, grassland, and a 20-year-old fallow site (after previous cereals). Species diversity was higher in *Miscanthus* than that in annual crops, more in line with grassland or long-term fallow with management intensity seen to be the most significant factor; the lower ground disturbance allowed earthworms from different ecological categories to develop a more heterogeneous soil structure. The highest number of species was found in the grassland sites (6.8) followed by fallow (6.4), *Miscanthus* (5.1), OSR (4.0), cereals (3.7), and maize (3.0) with total individual earthworm abundance ranging from 62 m^{-2} in maize sites to 355 m^{-2} in fallow with *Miscanthus* taking a medium position (132 m^{-2}), although differences in abundance were not found to be significant between land uses. There is some trade-off in this advantage for the earthworms however; the high-nitrogen-use efficiency and nutrient cycling which reduces the need for nitrogen fertilizer and its associated environmental harm means that, despite large volumes being available, *Miscanthus* leaf litter does not provide a particularly useful food resource due to its low-nitrogen, high-carbon nature (Ernst *et al.*, 2009; Heaton *et al.*, 2009) and earthworms feeding on this kind of low-nitrogen material have been found in other studies to lose overall mass (Abbott & Parker, 1981). In contrast, though, the extensive litter cover at ground level under *Miscanthus* compared to the bare soil under annual cereals was suggested to be a potentially significant advantage for earthworms in soil surface moisture retention and protection from predation.

Semere & Slater (2007a,b) sampled an exhaustive range of aboveground indicator species at five sites in Herefordshire, UK. They compared results between *Miscanthus*, reed canary grass, and switchgrass and found *Miscanthus* to contain high levels of diversity in comparison with the other energy grasses; particularly evident in terms of beetles, flies, and birds, with breeding skylarks and lapwings being recorded in the crop itself. It was pointed out by the authors, however, that although the overwinter vegetative structure provided an important cover and habitat resource, it was the noncrop weed species in and around the field sites that underpinned the food webs supporting the bird species. This link between crop density, weed content, and food resources for birds was again demonstrated by Dauber *et al.* (2015)

who recorded the abundances of ground fauna, beetles, spiders, etc., at 14 mature *Miscanthus* sites in SE Ireland. They found light penetration through the canopy directly related to within crop biodiversity, with *Miscanthus* planted on previous grassland showing higher levels of biodiversity compared to that planted on former arable.

This trade-off between crop success and within-crop biodiversity is to be expected. For an economic return, the most efficient capture of light by the crop canopy will inevitably reduce noncrop weed resources for other species. However, *Miscanthus* offers environmental benefits in structural heterogeneity, low chemical inputs, and overwinter ripening providing near continuous cover. Particularly in a landscape of high-input arable production, *Miscanthus* has the potential to offer a 10- to 15-year break crop allowing the soil and surrounding field margins time to recover from intensive production. Bellamy *et al.* (2009) looked at bird species and their food resources at six paired sites in Cambridgeshire comparing *Miscanthus* plantations up to 5 years old with winter wheat rotations in both the winter and summer breeding seasons. The authors found that *Miscanthus* offered a different ecological niche during each season; most of the frequently occurring species in the winter were woodland birds, whereas no woodland birds were found in the wheat; in summer, however, farmland birds were more numerous. More than half the species occurring across the sites were more numerous in the *Miscanthus*, 24 species recorded compared to 11 for wheat. During the breeding season, there was once again double the number of species found at the *Miscanthus* sites with individual abundances being higher for all species except skylark. Considering only birds whose breeding territories were either wholly or partially within crop boundaries, a total of seven species were found in the *Miscanthus* compared to five in the wheat with greater density of breeding pairs (1.8 vs. 0.59 species ha^{-1}) and also breeding species (0.92 vs. 0.28 species ha^{-1}). Two species were at statistically significant higher densities in the *Miscanthus* compared to wheat, and none were found at higher densities in the wheat compared to *Miscanthus*. As discussed, the structural heterogeneity, both spatially and temporally, plays an important role in determining within-crop biodiversity, autumn-sown winter wheat offers little overwinter shelter with ground cover averaging 0.08 m tall and very few noncrop plants, whereas the *Miscanthus*, at around 2 m, offered far more. In the breeding season, this difference between the crops remained evident; the wheat fields provided a uniform, dense crop cover throughout the breeding season with only tram lines producing breaks, whereas the *Miscanthus* had a low open structure early in the season rapidly increasing in height and density as the season progressed. Numbers of birds declined as the crop grew with two bird species in particular showing close (though opposite) correlation between abundance and crop height; red-legged partridge declined as the crop grew, whereas reed warblers increased, and these warblers were not found in the crop until it had passed 1 m in height, even though they were present in neighbouring OSR fields and vegetated ditches. In conclusion, the authors point out that, for all species combined, bird densities in *Miscanthus* were similar to those found in other studies looking at SRC willow and set-aside fields, all sites had greater bird densities than conventional arable crops.

It is through these added resources to an intensive agricultural landscape and reductions in chemical and mechanical pressure on field margins that *Miscanthus* can play an important role in supporting biodiversity but must be considered complementary to existing systems and the wildlife that has adapted to it. Clapham *et al.* (2008) reports, as do the other studies here, that in an agricultural landscape, it is in the field margins and interspersed woodland that the majority of the wildlife and their food resources are to be found, and the important role that *Miscanthus* can play in this landscape is the cessation of chemical leaching into these key habitats, the removal of annual ground disturbance and soil erosion, improved water quality, and the provision of heterogeneous structure and overwinter cover.

Summary statements

Based on the literature evidence reviewed above, we present here a number of summary statements addressing concerns and questions around the environmental sustainability of *Miscanthus* production in the UK.

Potential UK land availability

- By planting in appropriate locations, government targets of 0.35 Mha of dedicated energy crops could be sustainably met by *Miscanthus* production without impacting essential food production.
- 0.35 Mha (2007 Biomass Strategy) would provide the energy equivalent to 1.67 Mt of oil each year (1.17% of total energy).
- 0.72–2.8 Mha (2012 Bioenergy Strategy) would provide the equivalent of 3.44–13.38 Mt of oil yr^{-1} (2.41–9.39% of total energy).

Soil carbon

- Former arable land converted to *Miscanthus* is most likely to lead to no change or an accumula-

tion of soil organic carbon (SOC), becoming comparable to an agricultural grassland within the lifetime of the crop. *Miscanthus* contributes 0.98 ± 0.14 Mg C4-C ha^{-1} yr^{-1} through litter drop and root turnover.
- Converting semi-permanent grassland to *Miscanthus* by traditional establishment (spraying, ploughing, tilling, and planting) results in an initial short-term soil carbon loss which is recovered as the crop matures.

Chemical inputs

- Nitrogen fertilizer is unnecessary and can be detrimental to sustainability, unless planted into low-fertility soils where early establishment will benefit from additions of around 50 kg N ha^{-1}.
- Early season herbicide application for weed control is essential in the establishing years but becomes redundant as the crop matures, other pesticides are not needed.

GHG cost of energy production

- When considering the entire energy supply chain burning, *Miscanthus* produces far less GHG per MJ of energy than fossil fuels; 1.12 g CO_2-C eq. compared to coal (33), oil (22), gas (16–22).

Nitrous oxide (N_2O) emission

- N_2O emissions can be five times lower under unfertilized *Miscanthus* than annual crops, and up to 100 times lower than intensive pasture land.
- Inappropriate nitrogen fertilizer additions can result in significant increases in N_2O emission from *Miscanthus* plantations, exceeding IPCC emission factors although these are still offset by potential fossil fuel replacement.

Water balance

- Water-use efficiency is among the highest of any crop, in the range of 7.8–9.2 g DM (kg $H_2O)^{-1}$.
- Overall, water demand will increase due to high biomass productivity and increased evapotranspiration at the canopy level (e.g. ET up from wheat by 100–120 mm yr^{-1}).
- Improved soil structures mean greater water-holding capacity (e.g. up by 100–150 mm), although soils may still be drier in drought years.

- Reduced run-off in wetter years, aiding flood mitigation and reducing soil erosion.
- Drainage water quality is improved, and nitrate leaching is significantly lower than arable (e.g. 1.5–6.6 kg N ha^{-1} yr^{-1} Miscanthus, 34.2–45.9 maize/soya bean).

Biodiversity

- *Miscanthus* adds structural resources to agricultural landscapes, provides overwinter cover, and increases temporal variability which is accessed by different bird species in different seasons.
- Earthworm diversity and abundance is improved in arable soils and comparable to grassland soils although biomass may be reduced through poorer food quality.
- Reduced chemical inputs improve headland and field boundary quality for wildlife.
- Unpalatability of leaf litter and harvest residue removes the need for pest control but trade-off food resources for invertebrates within the cropped area are only provided by interspersed weed species are limited to weed species.

Conclusion

This study distils a large body of literature into simple statements around the environmental costs and benefits of producing *Miscanthus* in the UK, and while there is scope for further research, particularly around hydrology at a commercial scale, biodiversity in older plantations or higher frequency sampling for N$_2$O in land-use transitions to and from *Miscanthus*, clear indications of environmental sustainability do emerge. Any agricultural production is primarily based on human demand, and there will always be a trade-off between nature and humanity or one benefit and another; however, the literature suggests that *Miscanthus* can provide a range of benefits while minimizing environmental harm. Consideration must be given to appropriateness of plantation size and location, whether there will be enough water to sustain its production and the environmental cost of transportation to end-users; its role as a long-term perennial crop in a landscape of rotational agriculture must be understood so as not to interfere with essential food production. There is nothing new in these considerations, they lie at the heart of any agricultural policy, and decision-makers are familiar with these issues; the environmental evidence gathered here will help provide the scientific basis to underpin future agricultural policy. It is only through an understanding at the government level that uptake of *Miscanthus* will be able to fulfil its potential in the UK bioenergy sector. Despite clear environmental benefits and developing supply chains, uptake of *Miscanthus* production remains low among UK farmers and there is much inertia to overcome. Financial considerations are paramount in farmers' willingness to adopt novel crops and production practices and uncertainty around grant funding, establishment costs, potential yields, and sale price limits confidence (Sherrington et al., 2008; Adams et al., 2011). There is, however, growing awareness of the bigger picture of environmental stewardship and climate change mitigation (Glithero et al., 2013) and farmers stress the need for clear, unbiased information on all aspects of bioenergy; from the entire cycle of crop management and marketing to end use, biomass boilers for on-farm use and local energy supply. There is a problem of 'chicken and egg' in developing these markets; farmers need the incentive of a mature market to sell into to encourage their uptake of these crops, whereas energy producers need a large-scale, secure, predictable supply of biomass to invest in the technologies to utilize them. Without top-down intervention, and policy stability, it will be difficult for a 'critical mass' of growers to develop to provide confidence in energy crop supply.

Acknowledgements

This study was supported by the BBSRC (Sparking Impact Award and GIANT-LINK project LK0863), by Natural Environment Research Council (NERC) as part of the Carbo-BioCrop project (NE/H01067X/1) and by the MAGLUE project. Many thanks also to Jonathan Scurlock of the National Farmers' Union (NFU) and Jeremy Woods of Imperial College, London, for advice and feedback.

References

Abbott I, Parker C (1981) Interactions between earthworms and their soil environment. *Soil Biology and Biochemistry*, **13**, 191–197.

Adams PW, Hammond GP, McManus MC, Mezzullo WG (2011) Barriers to and drivers for UK bioenergy development. *Renewable and Sustainable Energy Reviews*, **15**, 1217–1227.

Agostini F, Gregory AS, Richter GM (2015) Carbon sequestration by perennial energy crops: is the jury still out? *BioEnergy Research*, 1–24.

Ahonsi MO, Agindotan BO, Williams DW, Arundale R, Gray ME, Voigt TB, Bradley CA (2010) First report of *Pithomyces chartarum* causing a leaf blight of *Miscanthus* × *giganteus* in Kentucky. *Plant Disease*, **94**, 480–480.

Alexander P, Moran D, Smith P et al. (2014) Estimating UK perennial energy crop supply using farm-scale models with spatially disaggregated data. *GCB-Bioenergy*, **6**, 142–155.

Anderson-Teixeira KJ, Davis SC, Masters MD, Delucia EH (2009) Changes in soil organic carbon under biofuel crops. *GCB Bioenergy*, **1**, 75–96.

Balesdent J, Mariotti A, Guillet B (1987) Natural 13 C abundance as a tracer for studies of soil organic matter dynamics. *Soil Biology and Biochemistry*, **19**, 25–30.

Bauen AW, Dunnett AJ, Richter GM, Dailey AG, Aylott M, Casella E, Taylor G (2010) Modelling supply and demand of bioenergy from short rotation coppice and *Miscanthus* in the UK. *Bioresource Technology*, **101**, 8132–8143.

Beale CV, Morison JI, Long SP (1999) Water use efficiency of C4 perennial grasses in a temperate climate. *Agricultural and Forest Meteorology*, **96**, 103–115.

Beccari G, Covarelli L, Balmas V, Tosi L (2010) First report of *Miscanthus* × giganteus rhizome rot caused by *Fusarium avenaceum*, *Fusarium oxysporum* and *Mucor hiemalis*. *Australasian Plant Disease Notes*, **5**, 28–29.

Behnke GD, David MB, Voigt TB (2012) Greenhouse gas emissions, nitrate leaching, and biomass yields from production of *Miscanthus* × giganteus in Illinois, USA. *BioEnergy Research*, **5**, 801–813.

Bellamy PE, Croxton PJ, Heard MS et al. (2009) The impact of growing *Miscanthus* for biomass on farmland bird populations. *Biomass and Bioenergy*, **33**, 191–199.

Blanco-Canqui H (2010) Energy crops and their implications on soil and environment. *Agronomy Journal*, **102**, 403–419.

Bond CE, Roberts J, Hastings AFSJ, Shipton Z, Joao E, Tabyldy K, Stephenson M (2014) ClimateXChange. Available at: http://www.climatexchange.org.uk/reducing-emissions/life-cycle-assessment-ghg-emissions-unconventional-gas1/ (accessed 2 February 2015).

Bosatta E, Ågren GI (1985) Theoretical analysis of decomposition of heterogeneous substrates. *Soil Biology and Biochemistry*, **17**, 601–610.

Brandao M, Milà i Canals L, Clift R (2011) Soil organic carbon changes in the cultivation of energy crops: Implications for GHG balances and soil quality for use in LCA. *Biomass and Bioenergy*, **35**, 2323–2336.

Cadoux S, Riche AB, Yates Nicola E, Machet JM (2012) Nutrient requirements of *Miscanthus* × giganteus: conclusions from a review of published studies. *Biomass and Bioenergy*, **38**, 14–22.

Cheng W (2009) Rhizosphere priming effect: its functional relationships with microbial turnover, evapotranspiration, and C-N budgets. *Soil Biology and Biochemistry*, **41**, 1795–1801.

Cherubini F, Bird ND, Cowie A, Jungmeier G, Schlamadinger B, Woess-Gallasch S (2009) Energy-and greenhouse gas-based LCA of biofuel and bioenergy systems: key issues, ranges and recommendations. *Resources, Conservation and Recycling*, **53**, 434–447.

Christian D, Riche A (1998) Nitrate leaching losses under *Miscanthus* grass planted on a silty clay loam soil. *Soil Use and management*, **14**, 131–135.

Christian DG, Lamptey JNL, Forde SMD, Plumb RT (1994) First report of barley yellow dwarf luteovirus on *Miscanthus* in the United Kingdom. *European Journal of Plant Pathology*, **100**, 167–170.

Christian DG, Poulton PR, Riche AB, Yates NE, Todd AD (2006) The recovery over several seasons of 15N-labelled fertilizer applied to *Miscanthus* × giganteus ranging from 1 to 3 years old. *Biomass and Bioenergy*, **30**, 125–133.

Christian D, Riche A, Yates N (2008) Growth, yield and mineral content of *Miscanthus* × giganteus grown as a biofuel for 14 successive harvests. *Industrial Crops and Products*, **28**, 320–327.

Clapham SJ, Slater FM, Booth E et al. (2008) The biodiversity of established biomass grass crops. *Aspects of Applied Biology*, **90**, 325–329.

Clayton H, McTaggart IP, Parker J, Swan L, Smith KA (1997) Nitrous oxide emissions from fertilised grassland: a 2-year study of the effects of N fertiliser form and environmental conditions. *Biology and Fertility of Soils*, **25**, 252–260.

Clifton-Brown JC, Lewondowski I (2000) Water use efficiency and biomass partitioning of three different *Miscanthus* genotypes with limited and unlimited water supply. *Annals of Botany*, **86**, 191–200.

Clifton-Brown JC, Stampfl PF, Jones MB (2004) *Miscanthus* biomass production for energy in Europe and its potential contribution to decreasing fossil fuel carbon emissions. *Global Change Biology*, **10**, 509–518.

Clifton-Brown JC, Breuer J, Jones MB (2007) Carbon mitigation by the energy crop, *Miscanthus*. *Global Change Biology*, **13**, 2296–2307.

Cox PM, Betts RA, Jones CD, Spall SA, Totterdell IJ (2011) Acceleration of global warming due to carbon-cycle feedbacks in a coupled climate model. *Nature*, **408**, 331.

Crutzen PJ, Mosier AR, Smith KA, Winiwarter W (2008) N 2 O release from agro-biofuel production negates global warming reduction by replacing fossil fuels. *Atmospheric Chemistry and Physics*, **8**, 389–395.

Curley EM, O'Flynn MG, McDonnell KP et al. (2009) Nitrate leaching losses from *Miscanthus* × giganteus impact on groundwater quality. *Journal of Agronomy*, **8**, 107–112.

Dauber J, Cass S, Gabriel D, Harte K, Astrom S, O'Rourke E, Stout JC (2015) Yield-biodiversity trade-off in patchy fields of *Miscanthus* × giganteus. *GCB Bioenergy*, **7**, 455–467.

Davis SC, Parton WJ, Dohleman FG, Smith CM, Del Grosso S, Kent AD, DeLucia EH (2010) Comparative biogeochemical cycles of bioenergy crops reveal nitrogen-fixation and low greenhouse gas emissions in a *Miscanthus* × giganteus agro-ecosystem. *Ecosystems*, **13**, 144–156.

DEFRA (2014a) UK non-food statistical summary. Available at: https://www.gov.uk/government/uploads/system/uploads/attachment_data/file/377944/non-food-statsnotice2012-25nov14.pdf (accessed 27 January 2015).

DEFRA (2014b) Change in the area and distribution of set-aside in the England and its environmental impact. Available at: http://archive.defra.gov.uk/evidence/statistics/foodfarm/enviro/observatory/research/documents/observatory08.pdf (accessed 27 January 2015).

DEFRA (2014c) Agriculture in the United Kingdom 2013. Available at: https://www.gov.uk/government/uploads/system/uploads/attachment_data/file/315103/auk-2013-29may14.pdf (accessed 27 January 2015).

DEFRA (2014d) Farming statistics – final crop areas, yields, livestock populations and agricultural workforce at 1 June 2013 – UK. Available at: https://www.gov.uk/government/statistics/farming-statistics-final-crop-areas-yields-livestock-populations-and-agricultural-workforce-at-1-June-2013-uk (accessed 27 January 2015).

Department of Energy and Climate Change (DECC) (2012) UK bioenergy strategy. Available at: http://www.gov.uk/government/uploads/system/uploads/attachment_data/file/48337/5142-bioenergy-strategy-.pdf (accessed 27 January 2015).

Department of Food and Rural Affairs (DEFRA) (2007) UK biomass strategy. Available at: http://www.biomassenergycentre.org.uk/pls/portal/docs/PAGE/RESOURCES/REF_LIB_RES/PUBLICATIONS/UKBIOMASSSTRATEGY.PDF (accessed 27 January 2015).

Dohleman FG, Heaton EA, Arundale RA, Long SP (2012) Seasonal dynamics of above-and below-ground biomass and nitrogen partitioning in *Miscanthus* × giganteus and *Panicum virgatum* across three growing seasons. *GCB Bioenergy*, **4**, 534–544.

Don A, Osborne B, Hastings A et al. (2012) Land-use change to bioenergy production in Europe: implications for the greenhouse gas balance and soil carbon. *GCB Bioenergy*, **4**, 372–391.

Dondini M, Hastings A, Saiz G, Jones MB, Smith P (2009a) The potential of *Miscanthus* to sequester carbon in soils: comparing field measurements in Carlow, Ireland to model predictions. *GCB Bioenergy*, **1**, 413–425.

Dondini M, Van Groenigen K-J, Del Galdo I, Jones MB (2009b) Carbon sequestration under *Miscanthus*: a study of 13C distribution in soil aggregates. *GCB Bioenergy*, **1**, 321–330.

Drewer J, Finch JW, Lloyd CR, Baggs EM, Skiba U (2012) How do soil emissions of N_2O, CH_4 and CO_2 from perennial bioenergy crops differ from arable annual crops? *GCB Bioenergy*, **4**, 408–419.

DUKES (2014a) Digest of UK energy statistics. Available at: https://www.gov.uk/government/collections/digest-of-uk-energy-statistics-dukes (accessed 27 January 2015).

DUKES (2014b) Digest of UK energy statistics, online summary figures. Available at: https://www.gov.uk/government/uploads/system/uploads/attachment_data/file/338768/DUKES_2014_internet_content.pdf (accessed 27 January 2015).

Eckert B, Weber OB, Kirchhof G et al. (2001) Azospirillum doebereinerae sp. nov., a nitrogen-fixing bacterium associated with the C4-grass *Miscanthus*. *International Journal of Systematic and Evolutionary Microbiology*, **51**, 17–26.

Ehdaie B, Waines JG (1993) Variation in water-use efficiency and its components in wheat: I. Well-watered pot experiment. *Crop Science*, **33**, 294–299.

Ernst G, Henseler I, Felten D, Emmerling C (2009) Decomposition and mineralization of energy crop residues governed by earthworms. *Soil Biology and Biochemistry*, **41**, 1548–1554.

ETI (2015) Bioenergy Insights into the future UK Bioenergy Sector, gained using the ETI's Bioenergy Value Chain Model (BVCM). Available at: http://www.eti.co.uk/wp-content/uploads/2015/03/Bioenergy-Insights-into-the-future-UK-Bioenergy-Sector-gained-using-the-ETIs-Bioenergy-Value-Chain-Model.pdf (accessed 14 April 2015).

Felten D, Emmerling C (2011) Effects of bioenergy crop cultivation on earthworm communities—a comparative study of perennial *Miscanthus* and annual crops with consideration of graded land-use intensity. *Applied Soil Ecology*, **49**, 167–177.

Felten D, Emmerling C (2012) Accumulation of *Miscanthus*-derived carbon in soils in relation to soil depth and duration of land use under commercial farming conditions. *Journal of Plant Nutrition and Soil Science*, **175**, 661–670.

Felten D, Fröba N, Fries J, Emmerling C (2013) Energy balances and greenhouse gas-mitigation potentials of bioenergy cropping systems (*Miscanthus*, rapeseed, and maize) based on farming conditions in Western Germany. *Renewable Energy*, **55**, 160–174.

Finch JW, Hall RL, Rosier PTW et al. (2004) *The Hydrological Impacts of Energy Crop Production in the UK-Final Report*. Department of Trade and Industry, London, 151.

Gams W, Klamer M, O'Donnell K (1999) Fusarium miscanthi sp. nov. from *Miscanthus* litter. *Mycologia*, **91**, 263–268.

Gauder M, Butterbach-Bahl K, Graeff-Hönninger S, Claupein W, Wiegel R (2012) Soil-derived trace gas fluxes from different energy crops-results from a field experiment in Southwest Germany. *GCB Bioenergy*, **4**, 289–301.

Gelfand I, Zenone T, Jasrotia P, Chen J, Hamilton SK, Robertson GP (2011) Carbon debt of Conservation Reserve Program (CRP) grasslands converted to bioenergy production. *Proceedings of the National Academy of Sciences of the United States of America*, **108**, 13864–13869.

Gifford RM, Roderick ML (2003) Soil carbon stocks and bulk density: spatial or cumulative mass coordinates as a basis of expression? *Global Change Biology*, **9**, 1507–1514.

Glithero NJ, Wilson P, Ramsden SJ (2013) Prospects for arable farm uptake of Short Rotation Coppice willow and miscanthus in England. *Applied Energy*, **107**, 209–218.

Godfray HCJ, Garnett T (2014) Food security and sustainable intensification. *Philosophical Transactions of the Royal Society B: Biological Sciences*, **369**, 20120273.

Goulding KWT, Bailey NJ, Bradbury NJ *et al.* (1998) Nitrogen deposition and its contribution to nitrogen cycling and associated soil processes. *New Phytologist*, **139**, 49–58.

Hansen EM, Christensen BT, Jensen LS, Kristensen K (2004) Carbon sequestration in soil beneath long-term *Miscanthus* plantations as determined by 13C abundance. *Biomass and Bioenergy*, **26**, 97–105.

Harvey J (2007) A versatile solution-growing *Miscanthus* for bioenergy. *Renewable Energy World*, **10**, 86.

Hastings A, Clifton-Brown J, Wattenbach M, Stampfl P, Paul Mitchell P, Pete Smith P (2008) Potential of *Miscanthus* grasses to provide energy and hence reduce greenhouse gas emissions. *Agronomy for Sustainable Development*, **28**, 465–472.

Hastings A, Clifton-Brown J, Wattenbach M, Mitchell CP, Stampfl P, Smith P (2009) Future energy potential of Miscanthus in Europe. *GCB-Bioenergy*, **1**, 180–196.

Hastings A, Yeluripati J, Hillier J, Smith P (2012) Chapter 12 – Biofuel crops and greenhouse gasses. In: *Biofuel Crop Sustainability* (ed. Singh BP), pp. 383–406. Wiley Blackwell, Danvers, MA. ISBN 978-0-470-96304-3

Hastings A, Tallis M, Casella E, Matthews R, Milner S, Smith P, Taylor G (2013) The technical potential of Great Britain to produce lingo-cellulosic biomass for bioenergy in current and future climates. *Global Change Biology – Bioenergy*, **6**, 108–122.

Heaton EA, Dohleman FG, Long SP (2008) Meeting US biofuel goals with less land: the potential of *Miscanthus*. *Global Change Biology*, **14**, 2000–2014.

Heaton EA, Dohleman FG, Long SP (2009) Seasonal nitrogen dynamics of *Miscanthus × giganteus* and *Panicum virgatum*. *GCB Bioenergy*, **1**, 297–307.

Heaton EA, Dohleman FG, Fernando Miguez A *et al.* (2010) *Miscanthus*: a promising biomass crop. *Advances in Botanical Research*, **56**, 75–136.

Hillier J, Whittaker C, Dailey G *et al.* (2009) Greenhouse gas emissions from four bioenergy crops in England and Wales: integrating spatial estimates of yield and soil carbon balance in life cycle analyses. *GCB Bioenergy*, **1**, 267–281.

Intergovernmental Panel on Climate Change (IPCC) (2007) *Climate Change 2007: The Physical Science Basis, Contribution of Working Group I, II and III to the Fourth Assessment Report of the IPCC* Chapter 2 (eds Solomon S, Qin D, Manning M *et al.*), pp. 1–996. Cambridge University Press, Cambridge.

Intergovernmental Panel on Climate Change (IPCC) (2014) Summary for policymakers. In: *Climate Change 2014: Impacts, Adaptation, and Vulnerability. Part A: Global and Sectoral Aspects. Contribution of Working Group II to the Fifth Assessment Report of the Intergovernmental Panel on Climate Change* (eds Field CB, Barros VR, Dokken DJ, Mach KJ, Mastrandrea MD, Bilir TE, Chatterjee M, Ebi KL, Estrada YO, Genova RC, Girma B, Kissel ES, Levy AN, MacCracken S, Mastrandrea PR, White LL), pp. 1–32. Cambridge University Press, Cambridge, UK and New York, NY.

Jones SK, Famulari D, Di Marco CF, Nemitz E, Skiba UM, Rees RM, Sutton MA (2011) Nitrous oxide emissions from managed grassland: a comparison of eddy covariance and static chamber measurements. *Atmospheric Measurement Techniques*, **4**, 2179–2194.

Jørgensen U (2011) Benefits versus risks of growing biofuel crops: the case of *Miscanthus*. *Current Opinion in Environmental Sustainability*, **3**, 24–30.

Jørgensen RN, Jørgensen BJ, Nielsen NE, Maag M, Lind AM (1997) N₂O emission from energy crop fields of *Miscanthus × Giganteus* and winter rye. *Atmospheric Environment*, **31**, 2899–2904.

Kahle P, Beuch S, Boelcke B, Leinweber P, Schulten HR (2001) Cropping of *Miscanthus* in Central Europe: biomass production and influence on nutrients and soil organic matter. *European Journal of Agronomy*, **15**, 171–184.

Keymer DP, Kent AD (2013) Contribution of nitrogen fixation to first year *Miscanthus × giganteus*. *GCB Bioenergy*, **6**, 577–586.

Kim S, Dale BE (2005) Life cycle assessment of various cropping systems utilized for producing biofuels: bioethanol and biodiesel. *Biomass and Bioenergy*, **29**, 426–439.

Kravchenko A, Robertson G (2011) Whole-profile soil carbon stocks: the danger of assuming too much from analyses of too little. *Soil Science Society of America Journal*, **75**, 235–240.

Kuzyakov Y (2010) Priming effects: interactions between living and dead organic matter. *Soil Biology and Biochemistry*, **42**, 1363–1371.

Lal R (2004) Soil carbon sequestration to mitigate climate change. *Geoderma*, **123**, 1–22.

Lamptey J, Plumb R, Shaw M (2003) Interactions between the Grasses Phalaris arundinacea, *Miscanthus* sinensis and Echinochloa crus-galli, and Barley and Cereal Yellow Dwarf Viruses. *Journal of Phytopathology*, **151**, 463–468.

Lange M (2011) The GHG balance of biofuels taking into account land use change. *Energy Policy*, **39**, 2373–2385.

Lewandowski I, Schmidt U (2006) Nitrogen, energy and land use efficiencies of *Miscanthus*, reed canary grass and triticale as determined by the boundary line approach. *Agriculture, Ecosystems & Environment*, **112**, 335–346.

Lewandowski I, Clifton-Brown JC, Scurlock JMO, Huisman W (2000) *Miscanthus*: European experience with a novel energy crop. *Biomass and Bioenergy*, **19**, 209–227.

Lewandowski I, Scurlock JMO, Lindvall E, Christou M (2003) The development and current status of perennial rhizomatous grasses as energy crops in the US and Europe. *Biomass and Bioenergy*, **25**, 335–361.

Lindroth A, Cienciala E (1996) Water use efficiency of short- rotation *Salix viminalis* at leaf, tree and stand scales. *Tree Physiology*, **16**, 257–262.

Lindroth A, Verwijst T, Halldin S (1994) Water-use efficiency of willow: variation with season, humidity, and biomass allocation. *Hydrology*, **156**, 1–19.

Long S (1983) C4 photosynthesis at low temperatures. *Plant Cell & Environment*, **6**, 345–363.

Lovett AA, Sünnenberg GM, Richter GM, Dailey AG, Riche AB, Karp A (2009) Land use implications of increased biomass production identified by GIS-based suitability and yield mapping for *Miscanthus* in England. *Bioenergy Research*, **2**, 17–28.

Lovett A, Sünnenberg G, Dockerty T (2014) The availability of land for perennial energy crops in Great Britain. *GCB Bioenergy*, **6**, 99–107.

Mackey B, Prentice IC, Steffen W *et al.* (2013) Untangling the confusion around land carbon science and climate change mitigation policy. *Nature Climate Change*, **3**, 552–557.

McIsaac GF, David MB, Mitchell CA (2010) *Miscanthus* and switchgrass production in central Illinois: impacts on hydrology and inorganic nitrogen leaching. *Journal of Environmental Quality*, **39**, 1790–1799.

Milner S, Lovett A, Holland R, Sunnenberg G, Hastings A, Smith P, Taylor G (2015) A preliminary assessment of the ecosystem service impacts of bioenergy in GB to 2050. *Global Change Biology-Bioenergy*, doi: 10.1111/gcbb.12263.

Miyamoto T, Kawahara M, Minamisawa K (2004) Novel endophytic nitrogen-fixing clostridia from the grass *Miscanthus* sinensis as revealed by terminal restriction fragment length polymorphism analysis. *Applied and Environmental Microbiology*, **70**, 6580–6586.

Naik SN, Goud VV, Rout PK, Dalai AK (2010) Production of first and second generation biofuels: a comprehensive review. *Renewable and Sustainable Energy Reviews*, **14**, 578–597.

Oenema O, Velthof GL, Yamulki S, Jarvis SC (1997) Nitrous oxide emissions from grazed grassland. *Soil Use and Management*, **13** (Suppl. 4), 288–295.

Ofgem (2014) UK biomass sustainability figures. Available at: https://www.ofgem.gov.uk/publications-and-updates/biomass-sustainability-dataset-2012-13 (accessed 27 January 2015).

O'Neill NR, Farr DF (1996) *Miscanthus* blight, a new foliar disease of ornamental grasses and sugarcane incited by Leptosphaeria sp. and its anamorphic state Stagonospora sp. *Plant Disease*, **80**, 980–987.

Poeplau C, Don A (2014) Soil carbon changes under *Miscanthus* driven by C4 accumulation and C3 decomposition-toward a default sequestration function. *GCB Bioenergy*, **6**, 327–338.

Post WM, Peng T-H, Emanuel WR *et al.* (1990) The global carbon cycle. *American Scientist*, **78**, 310–326.

Rafique R, Hennessy D, Kiely G (2011) Nitrous oxide emission from grazed grassland under different management systems. *Ecosystems*, **14**, 563–582.

Reay DS, Davidson EA, Smith KA *et al.* (2012) Global agriculture and nitrous oxide emissions. *Nature Climate Change*, **2**, 410–416.

Richter GM, Riche AB, Dailey AG, Gezan SA, Powlson DS (2008) Is UK biofuel supply from *Miscanthus* water-limited? *Soil Use and Management*, **24**, 235–245.

Richter GM, Agostini F, Redmile-Gordon M, White R, Goulding KWT (2015) Sequestration of C in soils under *Miscanthus* can be marginal and is affected by genotype-specific root distribution. *Agriculture Ecosystems & Environment*, **200**, 169–177.

Rochette P, Eriksen-Hamel NS (2008) Chamber measurements of soil nitrous oxide flux: are absolute values reliable? *Soil Science Society of America Journal*, **72**, 331–342.

Roth B, Jones M, Burke J, Williams M (2013) The effects of land-use change from grassland to *Miscanthus × giganteus* on soil N₂O emissions. *Land*, **2**, 437–451.

Roth B, Finnan JM, Jones MB, Burke JI, Williams ML (2015) Are the benefits of yield responses to nitrogen fertilizer application in the bioenergy crop *Miscanthus × giganteus* offset by increased soil emissions of nitrous oxide? *GCB Bioenergy*, **7**, 145–152.

Scharlemann J, Tanner EVJ, Hiederer R, Kapos V (2014) Global soil carbon: understanding and managing the largest terrestrial carbon pool. *Carbon Management*, **5**, 81–91.

Schneckenberger K, Kuzyakov Y (2007) Carbon sequestration under *Miscanthus* in sandy and loamy soils estimated by natural ^{13}C abundance. *Journal of Plant Nutrition and Soil Science*, **170**, 538–542.

Searchinger T, Heimlich R, Houghton RA *et al.* (2008) Use of US croplands for biofuels increases greenhouse gases through emissions from land-use change. *Science*, **319**, 1238–1240.

Semere T, Slater F (2007a) Ground flora, small mammal and bird species diversity in *Miscanthus* (*Miscanthus* × giganteus) and reed canary-grass (*Phalaris arundinacea*) fields. *Biomass and Bioenergy*, **31**, 20–29.

Semere T, Slater F (2007b) Invertebrate populations in *Miscanthus* (*Miscanthus* × giganteus) and reed canary-grass (*Phalaris arundinacea*) fields. *Biomass and Bioenergy*, **31**, 30–39.

Sherrington C, Bartley J, Moran D (2008) Farm-level constraints on the domestic supply of perennial energy crops in the UK. *Energy Policy*, **36**, 2504–2512.

Shield IF, Barraclough TJP, Riche AB, Yates NE (2014) The yield and quality response of the energy grass *Miscanthus* × giganteus to fertiliser applications of nitrogen, potassium and sulphur. *Biomass and Bioenergy*, **68**, 185–194.

Smith P, Powlson D, Glendining M, Smith JO (1997) Potential for carbon sequestration in European soils: preliminary estimates for five scenarios using results from long-term experiments. *Global Change Biology*, **3**, 67–79.

Stephens W, Hess TM, Knox JW (2001).Review of the effects of energy crops on hydrology. NF0416, Cranfield University MAFF.

Styles D, Jones MB (2007) Energy crops in Ireland: quantifying the potential life-cycle greenhouse gas reductions of energy-crop electricity. *Biomass and Bioenergy*, **31**, 759–772.

Styles D, Gibbons J, Williams AP *et al.* (2015) Consequential life cycle assessment of biogas, biofuel and biomass energy options within an arable crop rotation. *GCB Bioenergy*, doi: 10.1111/gcbb.12246

Tallis MJ, Casella E, Henshall PA, Aylott MJ, Randle TJ, Morison JIL, Taylor G (2013) Development and evaluation of ForestGrowth-SRC a process-based model for short rotation coppice yield and sopatial supply reveals poplar uses water more efficiently than willow. *GCB Bioenergy*, **5**, 53–66.

Tilman D, Socolow R, Foley JA *et al.* (2009) Beneficial biofuels—the food, energy, and environment trilemma. *Science*, **325**, 270.

Tilman D, Balzer C, Hill J, Befort BL (2011) Global food demand and the sustainable intensification of agriculture. *Proceedings of the National Academy of Sciences of the Unites States of America*, **108**, 20260–20264.

Tolk JA, Howell TA, Evett SR (1998) Evapotranspiration and yield of corn grown on three High Plains soils. *Agronomy Journal*, **90**, 447–454.

Tonini D, Hamelin L, Wenzel H, Astrup T (2012) Bioenergy production from perennial energy crops: a consequential LCA of 12 bioenergy scenarios including land use changes. *Environmental Science & Technology*, **46**, 13521–13530.

Valentine J, Clifton-Brown JC, Hastings A, Robson P, Allison G, Smith P (2012) Food vs. fuel: the use of land for lignocellulosic "next generation" energy crops that minimize competition with primary food production. *GCB Bioenergy*, **4**, 1–19.

Vanloocke A, Bernacchi CJ, Twine TE (2010) The impacts of *Miscanthus* × giganteus production on the Midwest US hydrologic cycle. *GCB Bioenergy*, **2**, 180–191.

Velthof G, Brader A, Oenema O (1996) Seasonal variations in nitrous oxide losses from managed grasslands in The Netherlands. *Plant and Soil*, **181**, 263–274.

Williams AG, Audsley E, Sandars DL (2010) Environmental burdens of producing bread wheat, oilseed rape and potatoes in England and Wales using simulation and system modelling. *The International Journal of Life Cycle Assessment*, **15**, 855–868.

Yaeger MA, Sivapalan M, McIsaac GF, Cai X (2013) Comparative analysis of hydrologic signatures in two agricultural watersheds in east-central Illinois: legacies of the past to inform the future. *Hydrology and Earth System Sciences*, **17**, 4607–4623.

Zatta A, Clifton-Brown JC, Robson P, Hastings A, Monti A (2014) Land use change from C3 grassland to C4 *Miscanthus*: effects on soil carbon content and estimated mitigation benefit after six years. *GCB Bioenergy*, **6**, 360–370.

Zenone T, Gelfand I, Chen J, Hamilton SK, Robertson GP (2013) From set-aside grassland to annual and perennial cellulosic biofuel crops: Effects of land use change on carbon balance. *Agricultural and Forest Meteorology*, **182**, 1–12.

Zhuang Q, Qin Z, Chen M (2013) Biofuel, land and water: maize, switchgrass or *Miscanthus*? *Environmental Research Letters*, **8**, 015020.

Zimmermann J, Dauber J, Jones MB (2012) Soil carbon sequestration during the establishment phase of *Miscanthus* × giganteus: a regional-scale study on commercial farms using 13C natural abundance. *GCB Bioenergy*, **4**, 453–461.

3

Spatial yield estimates of fast-growing willow plantations for energy based on climatic variables in northern Europe

BLAS MOLA-YUDEGO[1,2], JOHANNES RAHLF[1], RASMUS ASTRUP[1] and IOANNIS DIMITRIOU[3]

[1]*Norwegian Institute of Bioeconomy Research (NIBIO), P.O. Box 115, Ås 1431, Norway,* [2]*School of Forest Sciences, University of Eastern Finland (UEF), P.O. Box 111, Joensuu FI-80 101, Finland,* [3]*Department of Crop Production Ecology, Swedish University of Agricultural Sciences (SLU), P.O. Box 7016, Uppsala S-750 07, Sweden*

Abstract

Spatially accurate and reliable estimates from fast-growing plantations are a key factor for planning energy supply. This study aimed to estimate the yield of biomass from short rotation willow plantations in northern Europe. The data were based on harvesting records from 1790 commercial plantations in Sweden, grouped into three ad hoc categories: low, middle and high performance. The predictors included climatic variables, allowing the spatial extrapolation to nearby countries. The modeling and spatialization of the estimates used boosted regression trees, a method based on machine learning. The average RMSE for the final models selected was 0.33, 0.39 and 1.91 (corresponding to $R^2 = 0.77, 0.88$ and 0.45), for the low, medium and high performance categories, respectively. The models were then applied to obtain 1×1 km yield estimates in the rest of Sweden, as well as for Norway, Denmark, Finland, Estonia, Latvia, Lithuania and the Baltic coast of Germany and Poland. The results demonstrated a large regional variation. For the first rotation under high performance conditions, the country averages were as follows: >7 odt ha^{-1} yr^{-1} in the Baltic coast of Germany, >6 odt ha^{-1} yr^{-1} in Denmark, >5 odt ha^{-1} yr^{-1} in the Baltic coast of Poland and between 4–5 odt ha^{-1} yr^{-1} in the rest. The results of this approach indicate that they can provide faster and more accurate predictions than previous modeling approaches and can offer interesting possibilities in the field of yield modeling.

Keywords: bioenergy, biofuels, biomass, boosted regression trees, climatic restrictions, energy crops, predictive models, short rotation, willow, yield maps

Introduction

How much biomass can be produced from fast-growing plantations in a given area? Fast-growing plantations are an alternative feedstock of wood biomass for the emerging energy sector and related industries. Such plantations are based on fast-growing woody species (such as willow) generally established on agricultural land, being intensively managed. The expected life span of a willow plantation is considered to be about 25 years, and the same plantation can be harvested several times, with rotations (*cutting cycles*) from 3 to 6 years (e.g., Rahman *et al.*, 2014).

At present, Sweden provides a good basis for commercial experience: fast-growing willow plantations have been cultivated at commercial level since the 1980s, particularly willow, making Sweden the leader in

Europe both in long-term experience and total area planted, which entails c. 13 000–16 000 ha (Mola-Yudego & González-Olabarria, 2010). In addition, Denmark currently entails 5700 ha (Jørgensen *et al.*, 2014), England c. 2500 ha (DEFRA, 2014) and Germany c. 4000–5000 ha (Wühlisch, 2012), whereas there are ambitious goals for their expansion, for example: Poland, aiming at 170 000 ha to be planted with energy crops (Kunikowski *et al.*, 2005), or UK, planning 350 000 ha by 2020 (DEFRA, 2007), although these levels of ambition may be subject to changes, as the Swedish example shows (Mola-Yudego & González-Olabarria, 2010).

Planning the expansion of these plantations requires accurate and updated information concerning both current and potential yield, a prerequisite for a successful development of bioenergy markets based on energy crops (Mola-Yudego *et al.*, 2014). At local level, productivity estimates are required for planning wood fuel supply chains, for the location of bioenergy plants (e.g., Mola-Yudego & Pelkonen, 2011), for profitability

Correspondence: Blas Mola-Yudego
e-mails: blas.mola@nibio.no; blas.mola@uef.fi

analysis (e.g., Toivonen & Tahvanainen, 1998; Rosen-qvist & Dawson, 2005; Dimitriou & Rosenqvist, 2011) and for the general logistics and management associated to biofuels, among others. At national level, productivity estimates are needed for the construction of scenarios (e.g., Bauen *et al.*, 2010), for the implementation of policy incentives (e.g., Mola-Yudego & Pelkonen, 2008), or even for environmental assessment (e.g., Gonzalez-Garcia *et al.*, 2012), among others.

Methods to predict plantation productivity have been traditionally based on regression models, where yield was predicted as a function of different parameters, following pre-established equations. To address the spatial component necessary to predict potential yield in different areas, these parameters have been often related to climate. Indeed, temperature and precipitation have been considered the most important factors for the growth of willow plantations (Perttu, 1999), and several of the initial studies on willow plantation growth have modeled yields based on climatic variables (e.g., Nilsson & Eckersten, 1983; Perttu *et al.*, 1984). Also climate-based models were used in Sweden to model the potential productivity at spatial level of willow plantations (Lindroth & Båth, 1999), resulting in productivity maps based on the linear relationship between yield and precipitation during the growing season.

However, these methods present several modeling limitations, as in many cases, the relationships between the variables used as predictors are complex, presenting several interactions, and the use of predefined relations (i.e., linear regression) can lose predictive power. More recent approaches have considered the inherent spatial component of yield prediction, using more ambitious modeling techniques. Aylott *et al.* (2008) produced estimates based on partial least squares, aiming at mapping the plantations productivity in UK. In this case, the modeling approach required detailed data of each plantation's growth and management activities, which were obtained from a network of well-studied experimental trials. Mola-Yudego (2010) produced estimates based on *k-nn* imputation methods (see, e.g., Kilkki & Päivinen, 1987; Muinonen & Tokola, 1990; Tomppo, 1990; Tokola *et al.*, 1996), aiming at mapping the plantations' productivity in Sweden. Following this approach, the variables from a specific area are predicted as a weighted average of the spectrally closest plantations (which are defined as nearest neighbors, *nn*) and the feature spectrum is defined by a vector of climatic variables. In this case, no detailed data of each plantation's growth were needed, but relied on a large pool of plantations to get stable and accurate estimates. However, the nature of the method makes it difficult to model the specific relations between variables and yields and limits its potential for extrapolation to those areas outside the sampled data.

In this context, the aim of this study was to provide accurate estimates of productivity for fast-growing willow plantations, spatially extrapolating a large sample of Swedish plantations to nearby areas in northern Europe, using climatic data. The results of this approach can be applied, among others, to the policy and economic considerations associated with wood supply derived from energy crops, as well as their future development.

Materials and methods

Description of the data

The data set used in the calculations was based on harvesting records from the first rotation of 1790 willow plantations in central and south Sweden (see Mola-Yudego, 2010). The records therefore correspond to harvest aboveground leafless biomass. The field measurements were provided by Lantmännen Agroenergi AB. Data with inconsistent records (e.g., missing digits in the coordinates) or lacking information regarding the area planted or the location were excluded from the data set. All plots were georeferenced to a 1 km precision. The average yield was calculated by dividing the total harvested biomass by area planted and the number of years of the rotation. The plantations were cut after the first growing season after planting to promote sprouting (*cutback*). The data used included 7753 ha planted during the period 1986–2005 in the area defined from 55°20′N to 61°29′N and from 11°33′E to 18°56′E. The average size of the plantations was 4.3 ha (SD: 4.2 ha).

The climatic data were based on the climate layers calculated for northern Europe (Hijmans *et al.*, 2004; WorldClim database version 1.4). These data consist of a set of grid maps resulting from an interpolation process of temperature and precipitation averages (Hijmans *et al.*, 2005), based on the reference period 1960–1990, to assess the average climatic conditions of the area.

The maps used in this study had a 30-s spatial resolution (which provides ~1 km precision). To link the climatic variables with the ground data, the maps were projected from the originally projected datum into the same coordinate system as the yield data (UTM, zone 33N). The precision of the interpolated climatic variables was 0.1° C for temperature and 1 mm for precipitation. The monthly averages of maximum, mean and minimum temperatures and precipitation (referred as $Tmax$, $Tmean$, $Tmin$ and P, respectively) were obtained for each plantation.

Statistical modeling approach

The modeling approach was based on boosted regression trees (BRT). This approach combines statistical and machine learning techniques aiming at the improvement of the performance of a single model by fitting many models and combining them for prediction (Schapire, 2003). Besides the selection of the variables to be included in the models, common to any modeling approach, BRT requires additional parameters to be calibrated. BRT models are defined by different parameters: number of trees, learning rate (or *shrinkage*, related to the reduction of the impact of any additional tree), bag (random fraction of the residuals is selected to build the tree, per iteration) and number of interactions between variables.

The models and statistical analysis were developed in R version 3.2.0 (R development core team, 2014). The BRT models were based on the dismo extension of the GBM package (Ridgeway, 2006), developed by Elith *et al.* (2008).

The predicted variable was the mean annual growth per hectare, expressed as oven dry tonnes per hectare and year (odt ha^{-1} yr^{-1}). Due to the high variability of the yields resulting from different management practices (Mola-Yudego & Aronsson, 2008; Mola-Yudego, 2010), the plantations were classified according to their local performance, using municipalities as a unit of analysis. For each municipality, the plantations were ranked according to their measured yields and then grouped in three categories made of approximately an equal number of plantations (Fig. 1). There were 119 municipalities with data, with an average of 15 plantations each.

The categories were described as follows: *high* performance, *medium* performance and *low* performance (Table 1), corresponding to the upper third of yield level, the middle third and the lower third at each municipality. The average establishment year was 1994 for all three categories. Inside the same municipality, plantations of the category *medium* were established on average +0.09 years (P-value: 0.079) more recently than those of the *low*, and plantations of the category *high* were established +0.29 years (P-value: 0.622) more recently than those of the *medium* and +0.47 years, (P-value: 0.032) more recently than those of the *low*. There were not significant differences concerning the climatic profile between categories in the same municipality.

The climatic variables selected as predictors were chosen following Mola-Yudego (2011a). The main criteria were defined to reflect the influence of climatic characteristics on yield based on existing literature, to present minimal bias and minimal root-mean-square error (RMSE) and to avoid excessive multicorrelation. An additional criterion derived from the previous study was to remove variables whose empirical constant p was close to 0 when optimized for prediction (see Mola-Yudego, 2011a,b). A total of six climatic variables were included in the models: *Tmax2* (February), *Tmax7* (July), *Tmax8* (August), *Tmin10* (Oct), *Tmean5* (May) and *Psum* (sum of the precipitation from May to September).

The combination of different number of interactions between the selected variables, the required number of trees, learning rates and bags would result in a large number of potential models. Therefore, several models were tested sequentially, being the learning rates fixed at: 0.02, 0.01, 0.008, 0.005 and 0.001; the bag at: 0.2, 0.5 and 0.8; and the number of interactions at: 1, 2, 4, 6, 8, 10, 12, 14 and 16. The number of trees was obtained through optimization (see Ridgeway, 2006): first the BRT models were built with the default 10-fold cross-validation, and this procedure was used to estimate the optimal number of trees to be included. The final models were finally built on the full data set, using the number of trees identified as optimal. This resulted in 432 models. Among those, the model selection was based on a double criterion: the model should have a high predictive power and should present stable predictions (small variations in the parameters defining the model should not result in markedly different predictions). These two criteria were evaluated through the RMSE (root-mean-squared error of the predictions vs. the observed data) and the RMSD (root-mean-squared deviation), respectively. The RMSD was defined to assess the stability of the predictions: Each model was run in 6500 points generated on agricultural land in northern Europe. Therefore, the RMSD of those predictions was calculated for each model, being defined as follows:

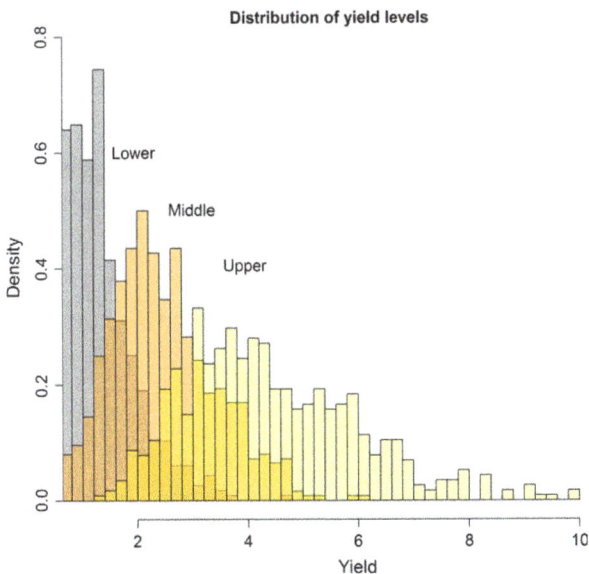

Fig. 1 Distribution of the commercial willow plantations (*n* = 1790) in Sweden used in the models. The groups correspond to a classification of local performance at municipality level: low, middle and high performance. Density refers to the probability density of the counts.

Table 1 Descriptors for the three categories defined. *N*: number of plantations. *N (municipality)*: average number of plantations per municipality. *Yield* refers to the harvested first rotation (cutting cycle). *Year Est.* refers to the year when the plantation was established. *Relative difference* refers to the differences in year of establishment among categories inside the same municipality. Numbers in parenthesis refers to the standard deviations

	Low performance	Medium performance	High performance
N	578	620	592
N (municipality)	4.9	5.2	5.0
Yield (odt ha^{-1} yr^{-1})	1.23 (0.57)	2.47 (0.95)	4.73 (2.45)
Year Est.	1994.1 (2.395)	1994.07 (2.398)	1994.33 (2.583)

$$RMSD = \sqrt{\frac{\sum(y_i - \bar{y}_i)^2}{n}} \qquad (1)$$

where, y_i is a prediction for a point using a given model, $\bar{y}_i$ is the mean of the predictions of all models for point I and n the total number of models considered for each performance category. After the evaluation, prediction maps were produced for northern Europe, in areas geographically and climatically similar to Sweden. The predictions were restricted to those areas defined as agricultural land using the Corine 2000 classification (EEA, 2000), 250 m resolution.

Finally, for each country, an estimate of the average annual productivity for the estimated life span of the plantations was calculated. The calculation was based on Mola-Yudego (2010) where yield estimates for the first cutting cycle are used as a reference value. The cutting cycle length was assumed to be 4 years, and it considered five rotations. The increments of productivity along the rotations were estimated based on Mola-Yudego & Aronsson (2008). The average annual yield was then calculated by dividing the accumulated production by the total number of years of the five rotations, plus one, to include the

initial year for cutback (in total: 21 years); these values were the basis for estimating of the potential energy produced under different percentages of available arable land.

Results

Of the 432 models constructed, 62 could not be calculated, particularly those with high learning rate (0.02), especially when there were many interactions and in the high performance category. In total, 142, 135 and 93 models were considered, for the low, middle and high performing categories (Fig. 2). The results showed that models with few interactions between the variables resulted in low RMSE (Fig. 3). The average RMSE for all the models presented was 0.40, 0.46 and 1.98 (corresponding to $R^2 = 0.64, 0.75$ and 0.36) for the categories of low, medium and high performance, respectively.

In general, the overall performance of the predictions was better for the medium and low productivity planta-

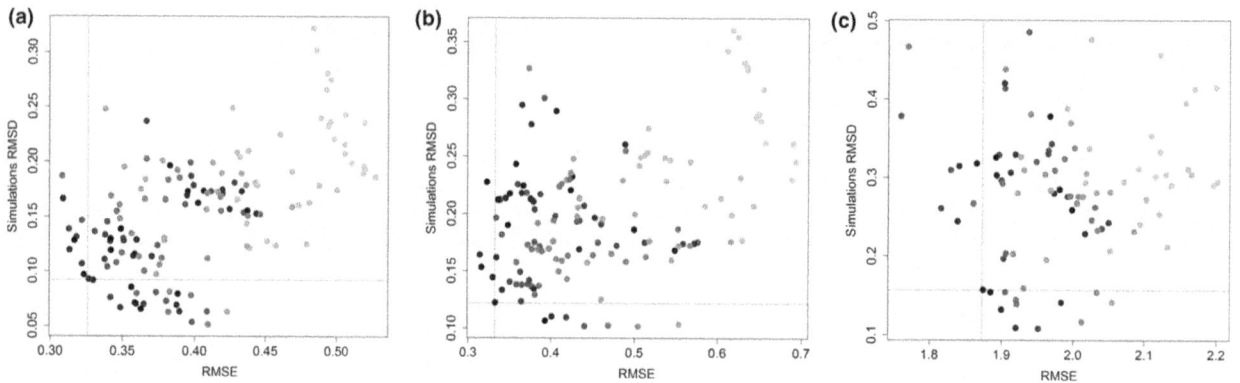

Fig. 2 Performance of the tested models (Root-mean-square error) and stability of the predictions (Root-mean-square deviation). Each point represents the overall results for one model. RMSD is calculated as the root mean squared of the differences in 6500 test points between predictions for a given model and the average predictions from all models. (a) Lowest performance ($n = 142$), (b) middle performance ($n = 135$), (c) highest performance category ($n = 93$). Color gradient: light = 1 interaction, darkest = 16 interactions.

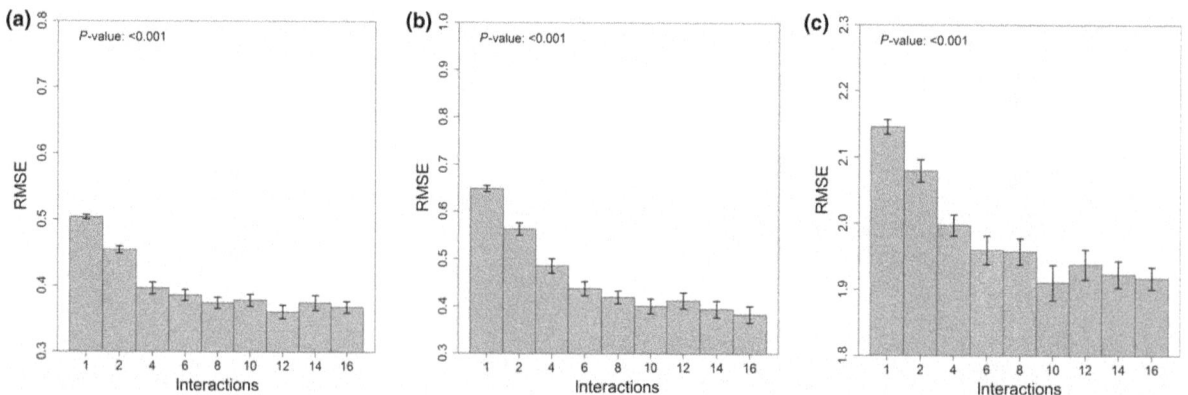

Fig. 3 Root-mean-square error (RMSE) as a function of the number of interactions. (a) Lowest performance, (b) middle performance, (c) highest performance category.

tions. The final models selected included the following: 14 interactions, 0.005 learning rate and 0.5 bag for the low performance; 16 interactions, 0.02 learning rate and 0.5 bag for the medium performance; and 16 interactions, 0.008 learning rate and 0.5 bag for the high performance. The coefficients of determination (R^2) were 0.77, 0.88 and 0.45, respectively (Table 2). The models failed to accurately predict the highest yields of the best performance group, especially above 10 odt ha^{-1} yr^{-1} (Fig. 4).

Psum and Tmax in February presented the highest weights in all categories (over 40%), whereas Tmax in August presented the lowest (lower than 10%) (Table 2). Psum presented strong interactions with all the variables related to temperature included in the models. The highest interactions for all categories were between Tmax in February and Psum, and Tmin in October and Psum. In the case of middle performance, an additional

strong interaction was between Tmax in July and Psum (Table 3).

The effect of the variables on yield was different for the different variables and performance categories. In general, precipitation had a positive effect on yield, for all performance groups. Tmax in February, Tmean in May and Tmin in October had a positive effect, whereas Tmax in July presented a negative effect. Tmax in August presented a positive effect for the middle performance group (Fig. 5). It must be taken into account

Table 2 Weights for each variable (%), estimates of model's Root-mean-squared error and coefficient of determination (R^2) of the predictions, for every performance group (Low: low performance, Medium: middle performance, High: upper performance)

Variable	Low	Medium	High
Tmax2	24.65	23.87	21.52
Tmax7	14.48	16.45	10.77
Tmax8	8.11	6.61	8.25
Tmin10	14.91	15.05	21.06
Tmean5	14.87	14.68	18.24
Psum	22.95	23.32	20.12
RMSE	0.33	0.33	1.87
R^2	0.77	0.88	0.45

Psum: aggregated precipitation May to September; T, temperature; Max, maximum; Min, minimum, numbers correspond to the calendar months.

Table 3 Variable interactions, for every performance

Variable	Tmax2	Tmax7	Tmax8	Tmin10	Tmean5	Psum
Low						
Tmax2	0	1.62	0.93	7.3	3.24	6.94
Tmax7	0	0	0.74	0.95	0.96	1.86
Tmax8	0	0	0	0.76	2.01	1.25
Tmin10	0	0	0	0	0.63	5.94
Tmean5	0	0	0	0	0	3.79
Psum	0	0	0	0	0	0
Medium						
Tmax2	0	3.02	1.83	8.74	28.62	30.64
Tmax7	0	0	1.4	2.39	1.83	21.02
Tmax8	0	0	0	5.2	0.82	5.36
Tmin10	0	0	0	0	2.86	6.51
Tmean5	0	0	0	0	0	15.72
Psum	0	0	0	0	0	0
High						
Tmax2	0	5.89	4.78	18.45	17.49	18.28
Tmax7	0	0	4.52	4.75	1.18	5.09
Tmax8	0	0	0	1.7	3.63	2.05
Tmin10	0	0	0	0	3.52	27.19
Tmean5	0	0	0	0	0	9.03
Psum	0	0	0	0	0	0

Psum: aggregated precipitation May to September; T, temperature; Max, maximum; Min, minimum, numbers correspond to the calendar months.

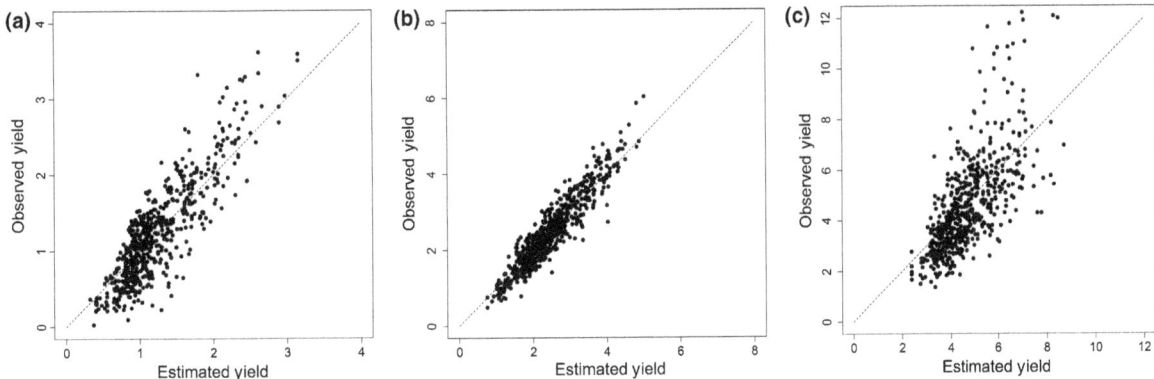

Fig. 4 Measured and predicted yield for the commercial fast growing willow plantations, according to the models. (a) Lowest performance, (b) middle performance, (c) highest performance group.

that the partial dependence is also defined by the inter-actions among variables (Table 3).

The resulting predictions were aggregated by region and country, to estimate the potential area under differ-ent productivity categories (Fig. 6). It must be taken into account that in the northernmost regions, there is scarce agricultural land, often located nearby the coastal areas in the best climatic conditions. This makes the regional averages of Lapland (Finland), Finnmark (Norway) and Upper and Middle Norrland (Sweden) to be high when aggregated at region level, compared to other more southern regions.

The analysis of the stability of the estimates was eval-uated through the standard deviation of the predictions of the models for each of 6500 random points (Fig. 7). The models show consistency concerning the predic-tions in Sweden, Finland, Estonia and most of Latvia. However, small changes in the model parameters resulted in larger differences in the predictions for the Western areas of Jutland (Denmark) as well as the west of Lithuania, and the Eastern parts of Northern Poland, especially concerning the best performance group.

Concerning the best performance category, the mean for the first rotation by country (Fig. 8) ranged from 4.1 odt ha^{-1} yr^{-1} (Finland) to 7.1 odt ha^{-1} yr^{-1} (Northern Germany). For the whole area, the average for the first rotation for this category was 4.8 odt ha^{-1} yr^{-1} (SD 0.98). However, it must be taken into account that there is a strong spatial variability, as most of the countries showed large regional differences (Fig. 9). When consid-ering the 10% agricultural land with the highest willow productivity (1.5 × 10^6 ha), the average yield for the area studied was 6.5 odt ha^{-1} yr^{-1} for the first rotation.

Discussion

This study aims at estimating the spatial distribution of production of biomass for energy from short rotation willow plantations by modeling their potential produc-tivity based on climatic variables. The data were based on harvest records for the first rotation from 1790 com-mercial plantations for the period 1989–2005, and it has been extensively used in the past for modeling purposes (e.g., Mola-Yudego & Aronsson, 2008; Mola-Yudego,

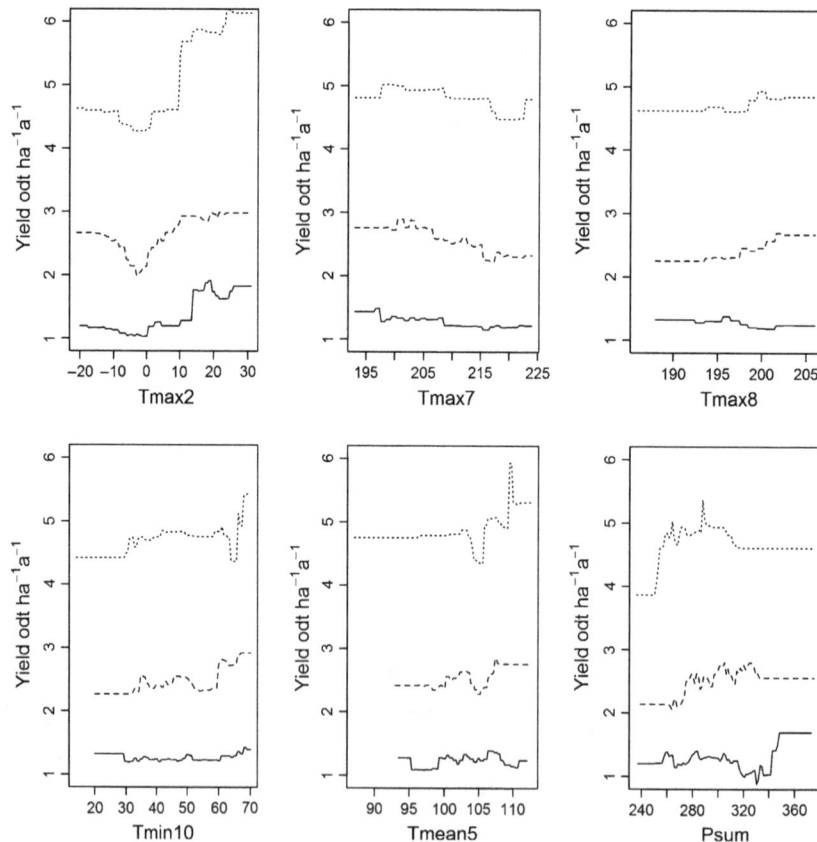

Fig. 5 Partial dependence plots of the predictors in the model, for the performance categories (lower, middle, upper). The tempera-tures are expressed in degrees Celsius (×10), and precipitation in mm of rainfall. Yield: average annual harvested yield of willow plantations. Psum: aggregated precipitation May to September, T: Temperature, Max: maximum, Min: minimum, numbers corre-spond to the calendar months.

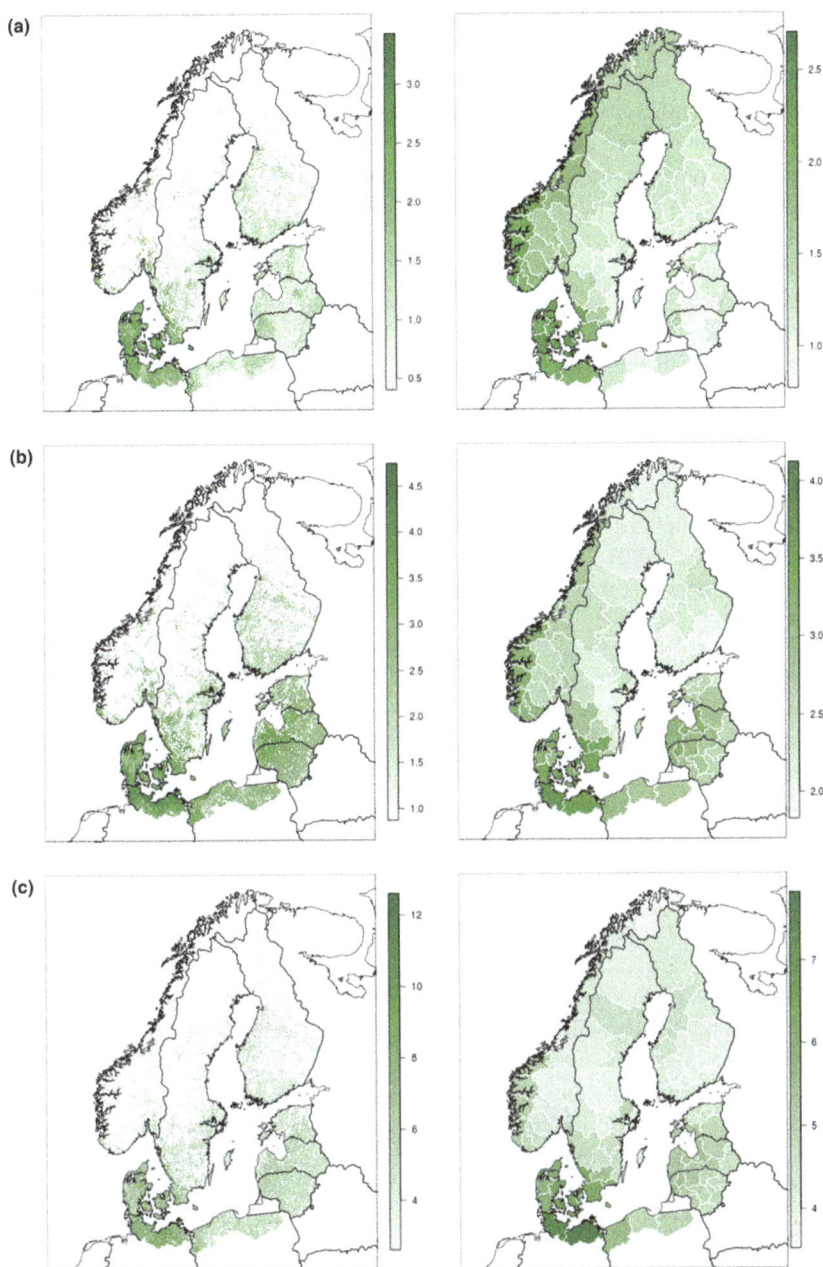

Fig. 6 Yield estimates for willow plantations in the agricultural areas of northern Europe. Darker colours indicate high productivity areas. (a) Lowest performance, (b) middle performance, (c) highest performance group. Left, estimates at 1 × 1 km resolution (see Supporting Information). Right, averages by regions. The average includes predictions only on agricultural land. Therefore, predictions in the northern latitudes are based on the limited amount of land available.

2011a,b). This data pool presents a realistic basis for yield estimates, as the predictions relate to the final amount of biomass that will be effectively utilized; for example, Sevel *et al.* (2012) estimated differences between nondestructive and harvested observations to be around 1.2 odt ha^{-1} yr^{-1}, and Searle & Malins (2014) and Mola-Yudego *et al.* (2015) observed that records from, for example, experimental trials tend to overesti-

mate yields especially when the plots are small. The large amount of data available allowed the inclusion of almost 60% of the whole area planted with willow for bioenergy in Sweden, which enhances the reliability of the estimates.

Nevertheless, a disadvantage of the data used was the lack of detailed information concerning the management practices performed by the farmers, as well as specific soil records from the plantations. Although some authors have considered that at large, temperature and precipitation are the most important factors for the growth of willow plantations (e.g., Perttu, 1983, 1999), several studies have been demonstrating the important role than soil type, in addition to climate, plays in yield performance (Aylott et al., 2008). In this line, Sevel et al. (2012) and Larsen et al. (2014) pointed out that soil, clone and its interaction are key variables to model the productivity of willow plantations. In this study, the explicit inclusion of soil variables presented several inconveniences as there is limited information available concerning the soil textures with the necessary spatial resolution and continuity to be included in the models for the studied countries. Similarly, the varieties used in the plantations play an important role in their yield performance (e.g., Lindegaard et al., 2001) but detailed information concerning the clones planted in most of the plantations was also not available.

The aggregation of the plantations in three performance categories, following Mola-Yudego (2010), was aimed at homogenizing the conditions inside the same group, thus incorporating to a certain extend management or local soil conditions. The municipality boundaries were used as a grouping factor, to allow a sufficient number of plantations to make the classification while at the same time assuring that the same climatic conditions would be shared between categories. In this sense, the high performing plantations inside a municipality would most likely be the result of better management, better clones and better soil conditions than the lower performing ones. It must be stressed that farmers decide the location of the plantations and the soil quality where they are established (Mola-Yudego & Aronsson, 2008). The working assumption proposed that the variability between categories may be attributed to variables related to the intensity of management (among others fertilization and rotation length), the local soil quality, the clone used and the interactions of these factors, whereas the climatic variables would rather explain the spatial variability of yields in plantations of the same category. In fact, the performance of the models was lower in the high performance category, underlining the fact that a large part of the variability between plantations in this category was explained by variables other than climate ($R^2 = 0.45$ for the final model), whereas for medium performance category, a

large part of the variability was effectively explained by the variables included ($R^2 = 0.88$). It must be taken into account that it is an *ad hoc* characterization based on final performance, and other factors may have affected yield (e.g., early frosts, heavy rains or moose browsing, among others).

Moreover, yield levels are not fixed and can be increased along time through new varieties and management techniques (Mola-Yudego, 2011b), thus to make a proper use of the predictions provided, they should be taken as a reference threshold defined by climate for the period studied, rather than final predictions. In this sense, the models presented can be recalibrated once new data are available or can be treated as a value based on climate that could incorporate both soil or clone factors by adding scenarios, correction factors in specific areas or an improvement factor based on trends (e.g., Mola-Yudego, 2010).

Several initial studies have modeled the yields in Sweden based on climatic variables (e.g., Nilsson & Eckersten, 1983). Lindroth & Båth (1999) proposed a semimechanistic model to calculate plantations' yield as a linear function of precipitation during the growing season. The application of the model provided maps of potential productivity for central and southern Sweden, but one disadvantage was that the models resulted in higher yield expectations than shown by empirical measurements based on the same commercial plantations (Mola-Yudego & Aronsson, 2008; Mola-Yudego, 2010) and that the method did not offer flexibility as the relationship between precipitation and yield was set to be linear.

The modeling approach taken in this study presents several advantages for modeling climatic data that may overcome these limitations. The use of BRT allows shaping the relationship between the variables and yield with almost no pre-assumptions concerning the shape of the relationship between the variables. This is a great advantage, as some of the variables and their interactions may have thresholds or limiting maxima. At the same time, BRT allows the inclusion of a large number of interactions between the variables and aims at finding those more statistically relevant, simplifying the modeling assumptions taken a priori. The approach has recently been used in productivity studies, to, for example, map the site index for different forest species (Aertsen et al., 2010) or the biomass yield of seminatural systems (Van Meerbeek et al., 2014).

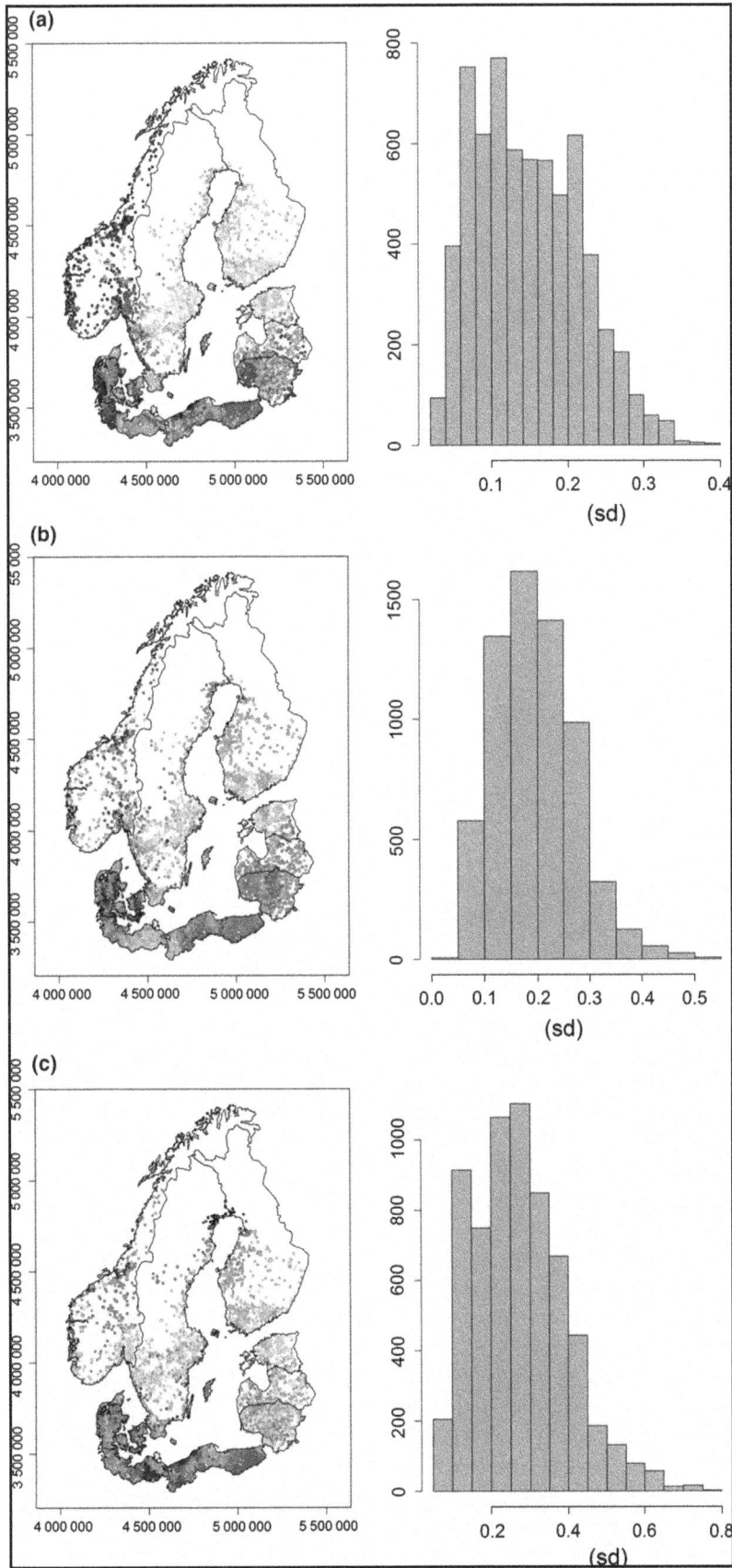

Fig. 7 Model sensitivity to changes in the calibration, spatially (left) and by frequency (right). Each point represents the standard deviation (SD) of the predictions resulting from the pool of models. Categories: (a) Lowest performance ($n = 142$ models), (b) middle performance ($n = 135$), (c) highest performance category ($n = 93$). There are 6500 predictions randomly distributed in agricultural land.

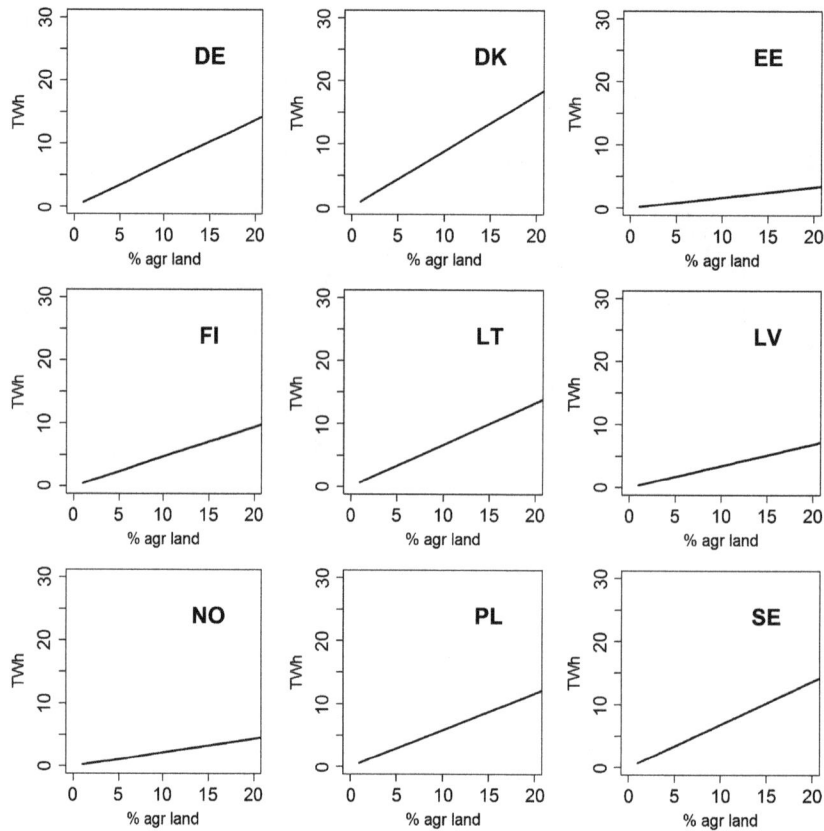

Fig. 8 Estimates for energy production assuming several percentages of available arable land (% agr land) for fast growing willow plantations, with regional average yields under high performance conditions. DE: northern regions (Schleswig-Holstein and Mecklenburg-Vorpommern), PL: northern regions (Warminsko Mazurskie, Województwo pomorskie and Zachodniopomorskie). DK: referred to 2012, the rest referred to 2013. NO: cultivated land, Statistics Norway (2013). The yield is estimated harvestable biomass during the whole lifespan of a plantation, including one initial year for cut-back and assuming high performance plantations established along the country.

In addition to its predictive power, the method is suitable for interpretation. The results showed that there was a strong effect of precipitation and temperature during early summer, which was expected and is in line with previous work (Lindroth & Båth, 1999). The minimum temperature in October had a strong and positive effect as it was found in Finland in Tahvanainen & Rytkönen (1999). The negative effects of the maximum temperatures in July may be related to higher evapotranspiration and shortage of rainfall, given the high water demands of willow (Linderson et al., 2007).

The models improved existing literature from commercial plantations, with higher prediction power than other studies with the same data (e.g., Mola-Yudego, 2010, 2011a,b), while at the same time being able to deliver high resolution estimates at 1 × 1 km, although it was observed than the BRT models had problems predicting values in the highest ranges, underestimating the most productive plantations. In general, the averages agreed with previous studies based on commercial

data: In Denmark, the averages found for the best performance category (6.5 odt ha^{-1} yr^{-1}) agree with the estimates of Sevel et al., 2012 (5.2–8.8 odt ha^{-1} yr^{-1}). In Sweden, the average and spatial estimates were matching the same spatial pattern Mola-Yudego (2011a,b) which used the same data set but with a kNN approach. In this case, the estimates are lower than in Mola-Yudego (2010) as the method could not incorporate the trends due to yield improvements along time. In Finland, the average was similar to Tahvanainen & Rytkönen (1999), although their study referred to standing biomass by nondestructive methods.

The predictions were based on the first harvest, and once plants have developed a root system the following harvests are substantially higher, which is confirmed in the literature (Hoffmann-Schielle, 1995, Labrecque & Teodorescu, 2003; Nordh, 2005). In this study, the average yield for the whole life span of the plantations was estimated using the predicted yield of the first rotation as reference value and then extrapolated using the

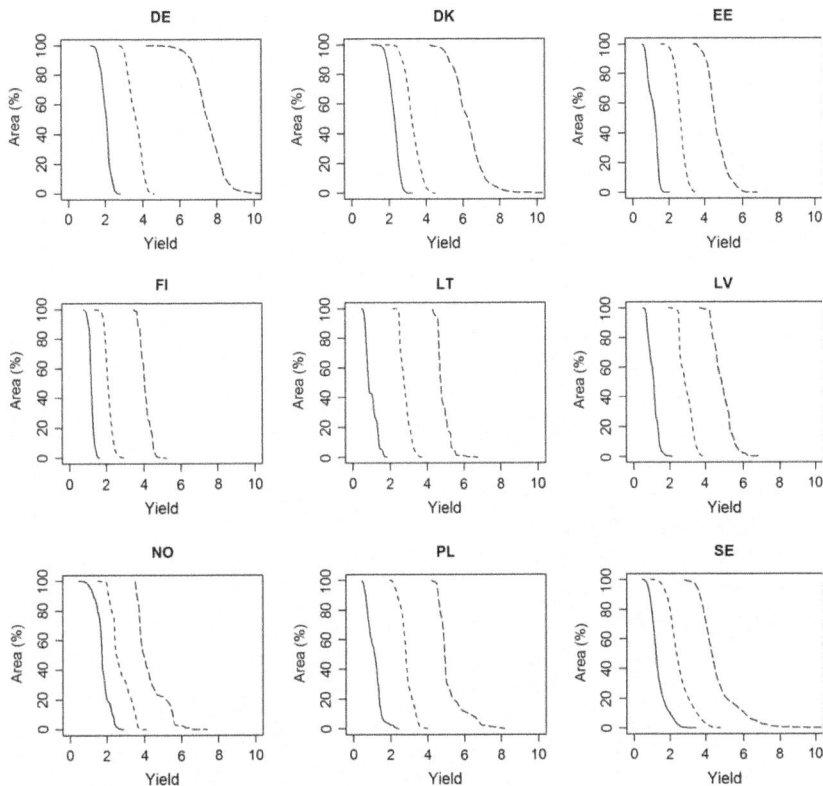

Fig. 9 Areas and potential productivity of short rotation willow plantations on agricultural land in the Nordic area. The areas are presented aggregated under different productivity values, for the low, medium and high performance groups, according to the estimations.

results of Mola-Yudego & Aronsson (2008) and Mola-Yudego (2011b). However, future research must be addressed to study the climatic effects during an extended period, including the specific effects that the climatic variables may have during the second and subsequent rotations. Also, it must be noticed that the final calculation of the whole life span included the first year (*cutback*) which slightly reduced the average yield.

Concerning the spatial accuracy of the predictions, the method applied can present the risk of overfitting when the models become too complex. One of the solutions was the inclusion of a source of stochasticity in the cross-validation used to assess and fit the parameters (Schonlau, 2005) and the use of random points helped to assess the stability of the predictions. The main goal was that, if the parameters used in the models are slightly changed, the resulting predictions should be consistent. The analysis of this spatial consistency showed that the west areas of Jutland in Denmark, as well as the west part of Lithuania and Latvia, and some areas in Poland are particularly sensitive to the calibration parameters and indicate that the yield extrapolation based solely on climatic variables and data from Swe-

den may need additional data for a proper calibration, perhaps related to soil conditions in the area.

The predictions were restricted to agricultural land, and the regional averages were therefore based on predictions for these areas (i.e., excluding forest lands, nonproductive land, lakes and rivers). This resulted in high regional values for northernmost areas (e.g., Lapland, Norrlands or Finnmark). As agriculture land is scarce on northern latitudes and located mainly in favorable conditions nearby the coast, the resulting averages for those counties are high as are based in a small sample distributed unevenly. Although those northern areas, in general, present the most adverse climatic conditions to willow establishment, attempts to develop highly productive clones with frost tolerance for those areas have been done since the late 1980s (Lumme & Törmälä, 1988), and relatively high yields cannot be excluded as a future scenario.

Finally, the country estimates considered the energy produced under different percentages of agricultural land up to 30%, for reference. It is difficult to estimate for the whole region the area that will eventually be dedicated to fast-growing plantations for energy. As a reference, however, Aust *et al.* (2014) suggested that considering all ecological, ethical, political and technical restrictions, as well as future climate predictions, 5.7% of cropland would be a suitable percentage for Germany.

The area studied shares many climatic and geographical features with the areas in Sweden already cultivated with fast-growing plantations. In fact, for most of Denmark and the west and southernmost cultivation zones in Finland, the Swedish experience concerning willow varieties and methods can probably be directly transferred. In the case of Norway, there is good potential in many parts of the country. Despite the country's overall limitations in agricultural land, areas located in the Østfold and Akerhus present good conditions for the establishment of plantation schemes: the results show high potential yields, share geographical proximity and climatic similarity with Sweden allowing the interchange of varieties, there is agricultural land available (c. 16% of the land, Statistics Norway, 2013), and there is a potential demand of wood chips for bioenergy due to the high density of population. In the Baltic countries and Baltic coast of Poland and Germany, the results show good potential yields that can translate into significant figures of energy production. Additionally, the enlargement of cultivated areas in those zones can result in rapid yield improvements once a critical market size is reached (Mola-Yudego *et al.*, 2014).

Indicatively, in the case of Sweden, estimates show that 1% and 5% of agricultural land could translate into 1.5% and 7.2% of the total annual heat consumption, respectively (annual heat estimated in 47 TWh, Swedish Energy Agency, 2013). However, there are several components of potential supply of biomass crops beyond yield estimation that can play an important role in developing the potential supply (e.g., the current agricultural systems, the role of relative prices, input costs and policy developments) and must be taken into account.

Acknowledgements

We are indebted to Gustav Melin and Stig Larsson at Lantmännen Agroenergi AB, for providing the data for the willow plantations, and to Robert J. Hijmans, Susan Cameron and Juan Parra, for providing the climatic data. We especially acknowledge the contribution made by the anonymous reviewers through their comments and suggestions.

References

Aertsen W, Kint V, Van Orshoven J, Özkan K, Muys B (2010) Comparison and ranking of different modelling techniques for prediction of site index in Mediterranean mountain forests. *Ecological Modelling*, **221**, 1119–1130.

Aust C, Schweier J, Brodbeck F, Sauter UH, Becker G, Schnitzler J-P (2014) Land availability and potential biomass production with poplar and willow short rotation coppices in Germany. *GCB Bioenergy*, **6**, 521–533.

Aylott MJ, Casella E, Tubby I, Street NR, Smith P, Taylor G (2008) Yield and spatial supply of bioenergy poplar and willow short-rotation coppice in the UK. *New Phytologist*, **178**, 358–370.

Bauen AW, Dunnett AJ, Richter GM, Dailey AG, Aylott M, Casella E, Taylor G (2010) Modelling supply and demand of bioenergy from short rotation coppice and Miscanthus in the UK. *Bioresource Technology*, **101**, 8132–8143.

DEFRA (2007) Department for Environment, Food and Rural Affairs. UK Biomass Strategy. Available at: http://www.defra.gov.uk/Environment/climatechange/uk/energy/renewablefuel/pdf/ukbiomassstrategy-0507.pdf (accessed 3 March 2008).

DEFRA (2014) Department for Environment, Food and Rural Affairs. Area of Crops Grown For Bioenergy in England and the UK: 2008 – 2013. Government Statistical Service. Department for Environment, Food & Rural Affairs. Available at: https://www.gov.uk/government/statistics (accessed 4 April 2015).

Dimitriou I, Rosenqvist H (2011) Sewage sludge and wastewater fertilisation of Short Rotation Coppice (SRC) for increased bioenergy production—biological and economic potential. *Biomass and Bioenergy*, **35**, 835–842.

EEA (2000) European Environmental Agency. Environment Image & Corine Land Cover 2000 for Denmark, Estonia, Finland, Germany, Latvia, Lithuania, Poland, Sweden. Available at: http://dataservice.eea.eu.int/dataservice/ (accessed 5 November 2005).

Elith J, Leathwick JR, Hastie T (2008) A working guide to boosted regression trees. *Journal of Animal Ecology*, **77**, 802–813.

Gonzalez-Garcia S, Mola-Yudego B, Dimitriou J, Aronsson P, Murphy RJ (2012) Environmental assessment of energy production based on long term commercial willow plantations in Sweden. *Science of the Total Environment*, **421**, 210–219.

Hijmans RJ, Cameron S, Parra J (2004) WorldClim, Version 1.4 (release 3). A square kilometer resolution database of global terrestrial surface climate.

Hijmans RJ, Cameron SE, Parra JL, Jones PG, Jarvis A (2005) Very high resolution interpolated climate surfaces for global land areas. *International Journal of Climatology*, **25**, 1965–1978.

Hoffmann-Schielle C, Jug A, Makeschin F, Rehfuess KE (1995) Short rotation plantations of balsam poplar, aspen and willows on former arable land in the Federal Republic of Germany. I. Site-growth relationships. *Forest Ecology and Management*, **121**, 41–55.

Jørgensen U, Sevel L, Georgiadis P, Larsen SU (2014) Hvordan skal energipil gødes?. Agrologisk. 1:8-10. Available at: http://www.agrologisk.dk/artikel?id=77064 (accessed 25 March 2015).

Kilkki P, Päivinen R (1987) Reference sample plots to combine field measurements and satellite data in forest inventory. In: Remote sensing-aided forest inventory. Seminars organized by SNS and Forest Mensurationist Club, Hyytiälä, Finland, December 10–12. 1996. University of Helsinki, Department of Forest Mensuration and Management, Research Notes 19: 209–215.

Kunikowski G, Ganko E, Pisarek M et al. (2005) Review – Regional energy crops experience. Renew Renewable fuels for advanced powertrains. SES6-CT-2003-502705. 63 pp.

Labrecque M, Teodorescu TI (2003) High biomass yield achieved by Salix clones in SRIC following two 3-year coppice rotations on abandoned farmland in southern Quebec, Canada. *Biomass and Bioenergy*, **25**, 135–146.

Larsen SU, Jørgensen U, Lærke PE (2014) Willow yield is highly dependent on clone and site. *BioEnergy Research*, **7**, 1280–1292.

Lindegaard KN, Parfitt RI, Donaldson G, Hunter T, Dawson WM, Forbes EGA, Larsson S (2001) Comparative trials of elite Swedish and UK biomass willow varieties. *Aspects of applied biology*, **65**, 183–192.

Linderson ML, Iritz Z, Lindroth A (2007) The effect of water availability on stand-level productivity, transpiration, water use efficiency and radiation use efficiency of field-grown willow clones. *Biomass and Bioenergy*, **31**, 460–468.

Lindroth A, Båth A (1999) Assessment of regional willow coppice yield in Sweden on basis of water availability. *Forest Ecology and Management*, **121**, 57–65.

Lumme I, Törmälä T (1988) Selection of fast-growing willow (Salix spp.) clones for short-rotation forestry on mined peatlands in northern Finland. *Silva Fennica*, **22**, 67–88.

Mola-Yudego B (2010) Regional potential yields of short rotation willow plantations on agricultural land in Northern Europe. *Silva Fennica*, **44**, 63–76.

Mola-Yudego B (2011a) Predicting and mapping productivity of short rotation willow plantations in Sweden based on climatic data using a non-parametric method. *Agricultural and Forest Meteorology*, **151**, 875–881.

Mola-Yudego B (2011b) Trends and productivity improvements from commercial willow plantations in Sweden during the period 1986–2000. *Biomass and Bioenergy*, **35**, 446–453.

Mola-Yudego B, Aronsson P (2008) Yield models from commercial willow biomass plantations in Sweden. *Biomass and Bioenergy*, **32**, 829–837.

Mola-Yudego B, González-Olabarria JR (2010) Mapping the expansion and distribution of willow plantations in Sweden: lessons to be learned about the spread of energy crops. *Biomass and Bioenergy*, **34**, 442–448.

Mola-Yudego B, Pelkonen P (2008) The effects of policy incentives in the adoption of willow short rotation coppice for bioenergy in Sweden. *Energy Policy*, **36**, 3062–3068.

Mola-Yudego B, Pelkonen P (2011) Pulling effects of district heating plants on the adoption and spread of willow plantations for biomass: the power plant In Enköping (Sweden). *Biomass and Bioenergy*, **35**, 2986–2992.

Mola-Yudego B, Dimitriou I, Gonzalez-Garcia S, Gritten D, Aronsson P (2014) A conceptual framework for the introduction of energy crops. *Renewable Energy*, **72**, 29–38.

Mola-Yudego B, Díaz-Yáñez O, Dimitriou I (2015) How much yield should we expect from fast-growing plantations for energy? Divergences between experiments and commercial willow plantations. *BioEnergy Research*, **8**, 1769–1777.

Muinonen E, Tokola T (1990) An application of remote sensing for communal forest inventory. In: The usability of remote sensing for forest inventory and planning.

Nilsson LO, Eckersten H (1983) Willow production as a function of radiation and temperature. *Agricultural Meteorology*, **30**, 49–57.

Nordh NE (2005) Long term changes in stand structure and biomass production in short rotation willow coppice. Faculty of Natural Resources and Agricultural Sciences. SLU, Uppsala, Sweden. Doctoral thesis No. 2005:120.

Perttu KL (1983) Temperature restraints on energy forestry in Sweden. *International Journal of Biometeorology*, **27**, 189–196.

Perttu KL (1999) Environmental and hygienic aspects of willow coppice in Sweden. *Biomass and Bioenergy*, **16**, 291–297.

Perttu K, Eckersten H, Kowalik P, Nilsson LO (1984) Modelling potential energy forest production. In Perttu, K. (Ed.) Ecology and management of forest biomass production systems. Swedish University of Agricultural Sciences. Report 15. Uppsala.

R Development Core Team (2014) R: A language and environment for statistical computing. R Foundation for Statistical Computing, Vienna, Austria. ISBN 3-900051-07-0, Available at: http://www.R-project.org.

Rahman MM, Mostafiz SB, Paatero JV, Lahdelma R (2014) Extension of energy crops on surplus agricultural lands: a potentially viable option in developing countries while fossil fuel reserves are diminishing. *Renewable and Sustainable Energy Reviews*, **29**, 108–119.

Ridgeway G (2006) Generalized boosted regression models. Documentation on the R package 'gbm', version 1.5-7. http://www.i-pensieri.com/gregr/gbm.shtml.

Rosenqvist H, Dawson M (2005) Economics of willow growing in Northern Ireland. *Biomass and Bioenergy*, **28**, 7–14.

Schapire RE (2003) *The Boosting Approach to Machine Learning: An Overview. In Nonlinear Estimation and Classification*. pp. 149–171. Springer, New York.

Schonlau M (2005) Boosted regression (boosting): an introductory tutorial and a Stata plugin. *Stata Journal*, **5**, 330.

Searle SY, Malins CJ (2014) Will energy crop yields meet expectations? *Biomass and Bioenergy*, **65**, 3–12.

Sevel L, Nord-Larsen T, Raulund-Rasmussen K (2012) Biomass production of four willow clones grown as short rotation coppice on two soil types in Denmark. *Biomass and Bioenergy*, **46**, 664–672.

Statistics Norway (2013) Table: 09595: Sub classes of land use and land cover (km²). Available at: www.ssb.no (accessed 4 April 2015).

Swedish Energy Agency (2013) Energy in Sweden 2013, Eskilstuna, Sweden: Swedish Energy Agency, 2014.

Tahvanainen L, Rytkönen VM (1999) Biomass production of Salix viminalis in southern Finland and the effect of soil properties and climate conditions on its production and survival. *Biomass and Bioenergy*, **16**, 103–117.

Toivonen RM, Tahvanainen LJ (1998) Profitability of willow cultivation for energy production in Finland. *Biomass and Bioenergy*, **15**, 27–37.

Tokola T, Pitkänen J, Partinen S, Muinonen E (1996) Point accuracy of a non-parametric method in estimation of forest characteristics with different satellite materials. *International Journal of Remote Sensing*, **17**, 2333–2351.

Tomppo E (1990) Designing a satellite image-aided national forest survey in Finland. In: The usability of remote sensing from forest inventory and planning. Proceedings from SNS/IUFRO workshop in Umea, 26–28 Feb 1990. Swedish University of Agricultural Sciences, Remote Sensing Laboratory, Report 4: 43–47.

Van Meerbeek K, Van Beek J, Bellings L, Aertsen W, Muys B, Hermy M (2014) Quantification and prediction of biomass yield of temperate low-input high-diversity ecosystems. *BioEnergy Research*, **7**, 1120–1130.

Wühlisch G (ed.) (2012) Poplars and Willows in Germany: Report of the National Poplar Commission, Johann-Heinrich von Thünen-Institut (vTI) Bundesforschungsinstitut für Ländliche Räume, Wald und Fischerei, Time period: 2008–2011, 26.

4

High-resolution spatial modelling of greenhouse gas emissions from land-use change to energy crops in the United Kingdom

MARK RICHARDS[1], MARK POGSON[1,2], MARTA DONDINI[1], EDWARD O. JONES[1], ASTLEY HASTINGS[1], DAGMAR N. HENNER[1], MATTHEW J. TALLIS[3,4], ERIC CASELLA[5], ROBERT W. MATTHEWS[5], PAUL A. HENSHALL[5], SUZANNE MILNER[3], GAIL TAYLOR[3], NIALL P. MCNAMARA[6], JO U. SMITH[1] and PETE SMITH[1]

[1]Institute of Biological and Environmental Sciences, University of Aberdeen, 23 St Machar Drive, Aberdeen, AB24 3UU, UK, [2]Academic Group of Engineering, Sports and Sciences, University of Bolton, Deane Road, Bolton, BL3 5AB, UK, [3]Centre for Biological Sciences, University of Southampton, Life Sciences Building, Southampton, SO17 1BJ, UK, [4]School of Biological Sciences, University of Portsmouth, King Henry Building, King Henry I Street, Portsmouth, PO1 2DY, UK, [5]Centre for Sustainable Forestry and Climate Change, Forest Research, Farnham, Surrey, GU10 4LH, UK, [6]Centre for Ecology and Hydrology, Lancaster Environment Centre, Library Avenue, Bailrigg, Lancaster, LA1 4AP, UK

Abstract

We implemented a spatial application of a previously evaluated model of soil GHG emissions, ECOSSE, in the United Kingdom to examine the impacts to 2050 of land-use transitions from existing land use, rotational cropland, permanent grassland or woodland, to six bioenergy crops; three 'first-generation' energy crops: oilseed rape, wheat and sugar beet, and three 'second-generation' energy crops: Miscanthus, short rotation coppice willow (SRC) and short rotation forestry poplar (SRF). Conversion of rotational crops to Miscanthus, SRC and SRF and conversion of permanent grass to SRF show beneficial changes in soil GHG balance over a significant area. Conversion of permanent grass to Miscanthus, permanent grass to SRF and forest to SRF shows detrimental changes in soil GHG balance over a significant area. Conversion of permanent grass to wheat, oilseed rape, sugar beet and SRC and all conversions from forest show large detrimental changes in soil GHG balance over most of the United Kingdom, largely due to moving from uncultivated soil to regular cultivation. Differences in net GHG emissions between climate scenarios to 2050 were not significant. Overall, SRF offers the greatest beneficial impact on soil GHG balance. These results provide one criterion for selection of bioenergy crops and do not consider GHG emission increases/decreases resulting from displaced food production, bio-physical factors (e.g. the energy density of the crop) and socio-economic factors (e.g. expenditure on harvesting equipment). Given that the soil GHG balance is dominated by change in soil organic carbon (SOC) with the difference among Miscanthus, SRC and SRF largely determined by yield, a target for management of perennial energy crops is to achieve the best possible yield using the most appropriate energy crop and cultivar for the local situation.

Keywords: bioenergy, carbon, greenhouse gas, land-use change, Miscanthus, short rotation coppice, short rotation forestry, soil

Introduction

Two of the greatest challenges facing humanity this century are climate change, and the need to produce enough energy to meet the demands of a growing and developing population (Edenhofer et al., 2014). Bioenergy has been proposed as a potential significant contributor to both issues; as a feedstock for delivering energy security (Sims et al., 2006), and as a contributor to climate mitigation, through substitution of fossil fuels, thereby reducing net greenhouse gas (GHG)

Correspondence: Pete Smith
e-mail: pete.smith@abdn.ac.uk

emissions from energy production (Creutzig et al., 2015). Further, if the carbon dioxide (CO_2) released on combustion for energy generation was pumped into long-term geological storage (bioenergy with carbon capture and storage: BECCS), it may also serve as a negative emission technologies, capable of removing CO_2 from the atmosphere (Fuss et al., 2014; Smith et al., 2016). Although bioenergy is not without its limitations (Creutzig et al., 2015; Smith et al., 2016), its potential role in contributing to climate mitigation and energy security has led to considerable attention over recent years (Creutzig et al., 2015) with some analyses suggesting that 20% of global energy demand could be met by biomass without impact on food supply (Beringer et al.,

2011; Slade *et al.*, 2014). In the United Kingdom, about 3% of primary energy was from bioenergy in 2015, with bioenergy contributing 72% of all renewable energy (DECC, 2015). However, much of the current biomass feedstock for bioenergy is imported and this is expected to continue up to 2050 (Howes *et al.*, 2011). The UK Bioenergy Strategy (DECC, 2012) suggested that sustainably sourced bioenergy (i.e. not from land with high C stocks such as peatland or forest, or land used for food production) could contribute ~8–11% to the UK's total primary energy demand by 2020 and ~8–21% by 2050. If the United Kingdom is to source a proportion of this bioenergy domestically, some land-use change (LUC) to bioenergy crops is required.

It is important to assess the GHG implications of land-use transitions to bioenergy crops because LUC entails change in soil organic carbon (SOC) stocks (Smith, 2008), and also potentially, emissions of other non-CO_2 GHGs (Smith *et al.*, 2008), namely nitrous oxide (N_2O) and methane (CH_4). This will better inform decisions about what energy crops to use, what current land uses to target (and avoid) for energy crop development, and where to best to grow each crop (Alexander *et al.*, 2014; Hastings *et al.*, 2014; Wang *et al.*, 2014). To these ends, the principal objective of the spatial modelling exercise described here was to estimate the effects of land-use change (LUC) to bioenergy crops, on SOC content and GHG emissions in the United Kingdom, in order assess the impact of potential bioenergy transitions. Eighteen land-use transitions were considered:

- Rotational crops (which includes rotations consisting entirely of arable crops and also those including rotational grass) to *Miscanthus (Miscanthus giganteus)*, short rotation coppice (SRC; here represented by willow genotype Joruun [*Salix viminalis* L. x *S. viminalis*], since this is the SRC species currently used in commercial plantations in the UK) and short rotation forestry (SRF; here represented by poplar [mixed cultivars; *Populus trichocarpa, deltoids, nigra*], since this generally shows the highest yield under UK conditions)
- Permanent grass and forest to wheat, oilseed rape (OSR), sugar beet, *Miscanthus*, SRC and SRF
- Three 'null' transitions for rotational crops, permanent grass and forest to provide results for unchanged land use as a baseline.

Conversion from rotational crops to OSR, sugar beet and wheat was not considered because the rotational crop land-use prior to transition is assumed to be similar to that following the transition, resulting in no change in net GHG balance.

Results from the spatial simulations to determine the effects of LUC transitions to bioenergy crops on SOC, GHG emissions and net soil GHG balance in the United Kingdom are presented. Net soil GHG balance from simulations carried out using data from low, medium and high emission climate scenarios is compared to determine the impact of climate uncertainty.

Materials and methods

The ECOSSE model

The ECOSSE (Estimation of Carbon in Organic Soils – Sequestration and Emissions) model simulates soil C and nitrogen (N) dynamics in both mineral and organic soils using meteorological, land use, land management and soil data and simulates changes in SOC and soil GHG emissions. The model is able to function at the field scale or at the national scale (using only the limited data available at this scale; Smith *et al.*, 2010a,b,c).

ECOSSE was developed from concepts originally derived for mineral soils in the RothC model (Jenkinson & Rayner, 1977; Jenkinson *et al.*, 1987; Coleman & Jenkinson, 1996) and SUNDIAL model (Bradbury *et al.*, 1993; Smith *et al.*, 1996). ECOSSE describes soil organic matter using 5 pools: inert organic matter, humus, biomass, resistant plant material (RPM) and decomposable plant material (DPM). All of the major processes of C and N turnover are included in the model, but each process is simulated using only simple equations driven by readily available inputs. This enables ECOSSE to be used for national scale simulations for which only limited input data are available.

ECOSSE simulates the soil profile to a depth of up to 3 m, dividing the soil into 5 cm layers to facilitate the accurate simulation of processes to depth. Plant C and N inputs are added monthly to the DPM and RPM pools. During the decomposition process, material is exchanged between the soil organic matter pools according to first-order equations, characterized by a specific decomposition rate for each pool. The decomposition rate of each pool is modified dependent on the temperature, water content, plant cover and pH of the soil (with additional modifiers dependent upon soil bulk density and inorganic N concentration in the case of anaerobic decomposition; Smith *et al.*, 2010c). The decomposition process results in gaseous losses of CO_2 and CH_4, with CO_2 losses dominating under aerobic conditions and CH_4 losses under anaerobic conditions. ECOSSE also simulates the oxidation of atmospheric CH_4, which, under aerobic conditions, can lead to the soil being a net consumer of CH_4 (Smith *et al.*, 2010c).

The nitrogen (N) content of the soil follows the decomposition of the soil organic matter, with a stable C:N ratio defined for each soil organic matter pool at a given pH, and N being either mineralized or immobilized to maintain that ratio. Nitrogen is released from decomposing soil organic matter as ammonium (NH_4^+) and may be then immobilized or nitrified to nitrate (NO_3^-). Carbon and N may be lost from the soil by the processes of leaching of NO_3^-, dissolved organic C, dissolved organic N, denitrification to nitric oxide (NO) and N_2O, volatilization of ammonia, or crop off-take of NO_3^- and NH_4^+. Carbon and N may be returned to the soil by plant input, application of inorganic fertilizers, atmospheric deposition or organic amendments (e.g. manure, crop residues).

ECOSSE simulates the soil water content of each layer using a 'tipping bucket' approach based on the SUNDIAL model (Bradbury *et al.*, 1993; Smith *et al.*, 1996). Water from precipitation entering the soil forces water in the soil deeper into the soil profile. Precipitation fills the uppermost soil layer with water until it reaches field capacity. Any remaining precipitation is then used to fill the next layer to field capacity. This process is repeated until no precipitation remains or the bottom of the profile is reached. Any precipitation water remaining after filling all layers to field capacity is partitioned between drainage (water leaving the soil profile) and excess, which is used to fill layers to saturation from the bottom of the profile upwards. This is performed using the observed depth of the water table, the available water at saturation and weather data to calculate the restriction to drainage (i.e. the fraction of the remaining water that becomes excess), that is required to achieve the observed water table depth. Water is also lost from the top of the profile as evapotranspiration, which is estimated using the Thornthwaite (1948) method.

The ECOSSE model has been thoroughly evaluated and shown to simulate SOC change, N_2O and CH_4 emissions reliably, for bioenergy crop transitions in the United Kingdom using field data from *Miscanthus*, SRC and SRF field sites, as described in Dondini *et al.*, 2015, 2016a,b).

Spatial application of the ECOSSE model

The spatial simulations of the United Kingdom are carried out on a 1 km grid basis giving a total of nearly 0.25 million grid cells. Grid cells which contain inappropriate land for growing bioenergy crops were excluded from the simulations based on the UKERC 7 land-use constraints (Lovett *et al.*, 2014). The UKERC 7 constraints mask excludes 100 m grid cells that meet one or more of the following criteria (Lovett *et al.*, 2014): slope $\geq 15\%$; peat (soil C $\geq 30\%$); designated areas; urban areas, roads, rivers; parks; scheduled monuments/world heritage sites; woodland and natural habitats from LCM2007 (including acid, neutral and calcarious grassland). We aggregated the UKERC 7 mask to 1 km, using the mode of the 100 m cells to determine exclusions at the 1 km scale. We also disaggregated the woodland category in order to permit use of woodland for transitions to SRF; we term this mask UKERC 7w. The land-use data that were used to initialize the ECOSSE 1 km grid were aggregated using the mode from the LCM2007 land cover from at 1 ha resolution in the UKERC mask (Lovett *et al.*, 2014).

The simulation of each LUC was carried out for up to 5 different soil types in each grid cell to capture soil heterogeneity at the subgrid cell level. All combinations of LUC from rotational crops, permanent grass and forest to wheat, OSR, sugar beet, *Miscanthus*, SRC and SRF were simulated, except for rotational crops to wheat, OSR and sugar beet which, being types of rotational crops, were considered to be equivalent to no LUC. Three 'null' transitions for rotational crops, permanent grass and forest were also simulated to provide results for unchanged land-use for comparison.

The rotational crop land-use category represents land used to grow arable crops and includes all-arable rotations and rotations that include rotational or temporary grassland for part of the rotation. The permanent grass land-use category represents permanent, uncultivated grassland only. Rotational grass is not a land-use and is part of rotational farming better represented by the rotational crops category (which may include periods of rotational grass).

Results were obtained using three different climate scenarios for a 35-year period running from 2015 to 2050. Prior to each simulation, the model was initialized to partition the SOC into the different SOC pools based on the assumption that the SOC in the soil column is at steady state under the land use given for the start of the simulation.

Following initialization, the main simulation was executed. This started with LUC from the initial land-use type to the bioenergy crop. Any soil cultivation carried out during LUC was simulated. As rotational cropland typically undergoes annual cultivation, the model assumes there is no additional cultivation required for the establishment of bioenergy crops. In contrast, the model simulates soil cultivation for LUC from permanent grass and forestry because these land-use types typically require ground preparation before bioenergy crops are planted. The model simulates physical fragmentation of soil organic matter resulting from cultivation by moving a proportion of the C and N in the humus pool, (which has a slow decomposition rate), to the DPM and RPM pools (which have faster decomposition rates; Smith *et al.*, 2010a). Redistribution of soil organic matter during cultivation is simulated by homogenizing the vertical distribution of the soil organic matter pools down to the cultivation depth – which might be expected with inversion ploughing followed by harrowing as ground preparation. The simulated cultivation depth for conversion from forest and permanent grass is 0.5 and 0.3 m, respectively.

After simulation of the cultivation associated with LUC, the model simulates soil dynamics under the bioenergy crop. The annual plant inputs of C and N to the soil are calculated from the annual yield of the crop (provided as an input to the model), using crop-specific ratios estimated from the literature.

For perennial bioenergy crops, the model simulates annual yield dynamics over the lifetime of the crop to account for reduced yields during establishment and peak yield later in the crop life cycle. Yield dynamics are modelled using the lifetime mean annual yield of the crop (as an input to the model) and five crop-specific parameters:

- $Y_{\text{peak-ratio}}$ – ratio of peak annual yield to lifetime mean annual yield, used to calculate peak annual yield.

- T_{peak} – time required for the crop to reach peak annual yield.

- T_0 – time spent at initial yield, before annual yield begins to increase towards peak annual yield. Used to approximate a sigmoidal growth curve.

- $Y_{\text{0-frac}}$ – initial yield as a fraction of lifetime mean annual yield. This parameter is calculated from the other parameters to ensure that the lifetime mean annual yield of the crop is preserved.

- Lifetime – the lifespan of the crop.

The parameter values for each perennial crop, which are based on expert opinion and informed by published studies

Table 1 Yield model parameters for *Miscanthus*, SRC and SRF. See text for an explanation of each parameter

Crop	$Y_{peak\text{-}ratio}$	T_{peak} (years)	T_0 (years)	$Y_{0\text{-}frac}$	Lifetime (years)
Miscanthus	1.1	5	0	0.299	20
SRC	1.1	6	0	0.433	20
SRF	1.6	15	4	0.267	20

such as Arundale *et al.* (2014), are given in Table 1. The simulated yield dynamics are characterized by 3 stages: a period spent at initial annual yield (SRF only), a period of linearly increasing annual yield and a period spent at peak annual yield. An example of the growth dynamics of each crop given by the parameter values in Table 1 is shown in Fig. 1. The lifetime mean annual yields used as an input to the model are taken from a number of sources described below.

The annual yield dynamics of perennial crops typically follow a sigmoidal curve. Here, we employed a simple linear-based approach to yield modelling to maintain model parsimony. *Miscanthus* and SRC establish quickly and do not have a very pronounced sigmoidal growth curve. Therefore, the linear increase during establishment will only result in a small error in the timing of plant inputs to the soil (and subsequent effects on the timing of changes in SOC and GHG emission). For SRF, which has a longer establishment time and a more pronounced sigmoidal growth curve, we introduced an additional flat growth phase at the start of establishment to better approximate the sigmoidal curve and minimize the error in the timing of plant inputs.

Each perennial bioenergy crop was re-established after a 20-year period (the estimated productive lifespan of the crop). It is assumed that re-establishment does not involve further cultivation. This assumption was made because perennial bioenergy crops can be re-established with only shallow soil disturbance

or very localized soil disturbance (McCalmont *et al.*, 2015). *Miscanthus* crops can be re-established by herbicide application of the existing crop followed by direct drilling of rhizomes, or planting plugs grown from seed (Clifton-Brown *et al.*, 2015). Ploughing of *Miscanthus* can be avoided by exposing the rhizomes on top of the soil so that they dehydrate and die (Caslin *et al.*, 2011a). The SRC can be removed by application of herbicide followed by mulching of the stools (using a bush-hogger), into the top 5–10 cm of the soil (Defra, 2004) and SRF may be re-established by planting between previous stumps (McKay, 2011). The impacts of soil disturbance during re-establishment of perennial bioenergy crops are poorly understood and require further research (Grogan & Matthews, 2002). However, as the re-establishment of these crops can be made with only shallow soil disturbance (the top 5–10 cm), or very localized disturbance (e.g. direct drilling of *Miscanthus* and replanting SRF between stumps), we expect the impacts on SOC to be small. Fertilizer was applied to *Miscanthus* and SRC at an annual rate of 30 and 60 kg N ha^{-1}, respectively, following recommended practice (Defra, 2010; Caslin *et al.*, 2011a). Fertilizer was applied to SRF at a rate of 45 kg N ha^{-1}. No fertilizer was applied to *Miscanthus*, SRC and SRF during the first 2 years after planting, again following best practice guidelines (Defra, 2010; Caslin *et al.*, 2011a).

Forest was assumed to be unfertilized. Rotational crops, permanent grass, wheat, OSR and sugar beet were assumed to be fertilized at a rate equal to the annual crop N demand. Crop N demand is a function of plant yield and the C:N ratio of the plant. Modelled crop N demand is high for wheat because it has a low C:N ratio and a relatively high yield. In contrast, modelled N demand for permanent grass was significantly lower because it has a higher C:N ratio.

For all land-use types, the changes in SOC and emissions of GHGs were calculated for the top metre of the soil profile. Only the top metre was considered because this is the depth to which soil parameters are provided by the Harmonised World Soil Database (HWSD; see below). Changes in SOC, CH$_4$ and

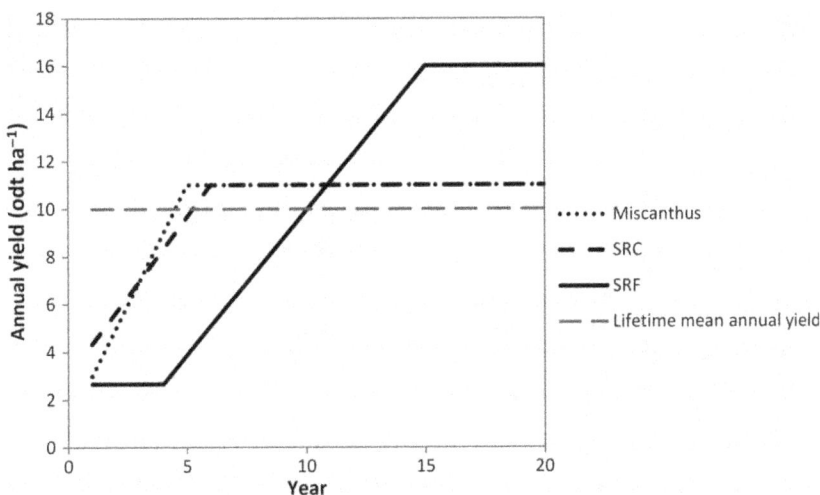

Fig. 1 Annual yield dynamics of *Miscanthus*, SRC and SRF over the 20-year lifespan of each crop, with a lifetime mean annual yield of 10 odt ha^{-1}. Lifetime mean annual yield is represented by the dashed grey line for comparison.

N_2O resulting from LUC were calculated by subtracting the results of the appropriate null transition from the LUC results, so that the change could be attributed solely to the LUC. For example, to calculate the impact of LUC from permanent grass to SRC, the results from the permanent grass null transition (i.e. grass remaining as grass) were subtracted from the permanent grass to SRC results. Each grid cell value in the model output represents the area-weighted mean of the simulations carried out for each soil type in the grid cell.

Results are given for the whole of the United Kingdom on a 1 km grid basis and express the area-weighted average obtained from simulations of the 5 most dominant soil types in each grid cell. For consistency and ease of comparison, all results (i.e. CH_4, N_2O, change in SOC and net GHG balance) are reported in terms of CO_2-equivalent values (CO_2e), using IPCC 100-year global warming potentials (GWPs; IPCC, 2001). More recent IPCC reports have provided updated GWPs from those given in the IPCC 2001 report, although, for consistency with national inventory GHG emission estimates, we have used the IPCC 2001 GWP values, following the recommended practice for national GHG inventories. Results show the cumulative total of each output variable and are relative to the value obtained if no transition had occurred (hence results directly show the effect of the transition).

Soil data

The HWSD version 1.2 was used to provide initial soil conditions in the model (FAO/IIASA/ISRIC/ISS-CAS/JRC, 2012). The HWSD provides soil data to a depth of 1 metre at a resolution of 30 arc s (approximately 1 km), for the dominant soil types in each grid cell; we reprojected this to the British National Grid using methods described by Ordnance Survey (Ordnance Survey, 2010). The soil properties used from this database to drive ECOSSE were as follows: organic C content, bulk density, pH, and sand, silt and clay faction. The HWSD does not include information on the water-holding capacities of soils so these were estimated using British Soil Survey pedotransfer functions (Hutson & Cass, 1987), which performed well in evaluations (Donatelli et al., 1996; Givi et al., 2004). The HWSD also provides the percentage of grid cell area covered by each soil type. The ECOSSE model is run for each dominant soil type in each grid cell and the output area-weighted by the percentage cover in each grid cell to calculate its mean response.

Climate data

ECOSSE requires precipitation and air temperature data which are used to drive the soil water model and to determine temperature-based rate modifiers for various soil processes. The meteorological driving data was taken from the UKCP09 Spatially Coherent Projections (Murphy et al., 2009). UKCP09 provides average monthly temperature and precipitation in a 25 km grid for overlapping 30-year periods centred on decades ranging from the 2020s to the 2080s, for high, medium and low emissions scenarios; again, we reprojected this to the British National Grid using methods described by Ordnance Survey (Ordnance Survey, 2010).

Yield data

ECOSSE requires yield data for each land-use type in order to estimate the monthly plant inputs to the soil. Yield data for the bioenergy crops have been obtained from a range of sources of varying spatial resolution. Baseline yields for first-generation crops were obtained from EUROSTAT (2014), which provides mean wheat and OSR yields across 12 NUTS (Nomenclature of Territorial Units for Statistics) regions in the United Kingdom, and a single mean national yield for sugar beet, based on Defra farm surveys. The baseline yield values for the rotational crop land-use category follow those of wheat.

Yield estimates for wheat, OSR and sugar beet under different climate scenarios were obtained by adjusting the baseline yields using the Miami model (Lieth, 1975). Miami is an empirical net primary production (NPP) model that estimates annual NPP from mean annual temperature and annual precipitation. The Miami estimate of NPP was calculated for each decade in each grid cell using the same UKCP09 climate data that was used for the ECOSSE simulations. The percentage change in NPP relative to the baseline Miami NPP was applied to the baseline yield data to adjust the yield for each climate scenario. Yield estimates for permanent grass and forest are obtained using NPP estimates from Miami, which are then linearly rescaled according to observed peak yields (Living Countryside, 2013) to reflect differences in grass and forest productivity.

Lifetime mean annual yield estimates for Miscanthus, SRC and SRF were obtained from simulations using the models MiscanFor (Hastings et al., 2009), ForestGrowth SRC (Tallis et al., 2013) and ESC-CARBINE (Thompson & Matthews, 1989; Pyatt et al., 2001), respectively. The yield predictions have been obtained using the same UKCP09 climate and HWSD soil data used as inputs to ECOSSE. These models were used due to their validated accuracy and use of compatible data. The lifetime mean annual yields were provided for each decade because the UKCP09 climate data provide long-term average climate values centred on each decade. As an ECOSSE simulation progresses, the annual yield for each year of the simulation is calculated from the lifetime mean annual yield for the current decade. Therefore, if the lifetime mean annual yield changes between decades, this is reflected by a change in the annual yield calculated within the model.

SRC was represented by willow. Although the yield modelling study of Hastings et al. (2014) found that SRC poplar outperformed SRC willow in almost all regions within Great Britain, willow remains the SRC species currently used in commercial plantations in the United Kingdom and thus, despite lower yields, is used to represent SRC here. SRF was represented by poplar, because Hastings et al. (2014) found that poplar outperformed all other SRF species included in the study except for Sitka spruce in the Scottish Highlands and Pennines (areas which are mostly excluded by the UKERC constraints mask). The other SRF species included in the study were as follows: aspen (Populus tremula L.), black alder (Alnus glutinosa L.), European ash (Fraxinus excelsior L.), sitka spruce (Picea sitchensis [Nong.] Carr.) and silver birch (Betula pendula Roth). The lifetime mean annual yields of SRF poplar across Great Britain were at least double those of other species. Given no clear commercial benefits of selecting other SRF species over poplar, we assumed the strong commercial incentive offered by the much higher yields will mean that poplar will be the dominant SRF species in the United Kingdom. Planting of lower yielding SRF species to avoid pest and disease outbreaks would result in consequent lower inputs to the soil.

The spatial distributions of yield used for each crop as driver for the ECOSSE model runs are presented in Fig. 2, showing mean annual yields in a single decade (2030s). Mean decadal yields changed by <2 oven dry tonnes (odt) ha^{-1} yr^{-1} over the 35-year study period for any single crop (data not shown).

If LUC leads to an increase in plant C inputs to the soil, the SOC content will gradually increase over time until a new equilibrium SOC content is reached (assuming all other factors remain equal). In ECOSSE, the quantity of new plant material entering the soil organic matter pools is determined by the amount of plant biomass (calculated from yield), minus the proportion of biomass that is removed during harvest.

Across the simulation area, SRF, Miscanthus and sugar beet are the highest yielding bioenergy crops with United Kingdom mean annual yields of 10–12 odt ha^{-1}. Based on reported harvest index values (Table 2), the model assumes that 75% of sugar beet biomass is removed during harvest, compared with 64% for Miscanthus and 60% for SRF. The low harvest index (relative to sugar beet) and high yields mean that SRF and Miscanthus have, on average, higher plant inputs to the soil than other bioenergy crops.

Conversion to bioenergy crops can also lead to a change in the quality of plant inputs to the soil. Plant residues from perennial grasses and woody plants such as Miscanthus, SRC and SRF are typically slower to decompose than residues from annual crops such as wheat, OSR and sugar beet (due in part to the residues having a higher C:N ratio). Slower decomposition rates reduce the rate of SOC loss. In the model, differences in crop residue decomposition rates are simulated through differential allocation of plant residues to two SOC pools: the DPM and RPM pools. The DPM pool has a faster decomposition rate than the RPM pool. To reflect the slower decomposition rates, residues from Miscanthus, SRC and SRF have a higher RPM:DPM ratio than residues from wheat, OSR and sugar beet.

Results

Differences in net GHG balance between the climate scenarios for land-use transitions to bioenergy crops are

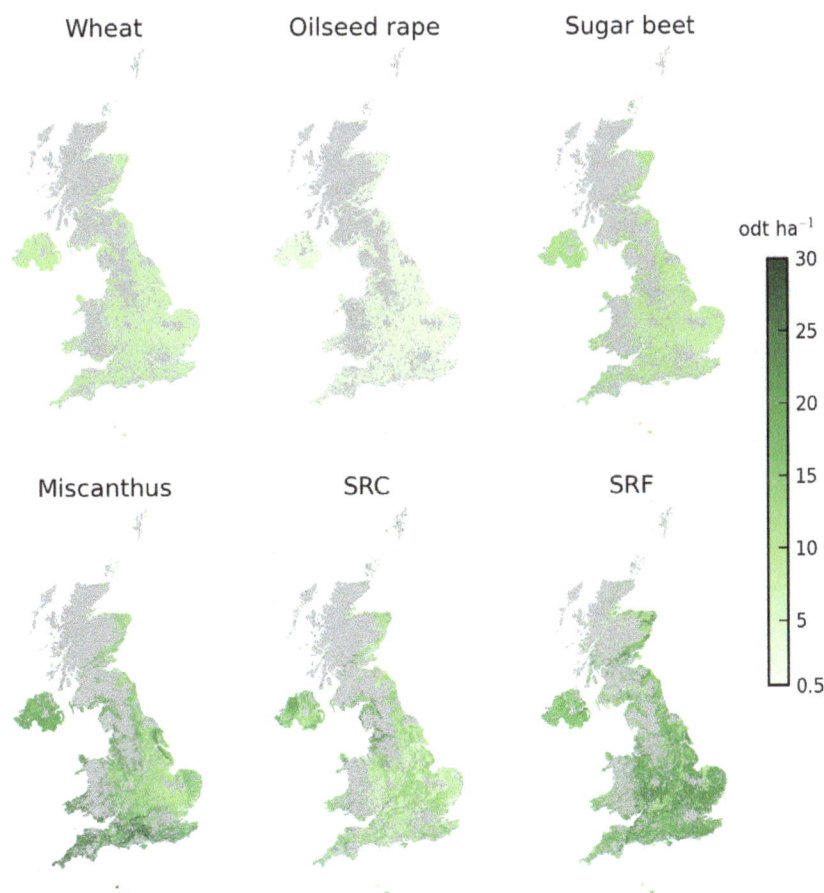

Fig. 2 Spatial distribution of modelled mean annual yield of bioenergy crops as odt ha^{-1} yr^{-1} (where odt is oven dry tonnes) in the 2030s under the UKCP09 medium emissions climate scenario. Miscanthus, SRC and SRF yields were obtained from simulations using the models MiscanFor, ForestGrowth SRC and ESC-CARBINE, respectively. Defra yield statistics from 2000 to 2008 were used to establish baseline yield values for wheat, OSR and sugar beet, which were then adjusted for future climate using the Miami model (see text).

Table 2 Harvest index parameter values of bioenergy crops. Note that the wheat harvest index includes the harvest of both grain and straw

Crop	Harvest index	Source
Miscanthus	0.64	Zhuang *et al.*(2013)
Oilseed rape	0.35	Kjellström & Kirchmann (1994), Dreccer *et al.*(2000), HGCA (2014)
SRC	0.6	Caslin *et al.*(2011b)
SRF	0.6	No data available so assumed to be the same as SRC
Sugar beet	0.75	Tsialtas & Karadimos (2003), Oritz *et al.*(2012)
Wheat	0.77	White & Wilson (2006), Stoddart & Watts (2012)

very small in comparison with the effects of LUC, being within ± 2 t CO_2e ha^{-1} over the 35-year period for any given transition. Given this very minor impact of climate to 2050, all results presented are for the medium climate scenario, and all results refer to the period 2015–2050 unless otherwise stated.

Net GHG balance represents the combined effects of changes in N_2O, CH_4 and SOC and is therefore the most comprehensive measure of bioenergy impacts; a negative GHG balance represents removals from the atmosphere (i.e. beneficial), and a positive GHG balance represents emissions to the atmosphere (i.e. detrimental). Only second-generation bioenergy crops (*Miscanthus*, SRC and SRF) showed any beneficial changes in soil GHG balance; all conversions to first-generation bioenergy crops (wheat, sugar beet and OSR) showed a detrimental impact on net GHG balance (Figs 3–5), except of course for 'conversion' from rotational crops, where a null transition is assumed (i.e. zero GHG balance).

Of the three initial land uses, conversion from rotational crops has the most favourable net GHG balance (Fig. 3). Conversion of rotational crops to SRF, SRC and *Miscanthus* each showed a beneficial response in almost all grid cells, with mean net GHG balance values of -126.9, -37.8 and -76.4 t CO_2e ha^{-1} over 35 years, respectively. In contrast, all conversions from permanent grass result in a detrimental change in net GHG balance in all grid cells except for SRF, which shows a small beneficial (>-21 t CO_2e ha^{-1}) change over large parts of the West Midlands, East Midlands and East Anglia (Fig. 4). Transitions from forest show detrimental soil net GHG balance in all grid cells with mean values of 88.7, 128.6 and 102.9 CO_2e ha^{-1} over 35 years for SRF, SRC and *Miscanthus*, respectively (Fig. 5). Overall, conversion of rotational crops to SRF is the most favourable conversion because it has the most beneficial net GHG balance over the largest area (Fig. 3). However, in some areas, most notably in parts of south-west England, southern England, south and west Wales, and in a narrow band north and south of the Humber, *Miscanthus* presents an equal or slightly better bioenergy opportunity than SRF (see the Discussion for further consideration of this). In contrast, SRC does not show a more beneficial net GHG balance than SRF or *Miscanthus* in any areas of significant size (Fig. 3).

The mean, minimum and maximum cumulative changes in SOC (expressed as loss of SOC, i.e. CO_2 emissions), in N_2O emissions and in CH_4 emissions from 2015 to 2050 following LUC from rotational crops, permanent grass and forest to all energy crops are shown in Table 3.

Conversion of land to bioenergy crops shows a large spatial and temporal variation in net GHG balance and its components: SOC, N_2O and CH_4. The impact of land-use change on net GHG balance depends upon the

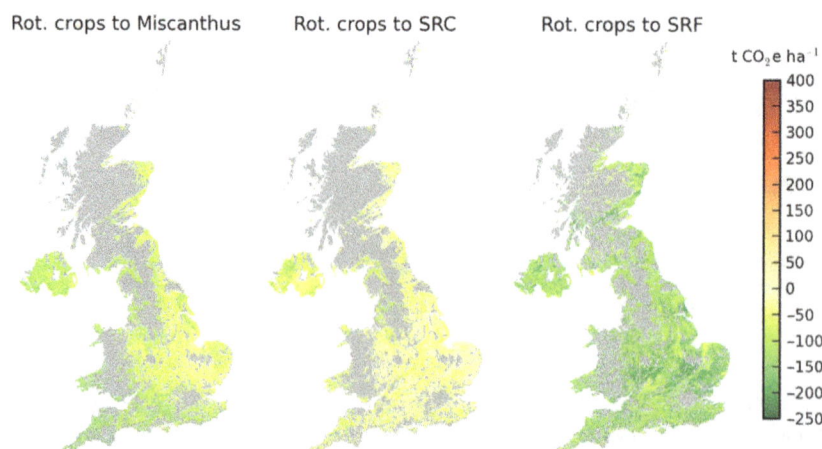

Fig. 3 Greenhouse gas emissions when rotational crops are converted to second-generation energy crops.

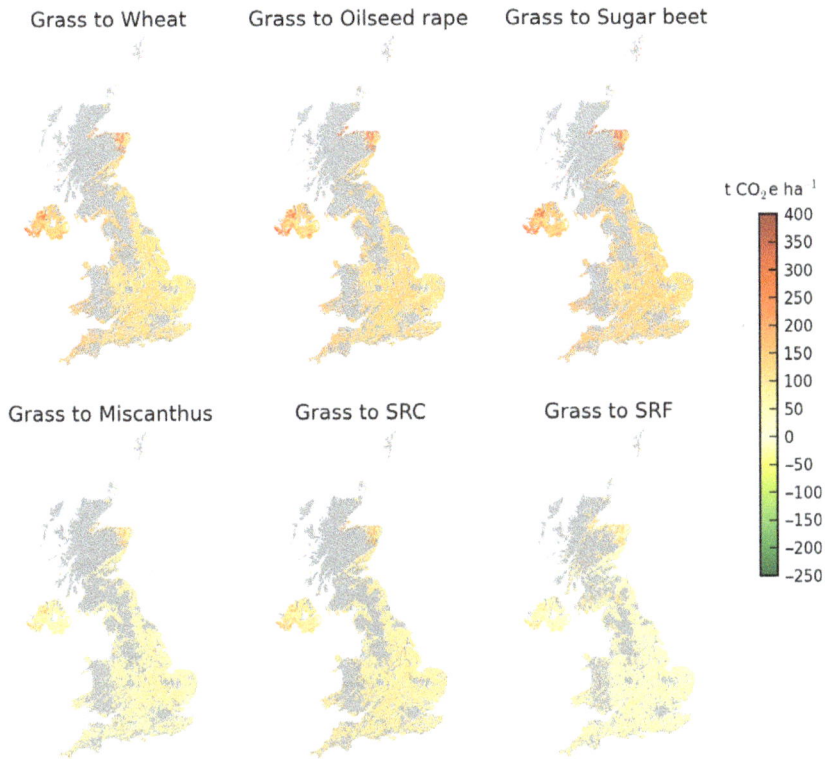

Fig. 4 Greenhouse gas emissions when permanent grass is converted to energy crops.

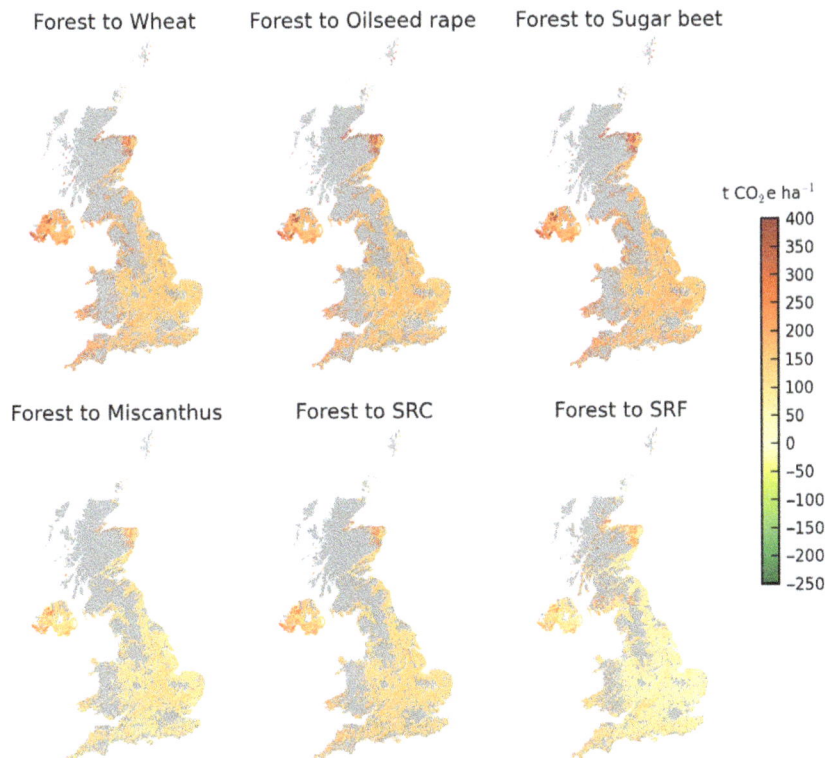

Fig. 5 Greenhouse gas emissions when forest is converted to energy crops.

Table 3 Mean (minimum and maximum range in parentheses; all t CO_2e ha^{-1}) change in SOC, N_2O emissions and CH_4 emissions from 2015 to 2050, following conversion to energy crops from rotational crops, rotational crops, permanent grass and forest. All positive values are emissions to the atmosphere, so an SOC gain (removal of CO_2 from the atmosphere) is shown as negative

Flux	Previous land use	Energy crop (all cumulative emissions in t CO_2e ha^{-1} for 2015–2050)					
		Wheat	OSR	Sugar beet	Miscanthus	SRC	SRF
SOC	Rotational crops	n/a	n/a	n/a	−55.4 (−121.8 to 0.7)	−18.7 (−114.4 to 27.4)	−102.9 (−205.4 to −9.9)
SOC	Grassland	85.4 (37.8 to 220.6)	119.9 (65.3 to 283.5)	118.6 (64.2 to −282.5)	44.7 (−1.7 to 147.8)	70.0 (33.0 to 186.3)	24.3 (−30.2 to 147.1)
SOC	Forest	117.9 (56.8 to 321.5)	149.7 (81.6 to 369.0)	148.5 (80.6 to 365.4)	78.2 (31.0 to 216.8)	102.0 (56.6 to 251.8)	64.0 (4.6 to 219.1)
N_2O	Rotational crops	n/a	n/a	n/a	−21.0 (−33.6 to −13.2)	−19.0 (−30.8 to −12.5)	−24.0 (−37.3 to −15.8)
N_2O	Grassland	35.9 (23.6 to 74.8)	10.9 (−0.6 to 54.3)	20.5 (10.3 to 62.1)	8.8 (0.2 to 43.9)	10.5 (1.6 to 49.2)	8.6 (−2.1 to 44.2)
N_2O	Forest	52.4 (39.0 to 88.0)	27.3 (16.8 to 56.4)	36.9 (25.4 to 66.5)	24.7 (13.6 to 42.9)	26.6 (17.9 to 47.7)	24.7 (16.5 to 42.3)
CH_4	Rotational crops	n/a	n/a	n/a	0.02 (−1.10 to 0.10)	−0.03 (−1.03 to 0.03)	−0.04 (−1.22 to 0.02)
CH_4	Grassland	−0.04 (−0.12 to 1.02)	−0.11 (−0.19 to 0.56)	−0.11 (−0.19 to 0.68)	−0.02 (−0.11 to 0.43)	−0.07 (−0.14 to 0.09)	−0.05 (−0.14 to 0.37)
CH_4	Forest	−0.02 (−0.10 to 1.78)	−0.09 (−0.18 to 0.74)	−0.09 (−0.18 to 0.89)	0.00 (−0.06 to 0.48)	−0.05 (−0.13 to 0.81)	−0.02 (−0.13 to 0.50)

type of land-use being converted, the type of bioenergy crop planted, the geographic location and the time since conversion. Overall, changes in SOC content have the largest impact on net GHG balance, followed by changes in N_2O, then CH_4 emissions (Table 3).

Figure 6 shows the SOC, N_2O and CH_4, respectively, for one transition, rotational crops to SRF, to show the relative influence of each component on the total net GHG balance (Fig. 6).

In general, most of the benefits to net GHG balance from favourable conversions are realized in the first 15–20 years following conversion; after this time, the rate of decrease in net GHG balance declines as SOC content approaches a new equilibrium (Fig. 7).

The changes in mean cumulative net GHG balance in soil emissions over time for each land-use transition are shown in Fig. 7. Conversion from rotational crops to *Miscanthus*, SRC and SRF show a decrease in net GHG balance in soil emissions over the 35-year simulation period, although there is little change during the first 5 years following conversion of rotational crops to SRC (Fig. 7). By contrast, all conversions from permanent grass and forest show a rapid increase in net GHG balance in soil emissions 5 years after LUC (Fig. 7). After the first 5 years, the net GHG balance of most LUCs continues to increase at a slower, broadly linear rate. However, in 2030 (15 years after conversion), the net GHG balance of permanent grass to SRF begins to decrease (Fig. 7).

Discussion

Land-use change emissions, such as those reported here, make a significant contribution to the overall GHG balance of energy crop transitions and are a relatively poorly constrained term in many bioenergy life cycle analyses. We discuss our findings further below.

Effects of land-use change

Conversion of land to bioenergy crops shows a large spatial and temporal variation in net GHG balance and its components; SOC, N_2O and CH_4. The impact of LUC on net GHG balance depends upon the type of land-use being converted, the type of bioenergy crop planted and the geographic location. Overall, changes in SOC have the largest impact on net GHG balance in soil emissions, followed by N_2O and then CH_4, accounting for the GWP of each flux.

Changes in soil organic carbon

Results for 2015 to 2050 show that both the initial and target land-use type have a very large impact on mean

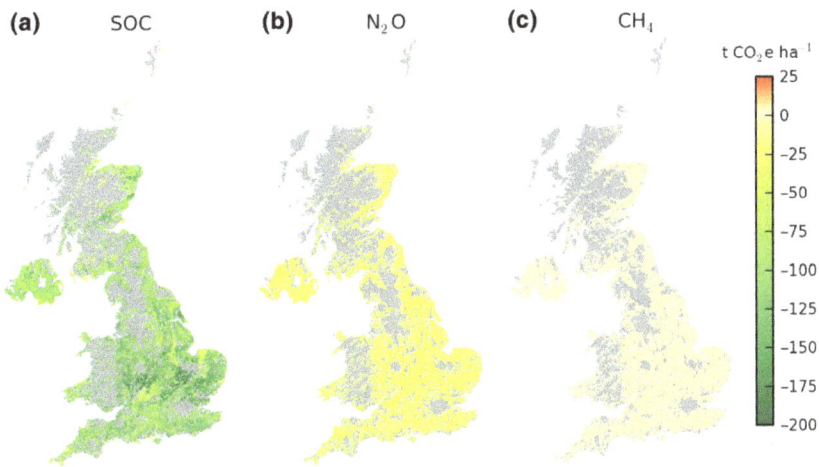

Fig. 6 All panels show the transition from rotational crops to SRF: panels a, b and c show maps of CO_2e for SOC (a), N_2O (b) and CH_4 (c).

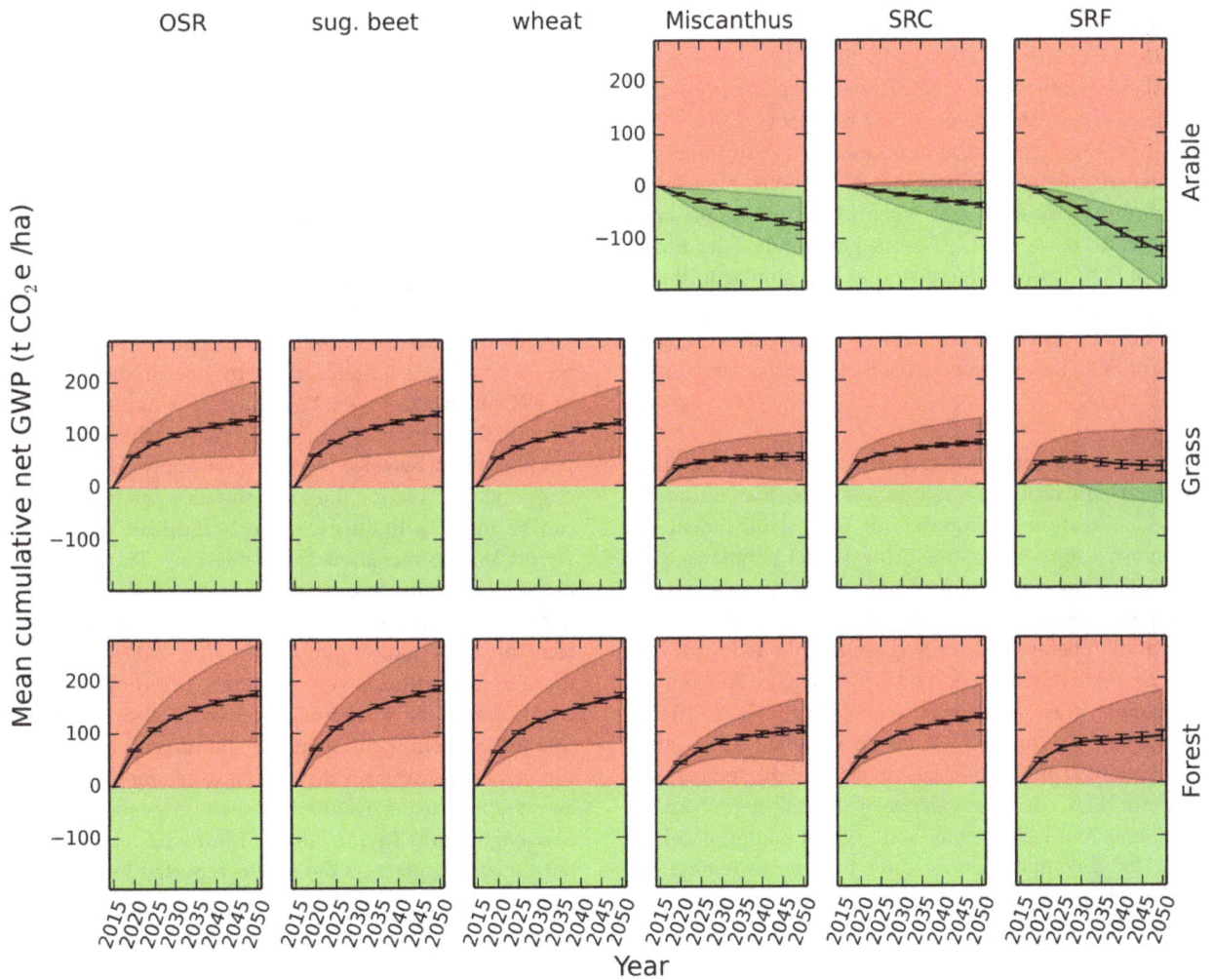

Fig. 7 Time series of mean cumulative net GHG balance resulting from land-use change to bioenergy crops in 2015 under the medium emissions climate scenario. Shaded areas show the 95% confidence interval of the distribution of modelled results (due to spatial variation) from the simulations across the United Kingdom. Error bars show the 95% confidence interval of estimated error based on the comparison of modelled and measured net GHG balance from site-level modelling studies (Dondini *et al.*, 2015, 2016a,b). The red portion of each panel shows a net GHG emission, the green portion shows a net GHG sink.

change in SOC. Conversion of rotational crops to *Miscanthus*, SRC and SRF and conversion of permanent grass to SRF are the only LUCs that lead to extensive beneficial changes in SOC. By contrast, all conversions from permanent grass to non-SRF bioenergy crops and all conversions from forest lead to mostly detrimental changes in SOC. These findings are broadly in-line with those of empirical and other modelling studies. Guo & Gifford's (2002) review of data from 74 LUC publications shows that conversion from arable land to plantation forest, secondary forest and pasture leads to significant increases in SOC, whereas conversion of forest and pasture to crop leads to large decreases. Murty *et al.* (2002) and Wei *et al.* (2014) also reported significant decreases in SOC following conversion of forest to cultivated agricultural land. For bioenergy crop transitions, the recent meta-analysis of Harris *et al.* (2015) that reported only empirical and not modelled data revealed in contrast to the findings here, broadly neutral SOC following LUC from grassland to SRC. For LUC from grassland to *Miscanthus*, the meta-analysis supports the findings here with SOC declining by 10.9 ($\pm$4.3) %. The literature survey of McCalmont *et al.* (2015), of the environmental impact of *Miscanthus* plantations, showed a significant increase in SOC when LUC from arable and a slight decrease for LUC from grasslands. The timescale of SOC losses in these studies was similar to those in our study, with most of the SOC loss occurring in the first 10–20 years after conversion. After this period, rates of SOC loss decline as the SOC approaches a new equilibrium.

Cultivation. As rotational cropland typically undergoes frequent cultivation, the model assumes that no additional cultivation is required for the establishment of bioenergy crops. By contrast, the model simulates soil cultivation for conversion of permanent grass and forestry because these land uses typically require ground preparation before bioenergy crops are planted. Cultivation of relatively undisturbed soil, such as soil under permanent grass and forest, usually has a large detrimental impact on SOC (Guo & Gifford, 2002), and McCalmont *et al.* (2016) also showed a slight decrease in SOC for LUC from grasslands. Cultivation physically fragments and redistributes soil organic matter, accelerating its decomposition, leading to a large release of CO_2 and subsequent decrease in SOC (Grandy & Robertson, 2006). As a rule, minimizing soil disturbance to the extent possible will minimize adverse impacts on SOC. The model captures this loss of SOC by simulating cultivation as described in the Methods section. This cultivation is responsible for the large detrimental change in SOC following LUC from permanent grass and forest (Figs 4 and 5). However, in the case of

perennial *Miscanthus*, SRC and SRF, this cultivation occurs only once for each 20-year crop cycle. It is possible in some cases for this detrimental effect to diminish slightly over time, where a higher yielding crop may produce greater SOC inputs than the previous land use, thus counterbalancing to some extent the effect of the initial soil disturbance, as evident in the grass to SRF transition in Fig. 7.

Plant inputs. The difference in quantity and quality of plant inputs is the principal reason behind the different SOC responses shown by each bioenergy crop type. As the quantity of plant inputs is partially based on yield, the spatial pattern of change in SOC broadly reflects the spatial pattern of yield. This is particularly apparent with *Miscanthus*, which shows a distinct area of high yield (as estimated by the MiscanFor model) in southern England and north and south of the Humber estuary (Figs 3–5), with a corresponding large increase in SOC in these areas following conversion from rotational crops (Fig. 3). The high yields in these two areas are due to the prevalence of chalky soils with high soil water-holding capacities, which the MiscanFor model predicts are favourable for the growth of *Miscanthus* (only MiscanFor treats these chalky soils differently). As SOC change is largely determined by yield (with higher yields giving higher C returns to the soil than lower yields), low yields can lead to a decline in SOC. The relatively detrimental impact of the permanent grass to SRC transition is largely driven by low predicted yields of SRC willow (see Fig. 2). A target for management of perennial energy crops is, therefore, to achieve the best possible yield by selecting the most appropriate energy crop and cultivar for the local situation, as long as this can be made without excessive N fertilizer use, which would lead to increased N_2O emissions. The difference between potential and reported yields in these second-generation bioenergy crops – the so-called *yield gap* remains large (Allwright and Taylor, 2015), reflecting the limited artificial selection and breeding in these crops, compared to annual food crops and suggesting that future yield increase may be dominated by the supply of new germplasm through next-generation molecular breeding using techniques such as genome editing (Allwright and Taylor, 2015). Improved yield would have a large impact on the results reported here.

Changes in N₂O emissions

Beneficial changes in N_2O emissions following conversion of rotational crops to *Miscanthus*, SRC and SRF occur because of reductions in N fertilizer inputs. In ECOSSE, reduced N fertilizer inputs lead to decreased N_2O emissions because: (a) the denitrification rate slows as the NO_3^- concentration in the soil decreases and (b) the proportion of denitrified N emitted as N_2O decreases as NO_3^- concentration in the soil decreases. In contrast to conversions from rotational crops, conversion of forest to wheat shows the greatest increase in N_2O because it involves a transition from a land-use that receives no fertilizer to a crop that receives a large amount of fertilizer (due to wheat's high N demand).

Beneficial changes in N_2O emissions following conversion from rotational crops show larger reductions in N_2O emissions in the west of the United Kingdom than the east (Fig. 6b). This is probably due to higher precipitation rates in the west leading to higher soil water contents. In the model, higher soil water content leads to two contrasting effects on N_2O emissions; firstly, the denitrification rate increases exponentially as the soil water content increases, and secondly, the proportion of denitrified N emitted as N_2O decreases linearly as soil water content increases. The exponential increase in the first process outweighs the linear decrease of the second process, leading to a simulated net increase in N_2O emissions as soil water content increases. This response reflects empirical evidence for N_2O emissions increasing as soil water content increases (e.g. Schindlbacher et al., 2004; Luo et al., 2013). The greater reductions in N_2O following conversion from rotational crops in the west of the country are, therefore, likely due to higher precipitation rates leading to higher soil water contents and in turn, higher N_2O emissions. Reductions in N fertilizer inputs in high precipitation grid cells will therefore lead to greater beneficial reductions in N_2O emissions.

The initial conversion of forest to pasture and cropland (Smith & Conen, 2004) and permanent grass to bioenergy crops (Gelfand et al., 2011; Nikièma et al., 2012; Palmer et al., 2013; Zenone et al., 2015) causes a large initial N_2O emission. Our results show a large emission of N_2O in the first 5 years after conversion from permanent grass and forest to all bioenergy crops. This arises due to the simulation of cultivation during LUC from permanent grass and forest due to soil disturbance increasing soil organic matter mineralization rates. After 5 years, the modelled rates of change in N_2O emissions decline. The large initial rates of N_2O emissions arise for similar reasons to the large SOC decreases that follow certain conversions; initial cultivation of land during LUC physically fragments and redistributes soil organic matter, accelerating its decomposition, releasing inorganic N that is used by denitrifying soil microbes leading to N_2O release (Grandy & Robertson, 2006). The subsequent slowing down of increases in N_2O emissions occurs as the rapidly decomposing soil organic matter resulting from cultivation becomes depleted and the N_2O emissions move towards the background rate. Changes in N_2O

emissions following conversions from permanent grass to OSR, *Miscanthus*, SRC and SRF start to level off and decrease after approximately 5 years. This occurs because the modelled N fertilizer inputs to OSR, *Miscanthus*, SRC and SRF are lower than for permanent grass. Recent work with *Miscanthus* demonstrates that the yield benefits of N fertilization were very small and in terms of GHG emissions did not offset increased soil N_2O emissions (Roth et al., 2015).

Changes in CH_4 emissions

The simulated CH_4 fluxes are very small for all land-use transitions throughout the simulation area. Owing to the absence of data for water table depth, we assumed that all soils in the simulations are freely drained, with no water table. This assumption could result in some uncertainty in the simulated CH_4 fluxes because CH_4 emissions are much higher from saturated than unsaturated soils (Segers, 1998). In the United Kingdom, observed CH_4 fluxes are much higher on organic soils (which are typically poorly drained in their natural state) than on mineral soils and are the main source of soil CH_4 emissions (Levy et al., 2012). Highly organic soils (and therefore the greatest sources of CH_4) were excluded by the UKERC constraints mask from the simulations because they are unsuitable for growing bioenergy crops.

Moreover, even if significant areas of poorly drained land with high CH_4 emissions are present within the simulated area, large changes in those CH_4 emission rates resulting from conversion to bioenergy crops are only likely to occur if the land is drained for bioenergy crops. We are not aware of any planned or actual drainage of extensive areas of land for bioenergy crops. Drainage is unlikely to take place on soils currently under rotational crops because the land will already have been drained (if it was necessary). Also, SRC willow and poplar are suitable for planting on soils with a shallow water table (1–2 m deep), with willow able to cope with water-logging, making it suitable for planting in areas with a high water table or areas prone to flooding (Hall, 2003). SRC therefore provides a bioenergy option that is unlikely to require the drainage of waterlogged land.

For the reasons described above, the uncertainty in the CH_4 emissions associated with the assumption of a freely draining soil is relatively small and simulated CH_4 fluxes are representative of the land suitable for bioenergy conversion. However, if extensive areas of water-logged land were to be drained for the establishment of bioenergy crops, it would be useful to explore the impacts on CH_4 fluxes (and changes in SOC and N_2O emissions) in more detail.

Effects of soil

The ECOSSE model requires input data for several soil properties: initial SOC content, pH, bulk density and clay content. These properties influence a range of processes within the model.

SOC content influences the amount of C lost as CO_2 during decomposition. All other factors being equal, soils with high organic C content will produce proportionally higher CO_2 emissions than a soil with low organic C content. We therefore expect soils with high organic C content to show greater sensitivity to changes in SOC resulting from LUC (e.g. due to cultivation). However, whilst the absolute loss of C due to cultivation is expected to be higher in soils with high organic C content, the relative loss of C may be lower if the clay content is higher.

SOC content increases as the clay content increases (Burke *et al.*, 1989). This increase occurs because clay particles strongly adhere to organic matter slowing down the decomposition process, and because clay forms aggregates that physically protect SOC from microbial decomposition (Rice, 2002). In ECOSSE, the effects of clay content on soil organic matter decomposition is modelled by altering the proportion of C released as CO_2 during decomposition (i.e. the efficiency of decomposition). As clay content increases, a smaller proportion of decomposed C is lost as CO_2 (i.e. the efficiency of decomposition increases), and a greater proportion is retained in the biomass and humus soil organic matter pools. Therefore, when clay-rich soils are cultivated during LUC (causing a large proportion of SOC to be moved from soil organic matter pools with a faster turnover rate to soil organic matter pools with a slower turnover rate), we would expect the modelled relative SOC losses to be lower than for soils with low clay content. This behaviour is in agreement with empirical evidence (e.g. Burke *et al.*, 1989).

A significant effect of soil pH on the rate of decomposition has been observed in many studies (e.g. Hall *et al.*, 1998; Andersson & Nilsson, 2001). In ECOSSE, the pH rate modifier for aerobic decomposition decreases linearly as pH drops below 4.5. For pH values >4.5, the rate modifier is set to 1 (i.e. has no effect upon the decomposition rate). Soils with a pH of <4.5 are typically highly organic. We therefore expect variations in pH between soil types to have very little impact on the model outputs because highly organic soils have been excluded from the simulation area.

In ECOSSE, bulk density affects the rate of CH_4 oxidation (i.e. consumption of CH_4). Empirical evidence shows that soils with a low bulk density tend to have higher rates of CH_4 oxidation (Borken & Brumme, 1997) because low bulk density soils are more permeable,

allowing atmospheric CH_4 and oxygen to diffuse more freely into the soil (Dörr *et al.*, 1993). Variation in bulk density in the simulated soils is very unlikely to have a significant effect on the results because (a) peat soils, which have a much lower bulk density (and therefore much higher potential oxidation rates than mineral soils), have been excluded from the simulation; (b) the simulated soil CH_4 production rates are very low so it is not possible for oxidation of CH_4 to significantly affect the net GHG balance.

Rotational grass

The permanent grass land-use type used in these simulations represents permanent, uncultivated grassland. Grassland, however, may also be temporary, used in rotation with arable crops, and in these circumstances can be regarded as a crop within an arable rotation. Permanent grassland is the most abundant type of grassland in the United Kingdom, covering 5.3 million ha in 2010, compared to 1.1 million ha of temporary (mostly rotational) grassland (Khan *et al.*, 2011) at any one time. Rotational grassland in any given year would be categorized as arable crops in different years, so the 1.1 million ha in any year represents a snapshot of the area of rotational grass. As such, rotational grass is not a land use; it is simply one component of rotational farming, which includes all-arable rotations as well as grass-arable rotations. Rotational grassland is usually represented as a crop within a rotation in most existing soil organic matter models and in ECOSSE is assumed to be a subset of arable rotational land. Permanent grassland represents a separate land-use transition as this land is only used for grass/livestock production. Rotational grass (by definition) occurs on the same land as is used for growing arable crops, so bioenergy conversion on rotational grass is equivalent to removal of land used for arable production. Rotational grassland can therefore be simulated in ECOSSE in the same way as arable-only rotations.

It is expected that rotational grassland would behave in a similar way to arable land in terms of net GHG response to LUC to bioenergy crops because a) it undergoes frequent cultivation and b) it typically receives more fertilizer than permanent grassland. This expectation is supported by empirical evidence. Long-term experiments at the Woburn Research Station (run by Rothamsted Research) in the United Kingdom found that conversion of continuous arable to rotational grassland (in this case either a 3-year grass or grass–clover ley followed by two arable crops in a 5-year cycle), resulted in only a 10–15% increase in SOC after 60 years (Johnston *et al.*, 2009). By contrast, the conversion of arable land to permanent grassland at the Rothamsted

Research Station resulted in a doubling of organic matter (indicated by total N), in 50 years (Johnston *et al.*, 2009). The small observed increase in SOC under rotational grassland suggests that the response of rotational grassland to LUC would fall between that of arable and permanent grass, but will be close to the all-arable rotations represented by our rotational crops category.

Uncertainty

There are a number of uncertainties associated with the modelled GHG balances. Uncertainty in national scale simulations has two components; uncertainty arising from the model and uncertainty arising from reduced detail and precision in data available at national scale compared to data available at the field scale. Uncertainty arising from the model was estimated as part of the site-specific modelling exercise reported in Dondini *et al.* (2015, 2016a,b). Here, we focus on uncertainties arising from the use of national scale data.

Uncertainty due to spatial and temporal resolution. The spatial and temporal resolutions of the driving data sets are given in Table 4. Due to the reduced detail of the inputs, the uncertainty in simulations at the national scale is likely to be greater than at the field scale. For example, in croplands, detailed management factors such as sowing date and timing and rate of fertilizer applications cannot usually be specified when the resolution of the simulations is larger than the size of the management unit. The resolution of the simulation here was a 1 km^2 grid cell, whereas the size of a management unit might be a 5 ha (0.05 km^2) field, so there will be many different values for the management factors within each 1 km^2 cell. For example, the rate of N

Table 4 Spatial and temporal resolution of driving datasets used in the spatial simulations

Input data	Spatial resolution	Temporal resolution
Harmonised World Soil Database	30 arc s (approx. 1 km grid cells)	N/A
UKCP09 climate projections	25 km grid cells	30-year averages
Crop yield	NUTS 1 regional averages for wheat, and oilseed rape, national average for sugar beet; 1 km grid cells for *Miscanthus*, SRC and SRF, 25 km grid cells for permanent grass and forest	Annual

fertilizer application to grassland varies considerably according to the clover concentration in the grass sward, livestock stocking density and soil N status (Defra, 2010).

Uncertainty in national scale simulations is also greater than at field scale due to the reduced precision of the input data. For example, the C content of the soil in a 5 ha field can be precisely measured and the error in the measurement defined using replicates, whereas at the national scale the soil C content for grid cells is estimated from typical or averaged soil C values for the major soil types identified in the cell (e.g. Batjes, 2009).

The uncertainty due to the reduced detail and precision of data available at the national scale can be quantified by evaluating the model at field scale, but using input drivers that are available at national scale (as performed for the error bars in Fig. 7).

Uncertainty due to soil. The uncertainty associated with the use of national scale soil data was quantified by simulating 40 paired land-use transition sites (Rowe *et al.*, 2016), using measured soil parameters and soil parameters obtained from the HWSD. A statistical analysis (data not shown) of simulations using the HWSD inputs across the 40 field sites shows that there was a good correlation between modelled and measured SOC (0–100 cm depth), when using the measured soil parameters ($r = 0.92$), and when using the HWSD parameters ($r = 0.79$). In both cases, there was no significant model error and no significant model bias.

Due to the nature of the HWSD data, where the locations of soils within each grid cell are unknown, it is not possible to define which HWSD soil type corresponds to a given field site, or whether the soil type of the field site is within the dominant soils reported in the HWSD. Despite this, there was a good correlation between modelled and measured values and a lack of model bias when using HWSD parameters as inputs. This suggests that uncertainty in model results arising from the use of HWSD data is fairly small.

A similar evaluation of national scale uncertainty using ECOSSE and National Soils Inventory of Scotland soil data to simulate SOC at 60 resampled field sites in Scotland was carried out by Smith *et al.* (2010b). That study found a very strong correlation between modelled and measured SOC ($r = 0.97$). The correlation was higher in the Smith *et al.* (2010b) study than the current study ($r = 0.97$ vs. $r = 0.79$). Smith *et al.* (2010b) obtained a higher correlation probably because the soil type at each field site could be matched to the corresponding soil type in the national soil database they used.

Initial land use was estimated from the LCM2007 data at 100 m resolution which was aggregated by mode to

1 km resolution to match the HSWD data. However, as the HSWD assigns up to 10 soil types and their proportion to each 1 km × km cell, it is not spatially explicit so that matching a soil type to land use could be ambiguous as we are only considering one initial land use per grid block and not considering the proportion of each land cover in the cell. In addition, land cover is usually associated with a soil type, for example podsols with forest and brown earths with arable land, and with the data available this level of detail and precision is not possible. In this way, a small number of unlikely combinations of soil type and initial land use may have been included in the average grid results. This will need future work to resolve.

Uncertainty due to yield. Climate variability and changes in the frequency and severity of extreme events can have significant, nonlinear impacts on crop yields because crops exhibit threshold responses to stress factors (Porter & Semenov, 2005; Trnka *et al.*, 2014). Therefore, the lack of short-term climate variation in the UKCP09 climate projections presents a potentially large source of uncertainty in the predicted yields and, subsequently, the bioenergy GHG balances (Hastings *et al.*, 2009).

None of the yield models used in this study explicitly account for the effects of atmospheric N deposition on productivity; all considered the yield with optimum N supply and the crops were not, therefore, N limited. Within the simulated area of the United Kingdom, N deposition typically adds between 10 and 30 kg N ha^{-1} yr^{-1} (Fowler *et al.*, 2004). However, we do not expect this level of N input to significantly affect the ECOSSE model outcomes for two reasons. Firstly, the yield models have been calibrated using UK field measurements of crops subjected to atmospheric N deposition, so the effects of N deposition are to some extent implicitly captured by the models. Secondly, farmers may adjust the rates of N fertilizer applied to crops according to the N deposition rate (Jones *et al.*, 2014). For example, UK wheat farmers are advised to increase their Soil N Supply index by 20 kg N ha^{-1} to allow for N deposition and the Defra Fertiliser Manual (Defra, 2010) factors in atmospheric N deposition (HGCA, 2009). Therefore, in fertilized cropping systems, the effects of N deposition may be largely mitigated by adaptation of fertilizer practices.

Levels of atmospheric N deposition in the United Kingdom are currently in decline due to reduced N emissions (Jones *et al.*, 2014), which could lead to reduced crop productivity. However, it is expected that fertilizer and other crop management practices will adjust to compensate for this reduction and so maintain the yields predicted by the models.

Further uncertainty arises because the crop yield projections are derived from several different sources which vary in spatial resolution, and, in the case of modelled values, the level of sophistication of the model. For example, the wheat and OSR yields are based on Defra average yield statistics for 12 regions in the United Kingdom (the NUTS level 1 regions), whereas sugar beet yields are based on a single national average yield value. Future wheat, OSR and sugar beet yields are obtained by modifying the baseline yield observations with a simple, empirical model, Miami (Lieth, 1975), whereas *Miscanthus* yield projections are obtained using a more complex, process-based model, MiscanFor (Hastings *et al.*, 2009).

The crop yield projections are based on models that are parameterized and calibrated for existing cultivars and current management practices. However, crop breeding and improvements in management practices will likely lead to increases in crop yield over time (Allwright and Taylor, 2015). In addition, the yield models do not consider the impact of pests and disease, nor extreme weather events.

These sources of uncertainty in yield forecasts are difficult to quantify, either due to lack of data (e.g. changes in the frequency of extreme climate events), or because they are inherently uncertain (e.g. impacts of future crop breeding), although it is likely that an increase in crop yield of 10% per decade would not be unreasonable for these largely unimproved crops and this could have a significant impact on model outputs (Allwright and Taylor, 2016). Because these uncertainties and their impact on GHG balance estimates we tested the sensitivity of the bioenergy GHG balances to changes in yields. The main findings from this sensitivity analysis were (a) for conversions from permanent grass and forest, yield increases of up to 50% were not sufficient to change a mean detrimental change in SOC to a mean beneficial change in SOC; (b) yield increases of up to 50% of any given bioenergy crop were generally insufficient to alter the crop's ranking in terms of changes in SOC, even when the yields of all other bioenergy crops were left unchanged; and (c) SRF and *Miscanthus* showed the greatest sensitivity to proportional changes in yield because they have the highest yields within the simulated area.

Although changes in estimated yields would certainly affect the total area of land favourable for conversion to bioenergy crops, our findings suggest that the broad conclusions inferred from the modelling results would remain unchanged.

Uncertainty due to fertilizer use. A large number of factors affect the amount of N fertilizer applied to a crop including the soil N status, expected crop N demand, weather, soil texture, regulations (e.g. in nitrogen vulnerable zones) and economic factors (e.g. cost of fertilizer). For grassland, additional factors may include the percentage of clover in the grass sward and stocking density. Many of these factors vary at a finer scale than the 1 km resolution of the simulations and are not described in any spatially defined databases. Therefore, the model makes assumptions about the amount of N fertilizer applied, which presents a source of uncertainty for the modelled changes in N_2O emissions.

To quantify this uncertainty, we conducted a sensitivity analysis to explore the impacts of a ±20% variation to the default N fertilizer application rate in a sample of the grid cells. The results of this analysis (data not shown) show that transitions to wheat were most sensitive to a proportional change in N fertilizer inputs: a 20% increase in N fertilizer led to a mean increase in N_2O emissions of about 5 t CO_2e ha^{-1} after 35 years (i.e. in 2050) and a 20% decrease reduced N_2O emissions by about 5.5 t CO_2e ha^{-1}. Other transitions showed mean deviations in N_2O emissions within ±2.5 t CO_2e ha^{-1}. The shifts in N_2O emissions resulting from a ±20% change in N fertilizer rates are modest, leading to a <5% change in the mean net GHG balance of each transition. Therefore, we do not expect uncertainty around N fertilization rates to be a source of large uncertainty in the modelling outcomes.

Future research needs

This study shows that future work should target second-generation bioenergy crops (*Miscanthus*, SRC and SRF), because these offer a much more favourable net GHG balance than first-generation bioenergy crops (wheat, sugar beet and OSR). It is noteworthy, however, that the overall GHG balance of bioenergy may still be positive, even if there are net preharvest soil GHG emissions due to land-use change, and the full value chain needs to be considered (Newton-Cross & Evans, 2015).

Whilst the type of land-use transition was the most important factor affecting net GHG balance, crop yield was found to be the most influential factor within each type of transition. However, a number of limitations of the yield data constrain the spatial accuracy of the soil GHG balance predictions and should be the focus of future research.

Firstly, Defra yield data for wheat (also used for the baseline rotational crop yield), sugar beet and OSR are spatially coarse, being available only at regional level, and only cover a short time span. The development of high-resolution spatial datasets of bioenergy crop yield would greatly improve the spatial accuracy of net GHG balance predictions.

Development of yield models is often hampered by lack of detailed soil and plant data from which to formulate process descriptions and evaluate the model. For example, only 11 UK experimental sites with sufficient data to validate the MiscanFor model were available (Hastings *et al.*, 2009). Future research should place an emphasis on detailed, long-term measurements of crop and soil attributes (yield, litter inputs, C and N contents of plant components and soil etc.), over the full life cycle of the crop, for the latest germplasm released from breeding programmes. Such data are required for the development of more robust and improved parameterizations of process-based models, critical for future predictions.

Models of future crop yield vary in the factors they take into account. For example (e.g. effects of elevated atmospheric CO_2 concentration), their level of sophistication and degree to which they have been calibrated for UK conditions. Moreover, where multiple models exist for a given crop, the yield estimates may differ considerably. For example, MiscanFor (Hastings *et al.*, 2009) predicts the highest *Miscanthus* yields to be in the south-west of England, whereas the empirical model of Richter *et al.* (2008) predicts relatively low yields in the south-west. Further work on model evaluation and model comparison is required to resolve these differences and reduce the uncertainty in model estimates. In the short-term, the uncertainty associated with choice of model could be quantified by modelling net GHG balance using yield forecasts produced from an ensemble of yield models for each crop.

Overall, the reliability and spatial accuracy of future net GHG balance modelling would benefit greatly from improvements in bioenergy yield modelling (or direct modelling of crop inputs of C to the soil). Standardization of yield models for different crops, such as the effect of soil type on soil water capacity, would enable more reliable comparison of different land-use transitions.

Finally, little is known about the impact of bioenergy crop re-establishment on SOC. Different re-establishment techniques involve different amounts of soil disturbance, which could lead to enhanced soil organic matter decomposition rates. Soil disturbance from re-establishment could have a significant effect on long-term C sequestration, with a proportion of the C sequestered during the previous planting cycle being lost again as CO_2 to the atmosphere (Grogan & Matthews, 2002). Research into the practicality of a range of potential re-establishment techniques and their impacts on soil C dynamics should be a high priority.

In conclusion, we have shown that increasing yield increases SOC so that in addition to optimizing the use of land, and obtaining the highest energy yield per unit area, research into improving the yield of SRC, SRF and *Miscanthus* genotypes will provide benefits to both energy security and GHG mitigation. When assessing the full GHG impacts of energy crops, all components of the bioenergy value chain (e.g. cultivation, management, harvest, transportation, processing, fossil fuel offset GHG impacts) need to be considered (Newton-Cross & Evans, 2015). The findings presented here fill the critical gap in preharvest GHG emission data, used to assess the full life cycle GHG emissions from energy crops in bioenergy value chains.

Acknowledgements

This work was part of the Ecosystem Land-Use Modelling (ELUM) project, which was commissioned and funded by the Energy Technologies Institute. The work also contributes to the EPSRC-Supergen funded MAGLUE project (EP/M013200/1). We are grateful to staff at the Energy Technologies Institute, particularly to Geraldine Newton-Cross, Geraint Evans and Hannah Evans for comments on previous versions of this manuscript and to Jonathan Oxley for project support.

References

Alexander P, Moran D, Smith P et al. (2014) Estimating UK perennial energy crop supply using farm scale models with spatially disaggregated data. *Global Change Biology Bioenergy*, 6, 142–155.

Allwright MR & Taylor G (2016) Molecular breeding for improved second generation bioenergy crops. *Trends in Plant Science*, 21, 43–54.

Andersson S, Nilsson SI (2001) Influence of pH and temperature on microbial activity, substrate availability of soil-solution bacteria and leaching of dissolved organic carbon in a mor humus. *Soil Biology and Biochemistry*, 33, 1181–1191.

Arundale RA, Dohleman FG, Heaton EA, Mcgrath JM, Voigt JB, Long SP (2014) Yields of *Miscanthus × giganteus* and *Panicum virgatum* decline with stand age in the Midwestern USA. *Global Change Biology Bioenergy*, 6, 1–13.

Batjes NH (2009) Harmonized soil profile data for applications at global and continental scales: updates to the WISE database. *Soil Use and Management*, 25, 124–127.

Beringer T, Lucht W, Schaphoff S et al. (2011) Bioenergy production potential of global biomass plantations under environmental and agricultural constraints. *Global Change Biology Bioenergy*, 3, 299–312.

Borken W, Brumme R (1997) Liming practice in temperate forest ecosystems and the effects on CO_2, N_2O and CH_4 fluxes. *Soil Use and Management*, 13, 251–257.

Bradbury NJ, Whitmore AP, Hart PBS, Jenkinson DS (1993) Modelling the fate of nitrogen in crop and soil in the years following application of ^{15}N-labelled fertilizer to winter wheat. *Journal of Agricultural Science*, 121, 363–379.

Burke IC, Yonker CM, Parton WJ, Cole CV, Schimel DS, Flach K (1989) Texture, climate, and cultivation effects on soil organic matter content in U.S. grassland soils. *Soil Science Society of America Journal*, 53, 800–805.

Caslin B, Finnan J, Easson L (Eds.) (2011a) Miscanthus best practice guidelines. Carlow & Hillsborough, Crops Research Centre & Agri-Food and Bioscience Institute. Available at: http://www.afbini.gov.uk/miscanthus-best-practice-guidelines.pdf (accessed 28 June 2014).

Caslin B, Finnan J, McCracken A (Eds.) (2011b) Short rotation coppice willow best practice guidelines. Carlow & Hillsborough, Crops Research Centre & Agri-Food and Bioscience Institute. Available at: http://http://www.afbini.gov.uk/willowbestpractice.pdf (accessed 28 June 2014).

Clifton-Brown J, Schwarz K, Hastings A (2015) History of the development of Miscanthus as a bioenergy crop: from small beginnings to potential realization. *Biology and Environment: Proceedings of the Royal Irish Academy*, 115, 1–12.

Coleman KW, Jenkinson DS (1996) RothC-26.3 - A model for the turnover of carbon in soil. In: *Evaluation of Soil Organic Matter Models Using Existing Ling-Term Datasets* (eds Powlson DS, Smith P, Smith J), pp. 237–246. Springer-Verlag, Heidelberg.

Creutzig F, Masera O, Ravindranath NH et al. (2015) Bioenergy and climate change mitigation: an assessment. *Global Change Biology Bioenergy*, 7, 916–944.

DECC (2012) *UK Bioenergy Strategy*. DECC, London.

DECC (2015) *Digest of UK Energy Statistics (DUKES) for 2015*. DECC, London. Available at: https://www.gov.uk/government/uploads/system/uploads/attachment_data/file/450302/DUKES_2015.pdf (accessed: 11 February 2016).

Defra (2004) *Growing Short Rotation Coppice: Best Practice Guidelines*. Defra, London.

Defra (2010) *Fertiliser Manual (RB2009)*, 8th edn. The Stationery Office, Norwich, UK.

Donatelli M, Acutis M, Laruccia N (1996) Pedotransfer functions: evaluation of methods to estimate soil water content at field capacity and wilting point. Available at: www.isci.it/mdon/research/bottom_modeling_cs.htm pp. 6–11 (accessed: 28 June 2014).

Dondini M, Jones EO, Richards M et al. (2015) Evaluation of the ECOSSE model for simulating soil carbon under short rotation forestry energy crops in Britain. *Global Change Biology Bioenergy*, 7, 527–540.

Dondini M, Richards M, Pogson M et al. (2016a) Evaluation of the ECOSSE model for simulating soil carbon under Miscanthus and short rotation coppice-willow crops in Britain. *Global Change Biology Bioenergy*. doi: 10.1111/gcbb.12286.

Dondini M, Richards M, Pogson M et al. (2016b) Simulation of greenhouse gases following land-use change to bioenergy crops using the ECOSSE model. A comparison between site measurements and model predictions. *Global Change Biology Bioenergy*. doi: 10.1111/gcbb.12298.

Dörr H, Katruff L, Levin I (1993) Soil texture parameterization of the methane uptake in aerated soils. *Chemosphere*, 26, 697–713.

Dreccer MF, Schapendonk AHCM, Slafer GA, Rabbinge R (2000) Comparative response of wheat and oilseed rape to nitrogen supply: absorption and utilisation efficiency of radiation and nitrogen during the reproductive stages determining yield. *Plant and Soil*, 220, 189–205.

Edenhofer O, Pichs-Madruga R, Sokona Y et al. (2014) (Eds) *Climate Change 2014: Mitigation of Climate Change. Contribution of Working Group III to the Fifth Assessment Report of the Intergovernmental Panel on Climate Change*. Cambridge University Press, Cambridge, UK and New York, NY, USA.

EUROSTAT (2014) Areas harvested, production by NUTS 2 regions. Available at: http://ec.europa.eu/eurostat (accessed: 14 July 2014).

FAO/IIASA/ISRIC/ISS-CAS/JRC (2012) *Harmonized World Soil Database (version 1.2)*. FAO, Rome, Italy and IIASA, Laxenburg, Austria.

Fowler D, O'Donoghue M, Muller JBA et al. (2004) A chronology of nitrogen deposition in the UK between 1900 and 2000. *Water, Air, and Pollution: Focus*, 4, 9–23.

Fuss S, Canadell JG, Peters GP et al. (2014) Betting on negative emissions. *Nature Climate Change*, 4, 850–853.

Gelfand I, Zenome T, Jasrotia P, Chen J, Hamilton SK, Robertson GP (2011) Carbon debt of conservation reserve program (CRP) grasslands converted to bioenergy production. *Proceedings of the National Academy of Sciences of the United States of America*, 108, 13864–13869.

Givi J, Prasher SO, Patel RM (2004) Evaluation of pedotransfer functions in predicting the soil water contents at field capacity and wilting point. *Agricultural Water Management*, 70, 83–96.

Grandy AS, Robertson GP (2006) Initial cultivation of a temperate-region soil immediately accelerates turnover and CO_2 and N_2O fluxes. *Global Change Biology*, 12, 1507–1520.

Grogan P, Matthews R (2002) A modelling analysis of the potential for soil carbon sequestration under short rotation coppice willow bioenergy plantations. *Soil Use and Management*, 18, 175–181.

Guo LB, Gifford M (2002) Soil carbon stocks and land use change: a meta analysis. *Global Change Biology*, 8, 345–360.

Hall RL (2003) Short rotation coppice for energy production hydrological guidelines [online]. Available at: http://www.berr.gov.uk/files/file14960.pdf (accessed 28 June 2014).

Hall JM, Paterson E, Killham K (1998) The effect of elevated CO_2 concentration and soil pH on the relationship between plant growth and rhizosphere denitrification potential. *Global Change Biology*, 4, 209–216.

Harris ZM, Spake R, Taylor G (2015) Land use change to bioenergy: a meta-analysis of soil carbon and GHG emissions. *Biomass & Bioenergy*, 82, 27–39.

Hastings A, Clifton-Brown J, Wattenbach M, Mitchell CP, Smith P (2009) Improved process descriptions for leaf area index development and photosynthesis rates in *Miscanthus x giganteus*: towards more robust yield predictions under different climatic and soil conditions. *Global Change Biology*, 1, 154–170.

Hastings A, Tallis MJ, Casella E et al. (2014) The technical potential of Great Britain to produce ligno-cellulosic biomass for bioenergy in current and future climates. *Global Change Biology Bioenergy*, 6, 108–122.

HGCA (2009) *Nitrogen for Winter Wheat – Management Guidelines*. HGCA, Stoneleigh, UK.

HGCA (2014) *Oilseed rape guide. January 2014. HGCA Guide 55*. Stoneleigh, UK.

Howes P, Bates J, Landy M, O'Brian S, Herbert R, Matthews R, Hogan G (2011) *UK and Global Bioenergy Resource – Final report*. AEA group, Harwell, UK.

Hutson JL, Cass A (1987) A retentivity function for use in soil water simulation models. *Journal of Soil Science*, **38**, 105–113.

IPCC (2001) *Climate Change 2001: Mitigation. Contribution of Working Group III to the Third Assessment Report. Intergovernmental Panel on Climate Change*. Cambridge University Press, Cambridge, UK and New York, USA.

Jenkinson DS, Rayner JH (1977) The turnover of organic matter in some of the Rothamsted classical experiments. *Soil Science*, **123**, 298–305.

Jenkinson DS, Hart PBS, Rayner JH, Parry LC (1987) Modelling the turnover of organic matter in long-term experiments at Rothamsted. *INTECOL Bulletin*, **15**, 1–8.

Johnston EA, Poulton PR, Coleman K (2009) Soil organic matter: its importance in sustainable agriculture and carbon dioxide fluxes. In: Sparks DL (Ed.). *Advances in Agronomy* **101**, 1–57.

Jones L, Provins A, Holland M *et al.* (2014) A review and application of the evidence for nitrogen impacts on ecosystem services. *Ecosystem Services*, **7**, 76–88.

Khan J, Powell T, Harwood A (2011) *Land Use in the UK*. Office for National Statistics, London, UK.

Kjellström CG, Kirchmann H (1994) Dry matter production of oilseed rape (*Brassica napus*) with special reference to the root system. *Journal of Agricultural Science*, **123**, 327–332.

Levy PE, Burden A, Cooper MDA *et al.* (2012) Methane emissions from soils: synthesis and analysis of a large UK data set. *Global Change Biology*, **18**, 1657–1669.

Lieth H (1975) Modeling the primary productivity of the world. In: *Primary Productivity of the Biosphere* (ed. Lieth H, Whittaker RH), pp. 237–263. Springer Berlin Heidelberg, Springer-Verlag, New York..

Living Countryside (2013) Available at: http://www.ukagriculture.com (accessed: 29 September 2013).

Lovett AA, Sünnenberg GM, Dockerty TL (2014) The availability of land for perennial energy crops in Great Britain. *Global Change Biology Bioenergy*, **6**, 99–107.

Luo GJ, Kiese R, Wolf B, Butterbach-Bahl K (2013) Effects of soil temperature and moisture on methane uptake and nitrous oxide emissions across three different ecosystem types. *Biogeosciences*, **10**, 3205–3219.

McCalmont JP *et al.* (2015) Environmental costs and benefits of growing *Miscanthus* for bioenergy in the UK. *Global Change Biology Bioenergy*, doi:10.1111/gcbb.12294.

McCalmont J, Hastings A, Robson P *et al.* (2016) The environmental credentials of *Miscanthus* as a bio-energy crop in the UK. *Global Change Biology Bioenergy*. doi: 10.1111/gcbb.12294

McKay H (ed.) (2011) Short rotation forestry: review of growth and environmental impacts. *Forest Research Monograph*, Forest Research, Surrey, **2**, 212.

Murphy JM, Sexton DMH, Jenkins GJ *et al.* (2009) *UK Climate Projections Science Report: Climate Change Projections*. Met Office Hadley Centre, Exeter.

Murty D, Kirschbaum MUF, McMurtrie RE, McGilvray H (2002) Does conversion of forest to agricultural land change soil carbon and nitrogen? A review of the literature. *Global Change Biology*, **8**, 105–123.

Newton-Cross G, Evans H (2015) *Bioenergy. Delivering Greenhouse Gas Emission Savings Through UK Bioenergy Value Chains*. Energy Technologies Institute, Loughborough, 2015. Available at: http://www.eti.co.uk/wp-content/uploads/2016/01/Delivering-greenhouse-gas-emission-savings-through-UK-bioenergy-value-chains.pdf (accessed 11 February 2016).

Nikièma P, Rothstein DE, Miller RO (2012) Initial greenhouse gas emissions and nitrogen leaching losses associated with converting pastureland to short-rotation woody bioenergy crops in northern Michigan, USA. *Biomass and Bioenergy*, **39**, 413–426.

Ordnance Survey (2010) *A guide to coordinate systems in Great Britain v2.1*. pp. 37–42.

Oritz JN, Tarjuelo M, de Juan A (2012) Effects of two types of sprinklers and height in the irrigation of sugar beet with a centre pivot. *Spanish Journal of Agricultural Research*, **10**, 251–263.

Palmer MM, Forrester JA, Rothstein DE, Mladenoff DJ (2013) Conversion of open lands to short-rotation woody biomass crops: site variability affects nitrogen cycling and N$_2$O fluxes in the US Northern Lake States. *Global Change Biology Bioenergy*, **6**, 450–464.

Porter JR, Semenov MA (2005) Crop responses to climatic variation. *Philosophical Transactions of the Royal Society B*, **360**, 2021–2035.

Pyatt G, Ray D, Fletcher J (2001) *An Ecological Site Classification for Forestry in Great Britain*. Bulletin 124. Forestry Commission, Edinburgh.

Rice CW (2002) Organic matter and nutrient dynamics. In: *Encyclopedia of Soil Science* (eds Lal R), pp. 925–928. Marcel Dekker Inc., New York, NY.

Richter GM, Riche AB, Dailey AG, Gezan SA, Powlson DS (2008) Is UK biofuel supply from *Miscanthus* water-limited? *Soil Use and Management*, **24**, 235–245.

Roth B, Finnan JM, Jones MB, Burke JI, Williams ML (2015) Are the benefits of yield responses to nitrogen fertilizer application in the bioenergy crop *Miscanthus x giganteus* offset by increased soil emissions of nitrous oxide? *Global Change Biology Bioenergy*, **7**, 145–152.

Rowe RL, Keith AM, Elias D, Dondini M, Smith P, Oxley J, McNamara NP (2016) Initial soil C and land use history determine soil C sequestration under perennial bioenergy crops. *Global Change Biology Bioenergy*. doi: 10.1111/gcbb.12311.

Schindlbacher A, Zechmeister-Boltenstern S, Butterbach-Bahl K (2004) Effects of soil moisture and temperature on NO, NO$_2$ and N$_2$O emissions from European forest soils. *Journal of Geophysical Research*, **109**, 1–12.

Segers R (1998) Methane production and methane consumption: a review of processes underlying wetland methane fluxes. *Biogeochemistry*, **41**, 23–51.

Sims REH, Hastings A, Schlamadinger B, Taylor G, Smith P (2006) Energy crops: current status and future prospects. *Global Change Biology*, **12**, 2054–2076.

Slade R, Bauen A, Gross R (2014) Global bioenergy resources. *Nature Climate Change*, **4**, 99–105.

Smith P (2008) Land use change and soil organic carbon dynamics. *Nutrient Cycling in Agroecosystems*, **81**, 169–178.

Smith KA, Conen F (2004) Impacts of land management on fluxes of trace greenhouse gases. *Soil Use and Management*, **20**, 255–263.

Smith JU, Bradbury NJ, Addiscott TM (1996) SUNDIAL: a PC-based system for simulating nitrogen dynamics in arable land. *Agronomy Journal*, **88**, 38–43.

Smith P, Martino D, Cai Z *et al.* (2008) Greenhouse gas mitigation in agriculture. *Philosophical Transactions of the Royal Society, B*, **363**, 789–813.

Smith JU, Gottschalk P, Bellarby J *et al.* (2010a) Estimating changes in national soil carbon stocks using ECOSSE – a new model that includes upland organic soils. Part I. Model description and uncertainty in national scale simulations of Scotland. *Climate Research*, **45**, 193–205.

Smith JU, Gottschalk P, Bellarby J *et al.* (2010b) Estimating changes in national soil carbon stocks using ECOSSE – a new model that includes upland organic soils. Part II. Application in Scotland. *Climate Research*, **45**, 193–205.

Smith JU, Gottschalk P, Bellarby J *et al.* (2010c) ECOSSE User Manual. Available at: https://www.abdn.ac.uk/staffpages/uploads/soi450/ECOSSE%20User%20manual%20310810.pdf (accessed 22 January 2015).

Smith P, Davis SJ, Creutzig F *et al.* (2016) Biophysical and economic limits to negative CO$_2$ emissions. *Nature Climate Change*, **6**, 42–50.

Stoddart H, Watts K (2012) *Biomass Feedstock, Residues and By-Products*. HGCA, Stoneleigh, UK. Available at: http://publications.hgca.com/publications/documents/HGCA_straw_paper_2012.pdf (accessed 3 February 2014).

Tallis MJ, Casella E, Henshall PA, Aylott MJ, Randle TJ, Morison JIL, Taylor G (2013) Development and evaluation of ForestGrowth-SRC a process-based model for short rotation coppice yield and spatial supply reveals poplar uses water more efficiently than willow. *Global Change Biology Bioenergy*, **5**, 53–66.

Thompson DA, Matthews RW (1989) *The Storage of Carbon in Trees and Timber*. Forestry Commission Research Information Note 160. Forestry Commission, Edinburgh, UK.

Thornthwaite CW (1948) An approach toward a rational classification of climate. *Geographic Review*, **38**, 55–94.

Trnka M, Rötter RP, Ruiz-Ramos M, Kersebaum KC, Olesen JE, Žalud Z, Semenov MA (2014) Adverse weather conditions for European wheat production will become more frequent with climate change. *Nature Climate Change*, **4**, 637–643.

Tsialtas JT, Karadimos DA (2003) Leaf carbon isotope discrimination and its relation with qualitative root traits and harvest index in sugar beet (*Beta vulgaris* L.). *Journal of Agronomy and Crop Science*, **189**, 286–290.

Wang S, Hastings A, Wang SC *et al.* (2014) The potential for bioenergy crops to contribute to meeting GB heat and electricity demands. *Global Change Biology Bioenergy*, **6**, 136–141.

Wei X, Shao M, Gale W, Li L (2014) Global pattern of soil carbon losses due to the conversion of forest to agricultural land. *Scientific Reports*, **4**, 4062.

White EM, Wilson FEA (2006) Responses of grain yield, biomass and harvest index and their rates of genetic progress to nitrogen availability in ten winter wheat varieties. *Irish Journal of Agricultural and Food Research*, **45**, 85–101.

Zenone T, Zona D, Gelfand I, Gielen B, Camino-Serrano M, Ceulemans R (2015) CO$_2$ uptake is offset by CH$_4$ and N$_2$O emissions in a poplar short-rotation coppice. *Global Change Biology Bioenergy*. doi:10.1111/gcbb.12269.

Zhuang Q, Qin Z, Chen M (2013) Biofuel, land and water: maize, switchgrass or *Miscanthus*? *Environmental Research Letters*, **8**, 015020 (6 pp).

5

A *Miscanthus* plantation can be carbon neutral without increasing soil carbon stocks

ANDY D. ROBERTSON[1,2,3,4], JEANETTE WHITAKER[1], ROSS MORRISON[5], CHRISTIAN A. DAVIES[2], PETE SMITH[3] and NIALL P. MCNAMARA[1]

[1]Centre for Ecology & Hydrology, Lancaster Environment Centre, Library Avenue, Bailrigg, Lancaster LA1 4AP, UK, [2]Shell International Exploration and Production, Shell Technology Center Houston, 3333 Highway 6 South, Houston, TX 77082-3101, USA, [3]Institute of Biological and Environmental Sciences, University of Aberdeen, 23 St Machar Drive, Aberdeen AB24 3UU, UK, [4]Department of Soil and Crop Sciences, Colorado State University, Fort Collins, CO 80523, USA, [5]Centre for Ecology & Hydrology, Maclean Building, Wallingford OX10 8BB, UK

Abstract

National governments and international organizations perceive bioenergy, from crops such as *Miscanthus*, to have an important role in mitigating greenhouse gas (GHG) emissions and combating climate change. In this research, we address three objectives aimed at reducing uncertainty regarding the climate change mitigation potential of commercial *Miscanthus* plantations in the United Kingdom: (i) to examine soil temperature and moisture as potential drivers of soil GHG emissions through four years of parallel measurements, (ii) to quantify carbon (C) dynamics associated with soil sequestration using regular measurements of topsoil (0–30 cm) C and the surface litter layer and (iii) to calculate a life cycle GHG budget using site-specific measurements, enabling the GHG intensity of *Miscanthus* used for electricity generation to be compared against coal and natural gas. Our results show that methane (CH_4) and nitrous oxide (N_2O) emissions contributed little to the overall GHG budget of *Miscanthus*, while soil respiration offset 30% of the crop's net aboveground C uptake. Temperature sensitivity of soil respiration was highest during crop growth and lowest during winter months. We observed no significant change in topsoil C or nitrogen stocks following 7 years of *Miscanthus* cultivation. The depth of litter did, however, increase significantly, stabilizing at approximately 7 tonnes dry biomass per hectare after 6 years. The cradle-to-farm gate GHG budget of this crop indicated a net removal of 24.5 t CO_2-eq ha^{-1} yr^{-1} from the atmosphere despite no detectable C sequestration in soils. When scaled up to consider the full life cycle, *Miscanthus* fared very well in comparison with coal and natural gas, suggesting considerable CO_2 offsetting per kWh generated. Although the comparison does not account for the land area requirements of the energy generated, *Miscanthus* used for electricity generation can make a significant contribution to climate change mitigation even when combusted in conventional steam turbine power plants.

Keywords: bioenergy, coal, decomposition, greenhouse gas, greenhouse gas intensity, life cycle assessment, litter, natural gas, net ecosystem exchange, soil C

Introduction

Climate change is unlikely to be solved with a short-term solution, but alternative renewable fuel sources, like bioenergy, can be a part of the long-term solution. Therefore, it is essential to ensure these bioenergy crops are helping to turn atmospheric carbon dioxide (CO_2) into stable long-lived carbon (C) forms, rather than the reverse. As alternative energy sources, bioenergy crops and lignocellulosic feedstocks often fare well against conventional fuels in both socio-economic (Paine *et al.*, 1996; Domac *et al.*, 2005; Remedio & Domac, 2003) and

environmental (Cherubini *et al.*, 2009; Smeets *et al.*, 2009;) comparisons. The bioenergy crop, *Miscanthus* x *giganteus* Greef et Deu (Hodkinson & Renvoize, 2001) (herein *Miscanthus*), has attracted attention in North America and Europe due to high yields (Christian *et al.*, 2008; Heaton *et al.*, 2008), low management requirements (Miguez *et al.*, 2008; Gopalakrishnan *et al.*, 2011; McCalmont *et al.*, 2015) and the potential for improved soil C stocks (Hansen *et al.*, 2004; Schneckenberger & Kuzyakov, 2007; Poeplau & Don, 2014). These characteristics make *Miscanthus* a particularly attractive crop in the light of climate change mitigation options (Hastings *et al.*, 2009; McBride *et al.*, 2011).

A key area of uncertainty when assessing the sustainability of bioenergy crops surrounds their potential to

Correspondence: Andy Robertson
e-mail: Andy.Robertson@colostate.edu

sequester more C in crop residues and soils than is emitted through production, transport and end-use processes of the harvested biomass. Quantifying the complete life cycle C budget of bioenergy plantations is therefore essential to accurately determine any potential GHG savings. This GHG mitigation potential is an important part of formal life cycle assessments (LCAs) for bioenergy crops that evaluate their environmental impact from cradle to grave (e.g. Adler *et al.*, 2007; Rowe *et al.*, 2011). To date, empirical measurements of the GHG balance of *Miscanthus* cultivation have produced inconsistent outcomes (Toma *et al.*, 2011; Drewer *et al.*, 2012; Zimmermann *et al.*, 2012; Poeplau & Don, 2014). As a consequence, GHG emissions data included in *Miscanthus* LCAs are often modelled (e.g. Hamelin *et al.*, 2012) or use IPCC default emission factors (e.g. Brandão *et al.*, 2011). To address this area of uncertainty, we focused on cultivation of *Miscanthus* from the cradle-to-farm gate to quantify the C sequestration potential of *Miscanthus*. For this, we measured four years of soil GHG emissions and net ecosystem exchange (NEE) from a 3- to 7-year-old commercial *Miscanthus* plantation in the United Kingdom, also measuring soil C stocks and accumulated plant litter.

Assessing the GHG budget of *Miscanthus* requires more than estimates of C assimilation through photosynthesis as soil C sequestration can offset a large proportion of GHG emissions from the field (Lal, 2004). Temperature (Kirschbaum, 1995) and water availability (Orchard & Cook, 1983; Wood *et al.*, 2013) are both major drivers of the microbial processes that incorporate C into soils. Further, the 'quality' of plant litter (quantified by C : N ratios or lignin : N ratios) can influence how quickly that C is decomposed (Taylor *et al.*, 1989; Donnelly *et al.*, 1990; Bonanomi *et al.*, 2013). Consequently, it is important to consider these factors when evaluating soil C sequestration. Senesced *Miscanthus* biomass is typically very low in N due to nutrient translocation. This results in low litter quality (Amougou *et al.*, 2011) which has a significant impact on the rate of C turnover from the litter layer into the topsoil (Cadoux *et al.*, 2012). Root decomposition also contributes to soil C sequestration, but *Miscanthus*-specific data are limited to a few studies (Rasse *et al.*, 2005; Agostini *et al.*, 2015). The majority (>50%) of belowground biomass is found in the top 30 cm (Neukirchen *et al.*, 1999; Amougou *et al.*, 2011), with C inputs from roots and rhizomes estimated to be as high as 0.86 tC ha^{-1} yr^{-1} and 2.66 tC ha^{-1} yr^{-1}, respectively (Agostini *et al.*, 2015). However, a recent study suggests that rhizosphere activity under *Miscanthus* may stimulate priming, causing a loss of native soil C and offsetting fresh C inputs (Zatta *et al.*, 2014). Long-term studies are therefore required to assess litter accumulation,

belowground biomass and soil C stock changes in *Miscanthus* plantations, in order to quantify its benefits for climate change mitigation (Poeplau & Don, 2014; Robertson *et al.*, 2015).

While C stocks in litter, standing biomass and soils are important 'pools' to quantify, their changes over time are relatively slow compared to the 'fluxes' of the system that include photosynthesis and respiration (Kuzyakov, 2011). These processes continually respond to environmental conditions and often follow diurnal patterns strongly influenced by crop physiology (Linn & Doran, 1984; Rochette *et al.*, 1999; Cheng *et al.*, 2003). At the ecosystem scale, the balance between C uptake and CO_2 efflux is described as the NEE, and within the C cycle, this is the largest flux between atmosphere and a bioenergy plantation. NEE is typically calculated using eddy covariance to continuously monitor changes in CO_2 concentration above the plantation canopy (Baldocchi, 2003). Although the C stored in aboveground biomass is often quantified for bioenergy crops when they are harvested, measurements of the NEE are required to ensure that the amount stored in pools is in excess of the amount emitted through fluxes.

In many agricultural systems, CO_2 is not the only GHG of importance with nitrous oxide (N_2O) emissions often contributing more to a crop's overall GHG balance than the NEE (Flessa *et al.*, 2002). Despite established measurement techniques, relatively few studies have measured soil GHG emissions from *Miscanthus* plantations. The limited data available show that emissions of both N_2O and methane (CH_4) from soils are low and CO_2 efflux dominates soil GHG emissions (Toma *et al.*, 2011; Drewer *et al.*, 2012; Gauder *et al.*, 2012). To accurately quantify an average annual efflux of these GHGs, data are required throughout the year and ideally over several years. In this study we measured GHG emissions and NEE in a *Miscanthus* plantation in Lincolnshire, UK, from 2009 to 2013 (growth years 3 to 7). We then used parallel measurements of climatic variables to explore the environmental controls on soil respiration (CO_2), CH_4 and N_2O emissions, including the temperature sensitivity of respiration at different stages in the crops growth cycle. The aims of the study were to quantify the relative contributions of each GHG towards the net GHG balance of the site, and to better understand their relationship to temperature and soil moisture as environmental drivers. CO_2 was expected to dominate site GHG fluxes, with warmer and wetter periods driving the greatest soil respiration rates. In addition, changes in soil C stocks and the litter layer were quantified over time, with the expectation that the dynamics of these C pools are largely responsible for sequestration rates reported for *Miscanthus* (e.g. Dondini *et al.*, 2009). These data were then used to calculate a life cycle GHG balance of *Miscanthus* cultivation in order to compare *Miscanthus* as a source of electricity to coal and gas.

Materials and methods

Study site

The field experiment was conducted in an 11.5-ha commercial *Miscanthus* plantation near Lincoln, Lincolnshire, UK. The soil type is a compacted loam that behaved like a heavy clay, with approximately 15 %, 36 % and 49 % of clay, silt and sand, respectively, in the top 30 cm of soil. The top 30 cm of soil had a mean total C and N concentration of 1.86 % and 0.18 %, respectively, with a soil pH ranging from 6.8 to 7.3. The bulk density of the soil was 1.46 ± 0.03 g cm^{-3} for the 0- to 15-cm layer and 1.53 ± 0.02 g cm^{-3} for the 15- to 30-cm soil layer. Root biomass (live and dead) was estimated at the end of the 7th growth year: 2.61 t dry mass ha^{-1} for 0–15 cm and 1.85 t dry mass ha^{-1} for 15–30 cm. Additional soil characteristics sampled monthly for two years within this study can be found in Table S1. The deeper soil profile showed an increasing bulk density (1.59 ± 0.20 g cm^{-3}, 30–50 cm; 1.62 ± 0.10 g cm^{-3}, 50–100 cm) and a clear B-horizon at the plough depth (30 cm). There was little evidence of root biomass propagation below 70 cm when trenches were dug in early 2009. The site had a mean annual precipitation of 605 mm and a mean annual temperature of 9.9 °C (30-year average 1980–2009). The *Miscanthus* was established in 2006 at a density of 10 000 rhizomes ha^{-1}. The crop was harvested annually in the spring, beginning in March 2008, but biomass was only removed from 2009 onwards; bale yields (20% moisture content) were recorded as 6.95, 10.28, 6.24, 7.58 and 6.87 dry t ha^{-1} for 2009 to 2013, inclusive. The only addition of fertilizer was in April 2010, when a phosphorus–potassium fertilizer was applied at a rate of 125 kg ha^{-1}. The land management prior to conversion to *Miscanthus* was a crop rotation of wheat and oilseed rape, with three years of wheat directly before conversion. Further site details can be found in Robertson *et al.*, 2016.

Sampling strategy and eddy covariance

In early May 2008 a meteorological tower was established in the north east corner of the *Miscanthus* plantation, along with a flux mast positioned to maximize CO$_2$ measurements given prevailing winds over the cropped area. The tower and mast were equipped with a number of devices to continuously (every 30 min) monitor a range of environmental conditions (Table S3), including an ultrasonic anemometer and infrared gas analyser (IRGA) to employ an eddy covariance (EC) system to examine NEE (more details can be found in S.1). Measurements were taken from 7 May 2008 until 10 March 2013 with some exceptions around the harvesting times where instrumentation was removed. NEE data were cumulated for each growth year (March to February) and an average taken over the four full years of measurements (March 2009 to February 2013), reported in g CO$_2$-C m^2.

Soil–atmosphere gas fluxes

Measurements of soil GHGs (CO$_2$, CH$_4$ and N$_2$O) were taken from October 2008 until March 2013 using the static chamber method described by Livingston & Hutchinson (1995), adapted to include the use of a pressure 'vent'. Five chambers made from PVC (40 cm diameter and 20 cm height) were inserted approximately 3 cm into the soil surface (exact volumes noted). This avoided severing many of the fine roots that were found very close to the soil surface (similar strategies have been recommended in different land uses by Heinemeyer *et al.*, 2011 and Mills *et al.*, 2011). All chambers remained in the soil except at harvest times. Chambers were replaced in the same approximate location after each harvest, with proximity to plants taken into consideration, aiming to represent the average spacing throughout the plantation. At the exact time of GHG sampling, and near the location of GHG sampling, volumetric soil moisture (0–6 cm depth) was measured using a ML2× Theta Probe and Meter HH2 (Delta T Devices, UK) as well as soil (0–7 cm depth) and air temperature measurements using a Tiny Tag temperature logger with integral stab probe (Gemini Data Loggers, UK). Measurements were not taken between December 2010 and April 2011 or in April 2012 due to funding constraints and harvest activities, respectively.

At times of sampling, chambers were closed with a reflective aluminium lid, which had a rubber seal around the edge to prevent leakage. Chambers were enclosed for 30 min with one 10-ml sample taken every 10 min for a total of four time points collected for each plot. At the time of sampling, gas samples were transferred from the chamber headspace into a 3-ml gastight exetainer (Labco Ltd, Lampeter, UK) via a needle and syringe inserted into the self-sealing septa in the chamber lid. The majority (>85%) of GHG measurements were taken between the hours of 10:30 and 14:30 with some exceptions due to field logistics. Exetainer gas samples were analysed on a Perkin-Elmer Autosystem XL Gas Chromatograph (GC) fitted with a flame ionization detector (FID) for CO$_2$ and CH$_4$ and an electron capture detector (ECD) for N$_2$O. All results were calibrated against certified gas standards (BOC, UK) (Case *et al.*, 2014) and converted to a total flux reported as mg CO$_2$-C m^{-2} h^{-1}, µg CH$_4$-C m^{-2} h^{-1} or µg N$_2$O-N m^{-2} h^{-1} in accordance with methods detailed in Holland *et al.* (1999).

Carbon and nitrogen in soil, vegetation and litter

In parallel with monthly GHG measurements, soil samples were collected using PVC pipes (5 cm internal diameter) hammered into the topsoil (0–15 cm) from five locations, one each within a 10 m radius of the static chambers. These cores were taken in March 2009 and March 2010 and then at monthly intervals from May 2011. Further, in October 2011, May 2012, October 2012 and March 2013 additional 30-cm-depth cores (split into 0- to 15-cm and 15- to 30-cm layers) were taken using a 2.5-cm-diameter gouge auger (Van Walt, Haslemere, UK). All soil collected was for destructive sampling and used for C and N determination. The routine monthly 0–15 cm cores were homogenized and freeze-dried (Alpha 1-4 LD, Martin Christ, Osterode am Harz, Germany) before being gently ground by hand to pass through a 2 mm sieve. The 0–30 cm cores were air-dried to constant weight at room temperature before being homogenized, ground and sieved. No differences in C or N concentration were seen between the freeze-dried and air-dried samples. All visible plant matter remains (e.g. roots and leaf litter) were removed before grinding. Small subsamples of the ground soil were taken for analysis of C and N concentration through combustion in an elemental analyser (Costech ECS 4010; Milan, Italy). C and N stocks were estimated by relation to fixed site bulk densities (1.46 for 0–15 cm and 1.53 for 15–30 cm) and the depth layer (Guo & Gifford, 2002). These bulk densities were taken from 15 replicates using a 4.8-cm-diameter, 40-cm-deep split-tube sampler (Eijkelkamp Agrisearch Equipment BV, Giesbeek, the Netherlands). Care was taken to avoid compaction during coring and, where necessary, bulk density was corrected for compression based on the depth of the hole. To ensure consistency when calculating C and N stocks, the resulting bulk density for 0–15 cm was checked against the PVC cores taken monthly.

In October 2011, an adjacent field was sampled to provide an estimate of soil conditions before the Miscanthus was planted (a paired-site approach). This allowed a comparison to be made where samples from the adjacent field represent time-zero reference values of soil C and N stocks. This field had followed the same land use as the Miscanthus field prior to planting in 2006, was seeded with oil seed rape in 2006 and 2010, and winter wheat all other years. Before sampling in 2011, it had recently been harvested for winter wheat before being ploughed and cultivated again. Three replicates at five random locations were cored using the same split-tube sampler (Eijkelkamp Agrisearch Equipment BV, Giesbeek, the Netherlands) and split into 0–15 cm and 15–30 cm ($n = 15$). The soil was then freeze-dried, sieved to 2 mm and analysed for C and N. The same procedure to remove plant matter remains from the soil samples was applied. Further, these cores were analysed for bulk density and corrected for compression through coring (0–15 cm, 1.13 ± 0.17 g cm^{-3}; 15–30 cm, 1.41 ± 0.15 g cm^{-3}). C and N stocks were calculated using the field-specific bulk density values. No carbonates were detected at either depth from either field.

Between October 2008 and March 2013 senesced above-ground biomass was collected using twenty five 2-m^2 litter traps. Traps were placed on top of the litter layer throughout the plantation, with senesced biomass collected and weighed on a monthly basis and values extrapolated to an average rate per hectare. Subsamples of the senesced biomass were weighed and returned to the laboratory for moisture content determination (oven-dried at 105 °C for 24 h). The resulting dried subsample was then ground by freeze-milling (6770 Freezer/Mill, SPEX SamplePrep, Stanmore, UK) before C and N concentrations were determined. The amount of biomass added to the litter layer after harvesting, termed harvesting inefficiency, was also quantified by measuring the size of the litter layer before and after harvest. This varied between years but was proportional to the aboveground yield. Using an average of the measurements taken, a standard value of 5% of the year's harvest was used in future calculations (this value was similar to that reported by Sanderson et al., 1997).

After harvesting (in May 2011, March 2012 and March 2013), the size (t ha^{-1}) of the litter layer was quantified by collecting all of the O-horizon (lightly raked from the soil surface) from 1.6 m^2 circles at 25 random locations throughout the plantation before extrapolating to a per area average for the site (after moisture content was determined by drying in an oven at 105 °C until constant weight, ~24 h). Additionally, the litter layer was quantified at 15 locations at six time points between March 2012 and March 2013 (May, June, August, September, October and January). Subsamples of the litter layer were dried, milled and analysed for C and N concentration. The decomposition rate of this litter layer was assessed assuming first-order decay rates as per Olson (1963), deriving a constant (k) to match a line of best fit through measured litter layer points. This constant was compared to two other studies for Miscanthus litter, Amougou et al. (2012) and Yamane & Sato (1975) who reported k values of 0.776 and 0.511, respectively.

Finally, standing biomass was partially harvested in October 2012 and March 2013 to assess C and N concentrations at the beginning and end of crop senescence. Nine stems were selected at random from different plants. Stems and leaves were separated, weighed and dried at 105 °C until constant weight (~24 h) to calculate moisture content. Dry biomass was then freeze-milled and analysed for C and N concentrations. All C and N concentrations were determined using the same elemental combustion analyser (Costech ECS 4010).

Site-specific life cycle GHG balance

To assess the contribution of site GHG emissions and changes in C stocks to the life cycle GHG balance of Miscanthus, an annual budget was calculated taking into account soil GHG fluxes, NEE and topsoil C stocks (0–30 cm). The mean annual NEE was used for net CO_2 emissions and cumulative annual CH_4 and N_2O emissions were derived from chamber fluxes using monthly data from the four years. CO_2 chamber data refer to soil emissions only and were not used in life cycle estimates. CH_4 and N_2O cumulative annual emissions were transformed using 100-year global warming potentials (GWPs), calculated as CO_2 equivalents (CO_2-eq) according to Myhre et al. (2013) ($CH_4 = 34$; $N_2O = 298$). The cradle-to-farm gate GHG balance was presented as an annual GHG balance per unit area extrapolated to the end of the plantation lifetime. This assumed an 18 year lifecycle of the plantation (DEFRA, 2007) and followed conventional cultivation routines (Table S4) including ploughing before planting as well as at the end of the plantation lifetime to prepare the site for the next crop (Styles & Jones, 2007; Thornley et al., 2009). Direct and indirect emissions associated with other site operations were considered according to Miscanthus-specific estimates of diesel requirements reported by Lewandowski et al. (1995), Smeets et al. (2009) and Thornley et al. (2009).

Applying an assumed 20% moisture content of Miscanthus biomass when harvested and combusted (Lewandowski et al., 2000) a realized calorific value (lower heating value (LHV)) of 14 MJ kg^{-1} (ECN, 2015) was used to estimate GHG intensity. Additionally, a lifetime harvested yield from the plantation was estimated to be 129.2 tonnes dry biomass ha^{-1} (Table S5).

In accordance with the common observation that productivity declines as a *Miscanthus* stand ages (Clifton-Brown *et al.*, 2007; Angelini *et al.*, 2009; Arundale *et al.*, 2014), the findings of Lesur *et al.* (2013) were applied to decrease yields proportional to stand age. Lesur *et al.* (2013) observed a maximum yield of 16.8 dry t ha^{-1} in year 8 and a decrease of 0.647 dry t ha^{-1} in each subsequent year. This reported maximum yield seems unrealistic at our site; therefore, the highest observed yield (10.28 dry t ha^{-1} in 2010) was assumed to be the site-specific maximum. Consequently, this is approximately 49% of that reported by Lesur *et al.* (2013) and so the rate of yield decline is scaled accordingly (0.396 dry t ha^{-1} yr^{-1}). The resulting life-time plantation yield (129.2 dry t ha^{-1}) compares well with the alternative approach (121.3 dry t ha^{-1}) to average measured yields of the first seven years and assume that average is stable over the plantation's lifetime (Table S5). It is important to note that in other areas of the world the harvested biomass may have a lower moisture content (Heaton, 2006), thereby incurring an increased LHV.

The final cradle-to-grave GHG balance was estimated for the *Miscanthus* plantation and reported using the standard notation of emissions per unit of energy generated (GHG intensity; g CO$_{2\text{-eq}}$ kWh^{-1}). This calculation was divided into three procedures (combustion, transportation and production) and was based on a number of informed assumptions. The *Miscanthus* biomass was assumed to be cofired for electricity generation in conventional steam turbine power stations where conversion efficiency of this solid biomass was 30% (1 MJ biomass = 0.30 MJ electricity) (Howes *et al.*, 2002). Although the conversion rate efficiency of biomass to energy can be considerably higher in combined heat and power (CHP) plants (~70%; Cannell, 2003), conventional electricity generation was employed to estimate the most realistic current scenario when comparing with traditional fossil fuels. This resulted in a GHG intensity associated with combustion as defined by Eqn (1).

$$GHG_{com} = \frac{Y \times C_{conc} \times \frac{CO_{2mol}}{C_{mol}}}{\left(\frac{Y \times Cal \times Eff}{E_{conv}}\right)} \qquad (1)$$

where GHG$_{com}$ is the GHG intensity of *Miscanthus* combustion for electricity generation in g CO$_{2\text{-eq}}$ kWh^{-1}; Y is the harvested yield in g biomass ha^{-1} at an assumed 20% moisture content (i.e. 129200000 over this plantation's lifetime); C$_{conc}$ is the carbon concentration of the harvested biomass as a fraction (0 to 1); CO$_{2\ mol}$ is the molecular mass of CO$_2$; C$_{mol}$ is the molecular mass of carbon; Cal is the calorific content of *Miscanthus* in MJ g biomass^{-1} (i.e. a LHV of 0.014 given an assumed 20% moisture content); Eff is the conversation rate efficiency in power stations as a fraction (i.e. 0.30); and E_{conv} is the energy conversion from MJ to kWh (i.e. 0.278 as 1 kWh = 3.6 MJ; Thompson & Taylor, 2008). The GHG intensity associated with transporting the biomass to a power plant assumed a 160-km round trip (based on the location of local power plants) using a vehicle averaging 2.44 km per litre of diesel while carrying the equivalent of 25 tonnes biomass (NAP, 2010). Total GHG emissions of using 1 l of diesel to transport over land were assumed to be 3644 g CO$_2$-eq (Smeets *et al.*, 2009). This resulted in Eqn (2).

$$GHG_{trans} = \frac{\left(\frac{PP_{dist}}{F_{eff}} \times F_{emi}\right)}{\left(\frac{L \times Cal \times Eff}{E_{conv}}\right)} \qquad (2)$$

where GHG$_{trans}$ is the GHG intensity of *Miscanthus* biomass being transported between the plantation and a power station in g CO$_{2\text{-eq}}$ kWh^{-1}; PP$_{dist}$ is the round trip distance to the power station (i.e. 160 km); F_{eff} is the fuel efficiency of the truck used in transportation (i.e. 0.41 l km^{-1}); F_{emi} is the truck emissions associated with 1 l of fuel used during transportation (i.e. 3644 g CO$_{2\text{-eq}}$ l^{-1}); and L is the truck load of biomass (i.e. 250 00 000 g). Finally, the GHG intensity of cradle-to-farm gate production was calculated using Eqn (3).

$$GHG_{prod} = \frac{(GHG_{site} \times P_{life})}{\left(\frac{Y \times Cal \times Eff}{E_{conv}}\right)} \qquad (3)$$

where GHG$_{prod}$ is the GHG intensity in g CO$_{2\text{-eq}}$ kWh^{-1} of *Miscanthus* biomass being grown and harvested including; GHG$_{site}$ is the GHG balance in g CO$_{2\text{-eq}}$ ha^{-1} yr^{-1} of all direct and indirect emissions, using NEE to estimate CO$_2$ exchange as well as CH$_4$ and N$_2$O measurements at the soil surface; and P_{life} is the plantation lifetime in years (i.e. 18). Ultimately, the sum of these three procedures were compared to full life cycle GHG budgets for coal and natural gas when used for electricity generation, as derived from MacKay & Stone (2013).

Statistical analysis

Outliers of GHG measurements were excluded when outside 2× standard deviation, as per Altman & Bland (1995), assuming normal distribution between all measurements of each gas at each time point, thereby retaining 95% of the data. All statistical analyses were performed with R version 3.0.2 (R Core Team, 2014). A global model was formed to define relationships between GHG fluxes and environmental variables (soil temperature, soil moisture, crop phase and a soil temperature * soil moisture interaction). User-defined growth phases of the crops were used to specify whether the crop was dormant (D), emerging (E) or growing (G). These each referred to four months of the year (November to February, March to June and July to October, respectively); the phases were found to be a significantly better predictor of CO$_2$ efflux than the traditional spring–summer–autumn–winter divisions.

Regression analysis was used to quantify the variance in GHG emissions explained by each of the environmental variables through use of the lme function as part of the NLME package (Pinheiro *et al.*, 2013) and the r.squaredGLMM function, part of the MUMIN package (Bartoń, 2012). To meet the assumptions of linear mixed effects (LME) models, log transformations to the flux data were required for soil CO$_2$ emissions and residuals were transformed using the varPower function (in NLME) for CH$_4$ and N$_2$O fluxes. Each chamber was used as the random effect to account for repeated sampling from the same location. This allowed estimates of how much variation in the measurements was explained by the different environmental factors.

Relationships of soil GHG emissions with soil temperature and soil moisture were explored in detail. The temperature sensitivity of CO_2 fluxes was determined as per Raich & Potter (1995) and Luo *et al.* (2001) to estimate a Q_{10} value associated with the relationship, defined as the relative change in CO_2 flux given a 10 °C rise in temperature. This followed a nonlinear (exponential) relationship and applied the nls function as part of the base stats package within R, reporting an associated *P* value to describe the closeness of the defined relationship and data points. Further, because the goodness-of-fit r^2 metric is not as statistically robust for nonlinear relationships (Spiess & Neumeyer, 2010), these are not reported and instead a Q_{10} value was calculated for each chamber individually, and therefore, a standard error could be applied to the average. These relationships were defined for both monthly averages and the full data set. This was done for two reasons: (i) to reduce bias where more measurements were taken in some certain crop phases and (ii) to assess how a few measurements at extreme temperatures influenced Q_{10} values. To test which relationship (monthly vs. all data) best described the temperature sensitivity a generalized additive model (GAM) approach was applied using the gam formula in the MGCV package (Wood, 2011). The resulting nls relationships were compared using the ANOVA function as part of the base stats package within R.

To compare the difference in chamber GHG measurements between temporal groups (days, months, phases or years), repeated-measures analysis of variance (ANOVA) was used applying the aov function as part of the base stats package in R. Where the assumptions of ANOVAs could not be met, residuals were transformed using either the varPower or varExp function as described earlier. The transformed (modelled) data were then analysed using the lme function with chamber as the random effect. This provided significance levels (i.e. *P*-values) to the tests performed.

Results

Climatic conditions and net ecosystem exchange

Continuous half-hourly measurements of air and soil temperature showed clear seasonal trends with annual means (9.60 and 9.55 °C, respectively) in line with 30-year averages (Fig. S1). While precipitation was distributed relatively evenly over the whole measurement period, on average March had the least rainfall (16.68 mm; 0.54 mm day^{-1}) and November had the most (70.60 mm; 2.35 mm day^{-1}). Both soil temperature and precipitation saw notable interannual variation with highs and lows in growth years 6 (9.86 °C) and 5 (8.91 °C) and in years 7 (818 mm) and 6 (405 mm), respectively (Table S2; Fig. S2). Mean NEE over the four full growing seasons was -678.08 ± 110.70 g CO_2-C m^{-2} yr^{-1} with more days between frosts in 2010 leading to the greatest uptake during this year. The large standard deviation reflects the notable interannual variation.

Soil GHG emissions and environmental controls on soil respiration

Soil fluxes of CH_4 and N_2O were largely negligible, with no discernible temporal trends and no clear relationships to environmental variables (Fig. 1). Using linear integration to cumulate average monthly fluxes to annual totals, CH_4 and N_2O emissions were found to be the same weight, totalling 0.38 kg CH_4-C ha^{-1} yr^{-1} and 0.38 kg N_2O-N ha^{-1} yr^{-1}, respectively. In the case of N_2O emissions, only the fluxes in June 2010 were significantly different from zero and therefore contributed largely to the cumulative annual average.

Soil CO_2 emissions were significantly higher than those of CH_4 and N_2O, contributing 3.00 ± 0.22 t CO_2-C ha^{-1} yr^{-1}. Emissions throughout the year followed a clear seasonal trend with highest emissions during the crops growth phase when soil temperatures were warmer; the lowest emissions were seen during the dormant crop phase when temperatures were cooler (Table 1). The climatic variables of temperature and precipitation explained the differences between years, with particularly warm and dry periods during measurements taken in June and September 2009 responsible for high cumulative totals in growth year 4. The highest single measurement (283 mg CO_2-C m^{-2} h^{-1}) was observed in September 2009 and the lowest (0.83 mg CO_2-C m^{-2} h^{-1}) in January 2013 (Fig. 2).

Using either all available data points or monthly averages, soil respiration correlated well with both soil temperature and soil moisture (GAM results for all correlations $P < 0.01$) (Fig. 3). Using nonlinear regressions for each block of chambers, mean Q_{10} values and standard errors were derived using both monthly average data and the full data set (Table 2). In all cases soil respiration was most sensitive to temperature during the crop growth phase and least sensitive during the dormant crop phase, when average temperatures were highest and lowest, respectively. ANOVA results showed the uncertainty of these Q_{10} values was lower ($P = 0.009$) when monthly averages were used in place of the full data sets.

Less than 5% of the variance observed for CH_4 or N_2O fluxes was explained by any of the environmental variables studied (Table 3). However, the same variables explained far more variation in chamber CO_2 fluxes; soil temperature explained more than half of the variance seen in soil respiration throughout the 4-year measurement period.

Carbon and nitrogen stocks

The paired-site proxy used as a 'time-zero' indicated that there was no temporal difference ($P > 0.05$) in soil

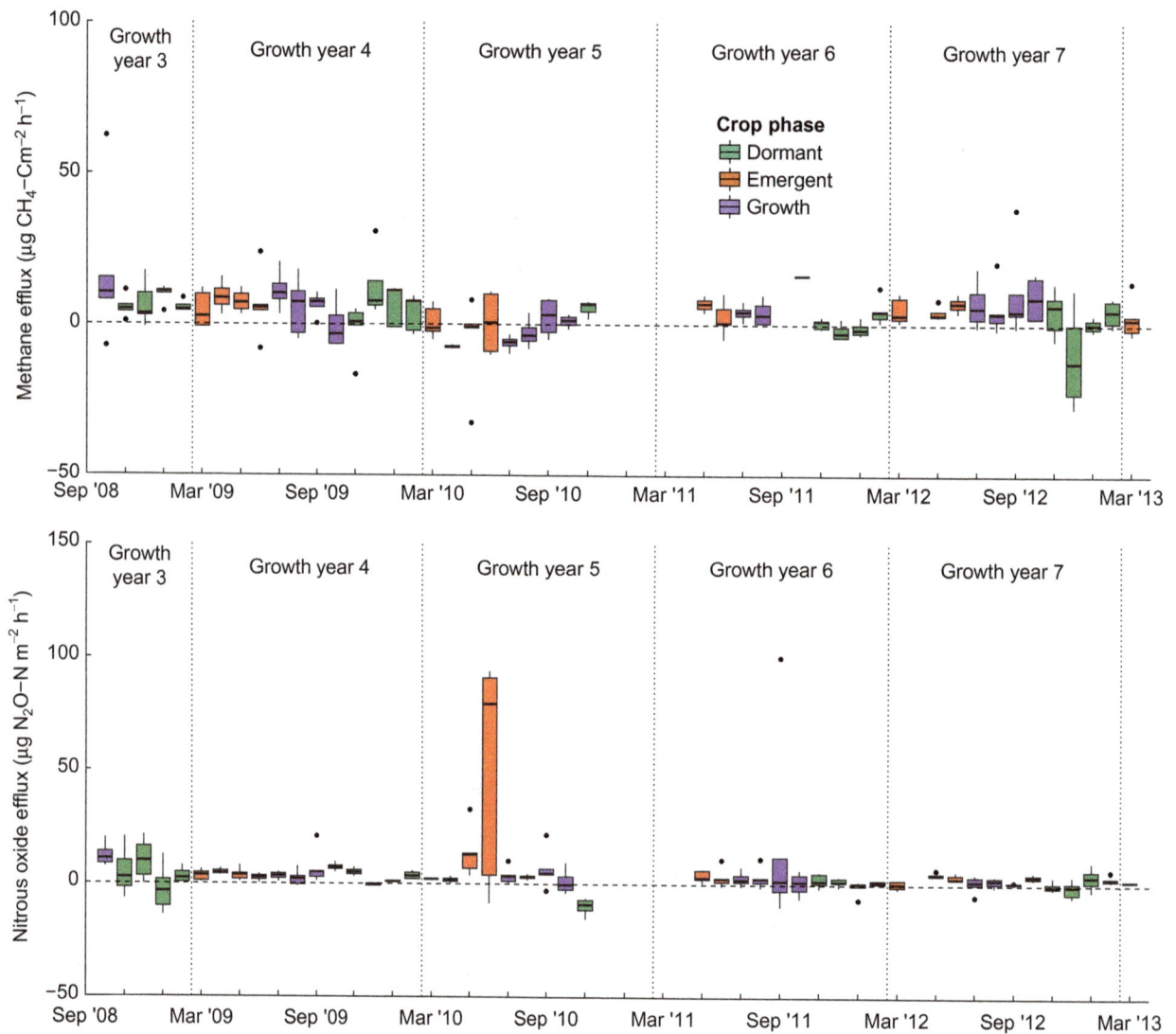

Fig. 1 Soil methane (CH_4) and nitrous oxide (N_2O) emissions in μg CH_4-C m^{-2} h^{-1} and μg N_2O-N m^{-2} h^{-1} calculated from static chambers ($n = 5$) within a *Miscanthus* plantation in Lincolnshire, UK. Measurements are grouped and coloured by crop phase (dormant, green; emergent, orange; growth, purple). The boxes represent the interquartile range (25% to 75%) and the line within is the median value; whiskers describe the highest and lowest data points still within 1.5× the interquartile range. Outliers of this 1.5× the interquartile range are shown by filled circles.

C or N stocks between 0- to 15-cm and 15- to 30-cm layers (Fig. 4). Soil C stocks were estimated to be 81.3 t ha^{-1} in the top 30 cm in March 2006 and, 7 years later, in March 2013, measured as 81.9 t ha^{-1} in the same soil layer. Similarly unchanging soil N stocks were observed with 8.2 t ha^{-1} in the top 30 cm in March 2006 and 8.1 t ha^{-1} in March 2013.

Annual inputs to the litter layer through crop senescence (not including harvesting inefficiency) decreased over time from 2.59 t dry biomass ha^{-1} in growth year 3 to 1.75 t dry biomass ha^{-1} in growth year 7. After heavily stunted growth during the first two years, all standing biomass was cut and left on the site in April 2008,

estimated to be 3 t biomass ha^{-1}. From this point, litter inputs comprised both senesced leaves (green bars; Fig. 5) and residues from harvesting inefficiency (grey bars; Fig. 5). Considerable litter accumulation was observed between 2009 and 2013 (blue points; Fig. 5), suggesting a decomposition rate (k) slower than the rate of inputs. Using our measurements of the litter layer, we estimated a decomposition rate between those reported by Amougou *et al.* (2012) and Yamane & Sato (1975): $k \sim 0.63$.

Both senesced and living *Miscanthus* biomass had similar C concentrations (Table 4). In contrast, N concentration in standing biomass almost halved between

Table 1 Soil respiration from four years of static chamber measurements under a *Miscanthus* plantation in Lincolnshire, UK. Measurements averaged and cumulated by crop phase (dormant, emergent, growth) within each growth year (March–February) between March 2009 and February 2013 ($\pm$ 1 SE)

Growth year	Crop phase	CO_2 efflux (mg CO_2-C m^{-2} h^{-1})	Cumulative CO_2 efflux (t CO_2-C ha^{-1})
4	Dormant	12.77 ± 1.86	0.64 ± 0.09
	Emergent	27.60 ± 2.26	0.75 ± 0.08
	Growth	106.89 ± 17.84	2.83 ± 0.45
	All	47.09 ± 7.70	4.22 ± 0.50
5*	Dormant	–	–
	Emergent	17.17 ± 2.39	0.33 ± 0.01
	Growth	56.86 ± 9.76	1.59 ± 0.39
	All	–	–
6	Dormant	16.12 ± 1.45	0.66 ± 0.11
	Emergent	30.47 ± 6.26	0.49 ± 0.05
	Growth	55.45 ± 5.41	1.33 ± 0.14
	All	34.30 ± 3.55	2.67 ± 0.19
7	Dormant	9.83 ± 1.75	0.31 ± 0.08
	Emergent	26.77 ± 2.60	0.61 ± 0.03
	Growth	42.07 ± 3.94	1.13 ± 0.07
	All	26.86 ± 2.58	2.24 ± 0.15

*denotes that the sensors were removed for too long to calculate average or cumulated emissions.

October (when senescence and nutrient translocation began) and March, and was reduced by a further 40 % in the litter layer (Table 4). Relatively little difference was seen in C concentration between stems and leaves, whereas N concentration was significantly different, resulting in C : N ratios of 206 and 56 for stems and leaves, respectively, in harvested biomass (Table 4). The mean oven-dried (0% moisture content) harvested yield was 6.07 t ha^{-1} yr^{-1} over the 5-year measurement period, equating to 2.85 t C ha^{-1} yr^{-1} (assuming 47% C concentration; Table 4); litter inputs were estimated as 2.69 t ha^{-1} yr^{-1} on average, equivalent to 1.24 t C ha^{-1} yr^{-1} (assuming 47% C concentration).

Life cycle GHG balance of Miscanthus vs. fossil fuels

When calculated over the predicted crop life cycle of 18 years, the total GHG balance from cradle to farm gate was a net removal of 441 t CO_2-eq ha^{-1} (Table 5). Soil C stocks were assumed to remain constant (as this creates the most cautious scenario and no empirical data at the site suggest otherwise) and the litter layer unchanged for the remainder of the crop's lifetime following the measurement period. Both CH_4 and N_2O emissions contributed very little to offsetting the net sequestration observed through NEE measurements.

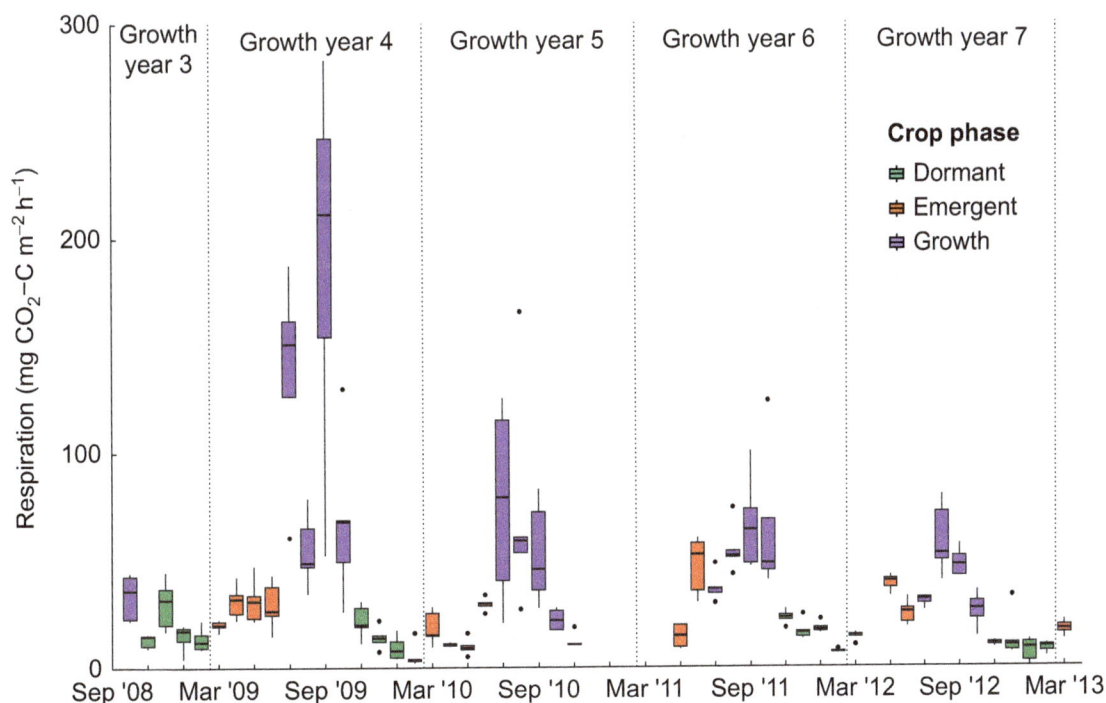

Fig. 2 Soil respiration (CO_2 emissions) in mg CO_2-C m^{-2} h^{-1} calculated from static chambers (*n* = 5) within a *Miscanthus* plantation in Lincolnshire, UK. Measurements are grouped and coloured by crop phase (dormant, green; emergent, orange; growth, purple). The boxes represent the interquartile range (25–75%) and the line within is the median value; whiskers describe the highest and lowest data points still within 1.5× the interquartile range. Outliers of this 1.5× the interquartile range are shown by filled circles.

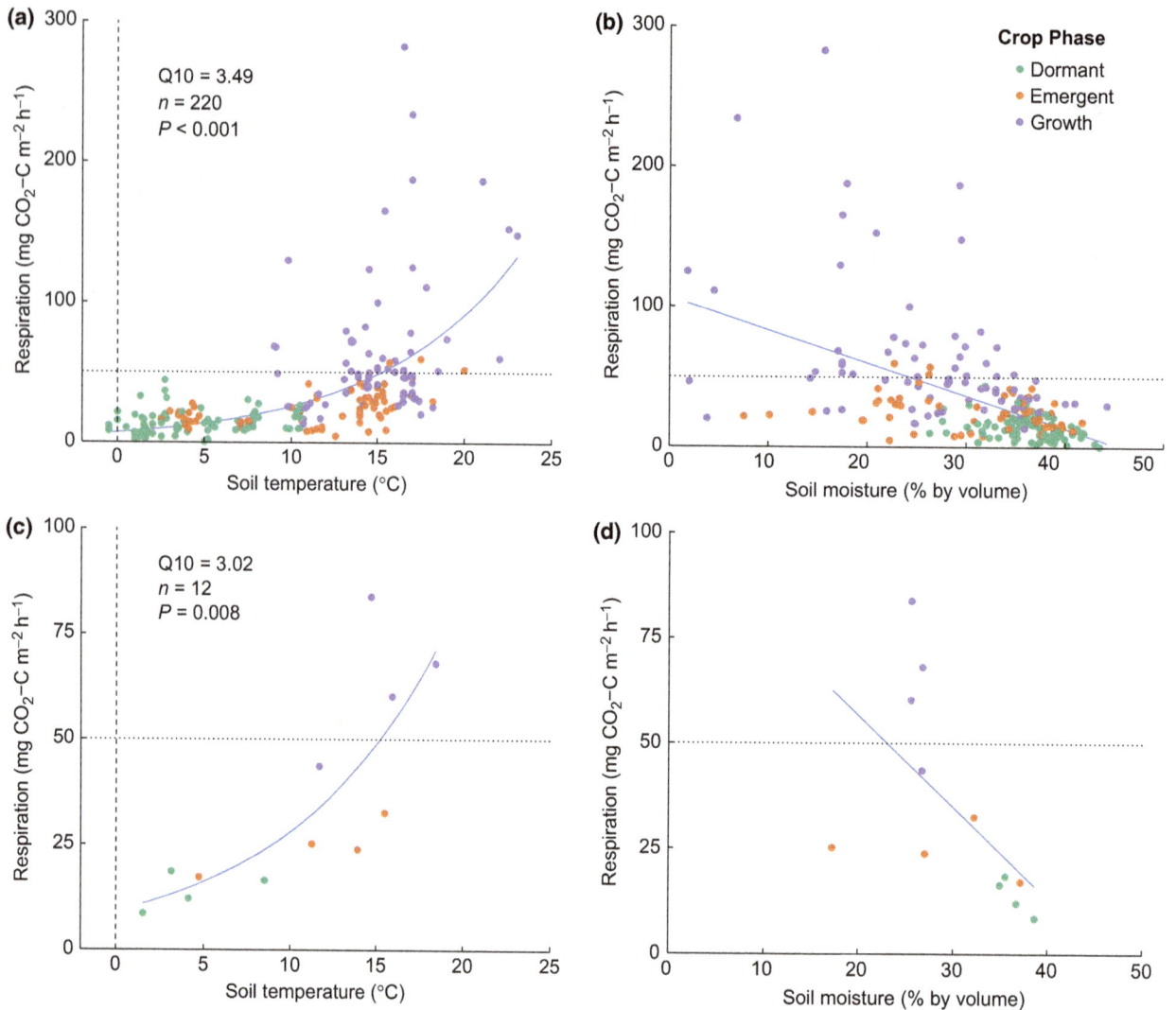

Fig. 3 Relationships between soil respiration and soil temperature (a and c) and soil moisture (b and d) using all available data points (a and b) and monthly average data (c and d) from measurements beneath a *Miscanthus* plantation in Lincolnshire, UK. Colours refer to crop phase: dormant (green), emergent (orange) and growth (purple). Regression analysis was used to fit an exponential relationship for soil temperature, reporting the associated *P*-values of how well the suggested relationship fit the measured data. Dashed vertical lines indicate 0 on plots where negative values were measured. Dotted horizontal lines are applied to aid comparison between top and bottom panels given that the scales differ.

Table 2 Temperature sensitivity of soil respiration calculated from monthly average data (± 1 SE) and the full data set of soil GHG emissions from a *Miscanthus* plantation in Lincolnshire, UK, between October 2008 and March 2013

Data set	Crop phase	Q_{10}	Mean soil temperature (range) (°C)
All chamber	All	4.39 ± 1.27	10.07 (−0.50 to 23.00)
All chamber	Dormant	1.64 ± 0.25	4.57 (−0.50 to 10.50)
All chamber	Emergent	2.03 ± 0.22	11.36 (2.50 to 20.00)
All chamber	Growth	3.18 ± 1.21	15.17 (9.00 to 23.00)
Monthly average chambers	All	3.03 ± 0.34	10.28 (1.52 to 18.41)

Table 3 Variance explained through regression analysis using linear mixed effects models on soil GHG emissions using all static chambers measurements between October 2008 and March 2013 under a *Miscanthus* plantation in Lincolnshire, UK

Factor	CO_2 (%)	CH_4 (%)	N_2O (%)
Soil temperature	48.48	1.62	0.90
Soil moisture	29.75	0.22	3.72
Crop phase	51.76	1.33	1.37
Temp × Moisture interaction	54.35	1.74	4.78

Cutting and baling the harvested biomass contributed the most to direct emissions but these were orders of magnitude lower than NEE measurements.

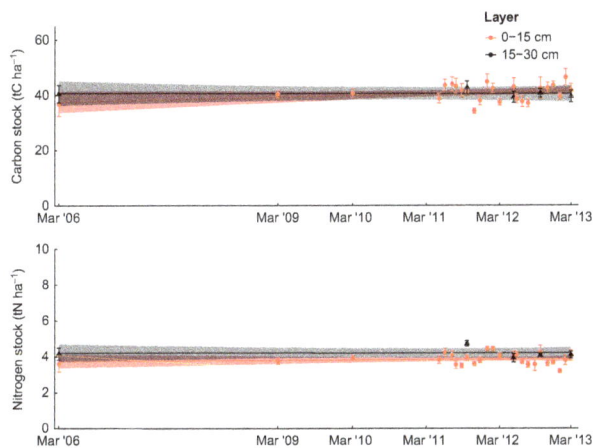

Fig. 4 Soil carbon and nitrogen stocks in tonnes per hectare measured from 2009 to 2013 at two depth intervals of the top-soil (red, 0–15 cm; grey, 15–30 cm) under a *Miscanthus* plantation in Lincolnshire, UK. Paired-site proxy measurements were used for 2006 data. Linear regression provided a relationship to time with colour-consistent shaded 95% confidence intervals.

Compared to the life cycles of coal and natural gas, *Miscanthus* had a substantially lower GHG intensity (Table 6). Further, the life cycle estimate of -1401 g $CO_{2\text{-eq}}$ kWh^{-1} suggests noteworthy sequestration beyond offsetting the known emissions. Any GHG intensity associated with cradle-to-farm gate 'production' below -1525.03 g $CO_{2\text{-eq}}$ kWh^{-1} would completely offset the emissions from transportation and combustion when using conventional power plants with conversion efficiency of 30% (Table 6). However, an important consideration in using GHG intensity as a comparison metric is that it does not account for the land area required to generate each unit of energy (kWh ha^{-1}). Consequently, a higher yield at this site, or an improved conversion efficiency (e.g. 70% achieved by CHP generators), would lead to lower emissions per kWh but would not necessarily increase net sequestration per kWh (Table 6). For reference, using 1 t of *Miscanthus* biomass (at 20% moisture content; LHV = 14 MJ kg^{-1}) for electricity generation produces 1167 kWh at 30% efficiency and 2722 kWh at 70% efficiency, while both emit 1722 kg CO_2-eq through combustion (assuming 47% C concentration) (Eqn 3).

Discussion

This study addressed three main objectives: i) to quantify GHG emissions from a *Miscanthus* plantation and examine the influence of soil temperature and moisture on these emissions, ii) to examine the dynamics of litter and soil C stocks that define long-term sequestration and iii) to estimate the life cycle GHG intensity of

electricity generation using *Miscanthus* harvested from this site, ultimately comparing this with conventional fossil fuels.

Net ecosystem exchange and soil GHG emissions

The annual net CO_2 flux, reported as NEE, was on average -24.85 t CO_2 ha^{-1} yr^{-1} (Table 5), despite low yields compared to other studies in similar climatic regions (Lewandowski *et al.*, 2000; Christian *et al.*, 2008). A trial in Illinois, USA, comparing *Miscanthus* with switchgrass (*Panicum virgatum*) and prairie grasslands reported a GHG balance of -20.31 t CO_2 ha^{-1} yr^{-1} for *Miscanthus* in its third year after establishment (Zeri *et al.*, 2011), 14% lower than switchgrass (-17.78), 88% lower than prairie (-10.82) and 18% higher than our reported NEE. This Illinois *Miscanthus* plantation produced approximately 16 t dry biomass ha^{-1} in October of the third growth season, more than double the spring yield at our Lincolnshire site. Both studies emphasize the large sequestration potential of *Miscanthus*, despite annual harvests removing all aboveground biomass. While the negative NEE at our site implied considerable sequestration, soil respiration (10.99 t CO_2 ha^{-1} yr^{-1}) offset a large portion and dominated the GHG flux at the soil surface. This value was within the same range as other *Miscanthus* plantations (Wanga *et al.*, 2005; Behnke *et al.*, 2012; Case *et al.*, 2014), as well as other bioenergy crops: switchgrass (*Panicum virgatum*) (Frank *et al.*, 2004; Lee *et al.*, 2012), maize (*Zea mays*) (Rochette *et al.*, 1999; Ding *et al.*, 2007) and short rotation coppice (SRC) poplar (*Populus* spp.) (Verlinden *et al.*, 2013).

In contrast to the CO_2 fluxes, both CH_4 and N_2O made a negligible contribution to the GHG budget of the plantation over 4 years. That said, in June 2010 N_2O emissions were an order of magnitude larger than all other months (Fig. 1). Soil N_2O efflux is often very sporadic (Parkin, 1987; Dalal *et al.*, 2003) and most commonly associated with rainfall events and rapid changes in water filled pore space (Dobbie *et al.*, 1999). Consequently, rainfall events that occurred prior to measuring are likely to have influenced the high flux measured in June 2010, although this is unlikely to be the sole cause. To elucidate the drivers of this lone peak, more regular flux measurements are required to gauge the influence of explanatory variables. If these events are short bursts and occur more often than detected by our measurement schedule, the contribution of N_2O to the overall GHG budget would be much larger due to the high GWP of N_2O.

The *Miscanthus* plantation was shown to be a small source of CH_4 contradicting two previous studies at other sites (Toma *et al.*, 2011; Gauder *et al.*, 2012); however, spatial heterogeneity in soils is likely to

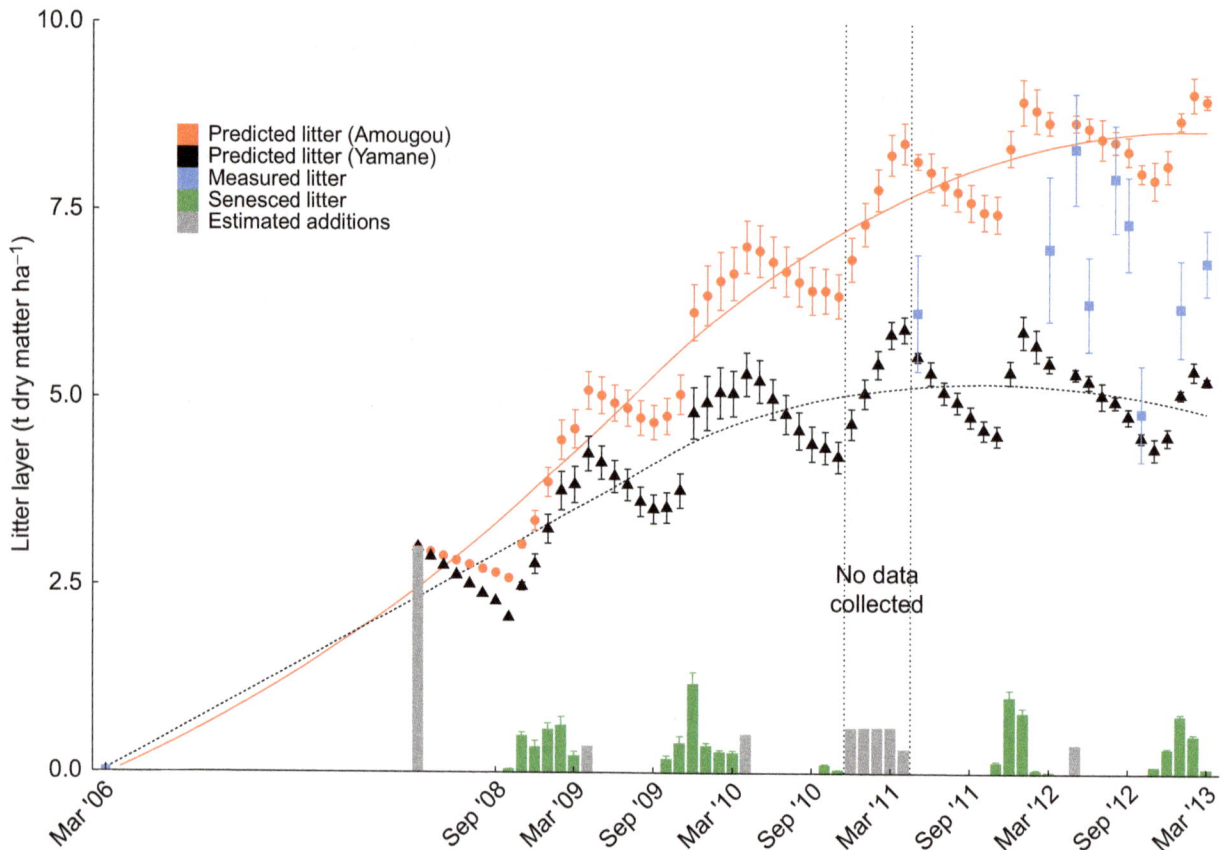

Fig. 5 Measured and modelled accumulation of *Miscanthus* plant litter on the soil surface over 7 years of growth. Two decomposition rates used (Amougou, $k = 0.776$, red points and line; Yamane, $k = 0.511$, black points and line) and smoothed loess regressions fitted through simulated data points. Senesced aboveground biomass (green bars) was measured through all months after September 2008 with the exception of December 2010 to April 2011 where senescence was estimated using an average from other years. Additions through harvesting inefficiency were estimated as 5% of total harvested biomass and occurred in April or May of each year (grey bars after September 2008).

Table 4 Average ($\pm$ 1 SE) carbon and nitrogen concentrations of *Miscanthus* biomass from a plantation in Lincolnshire, UK. Sampling occurred during the 7th growth year of the perennial crop; litter layer values refer to an average of all samples collected between November 2011 and March 2013

		Standing biomass		
		October 2012	March 2013	Litter layer*
Stems	C concentration (%)	46.25 ± 0.20	47.72 ± 0.22	–
	N concentration (%)	0.40 ± 0.03	0.24 ± 0.04	–
	C : N	118.16 ± 6.17	205.93 ± 30.49	–
Leaves	C concentration (%)	44.98 ± 0.19	45.85 ± 0.56	44.15 ± 0.30
	N concentration (%)	1.77 ± 0.07	0.98 ± 0.25	0.58 ± 0.06
	C : N	25.70 ± 1.06	55.78 ± 17.83	85.94 ± 7.41

*The litter layer consisted primarily of leaf litter but some stems were likely to be included. Standing biomass measurements of October represent the end of the growing season and March the end of senescence.

cause variation between sites (Smith *et al.*, 2000). While there are a number of factors which influence the processes that govern CH_4 and N_2O efflux (e.g. disturbance, Hütsch (2001); fertilizer, Mosier *et al.*

(1991); C : N of biomass, Gundersen *et al.* (2012)), the management intensity of *Miscanthus* plantations is typically low (no tillage, low fertilizer application) reducing the likelihood of high emissions. This may explain

Table 5 Life cycle greenhouse gas balance of *Miscanthus* cultivation from cradle-to-farm gate based on an 18-year life cycle and cultivation conditions of a *Miscanthus* plantation in Lincolnshire, UK

Process step	GER* (MJ diesel ha^{-1})	GCR† diesel (kg CO_2-eq. ha^{-1})	Times applied over life cycle	GHG balance (kg CO_2-eq. ha^{-1} yr^{-1})
Direct emissions				
Soil preparation				
Ploughing‡	744.0	63.77	2	7.09
Harrowing‡	310.3	26.60	1	1.48
Herbicide application‡	51.0	4.37	1	0.24
Planting‡	170.1	14.58	1.3	1.05
Rolling‡	340.1	29.15	1	1.62
Crop maintenance				
Fertilizer application‡	416.6	35.71	8	15.87
Harvesting				
Cutting¶	661.9	56.74	2	6.30
Cutting/baling¶	1486.3	127.39	16	113.24
Crop removal				
Herbicide application‡	51.0	4.37	1	0.24
Indirect emissions				
Rhizome propagation§ (10 000 ha^{-1})	2000	171.43	1	9.52
Herbicide production§		16.0	2	1.78
Measured field data				
Annual N_2O fluxes‖			18	176.53
Annual CH_4 fluxes‖			18	17.34
Annual NEE‖			18	−24 847.88
Annual total				**−24 495.81**
Life cycle total				**−440 925.66**

*GER: gross energy requirement conversion factor of 42.51 MJ l^{-1} diesel (Elsayed and Mortimer, 2001; DEFRA R-AEA, 2015).
†GCR: gross C requirement of diesel conversion factor 0.0857 kg CO_2 MJ^{-1} diesel (Smeets *et al.*, 2009).
‡Thornley *et al.* (2009).
§Smeets *et al.* (2009).
¶Styles & Jones (2007).
‖Data from site GHG budget.
Bold values refer to summed totals - both annually and over the full plantation lifetime.

Table 6 Greenhouse gas footprints of *Miscanthus* biomass used for electricity generation under two efficiency scenarios (30% and 70%) compared against coal and natural gas

Process step	*Miscanthus* (g CO_2-eq kWh^{-1})* 30% efficiency	70% efficiency	Coal (g CO_2-eq kWh^{-1}) 38% efficiency	Natural gas (g CO_2-eq kWh^{-1}) 47% efficiency
Production†	−2925.80	−1253.91		
Transportation‡	48.78	20.90		
Combustion§	1476.25	632.68		
Total	**−1400.77**	**−600.33**	837–1130¶	423–535¶

*1 kWh = 3.6 MJ.
†Data from site GHG budget (see method for details).
‡Smeets *et al.* (2009) (see method for details).
§Cannell (2003) (see method for details).
¶MacKay & Stone (2013).
Bold values refer to summed totals.

the low trace GHG emissions seen in this study and reported elsewhere (e.g. Toma *et al.*, 2011; Drewer *et al.*, 2012; Gauder *et al.*, 2012). It is worth noting that land use change to intensive management practices after *Miscanthus* propagation may stimulate rapid mineralization of labile nutrients (particularly C and N) that accumulated during the plantation's lifetime.

Environmental drivers of soil respiration

Due to very low CH_4 and N_2O fluxes, it is not possible to draw conclusions regarding the weak relationships observed between climatic variables and emissions. In contrast, soil respiration did vary significantly with season, closely following changes in soil temperature and crop phenology (Table 1; Fig. 2). This confirms results from other studies where largest CO_2 emissions were observed when temperatures and photosynthetic activity were greatest (Yazaki *et al.*, 2004; Wanga *et al.*, 2005; Gauder *et al.*, 2012) and follows conventional understanding of both heterotrophic and autotrophic soil respiration (Ryan & Law, 2005; Tang *et al.*, 2005). Soil respiration also varied interannually (4.22 to 2.24 t C ha^{-1} for growth years 4 and 7, respectively; Table 1) despite similar climatic conditions between years (Table S6). Yazaki *et al.* (2004) took similar measurements from a *Miscanthus sinensis* plantation in Japan, estimating much more consistent emissions between two years. While the average aboveground biomass was similar, annual soil respiration from the Japanese plantation was more than three times higher than ours (~14 t C ha^{-1}). Additionally, in the same study the temperature sensitivity (Q_{10}) of total soil respiration varied between 2.7 and 3.1. This agrees well with the average Q_{10} values calculated for our site (Table 2), despite the Japanese site having higher soil temperatures and not including Q_{10} estimates between December and April (when they are likely to be lowest). The relatively low soil temperatures at our site, and their impact on soil respiration, may explain why the low productivity still creates a lower NEE than that of the higher yielding site in Illinois (Zeri *et al.*, 2011); while C assimilation through photosynthesis in Illinois is considerably higher than in Lincolnshire, so is the annual mean air temperature (11.1 vs. 9.6 °C) and, in particular, temperatures during the growing season. Consequently, soil respiration is likely to greatly offset the increased C sequestration through photosynthesis; while biomass production in Illinois is larger than that in Lincolnshire, the overall GHG balance of the *Miscanthus* plantation may be more favourable in the cooler climate.

Carbon and nitrogen stocks

Soil C and N stocks did not change over 4 years and when compared with a proxy for before *Miscanthus* was planted, stocks were still unchanged (Fig. 4). While this is consistent with some studies of *Miscanthus* (Zatta *et al.*, 2014; Rowe *et al.*, 2015), many others report increases in topsoil (0–30 cm) C stocks of more than 1 t C ha^{-1} yr^{-1} with prior land use and management practices playing a key role in the direction of change (Kahle

et al., 2001; Dondini *et al.*, 2009; Zimmermann *et al.*, 2012; Poeplau & Don, 2014). There is a reasonable chance that topsoil C stocks were negatively impacted through disturbance of ploughing and planting, but were also enhanced by the addition of rhizomes and rapid fine root turnover as the plantation established itself. Indeed, Amougou *et al.* (2011) reported combined rhizome and root C input rates of 2.91 t ha^{-1} for the top 30 cm over the first three years after planting. These input rates are then expected to decline as the plantation ages; Richter *et al.* (2015) noted a combined C input rate of 1.43 t ha^{-1} for the top 100 cm over the first 14 years after planting (see Agostini *et al.* 2015 for a review of existing data on this topic). Aside from the lower yields noted at this Lincolnshire site, and therefore likely smaller belowground biomass pools, there is no clear reason why soil C stocks are not increasing over time. We hypothesize that at this site fresh C inputs may be stimulating (priming) the decomposition of existing soil C, therefore negating any C sequestration (Zatta *et al.*, 2014). Testing this hypothesis would require the use of stable isotopes to trace the fate of native soil C and fresh C inputs in these crops.

N deficiency in the soil may also explain low C sequestration rates through limitation of decomposition and microbial activity (Hu *et al.*, 2001; Craine *et al.*, 2007). The C : N ratio of senesced *Miscanthus* biomass was between 70 and 120, and soil C : N was around 10 (Table 4; Fig. 4). These are high values for an arable crop, and therefore, a lack of N fertilizer may be a limiting factor in microbial decomposition (Anderson & Domsch, 1989). That said, these C : N ratios are within a normal range for *Miscanthus* plantations where soil C sequestration has been noted (Dondini *et al.*, 2009; Amougou *et al.*, 2011) and therefore cannot alone explain the lack of sequestration at this site. Additionally, other studies have observed similar accumulation rates of senesced biomass (2 t ha^{-1} yr^{-1}; $k \sim 0.63$) while also reporting increased soil C stocks (Yamane & Sato, 1975; Amougou *et al.*, 2011, 2012).

In the absence of soil C sequestration at this site, the measured NEE of -6.78 t C ha^{-1} yr^{-1} is very low and requires an explanation for where C is being sequestered. Following biomass removal at harvest, C pools may remain in live belowground biomass, an increased O-horizon and in the soil organic matter (SOM) that was removed before calculating soil C stocks. When these additional pools are considered, -6.78 t C ha^{-1} yr^{-1} is not unrealistic: 2.85 t C ha^{-1} yr^{-1} was present in harvested biomass and 1.24 t C ha^{-1} yr^{-1} was added to the O-horizon through senescence and harvesting inefficiency (Fig. 5). This leaves 2.69 t C ha^{-1} yr^{-1} to be allocated to live belowground biomass, to soils below the measured topsoil (30 cm) and to SOM fractions, a realistic possibility given the recalcitrant nature of Miscanthus biomass (Amougou et al., 2011) and its characteristic deep-rooting (Neukirchen et al., 1999). Indeed, live and dead root biomass was estimated to be 4.46 t dry mass ha^{-1} in the top 30 cm of soils at this site and annual C inputs under Miscanthus can be substantial (Agostini et al., 2015). It is also important to note that dissolved organic carbon and carbon lost through root exudation may contribute to this unquantified sink of soil carbon (Hromadko et al., 2010).

Comparative life cycle GHG budgets of Miscanthus

Miscanthus was calculated to remove 441 t CO_2-eq ha^{-1} (over 18 years) from the atmosphere using a 'cradle-to-farm gate' analysis (Table 5). This compares well against a SRC willow plantation, grown for 23 years, removing 496 t CO_2-eq ha^{-1} without consideration of soil GHG emissions (Heller et al., 2003). It is worth noting that while our method of linear integration to cumulate soil CO_2 emissions is robust, it may be less appropriate for N_2O. Soil N_2O emissions are spatially and temporally heterogeneous and as a result chamber measurements may not capture the true site-scale emission rates (Williams et al., 1992; Bouwman et al., 2002; Stehfest & Bouwman, 2006). This may have contributed towards the favourable cradle-to-farm gate GHG balance in comparison with other studies, where soil GHG emissions were modelled rather than measured (Brandão et al., 2011; Hamelin et al., 2012). While we acknowledge that the low temporal resolution of measurements may limit our ability to accurately quantify the contribution of N_2O to the life cycle GHG budget, both this study and those previously published report low N_2O emissions under Miscanthus (Toma et al., 2011; Drewer et al., 2012; Gauder et al., 2012). Higher resolution (both temporally and spatially) N_2O measurements would reduce uncertainty and are needed to underpin the refinement of emission factors for use in LCAs. With respect to NEE, limiting gaps in NEE measurements would also improve the accuracy of field GHG emissions data for LCAs. The measurement gaps reported here were assumed to cause limited error because they occurred in winter when photosynthesis and GHG fluxes were low. Further, average annual values were derived from a full 48-month period. Ultimately, gaps during winter months are likely to have far smaller impact on annual NEE estimates than other factors such as interannual climatic variation (Massman & Lee, 2002; Baldocchi, 2014).

The life cycle GHG intensity of electricity generation using Miscanthus from this site is very low compared to that of electricity generated from coal or natural gas. While both fossil fuels are a net source of GHGs, the Miscanthus plantation was a noteworthy GHG sink,

offsetting between 0.6 and 1.4 kg CO_2-eq per kWh (Table 6). This range is very low compared to a similar study of Miscanthus grown in Canada (Sanscartier et al., 2014) where between 0.02 and 0.19 kg CO_2-eq was offset per kWh, including soil C sequestration. However, GHG intensity (emissions per unit energy generated) does not account for the land area required to generate each kWh – a major concern when determining the sustainability of bioenergy crops (Dornburg et al., 2003; Rowe et al., 2009). At this site, each hectare is capable of producing 8372 kWh of electricity, assuming a combustion efficiency of 30% and an average annual yield of 7.18 ha^{-1} (20% moisture content). A higher yielding site with similar environmental characteristics may increase C sequestration through NEE but not necessarily enough to improve the GHG balance per kWh produced, especially if these higher yields come at a cost of increased emissions during production and growth through intensive management (e.g. fertilizer application or precision planting). A recent study comparing Miscanthus with maize and switchgrass in North America (Qin et al., 2015) drew similar conclusions to those described here: Miscanthus has the potential to produce energy at low, or even C-negative, GHG intensities. It is also important to recall that soil C sequestration can offset a significant portion of the emissions derived from generating electricity. Given a 30% combustion efficiency and 129.2 t ha^{-1} yield (18 years at Lincolnshire), an increase of 1 t C ha^{-1} yr^{-1} in soils would offset 438 g CO_2-eq kWh^{-1} on a life cycle basis (Eqn 3, GHG$_{site}$ fixed at −3.66 t CO_2-eq ha^{-1} yr^{-1}). An increase of 1 t C ha^{-1} yr^{-1} in the top 30 cm is not unrealistic; at this site, Miscanthus inputs were previously shown to add 0.86 t C ha^{-1} yr^{-1} to the top 30 cm (Robertson et al., 2016) and Poeplau & Don (2014) saw an average increase of 1.68 ± 0.7 t C ha^{-1} yr^{-1} from a range of Miscanthus crops across Europe. The unchanged topsoil C stocks reported here, therefore, have important consequences for whether it is deemed a preferable alternative to conventional fossil fuels.

Due to minimal land management and fertilizer requirements (Cadoux et al., 2012), Miscanthus is often seen as an attractive option when land is unsuitable for conventional arable crops. However, policymakers still require more data to reliably assess its sustainability when used for bioenergy by combustion. As hypothesized, this study found CO_2 to dominate site GHG fluxes but noted substantially more sequestered than emitted over each year. Furthermore, despite relatively low yields and a lack of soil C sequestration, the crop studied here had a considerably lower GHG intensity than coal or natural gas when used for electricity generation. Additional research is required to elucidate why soil C stocks are not changing under this plantation

(Zatta *et al.*, 2014; Robertson *et al.*, 2016) and future bioenergy sustainability studies should prioritize land use efficiency over GHG intensity comparisons. Nevertheless, this study demonstrates that even when yields are lower than many other sites due to climate or establishment issues, GHG benefits can still outweigh costs and contribute to climate change mitigation through the provision of low C renewable energy.

Acknowledgements

The authors would like to thank the NERC Centre for Ecology & Hydrology and Shell for providing a joint PhD studentship grant award to Andy Robertson (CEH project number NEC04306). We are also grateful to Emily Clark, Simon Oakley and Rebecca Rowe at CEH Lancaster and Sean Case at the University of Copenhagen for help with fieldwork, laboratory work and manuscript comments.

References

Adler PR, Del Grosso SJ, Parton WJ (2007) Life-cycle assessment of net greenhouse-gas flux for bioenergy cropping systems. *Ecological Applications*, **17**, 675–691.

Agostini F, Gregory AS, Richter GM (2015) Carbon sequestration by perennial energy crops: is the jury still out? *Bioenergy research*, **8**, 1057–1080.

Altman DG, Bland JM (1995) Statistics notes: the normal distribution. *British Medical Journal*, **310**, 298.

Amougou N, Bertrand I, Machet J-M, Recous S (2011) Quality and decomposition in soil of rhizome, root and senescent leaf from Miscanthus x giganteus, as affected by harvest date and N fertilization. *Plant and Soil*, **338**, 83–97.

Amougou N, Bertrand I, Cadoux S, Recous S (2012) Miscanthus× giganteus leaf senescence, decomposition and C and N inputs to soil. *GCB Bioenergy*, **4**, 698–707.

Anderson TH, Domsch KH (1989) Ratios of microbial biomass carbon to total organic carbon in arable soils. *Soil Biology and Biochemistry*, **21**, 471–479.

Angelini LG, Ceccarini L, o Di Nasso NN, Bonari E (2009) Comparison of Arundo donax L. and Miscanthus x giganteus in a long-term field experiment in Central Italy: analysis of productive characteristics and energy balance. *Biomass and Bioenergy*, **33**, 635–643.

Arundale RA, Dohleman FG, Heaton EA, Mcgrath JM, Voigt TB, Long SP (2014) Yields of Miscanthus× giganteus and Panicum virgatum decline with stand age in the Midwestern USA. *GCB Bioenergy*, **6**, 1–13.

Baldocchi DD (2003) Assessing the eddy covariance technique for evaluating C dioxide exchange rates of ecosystems: past, present and future. *Global Change Biology*, **9**, 479–492.

Baldocchi D (2014) measuring fluxes of trace gases and energy between ecosystems and the atmosphere—the state and future of the Eddy covariance method. *Global Change Biology*, **20**, 3600–3609.

Bartoń K (2012) MuMIn: Multi-model inference. R package version 1.9.13.

Behnke GD, David MB, Voigt TB (2012) Greenhouse gas emissions, nitrate leaching, and biomass yields from production of Miscanthus x giganteus in Illinois, USA. *BioEnergy Research*, **5**, 801–813.

Bonanomi G, Incerti G, Giannino F, Mingo A, Lanzotti V, Mazzoleni S (2013) Litter quality assessed by solid state ^{13}C NMR spectroscopy predicts decay rate better than C/N and Lignin/N ratios. *Soil Biology and Biochemistry*, **56**, 40–48.

Bouwman AF, Boumans LMJ, Batjes NH (2002) Modeling global annual N$_2$O and NO emissions from fertilized fields. *Global Biogeochemical Cycles*, **16**, 1080.

Brandão M, Milà i Canals L, Clift R (2011) Soil organic C changes in the cultivation of energy crops: implications for GHG balances and soil quality for use in LCA. *Biomass and Bioenergy*, **35**, 2323–2336.

Cadoux S, Riche AB, Yates NE, Machet JM (2012) Nutrient requirements of Miscanthus x giganteus: conclusions from a review of published studies. *Biomass and Bioenergy*, **38**, 14–22.

Cannell MG (2003) C sequestration and biomass energy offset: theoretical, potential and achievable capacities globally, in Europe and the UK. *Biomass and Bioenergy*, **24**, 97–116.

Case SDC, McNamara NP, Reay DS, Whitaker J (2014) Can biochar reduce soil greenhouse gas emissions from a Miscanthus bioenergy crop? *Global Change Biology Bioenergy*, **6**, 76–89.

Cheng W, Johnson DW, Fu S (2003) Rhizosphere effects on decomposition. *Soil Science Society of America Journal*, **67**, 1418–1427.

Cherubini F, Bird ND, Cowie A, Jungmeier G, Schlamadinger B, Woess-Gallasch S (2009) Energy-and greenhouse gas-based LCA of biofuel and bioenergy systems: key issues, ranges and recommendations. *Resources, Conservation and Recycling*, **53**, 434–447.

Christian DG, Riche AB, Yates NE (2008) Growth, yield and mineral content of Miscanthus x giganteus grown as a biofuel for 14 successive harvests. *Industrial Crops and Products*, **28**, 320–327.

Clifton-Brown JC, Breuer J, Jones MB (2007) Carbon mitigation by the energy crop, Miscanthus. *Global Change Biology*, **13**, 2296–2307.

Craine JM, Morrow C, Fierer N (2007) Microbial nitrogen limitation increases decomposition. *Ecology*, **88**, 2105–2113.

Dalal RC, Wang W, Robertson GP, Parton WJ (2003) Nitrous oxide emission from Australian agricultural lands and mitigation options: a review. *Soil Research*, **41**, 165–195.

DEFRA (2007) Planting and growing Miscanthus: Best Practice Guidelines For Applicants to Defra's Energy Crops Scheme. Available at: http://webarchive.nationalarchives.gov.uk/20140605090108/http://www.naturalengland.org.uk/Images/Miscanthus-guide_tcm6-4263.pdf (accessed 18 September 2015).

DEFRA R-AEA (2015) Department for Environment, Food & Rural Affairs (DEFRA), Ricardo-AEA, Carbon Smart, 2015. Greenhouse gas conversion factor repository. Available at: http://www.ukconversionfactorscarbonsmart.co.uk (accessed 14 February 2016).

Ding W, Cai Y, Cai Z, Yagi K, Zheng X (2007) Soil respiration under maize crops: effects of water, temperature, and nitrogen fertilization. *Soil Science Society of America Journal*, **71**, 944–951.

Dobbie KE, McTaggart IP, Smith KA (1999) Nitrous oxide emissions from intensive agricultural systems: variations between crops and seasons; key driving variables; and mean emission factors. *Journal of Geophysical Research*, **104**, 26891–26899.

Domac J, Richards K, Risovic S (2005) Socio-economic drivers in implementing bioenergy projects. *Biomass and Bioenergy*, **28**, 97–106.

Dondini M, Hastings A, Saiz G, Jones MB, Smith P (2009) The potential of Miscanthus to sequester C in soils: comparing field measurements in Carlow, Ireland to model predictions. *Global Change Biology Bioenergy*, **1**, 413–425.

Donnelly PK, Entry JA, Crawford DL, Cromack Jr K (1990) Cellulose and lignin degradation in forest soils: response to moisture, temperature, and acidity. *Microbial Ecology*, **20**, 289–295.

Dornburg V, Lewandowski I, Patel M (2003) Comparing the land requirements, energy savings, and greenhouse gas emissions reduction of biobased polymers and bioenergy. *Journal of Industrial Ecology*, **7**, 93–116.

Drewer J, Finch JW, Lloyd CR, Baggs EM, Skiba U (2012) How do soil emissions of N$_2$O, CH$_4$ and CO$_2$ from perennial bioenergy crops differ from arable annual crops? *GCB Bioenergy*, **4**, 408–419.

ECN (2015) Energy research Centre of the Netherlands (ECN), Phyllis2 Database for biomass and waste. Available at: https://www.ecn.nl/phyllis2/ (accessed 14 February 2016).

Elsayed and Mortimer (2001) Carbon and energy modelling of biomass systems: conversion plant and data updates. DTI/Pub URN 01/1342. Available at: http://webarchive.nationalarchives.gov.uk/+/http://www.berr.gov.uk/files/file14926.pdf (accessed 14 February 2016).

Flessa H, Ruser R, Dörsch P, Kamp T, Jimenez MA, Munch JC, Beese F (2002) Integrated evaluation of greenhouse gas emissions (CO$_2$, CH$_4$, N$_2$O) from two farming systems in southern Germany. *Agriculture, Ecosystems and Environment*, **91**, 175–189.

Frank AB, Berdahl JD, Hanson JD, Liebig MA, Johnson HA (2004) Biomass and C partitioning in switchgrass. *Crop Science*, **44**, 1391–1396.

Gauder M, Butterbach-Bahl K, Graeff-Hönninger S, Claupein W, Wiegel R (2012) Soil-derived trace gas fluxes from different energy crops–results from a field experiment in Southwest Germany. *GCB Bioenergy*, **4**, 289–301.

Gopalakrishnan G, Cristina NM, Snyder SW (2011) A novel framework to classify marginal land for sustainable biomass feedstock production. *Journal of Environmental Quality*, **40**, 1593–1600.

Gundersen P, Christiansen JR, Alberti G et al. (2012) The greenhouse gas exchange responses of methane and nitrous oxide to forest change in Europe. *Biogeosciences Discussions*, **9**, 6129–6168.

Guo LB, Gifford RM (2002) Soil carbon stocks and land use change: a meta analysis. *Global Change Biology*, **8**, 345–360.

Hamelin L, Jørgensen U, Petersen BM, Olesen JE, Wenzel H (2012) Modelling the C and nitrogen balances of direct land use changes from energy crops in Denmark: a consequential life cycle inventory. *GCB Bioenergy*, **4**, 889–907.

Hansen EM, Christensen BT, Jensen LS, Kristensen K (2004) C sequestration in soil beneath long-term *Miscanthus* plantations as determined by ^{13}C abundance. *Biomass and Bioenergy*, **26**, 97–105.

Hastings A, Clifton-Brown J, Wattenbach M, Mitchell CP, Stampfl P, Smith P (2009) Future energy potential of *Miscanthus* in Europe. *Global Change Biology Bioenergy*, **1**, 180–196.

Heaton EA (2006) The Comparative Agronomic Potential of *Miscanthus* x *giganteus* and *Panicum virgatum* as Energy Crops in Illinois. University of Illinois, USA.

Heaton EA, Dohleman FG, Long SP (2008) Meeting US biofuel goals with less land: the potential of *Miscanthus*. *Global Change Biology*, **14**, 2000–2014.

Heinemeyer A, Di Bene C, Lloyd AR et al. (2011) Soil respiration: implications of the plant-soil continuum and respiration chamber collar-insertion depth on measurement and modelling of soil CO2 efflux rates in three ecosystems. *European Journal of Soil Science*, **62**, 82–94.

Heller MC, Keoleian GA, Volk TA (2003) Life cycle assessment of a willow bioenergy cropping system. *Biomass and Bioenergy*, **25**, 147–165.

Hodkinson TR, Renvoize SA (2001) Nomenclature of *Miscanthus* × *giganteus* (Poaceae). *Kew Bulletin*, **56**, 759–760.

Holland EA, Robertson GP, Greenberg J, Groffman PM, Boone RD, Gosz JR (1999) Soil CO2, N2O, and CH4 exchange. In: *Standard Soil Methods for Long-Term Ecological Research* (eds Robertson GP, Coleman DC, Bledsoe CS, Sollins P), pp. 185–201. Oxford University Press, Oxford.

Howes P, Barker N, Higham I et al. (2002) Review of power production from renewable and related sources. Environment Agency. Available at: https://www.gov.uk/government/uploads/system/uploads/attachment_data/file/290421/sp4-097-tr-e-e.pdf (accessed 14 February 2016).

Hromadko L, Vranova V, Techer D, Laval-Gilly P, Rejsek K, Formanek P, Falla J (2010) Composition of root exudates of *Miscanthus x Giganteus* Greef et Deu. Acta Universitatis Agriculturae et Silviculturae Mendelianae Brunensis (Czech Republic).

Hu S, Chapin FS, Firestone MK, Field CB, Chiariello NR (2001) Nitrogen limitation of microbial decomposition in a grassland under elevated CO2. *Nature*, **409**, 188–191.

Hütsch BW (2001) Methane oxidation in non-flooded soils as affect by crop production – invited paper. *European Journal of Agronomy*, **14**, 237–260.

Kahle P, Beuch S, Boelcke B, Leinweber P, Schulten HR (2001) Cropping of *Miscanthus* in Central Europe: biomass production and influence on nutrients and soil organic matter. *European Journal of Agronomy*, **15**, 171–184.

Kirschbaum MUF (1995) The temperature dependence of soil organic matter decomposition, and the effect of global warming on soil organic C storage. *Soil Biology and Biochemistry*, **27**, 753–760.

Kuzyakov Y (2011) How to link soil C pools with CO2 fluxes? *Biogeosciences*, **8**, 1523–1537.

Lal R (2004) Soil carbon sequestration to mitigate climate change. *Geoderma*, **123**, 1–22.

Lee J, Pedroso G, Linquist BA, Putnam D, van Kessel C, Six J (2012) Simulating switchgrass biomass production across ecoregions using the DAYCENT model. *Global Change Biology Bioenergy*, **4**, 521–533.

Lesur C, Jeuffroy MH, Makowski D et al. (2013) Modeling long-term yield trends of *Miscanthus×giganteus* using experimental data from across Europe. *Field Crops Research*, **149**, 252–260.

Lewandowski I, Kicherer A, Vonier P (1995) CO2-balance for the cultivation and combustion of *Miscanthus*. *Biomass and Bioenergy*, **8**, 81–90.

Lewandowski I, Clifton-Brown JC, Scurlock JMO, Huisman W (2000) *Miscanthus*: European experience with a novel energy crop. *Biomass and Bioenergy*, **19**, 209–227.

Linn DM, Doran JW (1984) Effect of water-filled pore space on C dioxide and nitrous oxide production in tilled and nontilled soils. *Soil Science Society of America Journal*, **48**, 1267–1272.

Livingston GP, Hutchinson GL (1995) Enclosure-based measurement of trace gas exchange: applications and sources of error. In: *Biogenic Trace Gases: Measuring Emissions from Soil and Water* (eds Matson PA, Harris RC), pp. 14–51. Marston Lindsey Ross International Ltd., Oxford.

Luo Y, Wan S, Hui D, Wallace LL (2001) Acclimatization of soil respiration to warming in a tall grass prairie. *Nature*, **413**, 622–625.

MacKay DJC, Stone TJ (2013) Department of Energy and Climate Change: Potential Greenhouse Gas Emissions Associated with Shale Gas Extraction and Use. Available at: https://www.gov.uk/government/uploads/system/uploads/attachment_data/file/237330/MacKay_Stone_shale_study_report_09092013.pdf (accessed 18 September 2015).

Massman WJ, Lee X (2002) Eddy covariance flux corrections and uncertainties in long-term studies of C and energy exchanges. *Agricultural and Forest Meteorology*, **113**, 121–144.

McBride AC, Dale VH, Baskaran LM et al. (2011) Indicators to support environmental sustainability of bioenergy systems. *Ecological Indicators*, **11**, 1277–1289.

McCalmont JP, Hastings A, McNamara NP, Richter GM, Robson P, Donnison IS, Clifton-Brown J (2015) Environmental costs and benefits of growing *Miscanthus* for bioenergy in the UK. *GCB Bioenergy*, doi:10.1111/gcbb.12294.

Miguez FE, Villamil MB, Long SP, Bollero GA (2008) Meta-analysis of the effects of management factors on *Miscanthus* x *giganteus* growth and biomass production. *Agricultural and Forest Meteorology*, **148**, 1280–1292.

Mills R, Glanville H, McGovern S, Emmett B, Jones DL (2011) Soil respiration across three contrasting ecosystem types: comparison of two portable IRGA systems. *Journal of Plant Nutrition and Soil Science*, **174**, 532–535.

Mosier A, Schimel D, Valentine D, Bronson K, Parton W (1991) Methane and nitrous oxide fluxes in native, fertilized and cultivated grasslands. *Nature*, **350**, 330–332.

Myhre G, Shindell D, Bréon F-M et al. (2013) Anthropogenic and natural radiative forcing. In: *Climate Change 2013: The Physical Science Basis. Contribution of Working Group I to the Fifth Assessment Report of the Intergovernmental Panel on Climate Change* (eds Stocker TF, Qin D, Plattner G-K, Tignor M, Allen SK, Boschung J, Nauels A, Xia Y, Bex V, Midgley PM), pp. 659–740. Cambridge University Press, Cambridge, UK and New York, NY, USA.

NAP (National Acadamies Press) (2010) Technologies and Approaches to Reducing the Fuel Consumption of Medium- and Heavy-Duty Vehicles. Chapter 2: Vehicle Fundamentals, Fuel Consumption, and Emissions. Available at: http://www.nap.edu/openbook.phprecord_id=12845&page=17. (accessed 18 September 2015).

Neukirchen D, Himken M, Lammel J, Czypionka-Krause U, Olfs HW (1999) Spatial and temporal distribution of the root system and root nutrient content of an established *Miscanthus* crop. *European Journal of Agronomy*, **11**, 301–309.

Olson JS (1963) Energy stores and the balance of producers and decomposers in ecological systems. *Ecology*, **44**, 322–331.

Orchard VA, Cook FJ (1983) Relationship between soil respiration and soil moisture. *Soil Biology and Biochemistry*, **15**, 447–453.

Paine LK, Peterson TL, Undersander DJ et al. (1996) Some ecological and socio-economic considerations for biomass energy crop production. *Biomass and Bioenergy*, **10**, 231–242.

Parkin TB (1987) Soil microsites as a source of denitrification variability. *Soil Science Society of America Journal*, **51**, 1194–1199.

Pinheiro J, Bates D, DebRoy S, Sarkar D, R Development Core Team (2013) nlme: Linear and Nonlinear Mixed Effects Models. R package version 3.1-113.

Poeplau C, Don A (2014) Soil C changes under *Miscanthus* driven by C4 accumulation and C3 decompostion – toward a default sequestration function. *Global Change Biology Bioenergy*, **6**, 327–338.

Qin Z, Zhuang Q, Zhu X (2015) C and nitrogen dynamics in bioenergy ecosystems: 2. Potential greenhouse gas emissions and global warming intensity in the conterminous United States. *GCB Bioenergy*, **7**, 25–39.

R Core Team (2014) R Foundation for Statistical Computing, Vienna, Austria.

Raich JW, Potter CS (1995) Global patterns of C dioxide emissions from soils. *Global Biogeochemical Cycles*, **9**, 23–36.

Rasse DP, Rumpel C, Dignac MF (2005) Is soil C mostly root C? Mechanisms for a specific stabilisation. *Plant and Soil*, **269**, 341–356.

Reichstein M, Falge E, Baldocchi D et al. (2005) On the separation of net ecosystem exchange into assimilation and ecosystem respiration: review and improved algorithm. *Global Change Biology*, **11**, 1424–1439.

Remedio EM, Domac JU (2003) Socio-economic analysis of bioenergy systems: A focus on employment. Rome: FAO.

Richter GM, Agostini F, Redmile-Gordon M, White R, Goulding KWT (2015) Sequestration of C in soils under Miscanthus can be marginal and is affected by genotype-specific root distribution. *Agriculture Ecosystems and Environment*, **200**, 169–177.

Robertson AD, Davies CA, Smith P, Dondini M, McNamara NP (2015) Modelling the C cycle of *Miscanthus* plantations: existing models and the potential for their improvement. *Global Change Biology Bioenergy*, **7**, 405–421.

Robertson AD, Davies CA, Smith P, Stott AW, Clark EL, McNamara NP (2016) Carbon inputs from *Miscanthus* displace older soil organic carbon without inducing priming. *Bioenergy Research*. doi:10.1007/s12155-016-9772-9.

Rochette P, Flanagan LB, Gregorich EG (1999) Separating soil respiration into plant and soil components using analyses of the natural abundance of C-13. *Soil Science Society of America Journal*, **63**, 1207–1213.

Rowe RL, Street NR, Taylor G (2009) Identifying potential environmental impacts of large-scale deployment of dedicated bioenergy crops in the UK. *Renewable and Sustainable Energy Reviews*, **13**, 271–290.

Rowe R, Whitaker J, Freer-Smith PH, Chapman J, Ludley KE, Howard DC, Taylor G (2011) Counting the cost of carbon in bioenergy systems: sources of variation and hidden pitfalls when comparing life cycle assessments. *Biofuels*, **2**, 693–707.

Rowe R, Keith A, Elias D, Dondini M, Smith P, Oxley J, McNamara N (2015) Initial soil carbon and land use history determine soil carbon sequestration under perennial bioenergy crops. *GCB Bioenergy*, **8**, 1047–1061.

Ryan MG, Law BE (2005) Interpreting, measuring, and modeling soil respiration. *Biogeochemistry*, **73**, 3–27.

Sanderson MA, Egg RP, Wiselogel AE (1997) Biomass losses during harvest and storage of switchgrass. *Biomass and Bioenergy*, **12**, 107–114.

Sanscartier D, Deen B, Dias G, MacLean HL, Dadfar H, McDonald I, Kludze H (2014) Implications of land class and environmental factors on life cycle GHG emissions of *Miscanthus* as a bioenergy feedstock. *GCB Bioenergy*, **6**, 401–413.

Schneckenberger K, Kuzyakov Y (2007) C sequestration under *Miscanthus* in sandy and loamy soils estimated by natural ^{13}C abundance. *Journal of Plant Nutrition and Soil Science*, **170**, 538–542.

Smeets EM, Lewandowski IM, Faaij AP (2009) The economical and environmental performance of *Miscanthus* and switchgrass production and supply chains in a European setting. *Renewable and Sustainable Energy Reviews*, **13**, 1230–1245.

Smith KA, Dobbie KE, Ball BC et al. (2000) Oxidation of atmospheric methane in Northern European soils, comparison with other ecosystems, and uncertainties in the global terrestrial sink. *Global Change Biology*, **6**, 791–803.

Spiess AN, Neumeyer N (2010) An evaluation of R2 as an inadequate measure for nonlinear models in pharmacological and biochemical research: a Monte Carlo approach. *BMC Pharmacology*, **10**, 6.

Stehfest E, Bouwman L (2006) N$_2$O and NO emission from agricultural fields and soils under natural vegetation: summarizing available measurement data and modeling of global annual emissions. *Nutrient Cycling in Agroecosystems*, **74**, 207–228.

Styles D, Jones MB (2007) Energy crops in Ireland: quantifying the potential life-cycle greenhouse gas reductions of energy-crop electricity. *Biomass and Bioenergy*, **31**, 759–772.

Tang J, Baldocchi DD, Xu L (2005) Tree photosynthesis modulates soil respiration on a diurnal time scale. *Global Change Biology*, **11**, 1298–1304.

Taylor BR, Parkinson D, Parsons WF (1989) Nitrogen and lignin content as predictors of litter decay rates: a microcosm test. *Ecology*, **70**, 97–104.

Thompson A, Taylor BN (2008) Guide for the Use of the International System of Units (SI) (Special publication 811). Gaithersburg, MD: National Institute of Stan-dards and Technology. Available at: http://physics.nist.gov/cuu/pdf/sp811.pdf (accessed 14 August 2016).

Thornley P, Upham P, Huang Y, Rezvani S, Brammer J, Rogers J (2009) Integrated assessment of bioelectricity technology options. *Energy Policy*, **37**, 890–903.

Toma Y, Fernández FG, Sato S et al. (2011) C budget and methane and nitrous oxide emissions over the growing season in a *Miscanthus* sinensis grassland in Tomako-mai, Hokkaido, Japan. *GCB Bioenergy*, **3**, 116–134.

Verlinden MS, Broeckx LS, Zona D et al. (2013) Net ecosystem production and C balance of an SRC poplar plantation during its first rotation. *Biomass and Bioenergy*, **56**, 412–422.

Wanga W, Ohse K, Liu J, Mo W, Oikawab T (2005) Contribution of root respiration to soil respiration in a C3/C4 mixed grassland. *Journal of Biosciences*, **30**, 507–514.

Williams EJ, Hutchinson GL, Fehsenfeld FC (1992) NOx and N$_2$O emissions from soil. *Global Biogeochemical Cycles*, **6**, 351–388.

Wood SN (2011) Fast stable restricted maximum likelihood and marginal likelihood estimation of semiparametric generalized linear models. *Journal of the Royal Statistical Society (B)*, **73**, 3–36.

Wood TE, Detto M, Silver WL (2013) Sensitivity of soil respiration to variability in soil moisture and temperature in a humid tropical forest. *PLoS ONE*, **8**, e80965.

Yamane I, Sato K (1975) Decomposition of litter of *Miscanthus* sinensis and *Sasa palmata* during five years under semi-natural condition. In: *Ecological Studies in Japanese Grasslands, With Special Reference to the IBP Area: Productivity of Terrestrial Communities* (ed. Numata M), pp. 212–215. University of Tokyo Press, Tokyo.

Yazaki Y, Mariko S, Koizumi H (2004) Carbon dynamics and budget in a *Miscanthus* sinensis grassland in Japan. *Ecological Research*, **19**, 511–520.

Zatta A, Clifton-Brown J, Robson P, Hastings A, Monti A (2014) Land use change from C3 grassland to C4 *Miscanthus*: effects on soil C content and estimated mitigation benefit after six years. *Global Change Biology Bioenergy*, **6**, 360–370.

Zeri M, Anderson-Teixeira K, Hickman G, Masters M, DeLucia E, Bernacchi CJ (2011) C exchange by establishing biofuel crops in Central Illinois. *Agriculture, Ecosystems and Environment*, **144**, 319–329.

Zimmermann J, Dauber J, Jones MB (2012) Soil C sequestration during the establishment phase of *Miscanthus* x *giganteus*: a regional-scale study on commercial farms using ^{13}C natural abundance. *Global Change Biology Bioenergy*, **4**, 453–461.

6

Bioenergy production and sustainable development: science base for policymaking remains limited

CARMENZA ROBLEDO-ABAD[1,2], HANS-JÖRG ALTHAUS[3,4], GÖRAN BERNDES[5],
SIMON BOLWIG[6], ESTEVE CORBERA[7], FELIX CREUTZIG[8], JOHN GARCIA-ULLOA[9],
ANNA GEDDES[10], JAY S. GREGG[6], HELMUT HABERL[11], SUSANNE HANGER[10,12],
RICHARD J. HARPER[13], CAROL HUNSBERGER[14], RASMUS K. LARSEN[15], CHRISTIAN
LAUK[11], STEFAN LEITNER[11], JOHAN LILLIESTAM[10], HERMANN LOTZE-CAMPEN[16,17],
BART MUYS[18], MARIA NORDBORG[5], MARIA ÖLUND[19], BORIS ORLOWSKY[20],
ALEXANDER POPP[16], JOANA PORTUGAL-PEREIRA[21], JÜRGEN REINHARD[22],
LENA SCHEIFFLE[16] and PETE SMITH[23]

[1]Department of Environmental Systems Science, USYS TdLab, ETH Zürich, Universitätstrasse 22, 8092 Zurich, Switzerland, [2]Helvetas Swiss Intercooperation, Maulbeerstr. 10, CH-3001 Bern, Switzerland, [3]Foundation for Global Sustainability (ffgs), Reitergasse 11, 8004 Zürich, Switzerland, [4]Lifecycle Consulting Althaus, Bruechstr. 132, 8706 Meilen, Switzerland, [5]Department of Energy and Environment, Chalmers University of Technology, SE 41296 Gothenburg, Sweden, [6]DTU Management Engineering, Technical University of Denmark, 4000 Roskilde, Denmark, [7]Institute of Environmental Science and Technology, and Department of Economics & Economic History, Universitat Autònoma de Barcelona, 08193 Barcelona, Spain, [8]Mercator Research Institute on Global Commons and Climate Change & Technical University Berlin, 10829 Berlin, Germany, [9]Institute of Terrestrial Ecosystems, ETH Zürich, Universitätstrasse 22, 8092 Zurich, Switzerland, [10]Institute for Environmental Decisions, ETH Zürich, Climate Policy Group, Universitätstrasse 22, 8092 Zurich, Switzerland, [11]Institute of Social Ecology Vienna (SEC), Alpen-Adria Universitaet (AAU), Schottenfeldgasse 29, 1070 Vienna, Austria, [12]International Institute for Applied Systems Analysis, Schlossplatz 1, Laxenburg, Austria, [13]School of Veterinary and Life Sciences, Murdoch University, South Street, Murdoch, WA 6150, Australia, [14]Department of Geography, University of Western Ontario, London, ON N6A 5C2, Canada, [15]Stockholm Environment Institute (SEI), Linnégatan 87D, 115 23 Stockholm, Postbox 24218, 104 51 Stockholm, Sweden, [16]Potsdam Institute for Climate Impact Research (PIK), PO Box 601203, 14412 Potsdam, Germany, [17]Humboldt-University zu Berlin, Unter den Linden 6, 10099 Berlin, Germany, [18]Division of Forest, Nature and Landscape, University of Leuven (KU Leuven), Celestijnenlaan 200E box 2411, BE- 3001 Leuven, Belgium, [19]Centre for Environment and Sustainability – GMV, University of Gothenburg, Aschebergsgatan 44, Göteborg, Sweden, [20]Climate-Babel, [21]Energy Planning Program, COPPE, Federal University of Rio de Janeiro, Centro de Tecnologia, Sala C-211, C.P. 68565, Cidade Universitária, Ilha do Fundão, 21941-972 Rio de Janeiro, RJ, Brazil, [22]Informatics and Sustainability Research Group, Swiss Federal Institute for Material Testing and Research, Empa, Ueberlandstrasse 129, 8600 Duebendorf, Switzerland, [23]Institute of Biological & Environmental Sciences, ClimateXChange and Scottish Food Security Alliance-Crops, University of Aberdeen, 23 St Machar Drive, Aberdeen AB24 3UU, UK

Abstract

The possibility of using bioenergy as a climate change mitigation measure has sparked a discussion of whether and how bioenergy production contributes to sustainable development. We undertook a systematic review of the scientific literature to illuminate this relationship and found a limited scientific basis for policymaking. Our results indicate that knowledge on the sustainable development impacts of bioenergy production is concentrated in a few well-studied countries, focuses on environmental and economic impacts, and mostly relates to dedicated agricultural biomass plantations. The scope and methodological approaches in studies differ widely and only a small share of the studies sufficiently reports on context and/or baseline conditions, which makes it difficult to get a general understanding of the attribution of impacts. Nevertheless, we identified regional patterns of positive or negative impacts for all categories – environmental, economic, institutional, social and technological. In general, economic and technological impacts were more frequently reported as positive, while social and environmental impacts were more frequently reported as negative (with the exception of impacts on direct substitution of GHG emission from fossil fuel). More focused and

Correspondence: Carmenza Robledo-Abad
e-mail: carmenza.robledo@usys.ethz.ch

transparent research is needed to validate these patterns and develop a strong science underpinning for establishing policies and governance agreements that prevent/mitigate negative and promote positive impacts from bioenergy production.

Keywords: agriculture, bioenergy, food security, forestry, mitigation, sustainable development

Introduction

During the last decades, developed and developing countries have introduced policies to encourage the use of bioenergy including *i.a.* the Brazilian National Alcohol Program (ProAlcool), the US Renewable Fuel Standard (RFS), the EU's Renewable Energy Directive (RED), the Alternative Energy Development Plan (AEDP) in Thailand, and the Indian National Policy on Biofuels (Sorda *et al.*, 2010). The promotion of bioenergy as a climate change mitigation measure has sparked a intensive discussion concerning potential impacts on sustainable development. Commonly mentioned positive impacts focus on opportunities for new uses of land, economic growth, climate change mitigation, increased energy security and employment (Smeets *et al.*, 2007; Nijsen *et al.*, 2012; Mendes Souza *et al.*, 2015). On the other hand, there are concerns about potential disruption to food security and rural livelihoods, direct and indirect greenhouse gas (GHG) emissions from land use change, enhanced water scarcity, ecological impacts, increased rural poverty, and displacement of small-scale farmers, pastoralists and forest users (Dauvergne & Neville, 2010; Delucchi, 2010; German *et al.*, 2011; Gamborg *et al.*, 2014; Hejazi *et al.*, 2015).

How bioenergy interacts with sustainable development has become a key scientific question as demand for bioenergy increases globally. The recent Intergovernmental Panel on Climate Change (IPCC) Working Group III contribution to the Fifth Assessment Report (WGIII AR5) highlights the relationship between context conditions, the use of bioenergy as a mitigation option and the impacts on sustainable development. Discussing impacts of bioenergy on sustainable development, the IPCC WGIII AR5 concludes that '...*the nature and extent of the impacts of implementing bioenergy depend on the specific system, the development context, and on the size of the intervention*' (Smith *et al.*, 2014).

Different case studies have documented that expanding production of the crops most commonly used to produce bioenergy can affect local incomes, food security, land tenure or health in positive and negative ways and that the outcomes of bioenergy production can be unequally distributed (Tilman *et al.*, 2009; Persson, 2014). Model-based assessments have tried to integrate sustain-

ability considerations, pointing out likely interactions between bioenergy and food prices as well as biodiversity and water use(Popp *et al.*, 2011; Lotze-Campen *et al.*, 2014; Scharlemann & Laurence, 2014). However, the effects of bioenergy on livelihoods and the role of governance agreements in promoting or mitigating specific types of impact have not yet been included in modelling exercises (Ackerman *et al.*, 2009; Lubowski & Rose, 2013; Creutzig *et al.*, 2014; Smith *et al.*, 2014). Furthermore, previous studies have concluded that more clarity about the relationships between bioenergy production, livelihoods and equity is still needed (Creutzig *et al.*, 2013; Hodbod & Tomei, 2013; Hunsberger *et al.*, 2014).

In the light of the urgent need for action on climate change (IPCC, 2014), persistent economic and social inequalities, and intensifying competition for land (Lambin & Meyfroidt, 2011; Haberl, 2015), there is a need for science-based policymaking with respect to the impacts of bioenergy on sustainable development. We have examined the scientific evidence base for such policymaking in a comprehensive systematic review using the scientific literature produced in the time period covered by the IPCC Fifth Assessment Report.

Methodology for reviewing impacts of bioenergy production on sustainable development

The aim of this systematic review was to analyse the state of knowledge about how the production of bioenergy resources affects sustainable development. This is a key for understanding to what extent the existent knowledge can provide advice for policymakers. The systematic review focuses on the following impact categories: social, economic, institutional, environmental and technological (including food security and human health as social). The review is based on the assumption that if production of a bioenergy resource impacts any of the focus categories, it also impacts sustainable development. Thus, analysing the reported impacts on these focus categories will facilitate an overview of the state of knowledge regarding the impacts from bioenergy production on sustainable development.

We followed the steps included in the methodological guidance for systematic reviews by (Petticrew & Roberts, 2008; Bartolucci & Hillegass, 2010). The review protocol that served as methodological basis included five steps: (i) definition of scope and aims, (ii) research questions, (iii) search for and selection of evidence, (iv) quality appraisal and (v) data extraction and synthesis (see detailed protocol of the systematic review in the supplementary material).

We investigated to what extent the scientific community has answered the following questions which are of high interest in various contexts, including policy, in which decisions on future implementation of bioenergy are decided upon: Where do sustainable development impacts from bioenergy production take place? What is the evidence for the purported impacts? How are impacts attributed and measured? Are there certain context conditions that enable the observed impacts? Are the reported impacts specific to particular biomass resources? These questions were motivated by the discussions addressed in AR5, WGIII (Smith *et al.*, 2014; annex on bioenergy). Although the AR5 considers impacts on sustainable development, it does not provide a geographically differentiated analysis or an understanding of the relation between context conditions and impacts. Several authors (Bustamante *et al.*, 2014; Creutzig *et al.*, 2014; Smith *et al.*, 2014; von Stechow *et al.*, 2015) explicitly highlight the need for improving the understanding of regional distribution of mitigation impacts on sustainable development, disaggregating by technologies and bioenergy inputs and under consideration of context conditions. The aim of this article was to make a first step in this direction through a stringent systematic review.

We used the same time frame for scientific publications as the Fifth IPCC Assessment report (AR5) (see supplementary information for the selection criteria and process) and went into a far more detailed analysis with regard to the questions reported above.

The AR5 defines bioenergy as *'energy derived from any form of biomass such as recently living organisms or their metabolic by-products'* (Allwood *et al.*, 2014). We include nine biomass resources in the review: forest residues, unutilized forest growth, dedicated biomass forest plantations, combined forest sources, agriculture residues, dedicated biomass agricultural plantations, organic waste, combined agricultural resources and combined forest and agricultural resources (see protocol in the supplementary information for specific definition of each biomass resource). As the focus of the research was to understand the impacts from production and collection of these biomass resources on development, we did not distinguish the technologies used for producing bioenergy from biomass (i.e. first or second generation) but considered the demand that both technologies can create on biomass resources.

We acknowledge that there is no general agreement on how to measure impacts on sustainable development (Sneddon *et al.*, 2006; Muys, 2013). Thus, we based the systematic review on the development impacts as outlined in the Agriculture, Forestry and Other Land Use (AFOLU) chapter of the IPCC WGIII AR5 (Smith *et al.*, 2014). We considered a set of 33 potential impacts on sustainable development structured into five impact categories: institutional, social and health-related, environmental, economic and technological (see Tables S3 and S4). We assumed that if production of a bioenergy resource affects any of these impact categories, it also affects sustainable development. Thus, analysing the reported impacts in a systematic manner provides an overview of the state of knowl-edge regarding how bioenergy production affects sustainable development as defined above.

Selection of studies and data extraction

The selection process was carried out in three steps: definition of search criteria, a search in two scientific collections and a quality appraisal. For the search criteria, we included thirty inclusion criteria covering all five development categories and two further criteria on bioenergy forms for a set of sixty inclusion criteria combinations; and we included 12 exclusion criteria (see 'article selection and data extraction' in the protocol included in the supplementary information for further details). We further refined the selection using 31 categories of Web of Science, including 12 research areas. We limited the search to articles in English. The search was conducted in the Web of Science and in Science Direct including all their databases. This procedure yielded a wide and inclusive sample of 1175 articles covering all five development categories. For the quality appraisal, we randomly selected a subset of articles ($n = 873$ or 74.3% of the original sample), which makes the subsample representative. Only 541 of these passed the quality appraisal (criteria and procedure for the appraisal is clarified in the 'quality appraisal' section in the protocol included in the supplementary information). A total of 408 articles of the 541 (75.4%) were randomly included in the data extraction, and the research team carefully reviewed all articles. During the data extraction, we removed 92 articles because none of the 33 potential impacts included in our list were discussed, although they did discuss issues belonging to the five categories (that explains why these articles passed the quality appraisal). Thus, the results presented below are based on the analysis of the detailed data extracted from 316 original research articles that discuss at least one of the 33 impacts included.

Data analysis

We analysed the data in three steps: (i) characterization of the study, (ii) consideration of the context conditions in the area of the study and (iii) reported impacts. Exploratory data analysis revealed a vast heterogeneity of how data were gathered, impacts attributed and results reported in the 316 analysed articles (see detailed counting of results in the supplementary information, file impacts trees). This heterogeneity combined with the number of variables mostly precluded the use of sophisticated statistical analysis methods, and our analysis is mainly based on descriptive tables and cross-tabulations, combining data from all three steps. The statistical significance of potentially interesting relations between context conditions and impacts was analysed using Fisher tests (R Core Team, 2014).

Results

Almost half of the articles in the systematic review analyse impacts from dedicated biomass plantations (agriculture and forestry), while few articles examine the sustainable development impacts from using agricultural or forestry residues (4 and 6%, respectively), or organic waste (2.5%) (see Table S10). Although several studies report that the use of organic waste as bioenergy feedstock can be associated with positive or low negative impacts, and hence considered an attractive bioenergy resource (Gregg & Smith, 2010; Haberl *et al.*, 2011; Odlare *et al.*, 2011), but the evidence in our review is insufficient to object or support this proposition as too few studies analyse this resource.

Different places, different state of knowledge

Our results show an uneven geographical distribution of the studies, with most articles focusing on developed regions: 26.7% on Europe and 26.3% on North America; compared to only 13.1% on Asia, 8.2% on Africa, 7.8% on Latin America (Central and South America), 2.2% on Oceania; 15.7% of the studies conduct global analyses (Fig. 2, Table S11). This distribution contrasts with the share of annual plant biomass production (approximated through Net Primary Production or NPP) of these regions: 16% in Europe, 12% in North America, 19% in Asia, 20% in Africa, 26% in Latin America and 6% in Oceania (Krausmann *et al.*, 2013). Although a multitude of socioeconomic and natural factors influences any region's technical or economic bioenergy potential, we consider NPP a useful proxy for its biophysical suitability for biomass production (Haberl *et al.*, 2013). Modelling and empirical data suggest that current NPP levels may underestimate achievable productivities in human-managed systems (DeLucia *et al.*, 2014), but should be viewed in the perspective of scales of cultivation required for bioenergy to make an important contribution to the future energy supply and also possible ecological impacts of high-input cultivation systems (Haberl, 2016).

Table 1 is divided into three categories of countries: (i) well-studied key countries, (section A in Table 1); (ii) potentially relevant but understudied countries, that is, countries with high NPP but few, if any, studies (section B in Table 1); and (iii) relatively overstudied countries, that is, countries with low NPP and hence a relatively minor global contribution to the global bioenergy potential but nevertheless with many studies associated with them (section C in Table 1).

The small share of studies considering impacts on sustainability in developing regions is surprising, as studies assessing global bioenergy potential commonly point to some of the countries in section B as possible large future suppliers of biomass and biofuels (Hoogwijk *et al.*, 2009; Smeets & Faaij, 2010; Beringer *et al.*, 2011; Haberl *et al.*, 2011; Nijsen *et al.*, 2012). For example, in Latin America, only Brazil (contributing 26 cases or 74% to the studies in countries of this region)

Table 1 Relation of studies and NPP values

Country	No. of studies	% of global NPP	Rank no. studies	Rank NPP
A. Countries with more than 1 study and more than 1% of global NPP				
United States	80	6.50	1	3
Brazil	25	12.10	2	1
China	13	5.60	4	5
India	13	2.30	5	10
Canada	9	6.00	10	4
Indonesia	9	3.20	12	8
United Republic of Tanzania	8	1.10	14	19
Australia	7	4.90	15	6
B. Countries with <5 studies and more than 1% of global NPP				
Russian Federation	3	11.30	27	2
Argentina	3	2.40	23	9
Dem. Rep. of the Congo	0	3.70	98	7
Colombia	0	1.90	89	11
Peru	1	1.60	51	12
Angola	0	1.50	65	13
Mexico	1	1.50	48	14
Venezuela	0	1.50	209	15
Bolivia	0	1.40	78	16
Sudan	0	1.30	192	17
Kazakhstan	0	1.20	131	18
C. Countries with 5 or more studies and <1% of global NPP				
Italy	14	0.24	3	63
Sweden	13	0.36	6	50
United Kingdom	12	0.23	7	65
Malaysia	10	0.56	8	32
South Africa	10	0.63	9	28
Germany	9	0.37	11	46
Thailand	9	0.51	13	35
Mozambique	6	0.91	16	22
Austria	5	0.08	17	97
Belgium	5	0.04	18	125
Spain	5	0.37	19	48
Denmark	4	0.05	20	119
France	4	0.58	21	31
the Netherlands	4	0.04	22	123

Net primary production (NPP) values calculated based on Haberl *et al.*, (2011). For this table, we counted studies specialized in one country and studies looking at multiple countries, which are considered otherwise as global or regional studies. 'Studies' refers to the articles included in this systematic review.

emerges as a focal point of the scientific literature, while the number of country-specific studies in other countries is small (three studies in Argentina and one study each in Costa Rica, Ecuador, Guatemala, Mexico and Peru). Hence, of the 20 countries in Latin America, only one country with a large NPP is well-studied, whereas six countries are under-studied despite

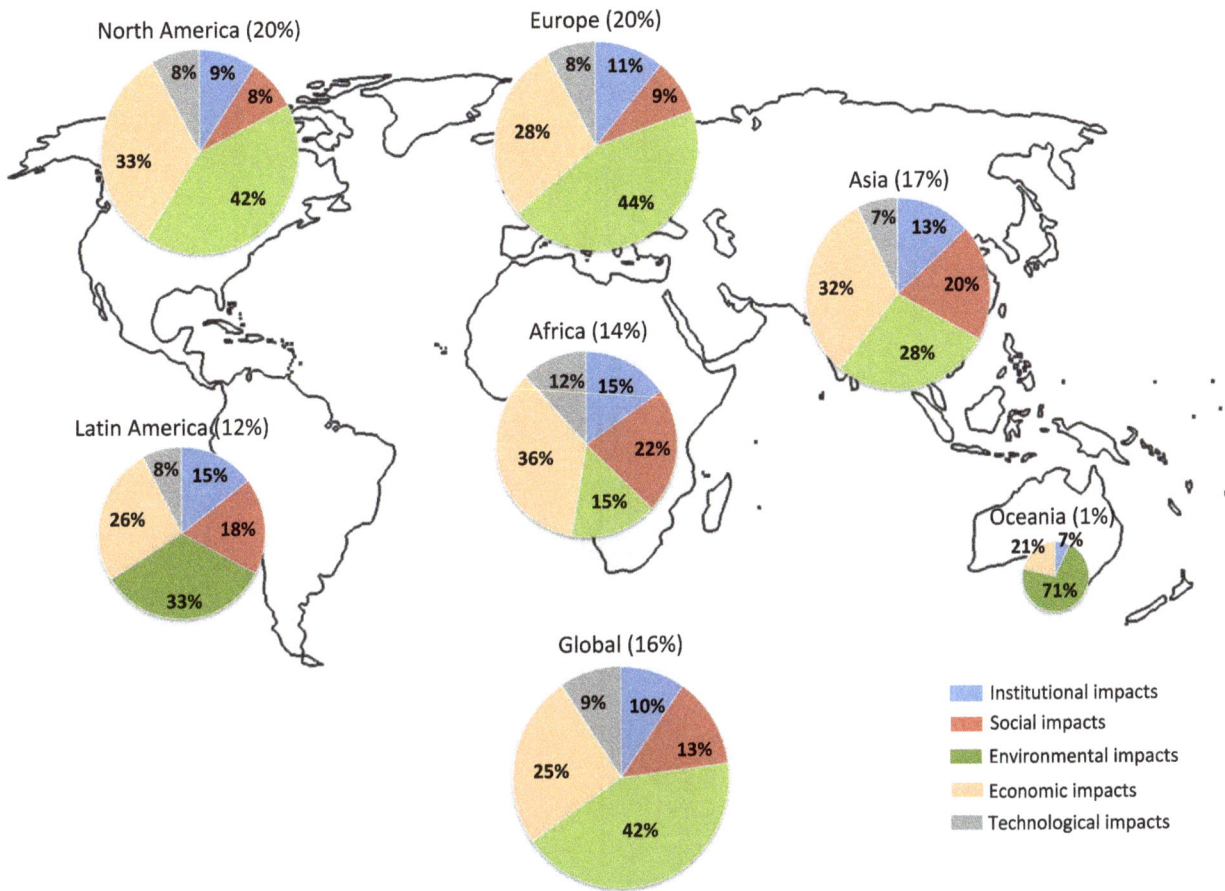

Fig 1 Regional distribution of the analysed impacts, reported as fraction of impacts within each category of all impacts analysed in each region. Percentage numbers after the region's name indicate the share of this region in the total of impacts considered and determine the size of the circle. Percentage numbers in the pies indicate the share of impacts each category contributes to the total number of impacts reported in the respective region. For all regions, the most reported social impact is food security; all other social impacts follow far behind. The outline map is from http://www.zonu.com/images/0X0/2009-11-05-10853/World-outline-map.png.

their large potential. Extrapolations of impacts from the local/national to the regional level are thus not yet possible.

When looking at which impacts have been considered and where, our results show that most regions focus on the environmental and economic categories and barely consider social impacts with the exception of food security (see Fig. 1 and Table 2). Only studies focusing on Asia and Africa show a more balanced interest across categories.

Only a small number of impacts have been studied across regions

Beyond the impact categories, we further analysed which specific impacts were most frequently considered in each region (see Table 3). Studies at the global level focus on impacts on displacement of activities, on deforestation or forest degradation, on soil and water, on food security and on GHG emissions. To a lesser extent,

but nevertheless important, global studies look at market opportunities, feedstock prices and technology development and transfer.

The regional distribution of the interest in specific impacts is uneven. In North America (mainly USA), impacts from the environmental category are included among the seven most frequent followed by impacts on prices of feedstock and on market opportunities from the economic category. The three most frequently analysed impacts in Europe and Latin America (mainly Brazil) are those on displacement of activities, on soil and water and on direct substitution of GHG emissions from fossil fuels. Studies from Oceania only consider six impacts: four of them in the environmental category with the most frequently analysed being impacts on soil and water.

The distribution of analysed impacts in Africa and Asia is more balanced. Most of the impacts have been considered in these two regions, suggesting a better engagement

Table 2 Distribution of analysed impacts per category

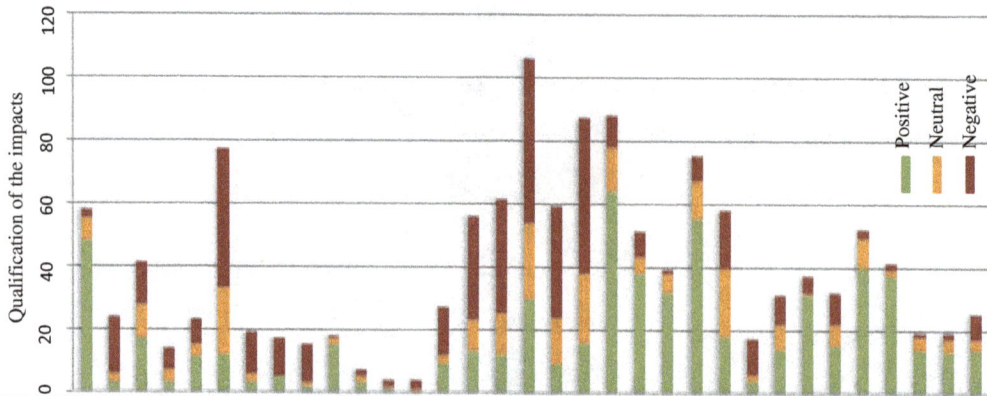

Impact category	No. Studies	Share of category in all studies	Impact on	No. Studies	Share of impact in all studies	Share of impact in the category
Institutional	105	33.23%	Energy independence	59	19%	56.2%
			Land tenure	25	8%	24.8%
			Cross-sectoral coordination(+)/conflicts (-)	41	13%	39.0%
			Labour rights	15	5%	14.3%
			Participative mechanisms	23	7%	21.9%
Social and health	105	33.23%	Food security	81	25.6%	74.3%
			Conflicts or social tension	19	6.0%	17.1%
			Traditional or indigenous practices	17	5.4%	16.2%
			Displacement of farmers	15	4.7%	13.3%
			Capacity building	18	5.7%	16.2%
			Women	7	2.2%	6.7%
			Elderly people	4	1.3%	3.8%
			Specific ethnic groups	4	1.3%	3.8%
			Health impacts	28	8.9%	26.7%
Environmental	222	70.25%	Deforestation or forest degradation	59	18.7%	26.1%
			Use of fertilizers with - impacts on soil and water	63	19.9%	27.9%
			Soil and water	115	36.4%	51.8%
			Biodiversity	64	20.3%	28.8%
			Displacement of activities or land uses	95	30.1%	41.9%
			Direct GHG emission substitution	96	30.4%	43.2%
Economic	165	52.22%	Economic activity	53	16.8%	32.1%
			Economic diversification	44	13.9%	26.1%
			Market opportunities	79	25.0%	47.9%
			Prices of feedstock	63	19.9%	37.0%
			Concentration of income	17	5.4%	10.3%
			Poverty	32	10.1%	18.8%
			Use of waste or residues creates socio-economic benefits	39	12.3%	23.6%
			Mid and long-term revenue's certainty	36	11.4%	21.2%
			Employment	55	17.4%	32.7%
Technological	72	22.78%	Technology development and transfer	44	13.9%	61.1%
			Infrastructure coverage	22	7.0%	30.6%
			Access to infrastructure	22	7.0%	30.6%
			Labour demand	28	8.9%	38.9%

As one single article can include several impacts, the percentage numbers in the columns 'share of impact in all studies' and 'share of impact in the category' do not add to 100% (see Tables S5 and S6).

Colours on the left side were used as codes for the five categories: institutional (blue), social (red), environmental (green), economic (orange) and technological (grey).

Table 3 Positive and negative impacts per region

Region / Impacts on	North America +	n	-	Europe +	n	-	Asia +	n	-	Africa +	n	-	Latin America +	n	-	Oceania +	n	-	Global +	n	-	n/a
Energy independency	9	2	0	10	3	1	8	0	0	8	2	1	6	0	0	0	0	0	6	0	1	2
Cross-sectoral coordination	4	1	3	2	1	6	6	2	0	2	0	0	1	2	2	0	1	0	2	4	2	
Land tenure	2	3	0	1	0	1	0	0	6	0	0	7	0	0	4	0	0	0	0	0	1	1
Food security	2	2	9	0	4	5	3	6	8	4	4	4	0	2	6	0	0	0	2	2	11	4
Health impacts	0	1	1	2	0	5	4	1	1	1	0	3	1	0	0	0	0	0	1	1	2	1
Conflicts or social tension	0	1	0	0	0	1	1	0	4	1	1	3	1	1	3	0	0	0	0	0	2	0
Soil and water	7	8	17	10	3	7	7	2	7	2	2	3	1	1	7	2	1	2	1	7	9	9
Direct substitution from GHG emissions	11	4	2	24	4	3	9	1	0	2	1	0	4	2	4	2	0	0	11	2	1	8
Displacement of activities or land uses	3	2	9	3	10	9	6	4	5	0	0	6	2	2	8	0	0	0	2	4	12	8
Biodiversity	3	5	9	5	5	6	1	0	6	0	0	1	0	0	7	0	0	1	0	5	5	5
Deforestation or forest degradation	3	3	4	2	2	4	2	0	4	2	0	4	1	0	6	0	0	0	2	3	11	5
Use of fertilizers with impact on soil & water	4	3	10	4	4	11	3	0	3	1	1	1	0	1	6	0	1	1	0	3	4	2
Market opportunities	18	1	2	15	3	1	5	2	1	3	2	1	5	0	2	1	1	1	8	3	0	4
Prices of feedstock	7	7	7	1	5	4	4	1	1	1	3	2	3	0	0	0	0	0	2	5	4	6
Employment	10	0	1	6	0	0	9	1	0	8	3	1	5	2	1	0	0	0	2	3	0	3
Economic activity	5	0	0	7	2	1	10	1	4	10	0	2	3	0	0	0	0	0	3	2	1	2
Economic diversification	7	1	0	8	2	0	5	1	1	5	1	0	5	0	0	0	0	0	2	1	0	5
Technology development and transfer	5	0	0	8	0	0	5	0	0	5	1	1	6	0	0	0	0	0	8	1	1	3
Labour demand	2	1	2	3	0	0	3	0	0	6	0	2	0	1	3	0	0	0	0	0	0	3
Infrastructure coverage	1	3	1	5	0	0	2	0	0	3	1	0	1	0	0	0	0	0	2	0	0	3

'+' indicates reported as positive; 'n' indicates reported as neutral; '-' indicates reported as negative. Colours indicated the proportion of positive, negative or neutral impacts per region, being green for positive, orange for neutral and red for negative. The colours present a weight related to the three numbers per impact and region and not only to the value in the specific cell. The criteria for including impacts in this table are (i) at least 40 articles considered the impact and/or (ii) the three more frequently reported impact per category. Colours in the left column identify the five categories of impacts and following the logic used in Fig. 1 (see supplementary information, file 'Regional distribution of impacts' for details on each impact).

Colours on the left side were used as codes for the five categories: institutional (blue), social (red), environmental (green), economic (orange) and technological (grey).

with the complexity of understanding sustainability impacts or an expectation that social impacts are relatively more important in these regions. The five impacts most often considered in Africa are impacts on food security, on energy independence, on economic activity, on employment and on poverty (in this order). In this region, impacts on *land tenure*, on *women* and on *capacity building*, are considered more often than in other regions. The five impacts most frequently considered in Asia are those on food security, on economic activity, on soil and water, on displacement of activities and on employment.

Unbalanced understanding about impacts on sustainable development

The perspective of whether impacts are positive, negative or neutral is also uneven across regions. Our analysis of a selection of impacts shows that mostly negative impacts are reported in Latin America and at the global level, while the other regions show a more balanced picture (see Tables 3). The more detailed analysis presented below shows interesting differences in the importance given to each category and on where specific impacts were assessed as positive or negative.

Institutional impacts are included in over 30% of the articles (see Table 2). Within this impact category, *energy independence* is the most frequently studied impact

across regions, especially in Europe and Africa, and biofuel deployment is reported mostly as having a positive impact on it. Other impacts in this category such as *cross-sectorial coordination* show mixed results for all regions, while *land tenure* was reported as negatively impacted in Africa, Asia and Latin America.

Social impacts are considered in over 30% of all studies, with *food security* being the most frequently addressed impact in this category (over 25% of the total studies and almost 75% of the articles considering social impacts). We undertook a detailed analysis of food security because it has been mentioned as one major concern for promoting deployment of bioenergy. Negative impacts on food security were reported twice as often as positive impacts. For all regions, impacts on food security are reported more often as negative than as positive, except in Africa where an equal number of studies report impacts as positive, negative or neutral (see Fig. 2 and Table 3).

In addition, we found that at the global level, the more often models are used for analysing impacts on food security, the higher the frequency of negative impacts (see Fig. 2). Although the small number of studies does not provide statistic robustness, this finding suggests a difference in the way impacts on food security are modelled or measured at the global level.

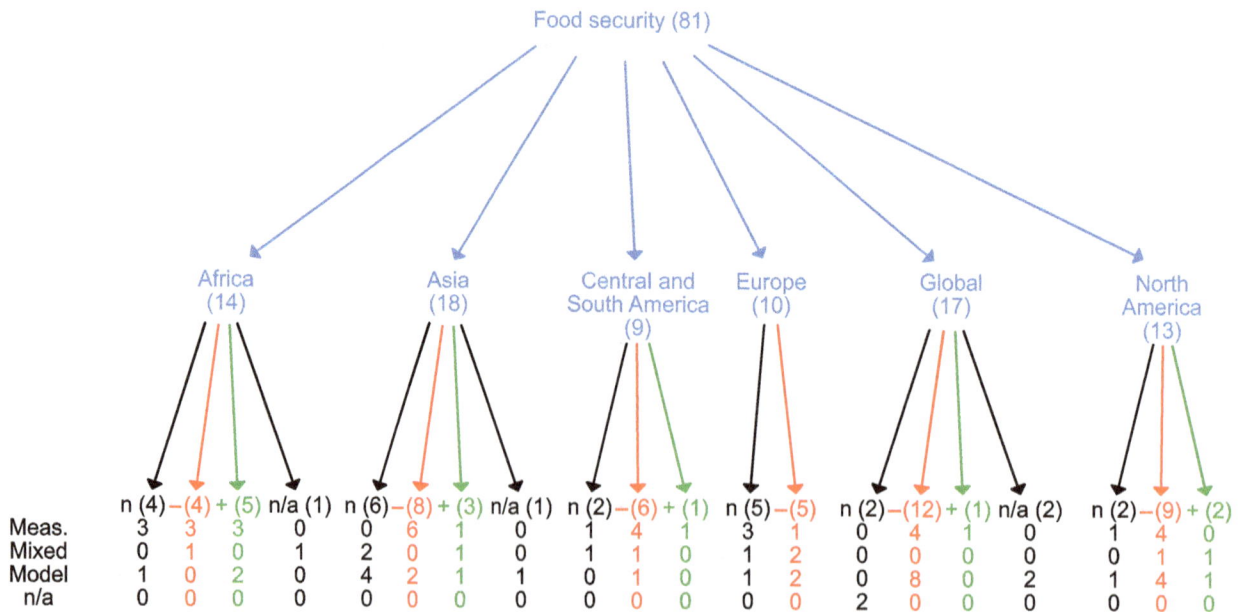

Food security (81)

	Africa (14)				Asia (18)				Central and South America (9)			Europe (10)		Global (17)				North America (13)		
	n (4)	– (4)	+ (5)	n/a (1)	n (6)	– (8)	+ (3)	n/a (1)	n (2)	– (6)	+ (1)	n (5)	– (5)	n (2)	– (12)	+ (1)	n/a (2)	n (2)	– (9)	+ (2)
Meas.	3	3	3	0	0	6	1	0	1	4	1	3	1	0	4	1	0	1	4	0
Mixed	0	1	0	1	2	0	1	0	1	1	0	1	2	0	0	0	0	0	1	1
Model	1	0	2	0	4	2	1	1	0	1	0	1	2	0	8	0	2	1	4	1
n/a	0	0	0	0	0	0	0	0	0	0	0	0	0	2	0	0	0	0	0	0

Fig 2 Impacts tree regarding food security. The blue arrows show the geographical distribution of the impacts on food security per regions as considered in the studies. In this case, there were no studies considering food security in Oceania. The first line indicates the number of positive (marked in green), negative (marked in red) or neutral impacts (marked in black). When the article did not specify the qualification of the impact, we considered it as nonavailable (n/a, marked in grey). From the second line downwards, we present how these impacts were identified, either using measurements, models or a combination (mixed). When the method was not clear in the article, we defined it as nonavailable (n/a). Impact trees for all other impacts considered in this systematic review are included in the supplementary information.

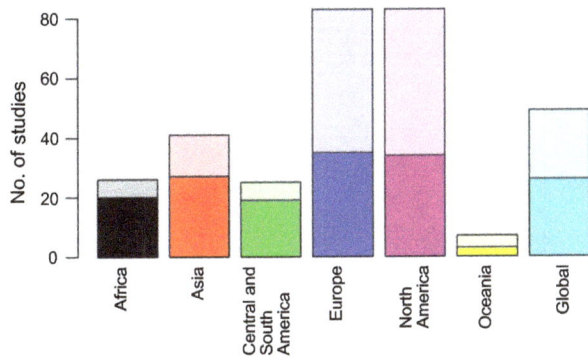

Fig 3 Geographical distribution of studies differentiating between studies considering or not considering context conditions. Solid colours indicate the number of studies with fully or partially matching context conditions. Transparent colours indicate the number of studies where context conditions were either not mentioned or do not correspond to the impact categories.

Other key social impacts – including gender and intragenerational impacts, social conflicts, displacement of farmers and impacts on traditional or indigenous practices – are insufficiently studied in all regions and practically not considered in global studies.

The environmental impact category is the most frequently considered category by the studies in the sample (over 70% of the total articles in the review, see Table 2), and each individual impact is addressed by at least a quarter of the studies. Across regions, all impacts in this category are reported as mostly negative or neutral, with the exception of *direct substitution of GHG emissions from fossil fuels,* which is considered positive or neutral in all geographical contexts. It is important to note, however, that over 65% of the studies used models for attributing direct substitution of GHG emissions from fossil fuels, and only 20% of these combined models with case study measurements. Thus, the qualification of this impact is highly dependent on the system boundaries and attribution criteria used. Negative impacts on the *displacement of activities or other land uses* are more frequently reported in Latin America, North America, Europe and at the global level (see Table 3). In Asia, slightly more positive impacts are reported compared to other regions.

Impacts on *biodiversity* are predominately reported as negative or neutral (see Table 3), except in a few studies from Europe and North America, whereas impacts on *deforestation or forest degradation* seem to be more negative for Latin America and at the global level. Further, impacts from the *use of fertilizers on soil and water* are reported as negative for Europe, North and Latin America, where these account for the majority of studies addressing this issue.

Economic impacts are considered in over half of all articles (see Table 2) and were predominantly positive

for most impacts assessed in this category. Positive effects on *market opportunities* are noticeably reported in studies for North America and Europe (see Table 3), whereas positive effects on *economic activity* were more frequently reported in Africa and Asia. Impacts on *prices of feedstock* show mixed results for all regions. As for other impacts where modelling was used far more often than case study measurements, the positive or negative character of the economic impacts category needs more analysis considering the system boundaries and attribution criteria used.

Over 20% of all articles consider technological impacts (see Table 2). *Technology development and transfer* is the most frequently considered impact, followed distantly by *impacts on labour demand, infrastructure coverage and access to infrastructure.* Impacts on technology development and transfer are seen mostly as positive in all regions with only two studies reporting negative impacts: one from Africa and one at the global level (see Table 3).

How context conditions influence development outcomes remains unclear

We analysed how impacts have been attributed by examining whether context conditions were explicitly reported. Context conditions describe the situation in the absence of additional biomass production and use for energy. Insight into these conditions is necessary for establishing a baseline or reference scenario and/or for attributing impacts on sustainable development from bioenergy production in a transparent manner. The systematic review includes 31 possible conditions that can describe the context in relation to the five impact categories (see supplementary information for a complete list of context conditions). We first analysed the extent to which impacts reported in the articles match to the corresponding context conditions at the level of category (i.e. whether context conditions were reported for those categories where impacts were identified).

The analysis shows that only 13.6% of the articles comprehensively describe the context conditions against the category of the reported impacts, whereas 23% do not report context conditions at all. For the remainder, conditions were partially or fully mismatched (i.e. context conditions are described but not for the category of impacts reported). This lack of clarity of the context conditions applies to articles dealing with developed and developing countries, as well as global analyses. However, we found that studies analysing bioenergy production in developing countries report context conditions more often than studies on Europe, North America or those with a global scope (see Fig. 3). The lack of information applies across all reported impacts. For instance, from

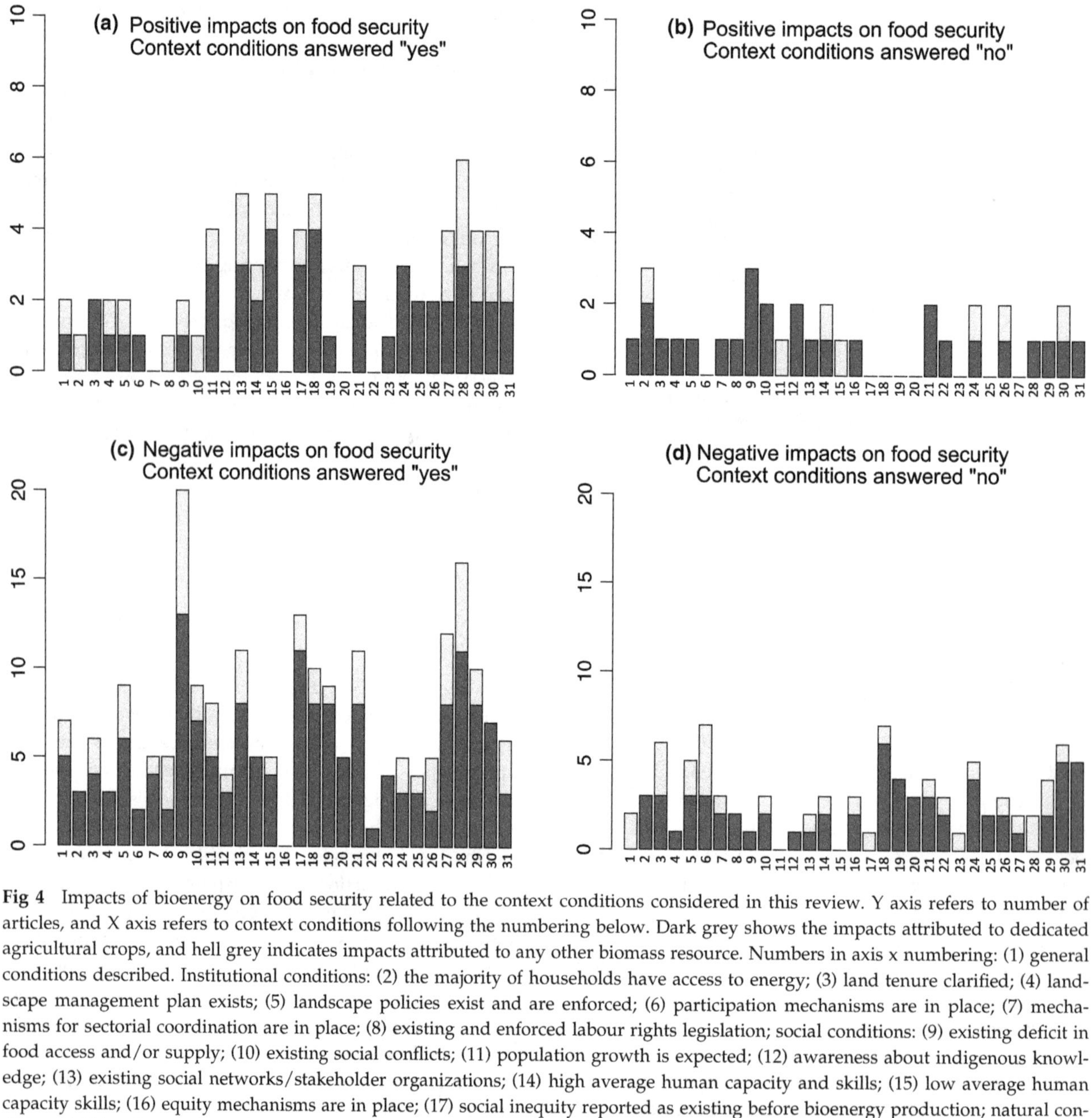

Fig 4 Impacts of bioenergy on food security related to the context conditions considered in this review. Y axis refers to number of articles, and X axis refers to context conditions following the numbering below. Dark grey shows the impacts attributed to dedicated agricultural crops, and hell grey indicates impacts attributed to any other biomass resource. Numbers in axis x numbering: (1) general conditions described. Institutional conditions: (2) the majority of households have access to energy; (3) land tenure clarified; (4) land-scape management plan exists; (5) landscape policies exist and are enforced; (6) participation mechanisms are in place; (7) mechanisms for sectorial coordination are in place; (8) existing and enforced labour rights legislation; social conditions: (9) existing deficit in food access and/or supply; (10) existing social conflicts; (11) population growth is expected; (12) awareness about indigenous knowledge; (13) existing social networks/stakeholder organizations; (14) high average human capacity and skills; (15) low average human capacity skills; (16) equity mechanisms are in place; (17) social inequity reported as existing before bioenergy production; natural conditions: (18) land is available for people living in the area; (19) water for agriculture/forestry is available for people living in the area; (20) drinking water is available to people living in the area; (21) land (use) competition previous any intervention is reported in the article; (22) air quality is reported as good; (23) high biodiversity index. Economic conditions: (24) availability of capital; (25) existing crediting mechanisms; (26) sharing mechanisms of economic benefits in place; conditions related to technology and infrastructure: (27) traditional technologies; (28) modern (industrial) technologies; (29) combination of modern and industrial technologies; (30) technology is available to major local stakeholders; (31) mechanisms for technology development and/or transfer given.

those articles quantifying impacts on food security, only 35% provide context conditions in the corresponding social category; concerning GHG emissions only 12% of articles provide corresponding baseline conditions. We recognize that for some standardized methodologies (e.g. LCA), and for most models, certain assumptions regarding context conditions are embedded in the proce-

dures used. However, when they are not reported and/or validated, which is often the case, it remains unclear how impacts were attributed.

We undertook a deeper analysis of the relationship between context conditions and several specific impacts. Initially, we conducted a descriptive analysis of impacts on food security, which is the most frequently reported

Table 4 Combinations of conditions and impacts with P-value below 5% in the Fisher test

Impact	Condition	P-value (Fisher test)	Combination condition/impact					
			Yes/+	Yes/−	Yes/n	No/+	No/−	No/n
Food security or food production (negative if reduced or positive if improved)	Existing deficit in food access and/or supply	0.00154111	2	20	3	3	1	4
Conflicts or social tension	Existing deficit in food access and/or supply	0.02222222	7	1	2	0	0	1
Direct substitution of GHG emissions reductions from fossil fuels	Sharing mechanisms of economic benefits in place	0.03571429	0	2	0	6	0	0
Prices of feedstock	Modern (industrial) technologies	0.04449388	11	4	13	1	2	0
Employment (being employment creation (+) or employment reduction (−))	Mechanisms for sectorial coordination are in place	0.04545455	7	0	0	2	1	2

Table 5 Regional distribution of relevant condition-impact combinations

Region/ Combination	'Existing deficit in Food access' and 'Food security'							'Sharing mechanisms in place' and 'Direct substitution of GHG emissions reductions'						
	Yes/+	Yes/−	Yes/n	No/+	No/−	No/n	Total	Yes/+	Yes/−	Yes/n	No/+	No/−	No/n	Total
Africa	1	2	2	1	0	0	6	0	0	0	0	0	0	0
Asia	1	6	1	0	0	2	10	0	0	0	2	0	0	2
Europe	0	2	0	0	0	1	3	0	0	0	1	0	0	1
North America	0	4	0	1	1	0	6	0	0	0	2	0	0	2
Oceania	0	0	0	0	0	0	0	0	0	0	0	0	0	0
Latin America	0	1	0	1	0	1	3	0	2	0	0	0	0	2
Global	0	5	0	0	0	0	5	0	0	0	1	0	0	1
Total	2	20	3	3	1	4	33	0	2	0	6	0	0	8

social impact, to determine whether it is possible to establish the context conditions that trigger positive or negative impacts on food security. About 80% of the articles mentioning impacts on food security include some description of the context conditions. We found that in articles reporting impact on food security, most context conditions are considered at least once (see Fig. 4) and that no particular context condition clearly stands out in relation to either positive or negative impacts (e.g. conditions that are most frequent in the food security analysis, such as the use of modern technologies, show up both for negative and positive impacts).

The general lack of correlation between context conditions and impact sign is also reflected in the P-values of Fisher tests, which we applied to all 1023 combinations of context conditions and impacts to check the influence of a particular context condition given or not given on the counts of impact signs. Table 4 displays that only 5 combinations have a P-value below 5% and reports their corresponding numbers of condition–impact combinations.

The Fisher test indicates whether the counts of impact signs in case of condition being 'yes' differ significantly from the counts of impact signs when the condition is 'no'. Thus, a low P-value does not represent strong evidence that the condition has an influence on the impact. This influence can only be postulated if the combination of conditions and impact also suggests its existence and direction. This is the case for only two combinations:

- Combination 1: context condition 'existing deficits in food access and/or food security' and impact on 'food security': when the context condition 'existing deficits in food access and/or supply' is given, then biomass production for bioenergy is almost exclusively reported to have a negative impact on food security. Studies reporting the absence of these deficits, on the other hand, report either a positive or a neutral impact on food security.

- Combination 2: context condition 'benefit sharing mechanism for economic benefits are in place' and impact on 'direct substitution of GHG emissions from fossil fuels': the impact on direct substitution of GHG emissions from fossil fuel is largely positive when no benefit-sharing mechanism for economic benefits is in place, while the presence of such mechanisms exclusively leads to this impact being negative.

For the other three combinations in Table 4, the number of impacts is very small if the condition is answered with 'no' and the distribution of impacts (positive, negative or neutral) is ambiguous. Thus, even if the condition being 'yes' suggests a positive impact sign in two of these cases, it is not known whether these conditions really influence the corresponding impacts.

The regional analysis for the two combinations that in total suggest a correlation between condition and impact are displayed in Table 5. Fisher tests showed no significant difference between 'yes' and 'no' answers for any region.

Patterns in the distribution of positive and negative impacts

The results show some general patterns that are worth highlighting (see especially Figs 2–4 and Table 3). Impacts on some economic and technological categories are persistently positive across studies and regions. Within these categories impacts on energy independence, direct substitution of GHG emissions from fossil fuels, market opportunities, economic activity and diversification, employment as well as different technological categories is far most often reported as positive. In contrast, most impacts in the social and environmental categories are reported largely as having negative impacts, especially on land tenure, food security, displacement of other activities, biodiversity loss, and conflict and social tension. These patterns indicate an important trade-off: that bioenergy projects may generate positive economic impacts but negative environmental and social impacts.

The incomplete information on context conditions (Fig. 3 and statistical analysis) makes it difficult to say anything conclusively across studies on what are the most relevant conditions triggering any specific impact. Yet, previous work has pointed to some reasons worth highlighting, notably that government institutions in countries targeted for bioenergy production often face severe constraints in implementing public policies and regulations intended to protect, for instance, land rights and food security (Ravnborg et al., 2013; Larsen et al., 2014). This is reinforced by our findings on context conditions related to food security and to some extent by the participation of governance-related conditions highlighted through the Fisher Test. It is also worth noting that because climate change mitigation has been an important motivator for promoting bioenergy, it has been a higher research priority than other goals such as those related to biodiversity or land tenure. The latest IPCC Assessment Report made a great advance in including ethics and sustainable development in its considerations and paves the way for a more systemic research approach towards understanding development

impacts from bioenergy production. More research is needed in the future to develop this approach, given the knowledge gaps identified in this review.

Conclusions and outlook

Understanding the impacts of bioenergy production on sustainable development has been an important research topic in recent years, but its coverage is uneven, both in terms of geographical coverage, feedstocks considered, and in the categories of impacts considered. Furthermore, results are hardly comparable because context conditions and attribution criteria are not properly reported in the majority of the studies.

In the following, we present our conclusions about the research questions in this review.

Where do sustainable development impacts from bioenergy production take place?

Geographically, we identified three distinct groups of countries, based on NPP as a proxy for biophysical biomass production potential, for considering bioenergy deployment in a given country. In the first group, we find countries with a high biophysical potential and a reasonable number of studies. These studies give good information about environmental and economic impacts, showing a tendency towards positive impacts from bioenergy production on direct substitution of GHG emissions from fossil fuels, market creation, technology development and transfer. However, social, institutional and technological impacts remain uncertain because they were far less often considered. The second group comprises countries with a high NPP but very few studies. Most of these are developing countries where there is a need for better understanding of possible sustainable development impacts of bioenergy implementation. For countries in this group, more research is needed to provide robust information for policymaking and governance agreements. The third group comprises countries with a relatively smaller NPP but many studies. This group consists mainly of developed countries and lessons on methodological issues from these studies can be used for future research in understudied countries.

What is the evidence for the purported impacts and how are impacts attributed and measured?

There is a lack of systematic reporting on criteria for attributing impacts. Despite the existing discussion on attribution of specific methodologies (e.g. Finkbeiner, 2013; Muñoz et al., 2015 on attribution of indirect land use change in LCA), this omission in the studies makes it impossible to pursue a consistent comparison of results. We found that the environmental and economic impact categories were more thoroughly studied, whereas far less is known about how bioenergy production will affect the social and institutional categories of sustainable development. Institutional and social impact categories are better considered in country-level studies than in global studies. Although there is an apparent indication of trade-offs between positive impacts on the economic category and negative impacts on the environmental and social categories, more clarity about what triggers the trade-offs could not be achieved due to the noncomparability of the results across the studies (lack of attribution criteria) and to the lack of information on context conditions in the majority of the studies.

Are there certain context conditions that enable the observed impacts?

We found that there is a gap on reporting the specific context conditions prior to any intervention aimed at producing biomass for bioenergy, with less than 15% of the studies providing a comprehensive presentation of the context conditions in the category on which they attributed impacts. The lack of consistency in reporting context conditions and their relation to the reported impacts prevents clear and definitive conclusions on how the context affects the development outcome. Previous assessments have highlighted the need for 'good governance' as a condition required for promoting positive impacts of bioenergy production (Creutzig et al., 2014; Hunsberger et al., 2014; Smith et al., 2014). The reported negative impacts on land tenure, food security and food production, or other social and institutional aspects bear witness that bioenergy deployment can result in undesirable consequences and on the importance of understanding the context conditions, especially existing governance of natural resources.

Are the reported impacts specific to particular biomass resources?

We found a concentration of studies dealing with dedicated biomass production, especially agricultural plantations. Other biomass resources have been less studied, and the use of waste as bioenergy feedstock has not received much systematic scrutiny. We conclude that analytical frameworks and methods that facilitate the analysis at a higher level of complexity, that is, including more categories or allowing aggregation from various studies, are still needed. Such frameworks need to ask for the inclusion and reporting of context conditions, explicitly and transparently, so that context-dependent differences can be identified. Future empirical research, especially case studies, should aim to

inform about the most effective governance arrangements – and identify situations where governance agreements have insufficient capacity to guarantee that bioenergy deployment consider international due diligence standards.

It is opportune to interpret our results in the context of the recent IPCC assessment of climate change. The IPCC author team concluded that:

One strand of literature highlights that bioenergy could contribute significantly to mitigating global GHG emissions via displacing fossil fuels, better management of natural resources, and possibly by deploying BECCS. Another strand of literature points to abundant risks in the large-scale development of bioenergy mainly from dedicated energy crops and particularly in reducing the land carbon stock, potentially resulting in net increases in GHG emissions (Smith et al., 2014)

One interpretation of this divergence is that the first strand of literature emphasizes technological opportunities, such as yield increases, to reduce land use impact, and reap economic opportunities, while the other strand of literature investigates environmental dimensions under risk of being harmed (Creutzig, 2014). The growing literature exploring sustainable landscape management systems for the provision of biomass and other ecosystem services might gradually come to bridge the gap between these two strands of literature. Not the least, the integration of bioenergy systems into agriculture landscapes has been recognized as a promising option for addressing environmental impacts associated with current agriculture systems (Clarke et al., 2014; Edenhofer et al., 2014; Smith et al., 2014).

The IPCC report annex on bioenergy also points out that environmental, social and economic consequences of bioenergy deployment are site specific, but remains inconclusive on weighting the consequences across case studies. This review goes beyond the IPCC assessment in providing a comprehensive meta-analysis, demonstrating that case studies evaluated so far tend to see increased economic and employment opportunities, GHG savings from fossil fuel displacement, and infrastructure development, but also risks related to land use change, in particular GHG emissions, food security, soil and water quality, biodiversity, and socially problematic outcomes.

Since the publication of the latest IPCC assessment report, further research on bioenergy has been published, which is in line with the main conclusions of our systematic review. The screening of this literature suggests that case studies mostly emphasize GHG emissions metrics and economic performance (e.g. (García et al., 2015; Mandaloufas et al., 2015)) and Dale et al. (2015) point out the importance of appropriate sustainability criteria and indicators. This observation suggests that the systematic bias observed in our survey of case studies can be interpreted as showing that social dimensions have been assigned a lower priority by scientists and policy processes than some environmental and economic dimensions.

There are limitations to the systematic review presented in this article. First, the complexity of the subject of analysis, such as the high number of potential interactions within the system boundaries and the lack of inclusion of criteria for analysing trans-boundary impacts or trade-offs between specific criteria and scale of the impacts, renders results of models and case studies partially inconclusive and subject to a priori values of investigators (Tribe et al., 1976). Second, most results in both cases depend on attributional accounting, which has been argued to be possibly misleading, while consequential accounting, being subject to higher uncertainties, might provide more policy-relevant information. This is especially relevant for studies using LCA methods (Brandao et al., 2013; Hertwich, 2014; Plevin et al., 2014a,b). Third, we focused on studies published in English only. These limitations should be considered in future studies and analysed using complementary assessment methods.

Overall, we find that comparatively assessing the impacts of bioenergy production on sustainable development using the available scientific literature is a considerable challenge, but we are able to propose four recommendations for future research: (i) pursue a more stringent use of frameworks and methodologies that attribute impacts of bioenergy production on all development categories, (ii) report context conditions and criteria for attributing development impacts transparently, (iii) improve understanding of impacts of bioenergy production in developing countries with potentially favourable biophysical conditions for bioenergy and (iv) improve understanding of potential sustainable development impacts in different regions of using other bioenergy feedstock than biomass from dedicated plantations (e.g. organic waste and/or agricultural/forestry residues). Addressing these issues is essential for providing a more solid scientific basis for policymaking and governance agreements in the field of bioenergy and sustainable development.

Acknowledgements

The authors gratefully acknowledge the participation Omar Masera, Richard Plevin, Roberto Schaeffer, Rainer Zah and Jacob Mulugetta during the literature appraisal. Carmenza Robledo-Abad acknowledges support from the Swiss State Secretary of Economic Affairs. Helmut Haberl gratefully acknowledges funding from the Austrian Provision Programme, the Austrian Academy of Sciences (Global Change Programme) and the EU-FP7 project VOLANTE.

Esteve Corbera acknowledges the support of the Spanish Research, Development and Innovation Secretariat through a 'Ramón y Cajal' research fellowship (RYC-2010-07183) and of a Marie Curie Career Integration Grant (PCIG09-GA-2011-294234). Simon Bolwig acknowledges the support of the Innovation Fond Denmark. Alexander Popp acknowledges the support from the European Union's Seventh Framework Program project LUC4C (grant agreement no. 603542). Bart Muys acknowledges support from the KLIMOS Acropolis research network on sustainable development funded by VLIR/ARES/DGD (Belgian Development Aid). Rasmus Kløcker Larsen acknowledges funding from the Swedish research council Formas. Carol Hunsberger acknowledges the support of a postdoctoral fellowship from Canada's Social Sciences and Humanities Research Council. John Garcia-Ulloa is supported by the Mercator Foundation Switzerland and the Zurich-Basel Plant Science Center. Johan Lilliestam, Anna Geddes and Susan Hanger acknowledge the support from the European Research Council (ERC) consolidator grant, contract number 313533. Joana Portugal-Pereira acknowledges the support of National Centre of Technological and Scientific Development (CNPq), under the Science Without Borders Programme (no 401164/2012-8). Richard Harper acknowledges funding from the Australian Department of Climate Change and Energy Efficiency.

References

Ackerman F, DeCanio SJ, Howarth RB, Sheeran K (2009) Limitations of integrated assessment models of climate change. Climatic Change, 95, 297–315.

Allwood JM, Bosetti V, Dubash NK, Gómez-Echeverri L, von Stechow C (2014) Glossary. In: Climate Change 2014: Mitigation of Climate Change. Contribution of Working Group III to the Fifth Assessment Report of the Intergovernmental Panel on Climate Change (eds Edenhofer O, Pichs-Madruga R, Sokona Y, Farahani E, Kadner S, Seyboth K, Adler A, Baum I, Brunner S, Eickemeier P, Kriemann B, Savolainen J, Schlömer S, von Stechow C, Zwickel T, Minx JC), pp. 1249–1279. Cambridge University Press, Cambridge, UK and New York, NY, USA.

Bartolucci AA, Hillegass WB (2010) Overview, strengths, and limitations of systematic reviews and meta-analyses. In: Evidence-Based Practice: Toward Optimizing Clinical Outcomes (ed. Chiappelli F), pp. 17–33. Springer, Berlin Heidelberg. (ISBN: 978-3-642-05024-4). Available at: http://link.springer.com/chapter/10.1007/978-3-642-05025-1_2 (accessed 05 March 2013).

Beringer T, Lucht W, Schaphoff S (2011) Bioenergy production potential of global biomass plantations under environmental and agricultural constraints. Global Change Biology Bioenergy, 3, 299–312.

Brandao M, Levasseur A, Kirschbaum MUF et al. (2013) Key issues and options in accounting for carbon sequestration and temporary storage in life cycle assessment and carbon footprinting. International Journal of Life Cycle Assessment, 18, 230–240.

Bustamante M, Robledo-Abad C, Harper R et al. (2014) Co-benefits, trade-offs, barriers and policies for greenhouse gas mitigation in the Agriculture, Forestry and Other Land Use (AFOLU) sector. Global Change Biology, 20, 3270–3290.

Clarke L, Jiang K, Akimoto K et al. (2014) Assessing transformation pathways. In: Climate Change 2014: Mitigation of Climate Change. Contribution of Working Group III to the Fifth Assessment Report of the Intergovernmental Panel on Climate Change (eds Edenhofer O, Pichs-Madruga R, Sokona Y, Farahani E, Kadner S, Seyboth K, Adler A, Baum I, Brunner S, Eickemeier P, Kriemann B, Savolainen J, Schlömer S, von Stechow C, Zwickel T, Minx JC), pp. 413–511. Cambridge University Press, Cambridge, UK and New York, NY, USA.

Creutzig F (2014) Economic and ecological views on climate change mitigation with bioenergy and negative emissions. GCB Bioenergy, 8, 4–10.

Creutzig F, Corbera E, Bolwig S, Hunsberger C (2013) Integrating place-specific livelihood and equity outcomes into global assessments of bioenergy deployment. Environmental Research Letters, 8, 035047.

Creutzig F, Ravindranath NH, Berndes G et al. (2014) Bioenergy and climate change mitigation: an assessment. GCB Bioenergy, 7, 916–944.

Dale VH, Efroymson RA, Kline KL, Davitt MS (2015) A framework for selecting indicators of bioenergy sustainability. Biofuels, Bioproducts and Biorefining, 9, 435–446.

Dauvergne P, Neville KJ (2010) Forests, food, and fuel in the tropics: the uneven social and ecological consequences of the emerging political economy of biofuels. Journal of Peasant Studies, 37, 631–660.

Delucchi MA (2010) Impacts of biofuels on climate change, water use, and land use. In: Year in Ecology and Conservation Biology 2010 (eds Ostfeld RS, Schlesinger WH), pp. 28–45, (ISBN: 978-1-57331-791-7). New York Academy of Sciences, Willey-Blackwell, New Jersey.

DeLucia EH, Gomez-Casanovas N, Greenberg JA et al. (2014) The theoretical limit to plant productivity. Environmental Science & Technology, 48, 9471–9477.

Edenhofer O, Pichs-Madruga R, Sokona Y et al. (2014) Technical summary. In: Climate Change 2014: Mitigation of Climate Change. Contribution of Working Group III to the Fifth Assessment Report of the Intergovernmental Panel on Climate Change (eds Edenhofer O, Pichs-Madruga R, Sokona Y, Farahani E, Kadner S, Seyboth K, Adler A, Baum I, Brunner S, Eickemeier P, Kriemann B, Savolainen J, Schlömer S, von Stechow C, Zwickel T, Minx JC), pp. 33–11. Cambridge University Press, Cambridge, United Kingdom and New York, NY, USA.

Finkbeiner M (2013) Indirect Land Use Change (iLUC) within Life Cycle Assessment (LCA) – Scientific Robustness and Consistency with International Standards. Publication of the Association of the German Biofuel Industry, Berlin, Germany. Available at: http://www.fediol.eu/data/RZ_VDB_0030_Vorstudie_ENG_Komplett.pdf (accessed 10 August 2015).

Gamborg C, Anker HT, Sandøe P (2014) Ethical and legal challenges in bioenergy governance: coping with value disagreement and regulatory complexity. Energy Policy, 69, 326–333.

García CA, Riegelhaupt E, Ghilardi A, Skutsch M, Islas J, Manzini F, Masera O (2015) Sustainable bioenergy options for Mexico: GHG mitigation and costs. Renewable and Sustainable Energy Reviews, 43, 545–552.

German L, Schoneveld GC, Gumbo D (2011) The Local Social and Environmental Impacts of Smallholder-Based Biofuel Investments in Zambia. Ecology and Society, 16, 12

Gregg JS, Smith SJ (2010) Global and regional potential for bioenergy from agricultural and forestry residue biomass. Mitigation and Adaptation Strategies for Global Change, 15, 241–262.

Haberl H (2015) Competition for land: a sociometabolic perspective. Ecological Economics, 111, 424–431.

Haberl H (2016) The growing role of biomass for future resource supply - prospects and pitfalls. In: Sustainability Assessment of Renewables-Based Products: Methods and Case Studies (eds Dewulf J, De Meester S, Alvarenga RAF), pp. 1–18. John Wiley and Sons, Ltd., New Jersey.

Haberl H, Erb K-H, Krausmann F et al. (2011) Global bioenergy potentials from agricultural land in 2050: sensitivity to climate change, diets and yields. Biomass and Bioenergy, 35, 4753–4769.

Haberl H, Erb K-H, Krausmann F, Running S, Searchinger TD, Smith WK (2013) Bioenergy: how much can we expect for 2050? Environmental Research Letters, 8, 031004.

Hejazi MI, Voisin N, Liu L et al. (2015) 21st century United States emissions mitigation could increase water stress more than the climate change it is mitigating. Proceedings of the National Academy of Sciences of the United States of America, 112, 10635–10640.

Hertwich E (2014) Understanding the climate mitigation benefits of product systems: comment on "using attributional life cycle assessment to estimate climate-change mitigation...". Journal of Industrial Ecology, 18, 464–465.

Hodbod J, Tomei J (2013) Demystifying the Social Impacts of Biofuels at Local Levels: where is the Evidence? Geography Compass, 7, 478–488.

Hoogwijk M, Faaij A, de Vries B, Turkenburg W (2009) Exploration of regional and global cost–supply curves of biomass energy from short-rotation crops at abandoned cropland and rest land under four IPCC SRES land-use scenarios. Biomass and Bioenergy, 33, 26–43.

Hunsberger C, Bolwig S, Corbera E, Creutzig F (2014) Livelihood impacts of biofuel crop production: implications for governance. Geoforum, 54, 248–260.

IPCC (2014) Climate Change 2014: Mitigation of Climate Change. Contribution of the Working Group III to the Fifth Assessment Report to the Intergovernmental Panel on Climate Change (eds Edenhofer O, Pichs-Madruga R, Sokona Y, Farahani E, Kadner S, Seyboth K, Adler A, Baum I, Brunner S, Eickemeier P, Kriemann B, Savolainen J, Schlömer S, von Stechow C, Zwickel T, Minx JC) Cambridge University Press, Cambridge, UK and New York, NY, USA.

Krausmann F, Erb K-H, Gingrich S et al. (2013) Global human appropriation of net primary production doubled in the 20th century. Proceedings of the National Academy of Sciences of the United States of America, 110, 10324–10329.

Lambin EF, Meyfroidt P (2011) Global land use change, economic globalization, and the looming land scarcity. Proceedings of the National Academy of Sciences of the United States of America, 108, 3465–3472.

Larsen RK, Jiwan N, Rompas A, Jenito J, Osbeck M, Tarigan A (2014) Towards "hybrid accountability" in EU biofuels policy? Community grievances and competing water claims in the Central Kalimantan oil palm sector. Geoforum, 54, 295–305.

Lotze-Campen H, von Lampe M, Kyle P et al. (2014) Impacts of increased bioenergy demand on global food markets: an AgMIP economic model intercomparison. Agricultural Economics, 45, 103–116.

Lubowski RN, Rose SK (2013) The potential for REDD+: key economic modeling insights and issues. Review of Environmental Economics and Policy, 7, 67–90.

Mandaloufas M, de Q, Lamas W, Brown S, Irizarry Quintero A (2015) Energy balance analysis of the Brazilian alcohol for flex fuel production. Renewable and Sustainable Energy Reviews, 43, 403–414.

Mendes Souza G, Victoria RL, Joly CA, Vedade L (Eds) (2015) Bioenergy & Sustainability: Bridging the Gaps. SCOPE Scientific Committee on Problems of the Environment, Paris Cedex, France.

Muñoz I, Schmidt JH, Brandão M, Weidema BP (2015) Rebuttal to "Indirect land use change (iLUC) within life cycle assessment (LCA) – scientific robustness and consistency with international standards". GCB Bioenergy, 7, 565–566.

Muys B (2013) Sustainable Development within Planetary Boundaries: A Functional Revision of the Definition Based on the Thermodynamics of Complex Social-Ecological Systems. Available at: http://librelloph.com/ojs/index.php/challengesinsustainability/article/view/22 (accessed 16 December 2014).

Nijsen M, Smeets E, Stehfest E, van Vuuren DP (2012) An evaluation of the global potential of bioenergy production on degraded lands. GCB Bioenergy, 4, 130–147.

Odlare M, Arthurson V, Pell M, Svensson K, Nehrenheim E, Abubaker J (2011) Land application of organic waste - effects on the soil ecosystem. Applied Energy, 88, 2210–2218.

Persson UM (2014) The impact of biofuel demand on agricultural commodity prices: a systematic review. Wiley Interdisciplinary Reviews: Energy and Environment, 4, 410–428.

Petticrew M, Roberts H (2008) Systematic Reviews in the Social Sciences: A Practical Guide. (ISBN: 978-1-4051-2110-1). Available at: http://onlinelibrary.wiley.com/book/10.1002/9780470754887 (accessed 05 March 2013).

Plevin RJ, Delucchi MA, Creutzig F (2014a) Using attributional life cycle assessment to estimate climate-change mitigation benefits misleads policy makers. Journal of Industrial Ecology, 18, 73–83.

Plevin R, Delucchi M, Creutzig F (2014b) Response to comments on "using attributional life cycle assessment to estimate climate-change mitigation...". Journal of Industrial Ecology, 18, 468–470.

Popp A, Lotze-Campen H, Leimbach M, Knopf B, Beringer T, Bauer N, Bodirsky B (2011) On sustainability of bioenergy production: integrating co-emissions from agricultural intensification. Biomass and Bioenergy, 35, 4770–4780.

R Core Team (2014) R: A Language and Environment for Statistical Computing. R Foundation for Statistical Computing, Vienna, Austria.

Ravnborg HM, Larsen RK, Vilsen JL, Funder M (2013) Environmental Governance and Development Cooperation – Achievements and Challenges. Danish Institute of International Studies, Copenhagen, Denmark. Available at: https://www.diis.dk/files/media/publications/import/extra/rp2013-15-environmental-governance_-web_1.pdf (accessed 10 December 2015).

Scharlemann JPW, Laurence WF (2014) How Green Are Biofuels? Animal Science Blogs. Available at: http://sites.psu.edu/tetherton/2008/02/28/how-green-are-biofuels/ (accessed 19 August 2015).

Smeets EMW, Faaij APC (2010) The impact of sustainability criteria on the costs and potentials of bioenergy production - Applied for case studies in Brazil and Ukraine. Biomass and Bioenergy, 34, 319–333.

Smeets EMW, Faaij APC, Lewandowski IM, Turkenburg WC (2007) A bottom-up assessment and review of global bio-energy potentials to 2050. Progress in Energy and Combustion Science, 33, 56–106.

Smith P, Bustamante M, Ahammad H et al. (2014) Agriculture, forestry and other land use (AFOLU). In: Climate Change 2014: Mitigation of Climate Change. Contribution of Working Group III to the Fifth Assessment Report of the Intergovernmental Panel on Climate Change (eds Edenhofer O, Pichs-Madruga R, Sokona Y, Farahani E, Kadner S, Seyboth K, Adler A, Baum I, Brunner S, Eickemeier P, Kriemann B, Savolainen J, Schlömer S, von Stechow C, Zwickel T, Minx JC), pp. 811–922. Chapter 11. Cambridge University Press, Cambridge, UK and New York, NY, USA.

Sneddon C, Howarth RB, Norgaard RB (2006) Sustainable development in a post-Brundtland world. Ecological Economics, 57, 253–268.

Sorda G, Banse M, Kemfert C (2010) An overview of biofuel policies across the world. Energy Policy, 38, 6977–6988.

von Stechow C, McCollum D, Riahi K et al. (2015) Integrating global climate change mitigation goals with other sustainability objectives: a synthesis. Annual Review of Environment and Resources, 40, 363–394.

Tilman D, Socolow R, Foley JA et al. (2009) Beneficial biofuels—the food, the energy, and environment trilemma. Science, 325, 270–271.

Tribe LH, Schelling CS, Voss J (Eds.) (1976) When Values Conflict. Essays on Environmental Analysis, Discourse, and Decision. published for the American Academy of Arts and Science by Ballinger Publishing Co, Cambridge, MA.

7

Initial soil C and land-use history determine soil C sequestration under perennial bioenergy crops

REBECCA L. ROWE[1,2], AIDAN M. KEITH[1], DAFYDD ELIAS[1], MARTA DONDINI[3], PETE SMITH[3], JONATHAN OXLEY[1] and NIALL P. MCNAMARA[1]

[1]Centre for Ecology & Hydrology, Lancaster Environment Centre, Library Avenue, Bailrigg, Lancaster, LA1 4AP, UK, [2]School of GeoSciences, University of Edinburgh, The King's Buildings, Alexander Crum Brown Road, Edinburgh, EH9 3FF, UK, [3]Institute of Biological and Environmental Sciences, University of Aberdeen, 23 St Machar Drive, Aberdeen, AB24 3UU, UK

Abstract

In the UK and other temperate regions, short rotation coppice (SRC) and *Miscanthus x giganteus* (*Miscanthus*) are two of the leading 'second-generation' bioenergy crops. Grown specifically as a low-carbon (C) fossil fuel replacement, calculations of the climate mitigation provided by these bioenergy crops rely on accurate data. There are concerns that uncertainty about impacts on soil C stocks of transitions from current agricultural land use to these bioenergy crops could lead to either an under- or overestimate of their climate mitigation potential. Here, for locations across mainland Great Britain (GB), a paired-site approach and a combination of 30-cm- and 1-m-deep soil sampling were used to quantify impacts of bioenergy land-use transitions on soil C stocks in 41 commercial land-use transitions; 12 arable to SRC, 9 grasslands to SRC, 11 arable to *Miscanthus* and 9 grasslands to *Miscanthus*. Mean soil C stocks were lower under both bioenergy crops than under the grassland controls but only significant at 0–30 cm. Mean soil C stocks at 0–30 cm were 33.55 ± 7.52 Mg C ha^{-1} and 26.83 ± 8.08 Mg C ha^{-1} lower under SRC ($P = 0.004$) and *Miscanthus* plantations ($P = 0.001$), respectively. Differences between bioenergy crops and arable controls were not significant in either the 30-cm or 1-m soil cores and smaller than for transitions from grassland. No correlation was detected between change in soil C stock and bioenergy crop age (time since establishment) or soil texture. Change in soil C stock was, however, negatively correlated with the soil C stock in the original land use. We suggest, therefore, that selection of sites for bioenergy crop establishment with lower soil C stocks, most often under arable land use, is the most likely to result in increased soil C stocks.

Keywords: bioenergy, Carbon Stocks, land-use change, *Miscanthus*, soil carbon, SRC willow

Introduction

Tackling climate change is one of the greatest challenges facing the world (IPCC, 2014). Along with other renewable energy sources and demand reduction, the use of biomass as a low-carbon (C) replacement for fossil fuels is seen as an essential part of the move towards a more sustainable energy system (Renewable Energy Road Map 2007; DECC *et al.*, 2012). Sources of biomass are diverse and include waste streams from food, forestry and conventional agricultural crops (Rowe *et al.*, 2009; DECC *et al.*, 2012). There is, however, increasing interest and utilisation of so-called second-generation (2G) bioenergy crops, especially in temperate developed nations such as Europe and the USA (Davis *et al.*, 2012; Don *et al.*, 2012). These 2G bioenergy crops, predominantly perennial grass and woody species, are grown

specifically to use as a renewable fuel source and are characterised by low input requirement and high growth rates. These traits result in a low energy requirement per unit of energy produced, limited management requirements, potentially higher C savings and reduced environmental impacts when compared to conventional food crops used for the production of first-generation biofuels (Fazio & Monti, 2011; Don *et al.*, 2012; Mohr & Raman, 2013; Walter *et al.*, 2014).

Assessing the C balance of 2G bioenergy crops presents a unique challenge as, in contrast to the use of conventional agricultural crops or waste streams, bioenergy crop production requires a major change in land use and management (Rowe *et al.*, 2009; Aylott & McDermott, 2012; Mohr & Raman, 2013). Land-use change (LUC) is known to be a primary factor affecting soil C stock (Guo and Gifford, 2002), and whilst impacts of harvesting and utilisation of these crops on the C balance are relatively well understood, impacts on soil C stocks are less well defined (Fazio & Monti,

Correspondence: Rebecca L. Rowe
e-mail: Rebrow@ceh.ac.uk

2011; Rowe *et al.*, 2011; Don *et al.*, 2012; Walter *et al.*, 2014).

In their meta-analysis, Don *et al.* (2012) highlighted the limited number of studies on the impacts of bioenergy crops on soil C stocks in temperate regions, and the highly variable and sometimes contradictory results reported across these. Even within single multi-site studies, impacts on soil C stock have been found to be variable between sites, with Walter *et al.* (2014), for example, reporting rates of change in soil C stocks across 21 SRC plantations in central Europe from -1.3 to 1.4 Mg C ha^{-1} yr^{-1} for transitions from arable land and -0.6 to 0.1 Mg C ha^{-1} yr^{-1} for transitions from grassland. Meanwhile, for *Miscanthus* transitions from arable land, Poeplau & Don (2014) found rates of change in soil C stocks within their study ranging from -0.17 to 1.54 Mg C ha^{-1} yr^{-1} and ranges in the literature of between -6.85 and 4.51 Mg C ha^{-1} yr^{-1}.

Some of the variations in the observed impact on soil C stocks, both between and within studies, have been related to differences in climatic conditions, original land use, soil types, management or crop genotype (Don *et al.*, 2012; Poeplau & Don, 2014; Richter *et al.*, 2015). These sources of variability can help to improve understanding of the mechanisms underlying changes in soil C stock, but comparison of studies can also be confounded by differences in quantification methods (Don *et al.*, 2012; Bárcena *et al.*, 2014). For example, LUC to SRC and *Miscanthus* can result in changes in soil C distributions within the soil profile and therefore sampling depth, which often differs between studies, can have a profound effect on the quantified impacts on soil C stocks (Poeplau & Don, 2014; Walter *et al.*, 2014). In their meta-analysis of impacts on soil C stocks of LUC to forestry, Bárcena *et al.* (2014) also highlighted the failure of many studies to adjust for change in soil bulk density (BD) that often co-occur with LUC. This results in an incorrect assessment of change in soil C stock and inflated between-study variability (Bárcena *et al.*, 2014). Apart from some notable exceptions (Walter *et al.*, 2014; Ferchaud *et al.*, 2015), few temperate bioenergy LUC studies have directly addressed the issue of changing BD (Don *et al.*, 2012).

In the context of mainland GB, and for the two dominant bioenergy crops in the UK, SRC willow and *Miscanthus* (Aylott & McDermott, 2012), we address these issues by providing a methodologically consistent data set of the impacts on soil C of land-use transitions to these crops, whilst incorporating variability in potential regulatory factors such as climate. This study aims both to assess within mainland GB the current impacts on soil C stocks of LUC to commercial plantations of either SRC or *Miscanthus*, and to provide insights and data on regulatory factors that can be incorporated into future

modelling activities (see Dondini *et al.*, 2015). To meet these aims, we undertook the assessment of soil C stocks under 20 *Miscanthus* and 21 SRC commercial plantations and their paired controls. Transitions were located across mainland GB and were purposefully selected to cover a wide range of climatic and soil conditions, including soil texture, pH, initial soil C stocks, a range of bioenergy crop ages and land-use transitions from both grassland and arable land uses, thus allowing the influence of these factors on changes in soil C stocks to be explored. Soil sampling utilised a combination of 0–30-cm and 0–1-m soil cores and soil C stocks were adjusted for changes in bulk density.

Materials and methods

Site selection

A database of potentially suitable commercial SRC and *Miscanthus* plantations was populated through liaising with bioenergy companies and individual growers. Data on soil C stocks prior to the land-use change were not available for these commercial sites, thus a paired-site approach was utilised, where impacts on soil C stock are assessed through a comparison between a target land use and an adjacent paired control representing the original land use (Davis & Condron, 2002; Laganière *et al.*, 2010). The paired-site method assumes no pre-existing differences between the control and bioenergy land uses that would confound changes in soil C stock (Wellock *et al.*, 2011; Hewitt *et al.*, 2012). Bioenergy plantations were therefore selected on the basis of the availability of a suitable paired control field in addition to the bioenergy crop age (time since establishment), geographical location and the type of LUC (i.e. from arable or from grassland). Selection aimed to provide the widest range of bioenergy crop age and geographical location, and a balance of transitions from arable and grassland to SRC and *Miscanthus* (Table 1). Each control and bioenergy plantation pair is referred to as a transition. In total, 41 transitions were assessed at 28 locations across mainland GB (Fig. 1).

The 41 transitions comprised 12 arable to SRC (all willow), 9 grasslands to SRC (8 willows, 1 poplar), 11 arable to *Miscanthus* and 9 grasslands to *Miscanthus* transitions (Table 1). Grassland was defined here using Defra definitions and includes both permanent pasture (>5 years old) and temporary grassland (5 years old and under), with the majority of sites being permanent pasture (Table 1). The lower number of grassland transitions reflects the greater difficulty experienced in locating bioenergy plantations established on former grassland.

Sampling method

Surface soil (0–30 cm). The surface soil of the cropped area of each bioenergy plantation or control field was sampled using a hierarchical design (Keith *et al.*, 2014), developed to capture variability across different spatial scales (Conant & Paustian, 2002; Conant *et al.*, 2003). Five sampling plots per field were

Table 1 Site details including transition location and type, current land use, duration of current land use, mean annual temperature (MAT), mean annual precipitation (MAP), soil texture (% clay) and C stocks at 30 cm and 100 cm

Site code	Transition number	Bioenergy crop	Control land use	Bioenergy planation age	Latitude, Longitude	MAP °C	MAT mm yr⁻¹	% Clay (0–30 cm; bioenergy crop)	Soil C stocks 0–30 cm (ESM reference mass of 3 Gg ha⁻¹) Bioenergy Mg C ha⁻¹ ± SD	Control Mg C ha⁻¹ ± SD	Soil C stocks 0–100 cm (ESM, reference mass of 13 Gg ha⁻¹) Bioenergy Mg C ha⁻¹ ± SD	Control Mg C ha⁻¹ ± SD
S1	1	SRC Willow	A	6	53.7, −0.8	9.63	603	8.08	54.65 ± 8.62	51.7 ± 6.92	127.72 ± 9.54	129.96 ± 10.79
S1	2	SRC Willow	A	13	53.7, −0.8	9.63	603	8.04	61.25 ± 11.08	51.7 ± 6.92	138.18 ± 8.29	129.96 ± 10.79
S2	3	SRC Willow	A	12	53.2, −0.8	9.77	580	6.73	45.62 ± 10.51	33.64 ± 3.19	85.62 ± 9.56	60.34 ± 0.98
S2	4	SRC Willow	A	8	53.2, −0.8	9.77	580	12.56	54.97 ± 7.41	33.64 ± 3.19	92.61 ± 10.05	60.34 ± 0.98
S3	6	SRC Willow	A	14	54.6, −2.7	7.64	1238	6.01	85.23 ± 10.86	73.93 ± 12.54	NA	NA
S5	9	SRC Willow	A	6	51.7, −0.9	10.04	625	5.75	57.44 ± 3.57	58.38 ± 13.00	110.75 ± 6.18	92.65 ± 8.76
S6	15	SRC Willow	A	7	51.5, −0.8	9.87	661	4.34	61.86 ± 11.30	54.38 ± 6.31	96.99 ± 6.62	99.22 ± 3.71
S7	18	SRC Willow	A	8	51.5, −1.6	9.95	663	9.86	115.79 ± 20.07	82.04 ± 12.13	143.28 ± 10.94	120.80 ± 19.95
S8	26	SRC Willow	A	5	50.7, −2.4	9.95	795	7.24	55.61 ± 9.62	47.9 ± 3.72	104.46 ± 11.12	99.95 ± 1.50
S9	33	SRC Willow	A	4	56.0, −3.6	8.36	946	4.25	87.75 ± 13.68	61.62 ± 6.41	161.83 ± 22.34	168.58 ± 13.43
S10	37	SRC Willow	A	6	54.8, −2.9	8.63	993	3.84	58.28 ± 10.15	64.36 ± 12.64	99.22 ± 16.21	71.85 ± 6.74
S11	41	SRC Willow	A	7	53.1, −0.3	9.95	582	6.76	45.70 ± 5.71	54.43 ± 8.32	132.64 ± NA	156.52 ± 26.22
S2	5	SRC Willow	PP	5	53.2, −0.7	9.77	580	9.39	124.52 ± 11.3	131.64 ± 9.78	252.05 ± 11.78	293.15 ± 15.25
S3	7	SRC Willow	PP	5	54.6, −2.6	7.64	1238	4.47	123.16 ± 19.48	127.16 ± 21.21	NA	NA
S4	8	SRC Willow	RG	5	50.9, −0.4	10.55	738	7.15	42.56 ± 6.78	61.24 ± 11.00	77.28 ± 13.31	63.84 ± 7.18
S7	17	SRC Willow	PP	23	51.5, −1.6	9.95	663	6.00	106.63 ± 8.63	184.66 ± 42.91	102.97 ± 3.51	218.08 ± 14.99
S12	20	SRC Willow	PP	10	52.2, −1.9	9.61	700	8.69	75.39 ± 12.64	95.84 ± 11.75	131.05 ± 0.80	148.76 ± 3.66
S12	21	SRC Poplar	PP	20	52.2, −1.9	9.61	700	8.82	65.59 ± 7.51	95.84 ± 11.75	105.70 ± 14.52	148.76 ± 3.66
S12	22	SRC Willow	PP	23	52.2, −1.9	9.61	700	5.98	69.00 ± 12.04	95.84 ± 11.75	99.27 ± 16.50	148.76 ± 3.66
S13	34	SRC Willow	TG	6	56.2, −3.2	8.58	810	4.69	62.57 ± 8.82	72.58 ± 16.10	106.53 ± 10.88	158.28 ± 6.03
S14	35	SRC Willow	PP	9	51.7, −4.7	10.34	882	6.65	76.49 ± 5.91	71.58 ± 12.07	103.97 ± 17.99	78.35 ± 5.58
S5	10	Miscanthus	A	6	51.7, −0.9	10.04	625	5.30	44.12 ± 6.58	58.38 ± 13.00	82.05 ± 5.82	92.65 ± 8.76
S15	11	Miscanthus	A	6	54.0, −1.2	9.17	634	4.12	39.84 ± 4.26	35.53 ± 3.87	97.56 ± 4.89	83.81 ± 6.10
S16	13	Miscanthus	A	3	53.4, −0.5	9.81	578	7.78	63.65 ± 7.3	59.83 ± 7.61	136.30 ± 9.22	124.95 ± 4.46
S17	16	Miscanthus	A	6	51.5, −1.3	10.14	633	7.05	63.98 ± 6.11	55.87 ± 5.52	102.05 ± 14.75	61.80 ± 7.21
S18	19	Miscanthus	A	6	51.8, −1.6	9.86	677	4.81	51.27 ± 5.82	99.38 ± 17.95	78.37 ± 6.34	146.64 ± 7.61
S19	27	Miscanthus	A	10	51.0, −3.1	10.22	832	8.69	45.35 ± 11.88	68.29 ± 8.61	78.57 ± 3.67	88.45 ± 4.15
S20	30	Miscanthus	A	8	50.4, −4.6	10.71	982	6.21	113.96 ± 23.65	87.44 ± 12.92	141.85 ± 27.88	105.30 ± 14.33
S14	36	Miscanthus	A	8	51.7, −4.8	10.34	882	6.57	90.89 ± 18.4	94.21 ± 16.94	139.87 ± 5.70	128.60 ± 12.53
S21	39	Miscanthus	A	6	52.6, 2.0	9.53	697	3.56	35.51 ± 5.68	44.37 ± 9.22	72.04 ± 9.80	90.35 ± 5.39
S22	40	Miscanthus	A	5	52.5, −0.5	9.78	584	9.95	82.98 ± 21.41	92.52 ± 23.4	197.89 ± 17.38	194.23 ± 7.89
S11	42	Miscanthus	A	7	53.1, −0.4	9.95	582	5.87	51.39 ± 9.77	54.43 ± 8.32	144.13 ± 21.54	156.52 ± 26.22
S23	14	Miscanthus	PP	8	53.2, 0.1	9.82	570	5.09	75.16 ± 6.6	95.89 ± 24.54	172.09 ± 13.18	241.41 ± 25.76

(continued)

Table 1 (continued)

Site code	Transition number	Bioenergy crop	Control land use	Bioenergy planation age	Latitude, Longitude	MAT °C	MAP mm yr⁻¹	% Clay (0–30 cm; bioenergy crop)	Soil C stocks 0–30 cm (ESM reference mass of 3 Gg ha⁻¹)		Soil C stocks 0–100 cm (ESM, reference mass of 13 Gg ha⁻¹)	
									Bioenergy Mg C ha⁻¹ ± SD	Control Mg C ha⁻¹ ± SD	Bioenergy Mg C ha⁻¹ ± SD	Control Mg C ha⁻¹ ± SD
S24	12	*Miscanthus*	TG	7	54.1, −1.1	9.17	634	7.06	51.99 ± 4.81	61.46 ± 7.36	118.03 ± 1.96	105.97 ± 4.88
S25	23	*Miscanthus*	PP	6	53.2, −3.7	8.18	1218	9.53	112.1 ± 12.64	96.89 ± 22.17	119.13 ± 7.73	68.74 ± 22.49
S26	24	*Miscanthus*	TG	1	52.4, −4.0	8.81	1502	10.51	136.3 ± 25.07	140 ± 30.68	131.86 ± NA	123.40 ± NA
S27	25	*Miscanthus*	PP	9	51.2, −2.8	10.31	765	7.55	152.7 ± 23.7	178.14 ± 39.45	232.40 ± 2.45	232.33 ± 40.35
S19	28	*Miscanthus*	PP	10	51.0, −3.1	10.22	832	6.88	49.91 ± 7.87	87.2 ± 9.71	81.30 ± 6.72	130.77 ± 11.46
S28	29	*Miscanthus*	PP	9	50.5, −4.8	10.00	1044	10.82	67.92 ± 7.78	85.95 ± 14.27	98.28 ± 9.31	108.49 ± 4.17
S20	31	*Miscanthus*	PP	7	50.4, −4.6	10.71	982	6.53	94.71 ± 14.44	146.8 ± 14.56	117.05 ± 12.65	182.47 ± 21.93
S21	38	*Miscanthus*	PP	6	52.6, 2.0	9.53	697	3.75	47.15 ± 0.51	91.65 ± 7.98	NA	NA

Control land-use classifications are based on Defra guidelines; A = arable land, PP = permanent pasture (defined as to land that is used to grow grasses or other herbaceous forage, either self-seeded or sown and has not been included in the crop rotation for 5 years or longer and has not been set aside during this 5-year period), RG = rough grazing (defined as to low-yielding permanent grassland, usually on low-quality soil, usually unimproved by fertiliser, cultivation, reseeding or drainage), TG = temporary grassland (defined as grass for grazing, hay or silage included as part of normal crop rotation, lasting at least one crop year and <5 years, sown with grass or grass mixture). With the exception of transitions 1, 2, 4, 5 and 41 and 42, none of the bioenergy crops received either inorganic or organic fertiliser; transitions 41 and 42 received wood waste and fibroflos applications, transitions 1–5 received a combination of inorganic fertiliser and treated sewage sludge. All arable fields were under conventional management receiving annual tillage and regular fertiliser applications.

Transition 32 was excluded as this was a short-rotation forestry plantation rather than a SRC planation.

Fig. 1 Map of sampling locations. Dark grey = SRC willow, light grey = *Miscanthus*; the data points of different bioenergy crops present at the same location are offset.

randomly selected from intersections of a grid overlaid on a map of the cropped area of field. The resolution of the grid was adjusted to ensure that there were a minimum of 50 grid intersections, with the condition that the resolution of the grid could not be <5 m. A 20-m perimeter buffer was also used to reduce potential edge effects. Within the five sampling plots, the three within-plot soil cores were taken using a split-tube soil sampler (Eijkelkamp Agrisearch Equipment BV, Giesbeek, The Netherlands) with an inner diameter of 4.8 cm to a depth of 30 cm. The first core was taken at the grid intersect, with two further cores taken at distances of 1 m and 1.5 m in random compass directions from the intersect. This gave a total of 15 spatially nested samples per field, accounting for both field-scale (between sampling plots) and plot-scale (cores within plots) variability. Before each core was taken, litter (L) and fermentation (L_f) horizons were collected from a 25 cm × 25 cm area centred on the coring location. Soil cores were divided in the field into 0–15 cm and 15–30 cm (measuring from the base of the core), individually bagged and returned to the laboratory. There was limited compression in some cores and this was allocated to the 0–15 cm section under the observation that most compression occurred in the upper layer of soil. The depth of the hole was always measured to ensure that the accurate core length was known.

Deep cores (0–100 cm). One of the five sampling plots was randomly selected and three 1-m cores were taken following the same spacing as the 30-cm cores, with the exact coring locations adjusted to avoid those of the 30-cm cores. Cores were taken using a window sampler system with a 4.4 cm cutting diameter (Eijkelkamp Agrisearch Equipment BV, Giesbeek, The Netherlands), allowing a full 1-m core to be extracted and subsequently transported in one section. If coring to the full depth was not possible, for example when large stones or bedrock were encountered, the precise depth of the cored hole was recorded.

Laboratory processing

Litter samples were dried at 80 °C for 24 h and dry mass of woody material (e.g. twigs, branches), leaves and undifferentiated material was recorded. Litter was assumed to have C concentration based on litter dry mass of 43% and 45% for *Miscanthus* leaves and stem, respectively (Beuch et al., 2000; Robertson et al., in preparation), 42% and 49% for willow leaves and stems, respectively (Chauvet, 1987; Heller et al., 2003), 46% for grass litter (Ross et al., 2002) and 41% for cereal litter (Aita et al., 1997).

Short cores (0–30 cm). The fresh mass of the 0–15 cm and 15–30 cm core sections was recorded and sections were then cut lengthways into quarters for separate subsequent analyses. One quarter was then set aside for processing for soil C and bulk density (BD, Table S1), together with the large stones and roots (>5 mm) hand-sorted from the remaining three sections. Another quarter was used to assess soil pH (Table S1) and the remaining sections were archived as a frozen sample (−20 °C).

For the assessment of soil pH, the fresh samples were bulked within each sampling plot but not across depths giving 10 composite samples per site (five each for the 0–15 cm and 15–30 cm depths). The fresh, bulked samples were sieved to 4 mm to remove stones and roots. 10 g of bulk soil was then mixed well with 25 ml of deionised water and allowed to stand for 30 min, before the pH of the liquid layer was recorded (Hanna pH210 Meter, Hanna Instruments Ltd., Befordshire, UK).

For BD, texture and soil C assessment, the fresh soil mass was recorded and then samples were air-dried at 25 °C for a minimum of 10 days. Air-dried samples were reweighed, sieved to 2 mm and the mass and volume of stones and roots remain on the sieve recorded. A subsample of the sieved soil (15–18 g) was oven-dried (105 °C for 12 h) and moisture-loss was recorded. The oven-dried subsample of soil was grounded in a ball mill (Fritsch Planetary Mill) and a 100-mg subsample was used for the assessment of C concentration using an elemental analyser (Leco Truspec CN, Milan, Italy). Prior to analysis using the elemental analyser, soil subsamples that were either from sites located on soil types known to contain inorganic C or which had pH values > 6.5 were tested for the presence of inorganic C using acid fumigation following Harris et al. (2001). All samples from sites which tested positive were treated to remove inorganic C following the same procedure.

A subsample of the sieved air-dried soil was also used to assess soil texture. As for pH measurement, samples were bulked across each field but not across depth, thus giving one value per field for each depth (0–15 cm and 15–30 cm). Analysis of the bulked samples was conducted by Macaulay Scientific Consulting Ltd. (Aberdeen, Scotland) with proportions of sand, silt and clay analysed by laser diffraction (Malvern Mastersizer 2000, Malvern Instruments Ltd., Worcestershire, UK). Analysis was conducted for both the bioenergy crops and the paired controls.

Bulk density of the whole core was calculated using values of moisture-loss from the air and oven-dried subsamples following methods in the GB Countryside Survey (Emmett et al., 2008; Reynolds et al., 2013). These calculations accounted for the measured mass and volume in the soil cores taken up by stones, and so are corrected to represent the fine earth proportion (Schrumpf et al., 2011). The Countryside Survey conducted a pilot study to compare different protocols to estimate BD in different soil types and found that the method used in this study was consistent with other protocols and within the ranges of typical values expected for each of the soil types (Emmett et al., 2008).

The soil C concentration and bulk density data were used to derive mass-based values of soil C stock to account for differences in bulk density across transitions. A soil C stock was calculated based on an equivalent soil mass approach (ESM), using a reference dry soil mass of 3 Gg ha^{-1}, following the method of Gifford & Roderick (2003).

Deep cores (0–1 m). On return to the laboratory, the 1-m cores were divided into three sections: 0–30, 30–50 and 50–100 cm. In cases where compression of the core had occurred during sampling, the length of the sections was reduced to account for the compression; a method also utilised by Walter et al. (2014). Depth increments of 0–30 cm, 30–50 cm and 50–100 cm were selected based on the common use of these increments in similar LUC studies (Laganière et al., 2010; Don et al., 2012).

Each 1-m core section was divided lengthways, one-half, and all root and stones (>5 mm) were processed for bulk density and C content as outlined for the 30-cm surface soil cores. The remaining half was retained as a frozen archive.

Soil C stocks were again calculated based on an equivalent soil mass approach (ESM), using a reference dry soil mass of 6 and 13 Gg ha^{-1} for the 0–50 cm and 0–1 m sections, respectively, following Gifford & Roderick (2003).

Treatment of under length core

In the ESM calculation, the length of the cores is not directly used to calculate soil C stocks (a reference mass is used and the deepest sections are used only to give C concentration). It is still necessary, however, to remove from the data set any cores that, due to the present of large stones or bedrock, do not reach a depth that provides a representative C concentration for the deeper soil layers. Therefore, based on inspection of the soil C profiles, cores <22.5 cm and 70 cm in length for the 30-cm and 1-m cores, respectively, were removed from

the data set prior to statistical analysis (see Table S2 for details).

Statistical analysis

The difference in soil C stock and litter variables between the land uses (SRC, *Miscanthus*, arable and grassland) was tested using linear mixed-effect models with the *nlme* package in the R statistical program (Pinheiro et al., 2014). Differences were observed in the control fields of the bioenergy crops with a higher overall mean soil C in the arable control sites of the *Miscanthus* transitions compared to the SRC transitions. The inclusion of site as a random factor was not sufficient to account for this underlying bias and, consequently, the SRC and *Miscanthus* transitions were analysed separately. Land use was entered as a fixed effect and field nested within site and plot nested within field entered as random effects in all models to ensure that appropriate comparisons of transition units were accounted for within site. The significance of the variable land use in the model was examined using a likelihood ratio test compared to the null model, including only random terms.

The significance of differences between the levels within 'land-use' was tested using Tukeys multiple comparison in the *glht* function in the *multcomp* package (Hothorn et al., 2008). Marginal (R_m^{2}) and conditional (R_c^2) R^2 values were calculated (Nakagawa & Schielzeth, 2013; Johnson & O'Hara, 2014) using the *r.squaredGLMM* function (Lefcheck, 2014) in the *MuMIn* package (Barton, 2015). Data on soil C for ESM at 0–30 cm and 0–1 m were log-transformed prior to testing to meet model assumptions. Litter data were x + 1 log-transformed due to high number of zero values in arable control fields. In all cases, means and standard errors given for land-use effects refer to model-estimated values, and therefore account for the random effect of site.

Difference in mean soil C stock between the controls and their paired bioenergy crops was divided by the age of the bioenergy plantation to estimate annual rates of change in soil C as Mg C ha^{-1} yr^{-1}. This procedure standardises differences in soil C stocks between the SRC and the *Miscanthus* control fields, allowing SRC and *Miscanthus* transitions to be combined into the same statistical test. Differences in annual rates of change between the 4 transitions (arable to SRC, grassland to SRC, arable to *Miscanthus* and grassland to *Miscanthus* transitions) were tested using a two-way ANOVA, with fixed factors of control land use (grassland or arable) and bioenergy crop (SRC and *Miscanthus*). Site was not included as a random factor as it was not found to improve the model fit.

Linear regression focused on the 0–30 cm depth where change was most likely and was used to explore variables influencing the impacts of transition to bioenergy crops on soil C stocks (clay content, soil pH, soil C stocks, bioenergy crop age, MAP and MAT). Data on percentage change from control were tested, again to standardise differences in soil C stocks between SRC and *Miscanthus* control fields, allowing SRC and *Miscanthus* transitions to be combined into the same statistical test.

The drivers of soil C changes were identified through model selection but the number of data points limited the complexity of candidate models. Therefore, $R_m^{2°}$ (Nakagawa & Schielzeth, 2013; Johnson & O'Hara, 2014) and Akaike's Information Criterion (AIC) were first used to assess the influence of each explanatory variable on the percentage change in soil C stock (Table S3). The explanatory variables were then added consecutively to the final models in the order indicated by greater $R_m^{2°}$ or lower AIC scores, provided the AIC of model continued to decrease. Site was included as a random variable in each model and calculation was performed in R using the *r.squaredGLMM* (Lef-

check, 2014) and *AIC* functions in the Lme4 and *MuMIn* package (Barton, 2015; Bates *et al.*, 2015).

Selection based on both the $R_m^{2°}$ and the AIC scores resulted in the selection of the same model which included the fixed factors control soil C stock and the bioenergy crop type and the random effect of site (Table S4). The significance of the explanatory variables within this model was examined using a likelihood ratio test.

Prior to this analysis, exploration of the soil texture data showed that in contrast to the percentage sand and silt, which

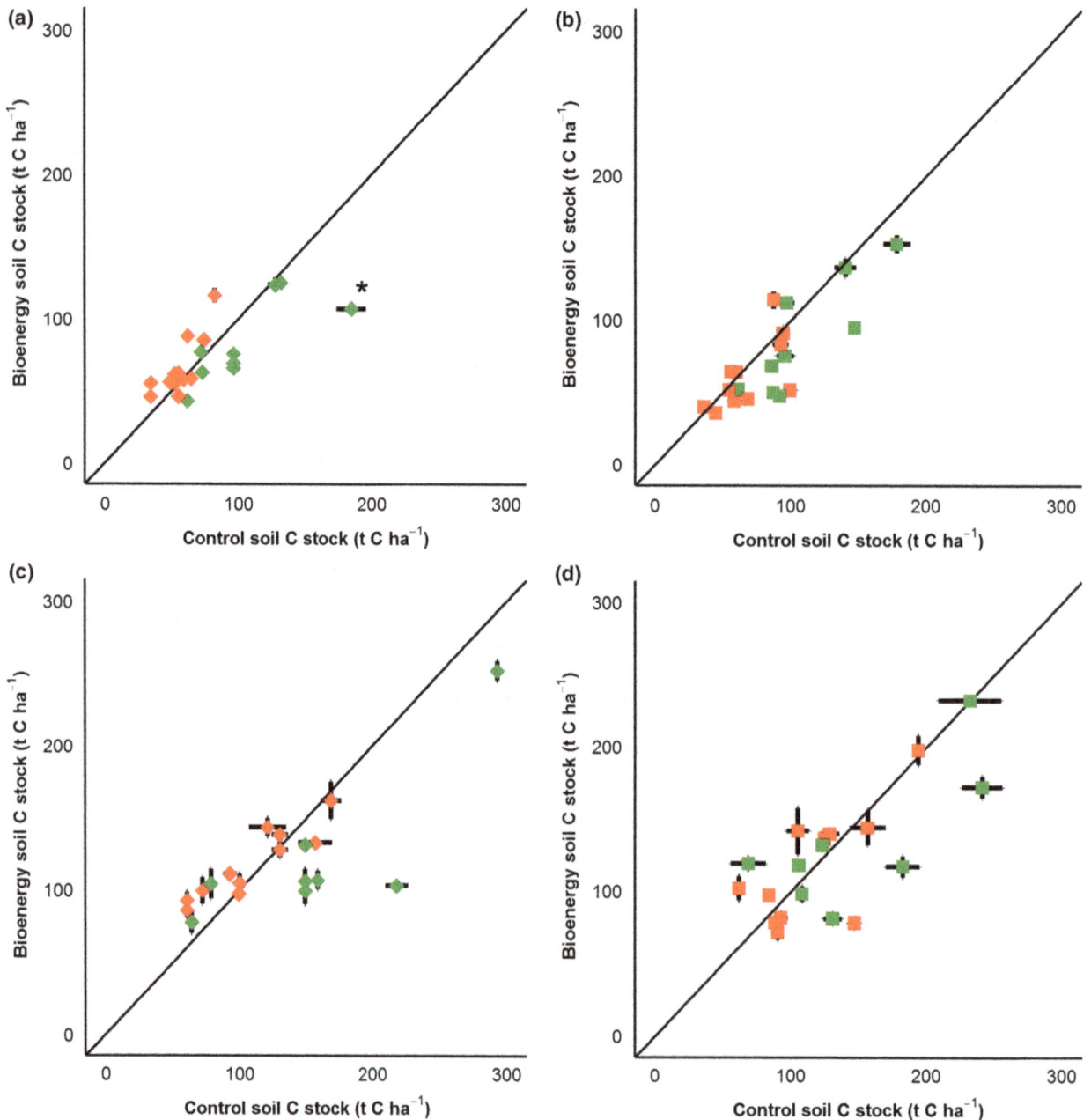

Fig. 2 Control versus bioenergy crops soil C stocks for the SRC transitions: 0–30 cm (a) 0–100 cm (c) depths, and the *Miscanthus* transitions 0–30 cm (b) 0–100 cm (d) depths; red symbols represent ex-arable transitions, and green symbols represent ex-grassland transitions. * indicates site 17 vs. 17C. Error bars give standard error.

showed a correlation between the bioenergy crops and paired control ($R^2 = 0.72$ and $R^2 = 0.71$, respectively; Fig. S1), the correlation for the percentage clay content was poor ($R^2 = 0.26$; Fig. S1). This poor correlation appeared to be related to high soil inorganic C in five of the sites (2, 5, 7, 17 and 19), a factor known to affect laser assessment of clay content (Kerry *et al.*, 2009). Removal of these sites resulted in an improvement to an R^2 of 0.62 but did not improve the explanatory power of the percentage clay in regard to the percentage change in soil C stocks (Table S3). Thus, this subset was not used in any subsequent analysis (Table S3).

Results

Soil C stocks 0–30 cm

Land use was found to affect surface (0–30 cm) soil C stock (Mg C ha^{-1}) in both the SRC ($\chi^2(3) = 15.30$, $P = 0.001$, $R_c^2 = 0.86$) and *Miscanthus* transitions ($\chi^2(3) = 13.71$, $P = 0.001$, $R_c^2 = 0.92$) (Fig. 2a,b). The greatest differences in soil C stocks were in the grassland transitions, with mean soil C stocks 33.55 ± 7.52 Mg C ha^{-1} and 26.83 ± 8.08 Mg C ha^{-1} lower under the SRC ($P = 0.004$) and *Miscanthus* plantations ($P = 0.001$), respectively (Fig. 2a,b, Table 2).

Differences between the arable controls and bioenergy crop were smaller than those seen in the grassland transitions, with greater variation between sites, and not significant ($P = 0.071$ and $P = 0.846$ for SRC and *Miscanthus* transitions, respectively) (Fig. 2). The nonsignificant differences in mean soil C stocks are being 16.27 ± 7.18 Mg C ha^{-1} higher under SRC, and 2.26 ± 8.18 Mg C ha^{-1} lower under *Miscanthus* plantations compared to arable controls.

Within the SRC data, the grassland control at site 17 had exceptionally high soil C compared to its paired bioenergy crop (Fig. 2a). This transition unit was located at a site with highly complex underlying geology and variable soil types. Removing this transition from the analysis of soil C stock reduced the difference between the SRC and the grassland control. The mean soil C stock under the SRC, however, was still significantly lower (-23.34 ± 8.37 Mg C ha^{-1}) than the grassland controls ($P = 0.047$).

Differences in soil C stocks between the bioenergy crops and the controls were reflected in the annual rates of change (Mg C ha^{-1} yr^{-1}) in the surface soil (0–30 cm) with effects of both the original land use ($F_{1,37} = 11.99$, $P = 0.001$) and also bioenergy crop type ($F_{1,37} = 6.59$, $P = 0.014$) but there was no interaction between these factors ($F_{1,37} = 0.326$, $P = 0.571$) (Table 3). Rates of change in the transitions from grassland, as would be expected by the differences in soil C stock, were consistently negative and significantly lower than observed in the arable transitions. Unlike the differences in soil C stock, annual rates of change also allowed the comparison of the two bioenergy crops and showed that the rates of change for the SRC transitions were more positive that those for the *Miscanthus* transitions (Table 3).

Soil C stocks 0–1 m

Over 0–1 m, soil C stocks (Mg C ha^{-1}) and annual rates of change followed a similar pattern to those seen in the surface soils (Fig. 1c,d, Tables 2 and 3). Unlike the surface soil, however, differences in soil C stocks between the controls and bioenergy crops were not significant in either the SRC ($\chi^2(3) = 1.93$, $P = 0.3813$, $R_c^2 = 0.92$) or *Miscanthus* transitions [$\chi^2(3) = 2.10$, $P = 0.350$, $R_c^2 = 0.90$)] (Table 2). Annual rates of change were not significantly different between the bioenergy crops ($F_{1,34} = 0.015$, $P = 0.902$), nor was there any impact of the original land use ($F_{1,34} = 2.432$, $P = 0.128$) or an interaction between these factors ($F_{1,34} = 1.166$, $P = 0.287$) (Table 3).

Over a shallower depth of 0–50 cm, there were differences in soil C stocks in the SRC transitions ($\chi^2(3) = 7.16$, $P = 0.028$, $R_c^2 = 0.91$) but not the *Miscanthus*

Table 2 Mean litter and soil C stocks (Mg C ha^{-1}) and standard error for the bioenergy crops (SRC and *Miscanthus*) and controls

Land use	C stock (Mg C ha^{-1})			
	Litter	0–30 cm	0–50 cm	0–100 cm
SRC	0.97 ± 0.18^a	70.31 ± 6.57^a	91.16 ± 8.98^a	116.91 ± 11.65^a
Arable	0.38 ± 0.19^b	54.04 ± 8.18^a	76.41 ± 12.14^a	107.22 ± 16.14^a
Grassland	0.21 ± 0.19^b	103.87 ± 9.5^b	$129.03 \pm 12.6^{b*}$	147.19 ± 16.56^a
Miscanthus	$2.09 \pm .24^a$	74.31 ± 7.84^a	108.77 ± 7.70^a	124.31 ± 11.39^a
Arable	0.78 ± 0.33^b	76.57 ± 8.53^a	94.41 ± 9.51^a	120.98 ± 12.69^a
Grassland	0.06 ± 0.36^b	101.14 ± 8.89^b	123.47 ± 10.14^a	140.49 ± 13.75^a

0–50 cm ESM and 0–100 cm ESM refer to soil C stock based on reference soil mass for these depths of 6 and 13 Gg ha^{-1}. Same litter indicates nonsignificant difference > P 0.05; * indicates that there was a near-significant difference ($P = 0.063$) between the grassland and the SRC. Test conducted on *Miscanthus* and SRC transitions separately and within each depth division.

Table 3 Annual rates of change in soil C stocks for 0–30 cm, 0–50 cm and 0–1 m soil cores based on ESM. Annual rates of change are estimated by dividing change in mean soil C compared to control by the years since transition. $n = 15$ and 3 for the 30-cm cores and 1-m cores, respectively

Land-use Change	Rate of change Mg C ha^{-1} yr^{-1} (SE)		
	0–30 cm	0–50 cm	0–1 m
SRC vs. Arable	1.54 ± 0.70	1.93 ± 1.37	1.26 ± 1.41
SRC vs. Grassland	−1.69 ± 0.81	−2.98 ± 1.61	−2.74 ± 1.65
Miscanthus vs. Arable	−0.93 ± 0.74	0.18 ± 1.37	0.05 ± 1.41
Miscanthus vs. Grassland	−3.17 ± 0.81	−2.11 ± 1.52	−0.69 ± 1.65

transitions (χ^2 (3) = 4.34, P = 0.114, R_c^2 = 0.85) (Table 2). The significant difference in the SRC transitions was, however, related to differences in soil C stocks between the grassland and the arable control (P = 0.008), although there was also nonsignificant trend for lower soil C stocks within the grassland controls compared to the SRC (P = 0.063).

Rates of change reflected the absence of a significant difference in soil C stock, which were similar in both bioenergy crops ($F_{1,35}$ = 0.188, P = 0.667). There was no interaction between the current land use and the control land use ($F_{1,35}$ = 0.761, P = 0.388) but rates of change were lower in the grassland compared to arable transitions ($F_{1,35}$ = 5.952, P = 0.019; Table 3), highlighting a difference that was less clear with soil C stock.

Driving factors determining changes in soil C

Based on the model selection, soil C stocks of the control field and the current land use (SRC, *Miscanthus*) were tested for their effect on the percentage change in soil C stocks resulting from the transition the bioenergy crops (Tables S3, S4). Soil C stocks was found to be negatively related to the percentage difference in soil C in bioenergy fields (χ^2 (1) = 8.70, P = 0.003, R_c^2 = 0.52). There was no interaction between current land-use type (SRC, *Miscanthus*) and soil C stock (χ^2 (1) = 2.138, P = 0.144, R_c^2 = 0.51), suggesting a similar relationship in both SRC and *Miscanthus*, but a near-significant effect of land-use type was observed (χ^2 (1) = 3.216, P = 0.073, R_c^2 = 0.22) likely resulting from the different intercepts of the linear relationships in the two bioenergy crops (Fig. 3 a & b). Examination of the residuals highlighted that three transitions (*Miscanthus* transitions 24 and 25,

and SRC willow transition 17) had a large influence on the results. Removal of these transitions influenced the slope of the linear relationships (Fig. 3 c & d), but did not change the overall significance of any of the factors.

Time since bioenergy establishment and the clay content of the bioenergy crop were the third most important factors influencing the change in soil C stock based on the marginal R^2 and AIC scores, respectively (Table S3). However, there was no clear relationship between the percentage change in soil C stock and either time since bioenergy establishment or clay content (Fig. 4).

Litter C stocks

Litter C stocks were different between the land uses in both the *Miscanthus* (χ^2 (2) = 25.42., P = 0.001, R_c^2 = 0.84) and the SRC plantations (χ^2 (2) = 43.68, P < 0.001, R_c^2 = 0.69), with *post hoc* testing showing that litter stock was higher in the bioenergy crops than in either the arable or grassland controls (Table 2). The addition of these relatively small litter C stocks to the surface soil C stocks (0–30 cm, Table 2) has little effect on the impact of the bioenergy crops on C stocks. C stocks remain lower in the bioenergy transitions than in grassland controls and are not significantly different to the arable controls.

Discussion

Soil C stocks in arable transitions

In this study, annual rates of change in the surface soil were more positive for the arable transitions than for the grassland transitions. Although, as soil C stocks in the SRC and *Miscanthus* plantations were not significantly different to the arable controls, the difference in the rates of changes is most likely related to the negative impacts on soil C stocks of transition from grassland, rather than any positive impacts of arable. This absence of a positive impact is contrary to a number of studies which have reported increases in topsoil C stocks following transitions from arable land uses to these bioenergy crops (Jug *et al.*, 1999; Dondini *et al.*, 2009; Schmitt *et al.*, 2010; Felten & Emmerling, 2012). These studies used a fixed depth method (FD) to calculate soil C stock which, unlike the ESM used in this study, makes no adjustment for changes in bulk density (BD) (Bárcena *et al.*, 2014). Applying FD methods to our data leads to a similar result to these studies with significantly lower surface soil C stock in the arable controls (Tables S5 and S6). The use of a FD method appears to inflate the differences between the arable control and the bioenergy crops, something that has been noted in a similar land-use change study (Bárcena *et al.*, 2014). The

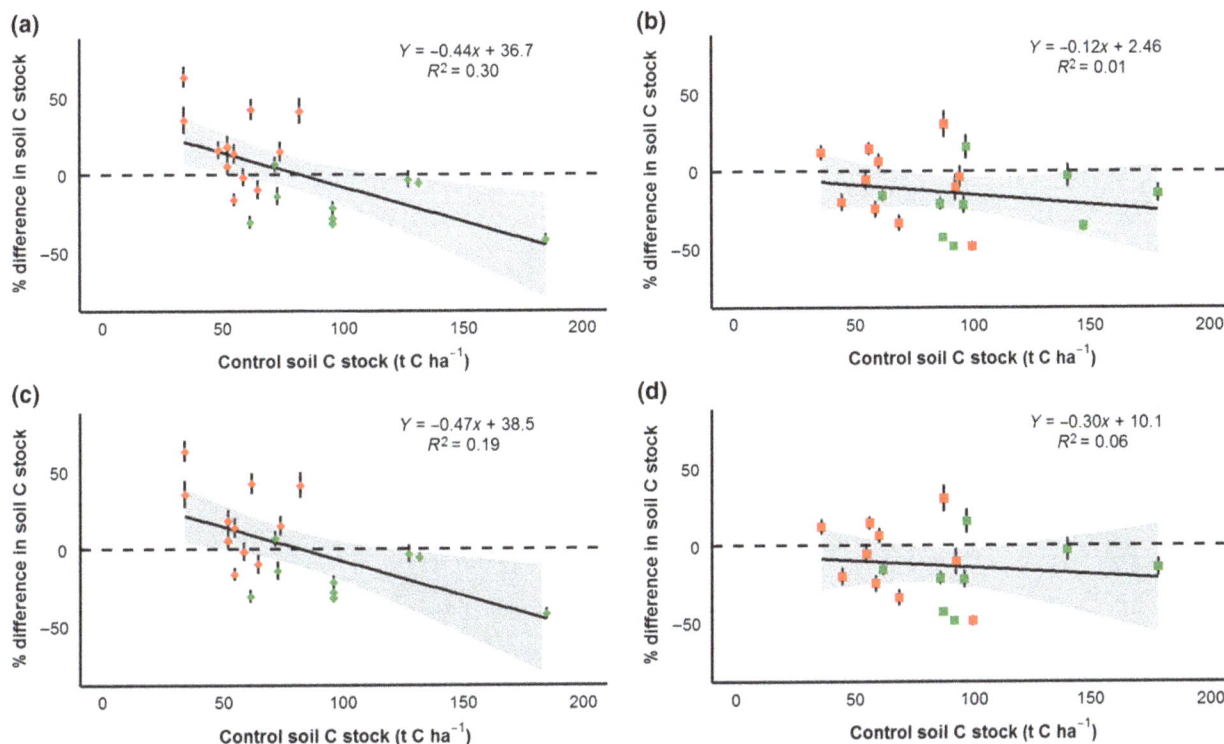

Fig. 3 Relationship between 0–30 cm soil C stocks in control crops and percentage differences for control in soil C resulting from land-use change for: SRC transitions (a), *Miscanthus* transitions (b), SRC transition without site 17 (c), *Miscanthus* transition without transitions 24 and 25 (d). Red markers indicate arable transition green grassland transition. The line shows linear regression of change in soil C stock with C stocks of the control fields; shaded area shows 95% of confidence interval, R^2 gives values for individual regression lines.

use of an ESM method is not widespread in bioenergy studies, and in the case of arable transitions, the only comparable study is that by Walter *et al.* (2014). Using an ESM method and a paired-site approach to assess impacts of arable to SRC transitions, Walter *et al.* (2014) also reported consistent changes in surface (0–30 cm) soil C stock.

Below the plough layer, BD is more consistent between land uses, and differences in C stock estimation due to method are less apparent. This is possibly reflected by the studies that have assessed soil C stock below 30 cm and reported no significant changes in transitions to either SRC (Coleman *et al.*, 2004; Lockwell *et al.*, 2012; Bonin & Lal, 2014; Walter *et al.*, 2014) or *Miscanthus* (Felten & Emmerling, 2012).

The age of the plantations studied may also have an impact on the soil C stock change. Hansen *et al.* (2004) reported higher soil C stocks under *Miscanthus* plantation compared to arable controls but only under the older of two plantations sampled (9 and 16 years old). A study of SRC by Dimitriou *et al.* (2012) also reported an increase in soil C stock compared to arable controls, but only one of the 14 sites sampled was under 15 years old. In addition, many were not in optimum condition

leading the authors to suggest that some of the increase in soil C concentration could be related to C inputs from decaying stools and roots.

Within this study, the difference in the mean age of the bioenergy crops may also explain the differences in the rate of change between the SRC and the *Miscanthus* transitions. The mean annual rate of change in the surface soil for the SRC to arable transitions (1.43 ± 0.71 Mg C ha^{-1} yr^{-1}), with a mean age of 8.5 years, was within the upper range of reported values from 0.38 to 1.59 Mg C ha^{-1} yr^{-1} (Kahle *et al.*, 2010, 2013; Chimento *et al.*, 2014). In contrast, the annual rate of change for transitions to *Miscanthus* from arable (-0.93 ± 0.74 Mg C ha^{-1} yr^{-1}), with a mean age of 6.4 years, was more negative than the mean reported values for topsoil changes of 0.28–2.24 Mg C ha^{-1} yr^{-1} (Clifton-Brown *et al.*, 2007; Dondini *et al.*, 2009; Zimmerman *et al.*, 2012; Chimento *et al.*, 2014). This possibly reflects the mature plantations in some of these studies (16 years and 14 years in Clifton-Brown *et al.*, 2007 and Dondini *et al.*, 2009; respectively) compared to this study. It is also clear that impacts on soil C vary greatly between sites, even within individual studies. For example, although mean rates of change in the study by Zim-

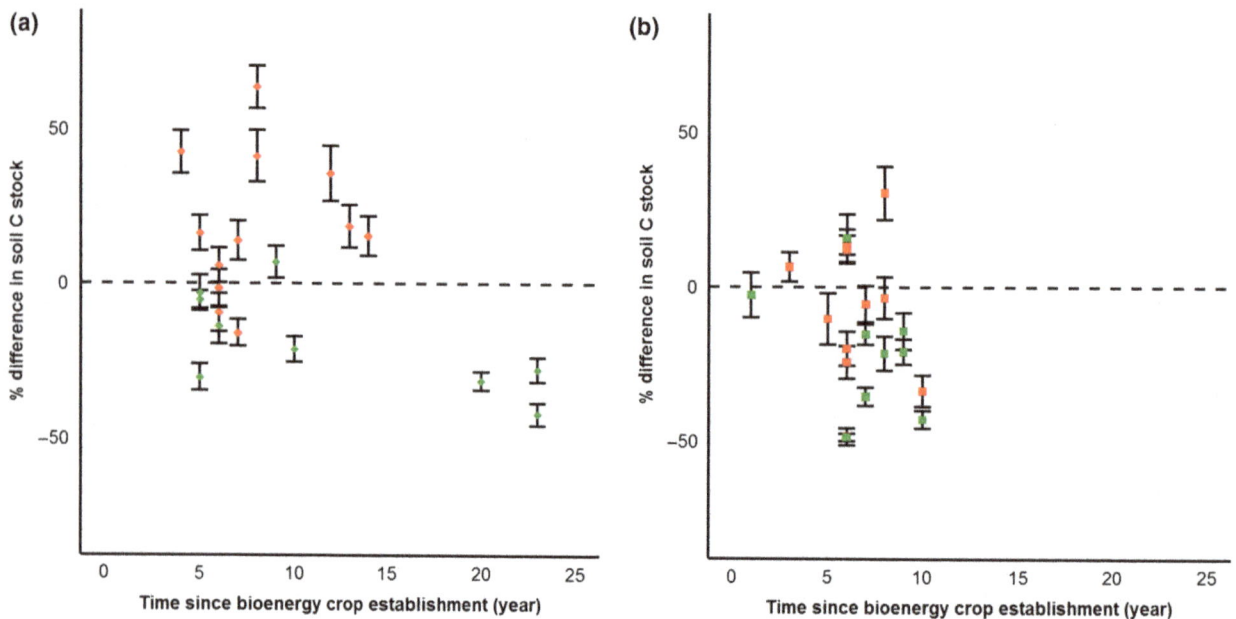

Fig. 4 Relationship between time since establishment (a) and bioenergy clay content, (b) and the percentage change in 0–30 cm soil C. Red markers indicate arable transition green grassland transition, diamond indicates SRC transitions and squares *Miscanthus* transitions. Error bars show pooled SE.

merman *et al.* (2012) were 1.79 Mg C ha^{-1} yr^{-1} from arable, the rates of change across sites within this study ranged from −6.85 to 7.7 Mg C ha^{-1} yr^{-1}. Site-specific factors clearly influence the impacts on soil C stocks, as reflected in the between-site variability observed within this study and reported in other multi-site studies (Coleman *et al.*, 2004; Dimitriou *et al.*, 2012; Don *et al.*, 2012; Walter *et al.*, 2014).

One possible additional source of variability between sites could be related to the willow clones selected. Nearly all the sites visited were planted by a single contractor whose records do not contain details of the clones planted at each site (F. Walters, Coppice Resources Ltd, Retford, pers.com.) but only that the mixed willow will contain 4–5 different clones. The lack of detailed information coupled with the practice of mixing clones throughout a single plantation (e.g. clones are not planted in uniform strips) for pest control purposes means that it is not possible within this study to examine differences between the influence of individual clones. However, any differences in soil C stock resulting from different clones are likely to be smaller than the impact resulting from the LUC from arable or grassland land uses.

Soil C stocks in grassland transitions

In contrast to the findings for arable soils, the lower soil C stocks in the topsoil (0–30 cm) and the negative rates of change of the SRC and *Miscanthus* plantations com-

pared to the grassland controls reflect findings in other studies (Don *et al.*, 2012; Rytter, 2012; Zimmerman *et al.*, 2012). The mean annual rate of change in the transition to SRC (−1.69 ± 0.82 Mg C ha^{-1} yr^{-1}, 0–30 cm) compares well, once again, with the values reported in a study of a 9-year-old SRC willow plantation by Lockwell *et al.* (2012) of −2.22 and −1.11 Mg C ha^{-1} yr^{-1} over 0–20 cm and 0–40 cm depths, respectively. The mean rate of change for transitions to *Miscanthus* from grassland, however, was again more negative (−3.17 ± 0.81 Mg C ha^{-1} yr^{-1}) than those reported from −1.66 and 0.83 Mg C ha^{-1} yr^{-1} by Zimmerman *et al.* (2012) and Zatta *et al.* (2014).

Over the greater depth of 1 m, the magnitude of differences in soil C stocks observed was similar to those seen in the surface soil, especially for the SRC transitions (−33.55 ± 7.52 Mg C ha^{-1} and −30.28 ± 10.96 Mg C ha^{-1} for 0–30 cm and 0–1 m, respectively) but differences were no longer significant. Fewer 1-m cores were taken compared to the 30-cm cores, resulting in reduced statistical power to detect impacts at greater depth. Walter *et al.* (2014) and Lockwell *et al.* (2012), however, reported similar findings in transitions from grassland to SRC, concluding that soil C losses in the surface soil were offset by increases lower in the soil profile, resulting in no significant changes in soil C stocks overall. *Miscanthus* shares the tendency of SRC to be deep rooting and *Miscanthus*-derived C inputs have been detected at depths of up 1.5 m (Felten & Emmerling, 2012), and thus, there is a mechanism by which both crops could alter soil C stocks at depth. In this study, mean difference in soil C stocks between both the SRC and *Miscanthus* and their grassland controls was less negative over 0–1 m than over 0–30 cm. Differences in sampling intensity between 0–30 cm and 0–1 m cores mean that it is not possible to directly attribute any redistribution of soil C within the soil profile.

Alternatively to a redistribution of soil C stocks, it is possible that changes in soil C stock were limited to the surface soil and that difficulties in detecting changes in soil C stock 0–1 m are instead due to the dilution of the impacts in the surface soil when including soil C stock at greater depths. This would agree with studies which report slower turnover times in the subsoil, with reported mean C resident times in soil layers below 20 cm of 2000–10 000 years (Fontaine *et al.*, 2007). Sampling subsoil is, however, still extremely valuable as although C stocks at depth may be characterised by long residence times, they have also been found to be susceptible to priming resulting from labile C inputs such as root exudates (Fontaine *et al.*, 2007; De Graaff *et al.*, 2014). Deep soil coring therefore provides a mechanism to detect both increase in soil C and any losses due to C priming.

Regardless, if losses in the surface soil are replaced with gains at depth or just diluted, any step taken to reduce surface soil C loss would be beneficial. Grassland soil C stocks have been shown to be negatively affected by tillage (Poeplau & Don, 2014). Thus, it has been suggested that the intensive cultivation undertaken prior to the bioenergy crop establishment may account for a substantial proportion of soil C losses observed (Don *et al.*, 2012; Walter *et al.*, 2014). A move to new, less intensive establishment methods may provide one option to reduce impacts on soil C stocks. However, it is unclear what role other factors, such as changes in the quality or quantity of inputs to soil, may play in addition to the effects of cultivation. For example, Poeplau & Don (2014) reported that transitions from grassland to forestry resulted not only in changes in soil C stocks but also a shift in soil C from stable to labile pools.

Factors influencing changes in soil C stock

Explaining variations in soil C stock changes within this study was explored through assessment of relationships between changes in soil C stock and selected factors. A negative relationship was found between changes in soil C stock and the soil C stock of the control field, suggesting that establishment of bioenergy crops on sites with low initial soil C provided the best opportunity to derive positive impacts on soil C stocks. Such a negative relationship was predicted for SRC poplar plantations

in modelling work by Garten *et al.* (2011) and generally agrees with the conclusions of Don *et al.* (2012) and Walter *et al.* (2014) that conversion of arable lands, which generally have low soil C stock, is preferable to conversion of grassland for bioenergy plantations. It is difficult to separate the impacts of original soil C stocks and original land use because they are highly correlated (e.g. higher soil C stocks are generally associated with grassland sites). As land use also affects soil C stability and turnover, as well as soil C stocks (Poeplau & Don, 2014), impacts of land-use transitions could be influenced by both the stability of the soil C and the total soil C stocks.

The relationships between control soil C and changes in soil C stock following bioenergy crop establishment are relatively weak, especially for *Miscanthus*. Transitions to SRC and *Miscanthus* have R^2 values of 0.30 and 0.01, respectively, which indicate considerable unexplained variability related to the impacts on soil C stocks at individual sites. Part of this unexplained variability may reflect the challenge of finding paired sites with no pre-existing differences in soil C stocks between the two land uses before conversion. In many cases, the bioenergy crops and paired sites were adjacent but the soil texture analysis does suggest, even for the more reliable sand and silt data, that in a few of the sites there may be some underlying differences between some of the transition pairs. In addition, whilst finding sites with generally similar land-use histories was relatively straightforward, the normal crop rotation practices (rotations wheat, barley, beans, etc.) and the variable nature of farming (fertiliser inputs, harvest times, etc.) combined with the limited nature of long-term data held by land owners meant that some variability between the bioenergy crop and the paired control was inevitable. A better understanding of between-site variability is also clearly needed. For example, in this study, the rates of change in soil C stock range from −3.75 to 0.58 Mg C ha^{-1} yr^{-1} for grassland to SRC transitions, and from −7.44 to 2.53 Mg C ha^{-1} yr^{-1} for grassland to *Miscanthus* transitions.

It is worth noting that whilst underlying differences between the paired sites could have influenced the analysis of the potential factors driving soil C stock change, and the rates of change, where the percentage change was calculated at the transitions level, in the assessment of soil C stocks individual core data rather than transition level mean were used. When using this core data, the mixed model is less sensitive to variation between the bioenergy crop and the control.

No relationship was found between bioenergy planation age or clay content and changes in soil C stock. In the case of clay content difficulties with both the analysis method and a limited range of clay content across the sites (3.50–12.56% Table 1 Fig. S1) may have reduced our ability to detect a relationship. However, the absence of any relationship between soil texture and changes in soil C stocks has also been reported for SRC (Walter *et al.*, 2014) and *Miscanthus* (Poeplau & Don, 2014). Clay content tends to be positively associated with soil C stock (Stockmann *et al.*, 2013) and the absorption of C compounds to clay minerals, together with occlusion into clay aggregates, has been shown to stabilise soil organic matter (Dungait *et al.*, 2012; Stockmann *et al.*, 2013). Therefore, there could be an expectation that higher clay content would protect soil C during LUC, and aid its accumulation post LUC (Laganière *et al.*, 2010). One possible reason why this is not seen could be that the current practice of intensively tilling prior to bioenergy crop planting could reduce the protection afforded by occlusion into clay aggregates (Stockmann *et al.*, 2013).

A relationship between changes in soil C stock and time since bioenergy crop establishment was also absent, something which has been reported in a number of other multi-site studies (Don *et al.*, 2012; Walter *et al.*, 2014). This is despite general agreement across a wide range of land-use transitions that time since LUC is an important factor in determining soil C stocks (Bárcena *et al.*, 2014; Poeplau & Don, 2014; Walter *et al.*, 2014). Bárcena *et al.* (2014) suggested that the time taken for soil C stocks to recover from any initial soil C loss following land-use transitions, and to reach a new equilibrium, may vary between sites. Thus, any assessment made between sites that are yet to near a new equilibrium will lead to highly variable results (Bárcena *et al.*, 2014). In case of transitions from arable to forestry, Bárcena *et al.* (2014) found that increases in soil C were only detectable in a chronosequence of independent sites after 30 years. The time required for soil C recovery in SRC and *Miscanthus* plantations is, as yet, unknown. Walter *et al.* (2014) did select older plantations (15–35 years) in their study of 21 SRC plantations, but were still unable to detect any relationship between plantation age and impacts on soil C stock. Therefore, it may be that the time period required to detect an effect of age on soil C under bioenergy crops will exceed the expected 25–30 year life span of these plantations.

The time taken to reach a new soil C equilibrium has potential to impact on the 'payback time' required for any decreases in soil C stocks within the soil to be replaced (Mello *et al.*, 2014). In contrast to the transitions from arable, where changes in soil C stock were not significant, there is not a soil C debt to be paid. A soil C debt was detected in the surface soil, at least in grassland transitions. To replace this debt through increases in the soil C stock, the bioenergy crops must

in theory reach a new soil C equilibrium that is equal to or greater that than of the grassland. The time it takes to reach this new equilibrium is also critical because, if it takes longer than the lifetime of the bioenergy crop, it may not be possible to repay the soil C debt through changes in soil C stock alone (Bárcena *et al.*, 2014; Mello *et al.*, 2014). Although it must be recognised that over greater depths this and other studies have found no significant negative impact of planting on grassland (Walter *et al.*, 2014). Although requiring a detailed life-cycle assessment to confirm, the C saving attributed to using biomass to offset fossil fuel use may be greater than any soil C loss as has been found to be the case for sugarcane planted on pasture in Brazil (Mello *et al.*, 2014).

It is possible that the difficulties in detecting a clear chronosequence may also result from different sites having different linear relationships between age and soil C and/or more complex nonlinear relationships. In addition, the C stock within the control field may not be in equilibrium, and for this reason, it is best to view controls as counterfactuals rather than a time zero. Long-term studies utilising both repeated sampling and the use of counterfactual paired sites, soil fractionation (Poeplau & Don, 2014) and process base modelling (Dondini *et al.*, 2015) are all methods which could help to provide a better understanding of the time it will take to reach a new equilibrium, and allow the comparison to other land-use options. The data collected in this study are highly suited for process models, which can be used to understand key drivers of soil C change, and such models can be used to predict impacts of future climate scenarios (Dondini *et al.*, 2015).

We conclude that where choices exist, the selection of arable land for bioenergy transitions to SRC and *Miscanthus* is likely to be more positive for soil C stocks than conversion from grassland, at least for soil C stocks within the surface soil. Whilst changes in soil C stocks at 0–1 m were not significant in any of the transitions types, the direction of changes mirrored those in the surface soil. Questions still remain as to why transitions from grassland can lead to negative changes in soil C, and work on soil C stability, especially during bioenergy crop establishment, would both address this question and potentially provide insight into management solutions that would maximise the soil C sequestration potential of these crops. Whilst these conclusions are valid for soil C, the findings also need to be considered in the wider context of other ecosystem services such as productivity, greenhouse gas regulation and water quality.

Acknowledgements

We are exceptionally grateful to all the land owners who have granted us access to sample their fields. Kate

Farrall, Jessica Adams, Neil Mullinger, Adam Dargan and Lou Walker for field and laboratory assistance. Pete Henrys (Centre for Ecology & Hydrology) for statistical guidance. This work was part of the Ecosystem Land-Use Modelling (ELUM) project, which was commissioned and funded by the Energy Technologies Institute.

References

Aita C, Recous S, Angers DA (1997) Short-term kinetics of residual wheat straw C and N under field conditions: characterisation by $^{13}C^{15}N$ tracing and soil particle size fractionation. *European Journal of Soil Sciences*, **48**, 283–294.

Aylott M, McDermott F (2012) *Domestic Energy Crops; Potential and Constraints Review*. NNFCC, Heslington, York, UK.

Bárcena TG, Kiær LP, Vesterdal L, Stefánsdóttir HM, Gundersen P, Sigurdsson BD (2014) Soil carbon stocks change following afforestation in Northern Europe: a meta-analysis. *GCB Bioenergy*, **20**, 2393–2405.

Barton K (2015) MuMIn: Multi-Model Inference R package version 1.15.1. Available at: http://CRAN.R-project.org/package=MuMIn (accessed 15 September 2015).

Bates D, Maechler M, Bolker BM, Walker S (2015) Fitting Linear Mixed-Effects Models using lme4. Journal of Statistical Software. Available at: http://arxiv.org/abs/1406.5823 (accessed 18 August 2015).

Beuch S, Boelcke B, Belau L (2000) Effect of organic residues of Miscanthus x giganteus on soil organic matter level of arable soils. *Journal of Agronomy and Crop Sciences*, **183**, 111–119.

Bonin CL, Lal R (2014) Aboveground productivity and soil carbon storage of biofuel crops in Ohio. *GCB Bioenergy*, **6**, 67–75.

Chauvet E (1987) Changes in the chemical composition of alder, poplar and willow leaves during decomposition in a river. *Hydrobiologia*, **148**, 35–44.

Chimento C, Almagro M, Amaducci S (2014) Carbon sequestration potential in perennial bioenergy crops: the importance of organic matter inputs and its physical protection. *GCB Bioenergy*. doi:10.1111/gcnn.12232.

Clifton-Brown JC, Breuer J, Jones MB (2007) Carbon mitigation by the energy crop, Miscanthus. *Global Change Biology Bioenergy*, **13**, 2296–2307.

Coleman MD, Isebrands JG, Tolsted DN, Tolbert VR (2004) Comparing soil carbon of short rotation poplar plantations with agricultural crops and woodlands in north central United States. *Environmental Management*, **33**, S299–S308.

Conant RT, Paustian K (2002) Spatial variability of soil organic carbon in grasslands: implications for detecting change at different scales. *Environmental Pollution*, **116**, S127–S135.

Conant RT, Smith GR, Paustian K (2003) Spatial variability of soil carbon in forested and cultivated sites. *Journal of Environmental Quality*, **32**, 278–286.

Davis MR, Condron LM (2002) Impact of grassland afforestation on soil carbon in New Zealand: a review of paired-site studies. *Australian Journal of Soil Research*, **40**, 675–690.

Davis SC, Parton WJ, Del Grosso SJ, Keough C, Marx E, Adler PR, DeLucia EH (2012) Impact of second-generation biofuel agriculture on greenhouse-gas emissions in the corn-growing regions of the US. *Frontiers in Ecology and the Environment*, **10**, 69–74.

De Graaff M-A, Jastrow JD, Gillette S, Johns A, Wullschleger SD (2014) Differential priming of soil carbon driven by soil depth and root impacts on carbon availability. *Soil Biology and Biochemistry*, **69**, 147–156.

DECC, DEFRA, DFT (2012) *UK Bioenergy Strategy*. Department of Energy & Climate Change, London. HMG.

Dimitriou I, Mola-Yudego B, Aronsson P, Eriksson J (2012) Changes in organic carbon and trace elements in the soil of willow short-rotation coppice plantations. *Bioenergy Research*, **5**, 563–572.

Don A, Osborne B, Hastings A et al. (2012) Land-use change to bioenergy production in Europe: implications for the greenhouse gas balance and soil carbon. *GCB Bioenergy*, **4**, 372–391.

Dondini M, Hastings A, Saiz G, Jones MB, Smith P (2009) The potential of Miscanthus to sequester carbon in soil; comparing field measurements in Carlow, Ireland to model predictions. *GCB Bioenergy*, **1**, 413–425.

Dondini M, Richards M, Pogson M et al. (2015) Evaluation of the ECOSSE model for simulating soil organic carbon under Miscanthus and short rotation coppice - willow crops in Britain. *GCB Bioenergy*. doi:10.1111/gcbb.12286.

Dungait JAJ, Hopkins DW, Gregory AS, Whitmore AP (2012) Soil organic matter turnover is governed by accessibility not recalcitrance. *Global Change Biology*, **18**, 1781–1796.

Emmett BA, Frogbrook ZL, Chamberlain PM et al. (2008) *Soils Manual. Countryside Survey Technical Report No.03/07. Centre for Ecology and Hydrology*, Natural Environment Research Council, Wallingford, UK.

Fazio S, Monti A (2011) Life cycle assessment of different bioenergy production systems including perennial and annual crops. *Biomass and Bioenergy*, **35**, 4868–4878.

Felten D, Emmerling C (2012) Accumulation of Miscanthus-derived carbon in soils in relation to soil depth and duration of land use under commercial farming conditions. *Journal of Plant Nutrition and Soil Science*, **175**, 661–670.

Ferchaud F, Vitte G, Mary B (2015) Changes in soil carbon stocks under perennial and annual bioenergy crops. *GCB Bioenergy*. doi:10.1111/gcbb.12249.

Fontaine S, Barot S, Barré P, Bdioui N, Mary B, Rumpel C (2007) Stability of organic carbon in deep soil layers controlled by fresh carbon supply. *Nature*, **450**, 277–280.

Garten CT Jr, Wullshleger SD, Classen AT (2011) Review and model-based analysis of factors influencing soil carbon sequestration under hybrid poplar. *Biomass and Bioenergy*, **35**, 214–226.

Gifford RM, Roderick ML (2003) Soil carbon stocks and bulk density: spatial or cumulative mass coordinates as a basis of expression? *Global Change Biology*, **9**, 1507–1514.

Guo LB, Gifford RM (2002) Soil carbon stocks and land use change: a meta-analysis. *Global Change Biology*, **8**, 345–360.

Hansen EM, Christensen BT, Jensen LS, Kristensen K (2004) Carbon sequestration in soil beneath long-term Miscanthus plantations as determined by ^{13}C abundance. *Biomass and Bioenergy*, **26**, 97–105.

Harris D, Horwath WR, van Kessel C (2001) Acid fumigation of soils to remove carbonates prior to total organic carbon or carbon-13 isotopic analysis. *Soil Science Society of America Journal*, **65**, 1853–1856.

Heller MC, Keoleian GA, Volk TA (2003) Life cycle assessment of willow bioenergy cropping system. *Biomass and Bioenergy*, **23**, 147–165.

Hewitt A, Forrester G, Fraser S, Hedley C, Lynn I, Payton P (2012) Afforestation effects on soil carbon stocks of low productivity grassland in New Zealand. *Soil Use and Management*, **28**, 508–516.

Hothorn T, Bretz F, Westfall P (2008) Simultaneous inference in general parametric models. *Biometrical Journal*, **50**, 346–363.

Intergovernmental Panel on Climate Change (IPCC) (2014) *Climate Change 2014: Mitigation of Climate Change. Contribution of Working Group III to the Fifth Assessment Report of the Intergovernmental Panel on Climate Change* (eds, Edenhofer OR, Pichs-Madruga Y, Sokona E et al.,). Cambridge University Press, Cambridge, United Kingdom and New York, NY, USA.

Johnson PC, O'Hara RB (2014) Extension of Nakagawa & Schielzeth's R^2 GLMM to random slopes models. *Methods in Ecology and Evolution*, **5**, 944–946.

Jug A, Makeschin F, Rehfuess KE, Hofmann-Schielle C (1999) Short-rotation plantations of balsam poplars, aspen and willows on former arable land in the Federal Republic of Germany. III. Soil ecological effects. *Forest Ecology and Management*, **121**, 85–99.

Kahle P, Baum C, Boelcke B, Kohl J, Ulrich R (2010) Vertical distribution of soil properties under short-rotation forestry in Northern Germany. *Journal of Plant Nutrition & Soil Science*, **173**, 737–746.

Kahle P, Möller J, Baum C, Gurgel A (2013) Tillage-induced changes in the distribution of soil organic matter and soil aggregate stability under former short rotation coppice. *Soil and Tillage Research*, **133**, 49–53.

Keith AM, Rowe RL, Parmar K, Perks MP, Mackie E, Dondini M, McNamara NP (2014) Implications of land-use change to Short-rotation Forestry in Great Britain for soil and biomass carbon. *GCB Bioenergy*, **7**, 541–552.

Kerry R, Rawlins BG, Oliver MA, Lacinska AM (2009) Problems with determining the particle size distribution of chalk soil and some of their implications. *Geoderma*, **152**, 324–337.

Laganière J, Angers DA, Parè D (2010) Carbon accumulation in agricultural soils after afforestation: a meta-analysis. *Global Change Biology*, **16**, 439–453.

Lefcheck J (2014) R-squared for generalized linear mixed-effects models. Available at: https://github.com/jslefche/rsquared.glmm, GitHub (accessed 10 June 2015).

Lockwell J, Guidi W, Labrecque M (2012) Soil carbon sequestration potential of willows in short-rotation coppice established on abandoned farm lands. *Plant and Soil*, **360**, 299–318.

Mello FFC, Cerri CEP, Davies CA et al. (2014) Payback time for soil carbon and sugar-cane ethanol. *Nature Climate Change*, **4**, 605–609.

Mohr A, Raman S (2013) Lessons from first generation biofuels and implications for the sustainability appraisal of second generation biofuels. *Energy policy*, **63**, 114–122.

Nakagawa S, Schielzeth H (2013) A general and simple method for obtaining R^2 from generalized linear mixed-effects models. *Methods in Ecology and Evolution*, **4**, 133–142.

Pinheiro J, Bates D, DebRoy S, Sarkar D, R Development Core Team (2014) nlme: Linear and Nonlinear Mixed Effects Models. R package version 3.1-118.

Poeplau C, Don A (2014) Soil carbon changes under Miscanthus driven by C_4 accumulation and C_3 decomposition – toward a default sequestration rates. *GCB Bioenergy*, **6**, 327–338.

Renewable Energy Road Map (2007) Renewable energies in the 21st century: building a more sustainable future. *Communication from the commission to the council and the European Parliament*. Brussels, 10.1.2007 COM(206) 848 final.

Reynolds B, Chamberlain PM, Poskitt J *et al.* (2013) Countryside survey: national "soil change" 1978–2007 for Topsoils in Great Britain—Acidity, Carbon, and Total Nitrogen Status. *Vadose Zone Journal*, **12**. doi:10.2136/vzj2012.0114.

Richter GM, Agostini F, Redmile-Gordon M, White R, Goulding KWT (2015) Sequestration of C in soils under Miscanthus can be marginal and is affected by genotype-specific root distribution. *Agriculture, Ecosystems and Environment*, **200**, 169–177.

Ross DJ, Tate KR, Newton PCD, Clark H (2002) Decomposability of C_3 and C_4 grass litter sampled under different concentrations of atmospheric carbon dioxide at a natural CO_2 spring. *Plant and Soil*, **240**, 275–286.

Rowe RL, Street NR, Taylor G (2009) Identifying potential environmental impacts of large-scale deployment of dedicated bioenergy crops in the UK. *Renewable and Sustainable Energy Reviews*, **13**, 271–290.

Rowe R, Whitaker J, Freer-Smith PH, Chapman J, Ludley KE, Howard DC, Taylor G (2011) Counting the cost of carbon in bioenergy systems: sources of variation and hidden pitfalls when comparing life cycle assessments. *Biofuels*, **2**, 693–707.

Rytter R-M (2012) The potential of willow and poplar plantations as carbon sinks in Sweden. *Biomass and Bioenergy*, **36**, 86–95.

Schmitt AK, Tischer S, Elste B, Hofmann B, Christen O (2010) Effect of energy forestry on physical, chemical and biological soil properties on a chernozem in continental dry climate conditions in central Germany. *für Kulturpflanzen*, **62**, 189–199.

Schrumpf M, Schulze ED, Kaiser K, Schumacher J (2011) How accurately can soil organic carbon stocks and stock changes be quantified by soil inventories? *Biogeosciences*, **8**, 1193–1212.

Stockmann U, Adams MA, Crawford JW *et al.* (2013) The knowns, known unknowns and unknowns of sequestration of soil organic carbon. *Agriculture, Ecosystems & Environment*, **164**, 80–99.

Walter K, Don A, Flessa H (2014) No general soil carbon sequestration under central European short rotation coppices. *GCB Bioenergy*, **7**, 727–740.

Wellock ML, LaPerle CM, Kiely G (2011) What is the impact of afforestation on the carbon stocks of Irish mineral soils? *Forest Ecology and Management*, **262**, 1589–1596.

Zatta A, Clifton-Brown J, Robson P, Hastings A, Monti A (2014) Land use change from C3 grassland to C4 *Miscanthus*: effects on soil carbon content and estimated mitigation benefit after six years. *GCB Bioenergy*, **6**, 360–370.

Zimmerman J, Dauber J, Jones MB (2012) Soil carbon sequestration during the establishment phase of *Miscanthus x giganteus*: a regional-scale study on commercial farms using 13C natural abundance. *GCB Bioenergy*, **4**, 453–461.

8

Investment risk in bioenergy crops

THEODOROS SKEVAS[1], SCOTT M. SWINTON[2], SOPHIA TANNER[2], GREGG SANFORD[3]
and KURT D. THELEN[4]

[1]*Gulf Coast Research and Education Center, University of Florida, 14625 County Road 672, Wimauma, FL 33598, USA,*
[2]*Department of Agricultural, Food, and Resource Economics, Michigan State University, Justin S. Morrill Hall of Agriculture 446 West Circle Dr., East Lansing, MI 48824-1039, USA,* [3]*Department of Agronomy, University of Wisconsin-Madison, 1575 Linden Dr, Madison, WI 53706, USA,* [4]*Department of Plant, Soil and Microbial Sciences, Michigan State University, 1066 Bogue St A286, East Lansing, MI 48824, USA*

Abstract

Perennial, cellulosic bioenergy crops represent a risky investment. The potential for adoption of these crops depends not only on mean net returns, but also on the associated probability distributions and on the risk preferences of farmers. Using 6-year observed crop yield data from highly productive and marginally productive sites in the southern Great Lakes region and assuming risk neutrality, we calculate expected breakeven biomass yields and prices compared to corn (*Zea mays* L.) as a benchmark. Next we develop Monte Carlo budget simulations based on stochastic crop prices and yields. The crop yield simulations decompose yield risk into three components: crop establishment survival, time to maturity, and mature yield variability. Results reveal that corn with harvest of grain and 38% of stover (as cellulosic bioenergy feedstock) is both the most profitable and the least risky investment option. It dominates all perennial systems considered across a wide range of farmer risk preferences. Although not currently attractive for profit-oriented farmers who are risk neutral or risk averse, perennial bioenergy crops have a higher potential to successfully compete with corn under marginal crop production conditions.

Keywords: bioenergy, cellulosic biomass, energy crops, investment analysis, Monte Carlo simulation, risk, stochastic budgeting

Introduction

Although annual corn is currently the most important bioenergy crop in the United States, perennial crops such as giant miscanthus (*Miscanthus × giganteus* Greef & Deuter ex Hodkinson & Renvoize) and switchgrass (*Panicum virgatum* L.) have shown the potential systematically to produce higher biomass yields (Heaton *et al.*, 2008; Dohleman & Long, 2009). Perennial crops represent long-term investments, due to the initial cost of crop establishment and the delay before harvestable biomass is available. While production costs may be predicted with some confidence, farmers are exposed to potentially large variability in biomass yield and price (Bocquého & Jacquet, 2010). To understand the potential for adoption of bioenergy crops, there is a need to analyze profitability risk associated with investments in the production of perennial bioenergy crops relative to crops that farmers already choose to grow.

A critical factor in adopting new crops, such as bioenergy crops, is their profitability relative to that of

existing cropping systems. Most farmers will allocate land to bioenergy crops only if the economic returns from these crops are at least equal to returns from the most profitable conventional alternatives (Jain *et al.*, 2010; James *et al.*, 2010; Kells & Swinton, 2014). The adoption of new agricultural technologies is also affected by risk (Ghadim *et al.*, 2005; Marra *et al.*, 2003; Chavas, *et al.*, 2009). Farmers' risk attitudes (Just & Zilberman, 1983) and perception about the distribution of future payoffs from the new technology (Marra *et al.*, 2003), potential sunk costs (Chavas *et al.*, 1994), and the opportunity cost of switching to a relatively unknown production system do affect the uptake of emerging agricultural technologies. An extensive literature models the investment uncertainty associated with adopting new agricultural technologies (Price & Wetzstein, 1999; Khanna *et al.*, 2000; Pietola & Myers, 2000; Carey & Zilberman, 2002; Isik & Yang, 2004; Odening *et al.*, 2005; Koundouri *et al.*, 2006; Tozer, 2009; Schoengold & Sunding, 2014; Anderson & Weersink, 2014). Yet scant empirical evidence is available on how investment uncertainty affects the adoption of bioenergy perennials. A notable exception is the study by Song *et al.* (2011) who model land conversion decisions between tradi-

Correspondence: Theodoros Skevas
e-mail: skevast@ufl.edu

tional crops and switchgrass under costly reversibility, and revenue uncertainty. However, these authors rely on secondary data and fail to account explicitly for the effects of crop failure and variable yield trajectories on investment returns from perennial bioenergy crops.

The agronomic and economic characteristics of bioenergy perennials make them risky choices. Investment in perennial energy crops is characterized by high establishment cost (Lewandowski et al., 2003), establishment problems related to extreme climatic and pest events (Thinggaard, 1997; Clifton-Brown & Lewandowski, 2000), foregone income while awaiting mature yield (Song et al., 2011), and considerable removal costs to make land available for a new crop. Moreover, the risk of investing in perennial bioenergy crops is aggravated by the absence of commodity markets or crop insurance for these crops, as well as limited farming experience with them.

Breakeven budgeting addresses profitability risk by establishing a lower bound for price or quantity that is required to cover costs. Various studies have calculated the average profitability of different biomass feedstock crops (e.g., Lewandowski et al., 2003; Heaton et al., 2004). Simple breakeven analysis studies have calculated the yields and prices at which a producer would cover costs of production (Mooney et al., 2009). One step more advanced are comparative breakeven analyses that calculate the yield or price required for a producer to earn profit at least equal to the return on a reference crop (Jain et al., 2010; Landers et al., 2012; DeLaporte et al., 2014; James et al., 2010). These studies rely mostly on secondary data, and they fail to account explicitly for risk. All of these studies ignore crop establishment risk and the temporal distribution of crop yield. Yet the highest biomass yielding bioenergy crop—giant miscanthus —has demonstrated susceptibility to winterkill during its first year (Kucharik et al., 2013), making establishment risk a serious concern. Moreover, risk associated with the time delay for perennial crops like giant miscanthus and switchgrass to reach harvestable yield may be substantial (Heaton et al., 2004). Both of these risk factors supplement conventional year-to-year yield variability of mature crops in ways that could significantly affect their profitability appeal to potential adopters.

Past stochastic simulation studies that have calculated probability distributions of net returns from bioenergy crops have taken two approaches to the crucial step of simulating crop yields. In the absence of adequate data on bioenergy crop yields, one group has relied upon general crop growth simulation models, such as ALMANAC and DayCENT (Dolginow et al., 2014; Miao & Khanna, 2014). These models have the advantage of being able to simulate crop yield over large regions. However, they have typically been validated at just a few individual sites, which may be problematic given

that they lack well-developed parameters for perennial bioenergy crops. One study (Clancy et al., 2012) statistically estimates yields of bioenergy crops across time, using a one-period-lagged, linear and plateau function and using residuals to simulate the probability distribution of random variability around expected yields. The Clancy et al. (2012) study is unique in recognizing the relevance of winter survival risk in giant miscanthus, which they assume to be ten percent. Finally, Bocquého & Jacquet (2010) relied on interview responses and recorded secondary data for short-term empirical distributions of bioenergy crop yields.

Our research draws on new bioenergy crop yield data to construct more nuanced, probabilistic, biomass yield functions for six bioenergy crop systems, linking those functions to stochastic price predictions through a stochastic investment budget model. Specifically, this study makes three contributions to the literature on economic risk of bioenergy crop production. First, it uses new multiyear field data on cellulosic biomass production to inform comparative breakeven analysis of perennial bioenergy crops relative to corn with grain and stover removal. Second, it explicitly considers three stochastic elements when evaluating bioenergy investment projects: (i) crop failure risk, (ii) time to maturity risk, and (iii) variability in mature yields. Third, it evaluates the economic performance of a broad range of bioenergy crops that includes not only corn, giant miscanthus, and switchgrass, but also restored prairie, native grasses, and early successional vegetation (long-term fallow). Using data from southern Michigan and Wisconsin, the modeling approach offers broader insights about the comparative riskiness of these bioenergy crops and what drives that risk.

Materials and methods

Conceptual framework

Rational economic decision-makers are assumed to make crop production choices by choosing crop j to maximize a utility function (U) that includes the value of net returns across a range of possible states of nature (i) in light of the decision-maker's risk preferences:

$$\text{Max}_j U(\text{NPV}_{ij}, \lambda) = \int_{i=1}^{N} U(\text{NPV}_{ij}, \lambda) f(\text{NPV}_{ij}) \text{dNPV}_{ij} \quad (1)$$

where NPV is the net present value of crop j, $j = 1, 2, \ldots, M$, λ is a measure of risk aversion, t is year, and T is the final year of the planning horizon. Location matters as well, but we suppress that factor to simplify notation.

When the model in Eqn (1) is applied to the case of growing bioenergy crops, an individual decision-maker makes crop production choices based on cash flows over the time horizon (T)

for the crop investment. The NPV for cropping system j over a period of T years is defined as follows:

$$\text{NPV}_{ij} = \sum_{t=1}^{T} \delta^t G_{ijt} \qquad (2)$$

where δ is the discount factor, and G_{ijt} denotes the gross margin (cash flow) of crop j cultivated in year t under state of nature i. Eqn (2) provides the discounted value of annual gross margins. Because crop prices and yields are stochastic, each time NPV_{ij} is a random draw representing state i from the probability distribution of possible discounted investment net returns.

The appropriate ranking of biomass investment projects will depend on the investor's risk preference. For a risk neutral decision-maker ($\lambda = 0$), maximizing Eqn (1) is equivalent to maximizing the expected net present value. However, most investors are not indifferent to risk. We adopt an expected utility theory approach to decision-making under risk (Hernstein & Milnor, 1953; Mongin, 1997). Following a substantial body of empirical evidence that farmers are risk averse (Pope & Just, 1991, Pannell et al., 2000; Hardaker, 2006), we assume that the decision-maker exhibits constant absolute risk aversion (CARA; Pratt, 1964) and that risk preference is embodied in the CARA function coefficient aversion, λ, that can vary over a range from risk neutral to highly risk averse.

Crop gross margin risk in the term, G_{ijt}, in Eqn (2) can be decomposed into three yield quantity factors and one price element drive: (i) survival risk, (ii) maturation risk, (iii) yield fluctuation risk in mature crops, and (iv) price risk. Survival risk in

bioenergy perennials refers to mortality losses following the first season after planting. Extreme climatic conditions and pest infestations are common causes. In particular, giant miscanthus rhizomes have failed to survive the winter when soil temperatures fall below −3.5 °C for a period of 3 days or more (Clifton-Brown & Lewandowski, 2000; Kucharik et al., 2013). Figure 1 depicts the effect of establishment failure, and delayed maturity on the NPV of an investment project of perennial biomass crops. Figure 1 (Panel a) illustrates the effect of crop failure risk on the NPV of a biomass investment project. The top graph shows how establishment failure delays the flow of biomass yield (Y), while the bottom graph shows the consequences for NPV. At t_0, crop establishment costs ($-I$) have been incurred and therefore the NPV (bottom of both panels) of a biomass cropping system is negative. In the subsequent period, the NPV continues to decrease due to a lack of harvestable biomass (and, thus absence of revenues), alongside rising crop variable costs, such as fertilization and crop protection. Following this period, as harvestable biomass becomes available, the NPV increases, potentially breaking even. Establishment failure is especially problematic in a crop like giant miscanthus that is costly to plant.

Maturation risk refers to variability in both the time required for a perennial crop to reach a plateau of mature yield and the level of the plateau that is reached. Figure 1 (Panel b) displays the effect of a delay in achieving a full yield potential on firm's returns. The top graph illustrates how random factors may delay the maturation of a perennial crop causing the biomass

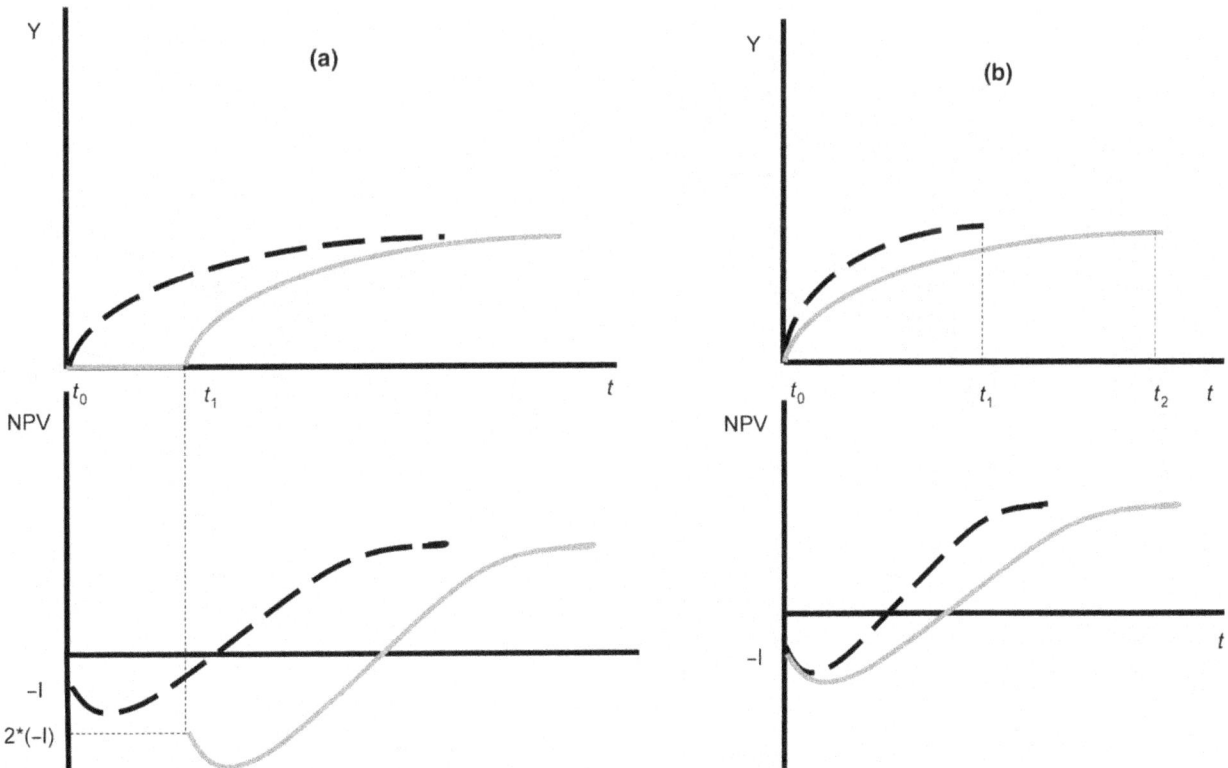

Fig. 1 Establishment failure (left) and delayed maturity (right) implications on the NPV of an investment project of bioenergy perennials.

yield trajectory to shift from the dashed black linen to the solid gray line. This delay shifts the NPV accumulation trajectory in the bottom graph to one that takes longer to break even. Delayed maturity permanently reduces investment return because early revenues have higher present value. Maturation risk can increase both the variance and skewness of the distribution of gross margins.

Finally, as with annual crops, revenue risk is also driven by regular fluctuations in mature yield and in crop prices. Mature yields vary due to factors such as climate (Parry and Carter, 1985; Nuñez & Trujillo-Barrera, 2014), soil type (Dinkins & Jones, 2008), and pests (Skevas et al., 2013). Agricultural prices vary due to changes in markets, which vary spatially from local to global (Harwood et al., 1999). We next present the empirical methods used to analyze how these four sources of risk are likely to affect farmer decisions about adopting bioenergy crops.

Empirical model

To examine the effect of risk on likely farmer adoption choices, we compare results from comparative breakeven budgeting to those from stochastic simulation of investment analysis. Comparative breakeven budgeting is a simple, widely used method that identifies the minimum price or yield needed for revenues to cover costs (Dillon, 1993). The version used here is adapted for investment analysis using the NPV method, so it incorporates discounting of future cash flows to adjust all values to initial year 'present' values (Kells & Swinton, 2014). To calculate a comparative breakeven price, crop yield and opportunity of not adopting the best alternative crop must be known; to calculate a breakeven yield, crop price and opportunity cost must be known. To accommodate policy incentives to encourage adoption of bioenergy crops, we do sensitivity analysis with both direct subsidies and crop insurance.

Stochastic simulation for investment analysis allows developing probability distributions of NPVs that allow comparison of bioenergy investment alternatives over a broad range of yield and price conditions and for decision-makers with different levels of risk aversion. We first describe methods for comparative breakeven analysis, including incentive policy scenarios; then, we move to methods for stochastic simulation and comparison of probability distributions of NPVs.

Risk neutral case: Comparative breakeven investment analysis

Comparative breakeven investment analysis is used to compute the economic performance of cellulosic biomass feedstock investment projects. The six biomass investment alternatives are corn, giant miscanthus, switchgrass, native grasses, restored prairie, and early successional vegetation (fallow). Revenues and expenditures are used to calculate annual cash flows for each cropping system. For convenience in comparing results between annual and perennial crops, we present all results as annualized values using the following annuity formula to convert NPVs to annual equivalents (Weston & Copeland, 1986):

$$A = \left[\frac{r\text{NPV}}{1 - 1/(1+r)^T} \right] \quad (3)$$

where A is the annual payment, and r is the discount rate. The time horizon is 6 years, a time horizon sufficient for most perennial crops to have attained mature yield for 3–4 years and hence for farmers to judge the appeal of adopting them. However, there is evidence that the optimal replacement interval of bioenergy perennials such as miscanthus and switchgrass can exceed 10 years (Pyter et al., 2007). We assume a real discount rate of 5%, following Erickson et al. (2004). Each cropping system has a different production cycle, with corn resulting in harvestable yield each year of the 6-year time horizon, while the perennial cropping systems experience delays of 1–2 years before producing harvestable yield.

The appeal of comparative breakeven budgeting for predicting adoption of new crops is that it builds in the opportunity cost of foregoing new income from the best benchmark crop. Given that corn is the most widely grown field crop in the United States, we treat it as the benchmark crop—the basis for comparison. We conduct the comparative breakeven price and yield analyses to identify the cellulosic biomass prices and yields that would make perennial crops equally profitable with corn. The breakeven price analysis takes into account the direct costs of production, expected yields, and the opportunity cost of replacing the existing cropping system. Net returns from corn are assumed to come from harvesting all grain plus 38% of stover (Brechbill & Tyner, 2008), a level of stover harvest consistent with maintaining soil organic matter. Following Kells & Swinton (2014), the comparative breakeven price of a cellulosic perennial crop to replace corn is as follows:

$$\text{BE}_{\text{pr}} = \frac{\text{NPV}_D + \sum_t \left(\frac{c_t}{(1+r)^t} \right)}{\sum_t \left(\frac{y_{C_t} - y_{D_t}}{(1+r)^t} \right)} \quad (4)$$

where BE_{pr} is the comparative breakeven price, NPV_D is the expected NPV of the 'defender' crop (corn), c_t the expected cost of producing the new biomass crop, y_{C_t} is the expected biomass yield achieved by the 'challenger' bioenergy crop, y_{D_t} is the expected biomass yield of the defender crop, and r and T as previously defined. The denominator represents the biomass yield gain of the challenger crop over the defender cropping system and implies that a new bioenergy crop breaks even in the comparative sense only if its biomass yield exceeds that of corn stover.

Comparative breakeven yield identifies the minimum yield of cellulosic biomass required for a producer to attain annualized investment returns earn equal to corn, given an expected biomass price. Using the same notation as above, the breakeven yield Y_{BE} is computed as follows:

$$Y_{\text{BE}} = \frac{\text{NPV}_D + \sum_t \left(\frac{\text{adc}_t}{(1+r)^t} \right)}{\sum_t \left(\frac{P_t - \text{ydc}_t}{(1+r)^t} \right)} \quad (5)$$

where adc_t is acreage dependent costs (i.e., cost of planting material, agrochemicals, and machinery–labor), P_t is the expected biomass price, and ydc is yield dependent costs (e.g., baling, storage, and bale transportation).

Policy incentives for bioenergy crops: Subsidies and insurance

As variants of the comparative breakeven investment analysis, we consider two sets of policy incentives to encourage adoption of perennial bioenergy crops. The first set already exists in the form of the U.S. Department of Agriculture's Biomass Crop Assistance Program (BCAP) (USDA, 2014). The second policy is based on existing crop revenue insurance that has not so far been extended to perennial bioenergy crops.

Under BCAP, we examine the impact of three BCAP payment forms on the investment returns from the bioenergy crop alternatives. The BCAP payments include the following: (i) establishment payments, (ii) annual rental payments, and (iii) matching payments. Establishment payments cover 50 percent of the costs of establishing dedicated energy crops and the total payments per acre are capped at $500. Annual rental payments include a payment (for a maximum of 5 years) based on typical rental rates for cropland, marginal land or forest land. They are used to cover the foregone income from the land during the establishment phase (before the crop reaches economically harvestable levels). Matching payments of $20 per ton (for a maximum of 2 years) are used to mitigate the cost of harvesting and transporting biomass to a biorefinery. The annual payment is reduced when a matching payment has been earned.

A second potential type of policy would allow growers of bioenergy crops to purchase revenue insurance to offset some of the risk associated with production variability. This study calculates insurance premiums that would support a policy that would pay off whenever the NPV did not reach the zero threshold. Based on the insurance premium approach presented in Goodwin (1994), a premium that is free of distributional assumptions and accounts for the time that net revenues cross the zero threshold can be calculated as follows:

$$\text{Premium}_j = \sum_{i=1}^{n} \theta_{ij}/n \qquad (6)$$

where $\theta_{ij} = 0 - \text{NPV}_{ij}$ if $\text{NPV}_{ij} < 0$, and 0 otherwise. The calculation of insurance premiums can indicate the cost of reducing net revenue risk exposure to potential adopters of bioenergy crops.

Risk averse case: stochastic capital budgeting

The stochastic capital budgeting model introduces the three forms of yield risk plus price risk into simulation of probability distributions of NPVs for each bioenergy crop. It also enables calculation of the monetary value of the certainty equivalent of each NPV distribution for a range of decision-makers with CARA risk preferences. The steps involved in building the stochastic (Monte Carlo) investment analysis model are detailed below. They include (i) statistical estimation of the equations for the three forms of biomass yield risk using appropriate functional forms, (ii) retention of coefficient standard errors to simulate random coefficient models, (iii)

fitting of parameters to appropriate probability distributions for additive random errors, (iv) collection of suitable random price data, (v) synthesis of these components into a stochastic simulation of NPV distributions by crop, and (vi) analysis of results as certainty equivalents for risk neutral and risk averse decision-makers.

Estimation of stochastic biomass yields was performed in three parts: first, estimation of the chance of crop establishment failure at each site (giant miscanthus only); second, estimation of time-to-maturity trajectories for each crop; and third, fitting of probability distributions for additive random errors. Estimations of time-to-maturity risk and risk in mature yields were based on 6 years of field experiments from 2008 to 2013 at Arlington (ARL) in south-central Wisconsin and the Kellogg Biological Station (KBS) in southwest Michigan. At each site, there were five plots each of corn, switchgrass, giant miscanthus, restored prairie, mixed native grasses, and early successional vegetation treatments. At ARL, there was winter kill of giant miscanthus in 2008/2009, and it was not replanted until 2010. In addition, at KBS, switchgrass, native grasses, and restored prairie all experienced crop failure in 2008 due to heavy rains and were replanted in 2009. As a result, these crops have fewer years of data.

Simulation of the probability of winterkill was conducted for giant miscanthus, based on evidence of plant mortality when soil temperatures at a depth of 10 cm fall below −3.5 degrees C. for a duration of three or more days (Kucharik et al., 2013). Soil temperature data from the University of Wisconsin Extension Ag Weather network spanning 20 years (August 1994-June 2014) revealed that 9 of 20 years exceeded that threshold at ARL, for a 45% chance of rhizome winterkill. Soil temperature data from KBS were not available; instead, data from Michigan State University's Enviroweather series collected in East Lansing between January 1996 and December 2014 were used. Because average soil temperature at 10 cm was not available, the 10 cm minimum and maximum temperatures were averaged and 3-day running means were calculated. Two of nineteen years of data (including 1996) saw soil temperatures fell below the −3.5 degree threshold, for a 10.5% probability of winterkill at KBS.

Data from the two sites were used to estimate the trajectory of biomass yield over the first 6 years, using a set of theoretically consistent functional forms. The functions evaluated included Spillman and Mitscherlich, as both increase to a plateau or upper asymptote, as well as linear and step-to-plateau functions. The Mitscherlich function and simpler linear functions performed well for crops that take time to reach mature yields such as switchgrass, giant miscanthus, and native grasses. For the crop yield trajectories that were modeled using the Mitscherlich function, coefficients were estimated using nonlinear least squares. Table 1 shows the functional forms and parameter estimates for yield trajectories of perennial crops at ARL and KBS. The Mitscherlich function has a marginal product that is unrestricted but nonswitching in sign, the linear function has a marginal product that is unrestricted in sign but constant in value. For further background, Griffin et al. (1987) review the properties of these functional forms and their optimality conditions. For these crops that

Table 1 Yield trajectories of perennial crops at ARL and KBS: functional forms and parameter estimates (explanatory variable $t = 0$–5 is years since planting)

Crop (location)	Functional form	Maximum (α)	Slope (β, m)	Intercept (b)	Mean (a)
Switchgrass (ARL)	Mitscherlich $y = \alpha(1-\exp(-\beta t))$	9.0392*** (.7983)	0.4521*** (0.0923)	n/a	n/a
Switchgrass (KBS)	Linear $y = 0$ if $t = 0$ $y = mt + b$ if $t > 0$	n/a	1.6358*** (0.3018)	3.5848*** (0.5647)	n/a
Giant miscanthus (ARL)	Mitscherlich $y = \alpha(1-\exp(-\beta t))$	15.0085*** (2.7872)	0.8912* (0.4503)	n/a	n/a
Giant miscanthus (KBS)	Mitscherlich† $y = \alpha(1-\exp(-\beta t))$	28.8517* (14.2383)	0.2182** (0.1661)	n/a	n/a
Native grasses (ARL)	Linear $y = 0$ if $t = 0$ $y = mt + b$ if $t > 0$	n/a	0.3300* (0.1710)	4.4244*** (0.4189)	n/a
Native grasses (KBS)	Linear $y = 0$ if $t = 0$ $y = mt + b$ if $t > 0$	n/a	0.7876* (0.4311)	3.2506*** (0.8065)	n/a
Early successional (ARL)	Mean value $y = a$	n/a	n/a	n/a	2.9843
Early successional (KBS)	Mean value $y = a$	n/a	n/a	n/a	2.365
Restored prairie (KBS)	Step to mean value $y = 0$ if $t = 0$ $y = a$ if $t > 0$	n/a	n/a	n/a	2.8925
Restored prairie (ARL)	Step to mean value $y = 0$ if $t = 0$ $y = a$ if $t > 0$	n/a	n/a	n/a	4.1296

Note: Numbers in parenthesis are standard errors of parameter estimates. ***Significant at 1% level, **significant at 5% level, *significant at 10% level.
†Davidson–MacKinnon test was inconclusive.

exhibited time-to-maturity risk, that risk was simulated using random slope coefficients, where the coefficients were drawn from normal distributions with mean at the estimated parameter and standard deviation equal to the estimated coefficient standard error. For early successional vegetation and restored prairie, yields showed no trend over time, so mean values suffice. Choices of functional form were based chiefly on theoretical consistency and supported by Davidson–MacKinnon tests (this test is not valid when comparing linear functions vs. mean values because of collinearity). Although linear yield functions fail to exhibit the expected diminishing marginal product over time, these functions were selected as the best fit for the native grasses, and they are acceptable for simulations that are limit to a 6-year time horizon.

In addition to time to maturity risk, yields were assumed to have an additive random error to account for yearly fluctuations on yield. Table 2 presents the probability distributions of random additive annual yield disturbance terms that were drawn from continuous distributions fitted from regression residuals using the @Risk add-in to Microsoft Excel.

To abstract from current market conditions, biomass prices were drawn at random from stochastic simulations of corn and warm season grass prices projected to 2018 that were prepared for the March 2014 outlook report by Food and Agricultural Policy Research Institute at University of Missouri (FAPRI-MO) (Personal communication by Wyatt Thompson to Scott Swinton by email, Dec. 13, 2014).

The stochastic budgeting model was programmed in Microsoft Excel and simulated using @Risk. Latin hypercube

Table 2 Probability distributions of additive random annual crop biomass yield disturbance terms that were drawn using @Risk

Crop	Site	Distribution
Giant miscanthus	ARL	Logistic (−0.1015, 1.8406)
	KBS	Normal (0.0702, 4.8657)
Switchgrass	ARL	Logistic (0.0212, 0.4584)
	KBS	ExtValueMin (0.7540, 1.2934)
Restored prairie	ARL	Weibull (2.8858, 4.4978) −4.0046*
	KBS	ExtValue (−0.5164, 0.9432)
Native Grasses	ARL	ExtValue (−0.5765, 0.9889)
	KBS	ExtValueMin (1.0837, 1.8946)
Early successional	ARL	Weibull (1.8545, 2.4424) −2.1731*
	KBS	ExtValueMin (0.5234, 0.9372)

*Weibull distribution shifted down by value of this constant (RiskShift parameter in @Risk).

sampling with a sample size of 1000 was used to estimate the distribution of the stochastic variables for each risky investment.

The flowchart of the steps performed in implementing the stochastic capital budgeting analysis appears in Fig. 2. The stochastic simulation cycles differed between corn, an annual, and the five perennial bioenergy crops. As shown on the left side of Fig. 2, each 6-year corn simulation cycle begins with drawing six corn grain prices and six biomass prices. For each

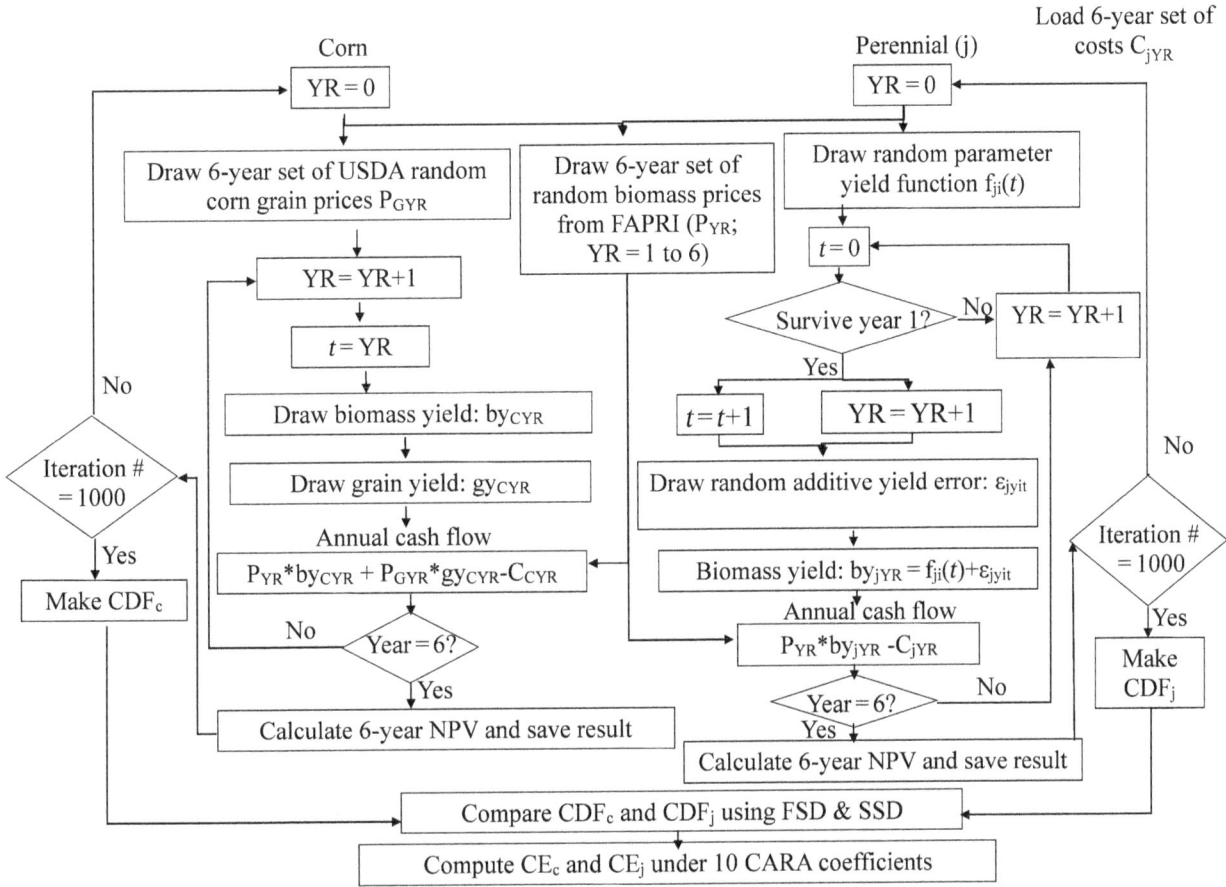

Fig. 2 Flowchart of stochastic simulation of 6-year net present values of investment returns.

year (1–6), the model first draws biomass and grain yields; then with random price and production cost, it calculates annual cash flow. After 6 years, it calculates the NPV for that period. The simulation cycle is repeated 1000 times. The simulation process for perennial bioenergy crops appears on the right side of Fig. 2. Each 6-year simulation cycle begins with drawing random biomass prices and coefficients for the random parameters yield function. If the crop fails in Year 1, it is replanted. If it survives, the biomass yield for that year is computed from the yield function plus an additive random error. Annual cash flow is the product of random biomass price and the computed yield, minus expected production cost. As with the corn model, NPV is calculated after 6 years. Upon completion of the 1000 simulation runs, cumulative distributions are constructed by ordering outcomes from smallest to largest.

Comparison of the alternative bioenergy crop NPV cumulative distributions for decision-makers who may be risk averse is performed using stochastic dominance criteria. These criteria allow ranking of investment prospects by comparing the empirical distributions of investment returns without requiring explicit knowledge of individual risk preferences. Common stochastic dominance criteria are first-degree (FSD) and second-degree stochastic dominance (SSD). FSD requires only the assumption that the decision-maker prefers higher returns to lower returns, and it covers all risk preferences. SSD requires

the added assumption that the decision-maker is risk averse, so it omits risk-preferring individuals. Both approaches involve pairwise comparison of the cumulative distribution functions (CDF) of NPVs from alternative investment options. When FSD and SSD cannot identify preferred alternatives, an approach with more restrictive assumptions but stronger discriminating power is stochastic efficiency with respect to a function (SERF) (Hardaker et al., 2006). Under the assumption that a decision-maker's risk preferences are known (as CARA with assumed coefficients, in this case), certainty equivalent (CE) values can be calculated as the monetary value that would leave the decision-maker indifferent between receiving the CE and the entire CDF from the risky investment. SERF ranks a set of risky alternatives in terms of CEs. Following Pratt (1964), we use the negative exponential constant absolute risk aversion (CARA) utility function: $U_{CARA}(G) = -e^{-\lambda G}$. Using this function, the CE is computed as follows:

$$CE_{CARA}(G, \lambda) = -\ln\left(-\frac{1}{n}\sum_i^n e^{-\lambda G}\right)\Big/\lambda \quad (7)$$

The CE represents the amount of money a decision-maker would require to be indifferent between receiving that amount for certain and receiving a potential result from the risky investment. When using agronomic experimental data, CARA is an appropriate utility function because there is no need to account for heterogeneity in decision-maker wealth levels. Following King & Robison (1981) and Cochran *et al.* (1985), the risk aversion coefficients used in this analysis range from 0 (risk neutral) to 0.001 (highly risk averse).

Data

The analyses reported here draw bioenergy crop management practices and yields from the 6-year period 2008–2013 from the Great Lakes Bioenergy Research Center (GLBRC) Biofuel Cropping System Experiment established at the Kellogg Biological Station (KBS) at Hickory corners, MI, and at the Arlington (ARL) Agricultural Research Station in Arlington, WI (see details at http://data.sustainability.glbrc.org/pages/1.html, and in Sanford *et al.*, 2016). The cropping system treatments discussed here include corn (with stover removal), giant miscanthus, switchgrass (Cave-in-Rock variety), native grasses, restored prairie, and early successional. Yield data and output prices are presented in Table 3. For the breakeven investment analysis, 2018 FAPRI price forecasts for corn are used, while cellulosic feedstock price is assumed to be \$50 mg^{-1}. At \$159 Mg^{-1} (=\$4 $^{-1}$), the simulated mean corn grain price is lower than the

Table 3 Crop yields and prices in the southern Great Lakes area in 2008–2013 (basis for comparative breakeven investment analysis)

Crop	Location	Yield* (Mg ha^{-1}) Mean	SD	Output price† (\$ Mg^{-1})
Corn grain	ARL	12.65	1.61	159
	KBS	9.82	3.19	
Corn stover	ARL	5.88	1.40	50
	KBS	2.62	1.58	
Switchgrass	ARL	4.88	3.21	50
	KBS	4.08	3.46	
Giant miscanthus	ARL	5.93	6.75	50
	KBS	11.02	8.16	
Native grasses	ARL	4.24	2.30	50
	KBS	2.95	2.93	
Early successional	ARL	2.99	1.23	50
	KBS	2.37	1.10	
Restored prairie	ARL	3.44	2.11	50
	KBS	1.89	1.61	

*Yield data are from field trials at the Great Lakes Bioenergy Research Center (GLBRC), intensive research sites at the University of Wisconsin agronomic research station at Arlington (ARL) in south-central Wisconsin and at the Kellogg Biological Station (KBS) in Hickory Corners, Southwest Michigan.
†Corn grain price is the average (FAPRI) 2018 price forecast for corn. The respective corn grain price in \$ bu^{-1} is 4. The biomass price is derived from rounding to the nearest \$5 Mg^{-1} both the average (FAPRI) 2018 price forecast for dry biomass from warm season grass and the Michigan State University T.B. Simon power plant purchases of switchgrass and restored prairie biomass from GLBRC in 2013 (based on coal-equivalent BTU content).

observed price during 2008–2013 (\$196 Mg^{-1} (=\$5 bu^{-1}) (National Agricultural Statistics Service). The \$159 price was chosen for this analysis because 1) the observed price is an historic high that appears not to be indicative of likely future values and 2) using the same price as for the stochastic simulation analysis later in the paper allows direct comparison of results. The cellulosic feedstock price was selected because it is close to the rounded average of the 2018 Food and Agricultural Policy Research Institute (FAPRI) price forecasts for warm season grass (i.e., \$50.79 mg^{-1}) and the Michigan State University T.B. Simon power plant energy biomass purchases (of switchgrass and restored prairie) from GLBRC in 2013 (i.e., \$51.14 Mg^{-1}).

The Simon power plant payments are meaningful, because they are based on the energy equivalent of coal, and thus indicative of what commercial power plants would pay for delivered biomass for co-firing with coal. For the stochastic capital budgeting, 2018 FAPRI price forecasts for corn and warm season grass were used. These prices are calculated from 500 simulated iterations. The average FAPRI price for corn and warm season grass was \$159 Mg^{-1} (i.e., \$4 bu^{-1}) and \$51 Mg^{-1}, respectively. Tables 1 and 2 in the appendix present the costs of the main inputs used in crop production for each cropping system and location. These costs include planting materials, agrochemicals, machinery–labor, and postharvest. Input cost data come from secondary sources, and when there was a lack of cost data for Wisconsin or Michigan, cost data from neighboring states were used. The input cost data used in the current study represent 2008–2013 production conditions in the southern Great Lakes region.

Results

Profitability by cropping system at 2008–2013 prices and yields

The mean profitability of the bioenergy cropping systems at KBS in southwest Michigan and ARL in south-central Wisconsin is presented as annualized NPV in Figs 3 and 4. In both locations, the profitability of corn far exceeded that of any of the perennial crop systems for two primary reasons. First, corn revenues benefit from two components: the valuable grain product plus the less valuable cellulosic biomass product. Second, predicted corn prices at \$4 bushel^{-1} are strong compared to historic levels, despite being below the high levels of 2008–2013. Although agrochemicals are more costly in corn than any of the other cropping systems, revenues offset those costs. By contrast, the high cost of giant miscanthus planting material (rhizomes) is not fully compensated at current prices, despite the high biomass yield of giant miscanthus. Due to better soils at ARL than KBS, all crops except giant miscanthus yielded better at ARL. However, the relative benefit of good soils was greater for corn yield than for the biomass yield of giant miscanthus, switchgrass, and early successional vegetation—indicating that lower

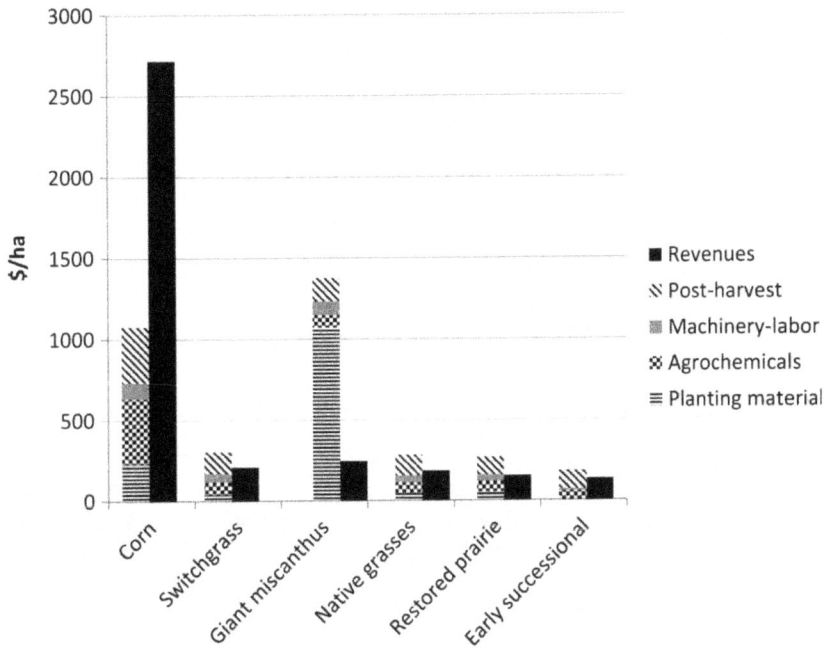

Fig. 3 Revenues and production costs (annualized NPV in $ ha^{-1}) of biomass crops, ARL, WI.

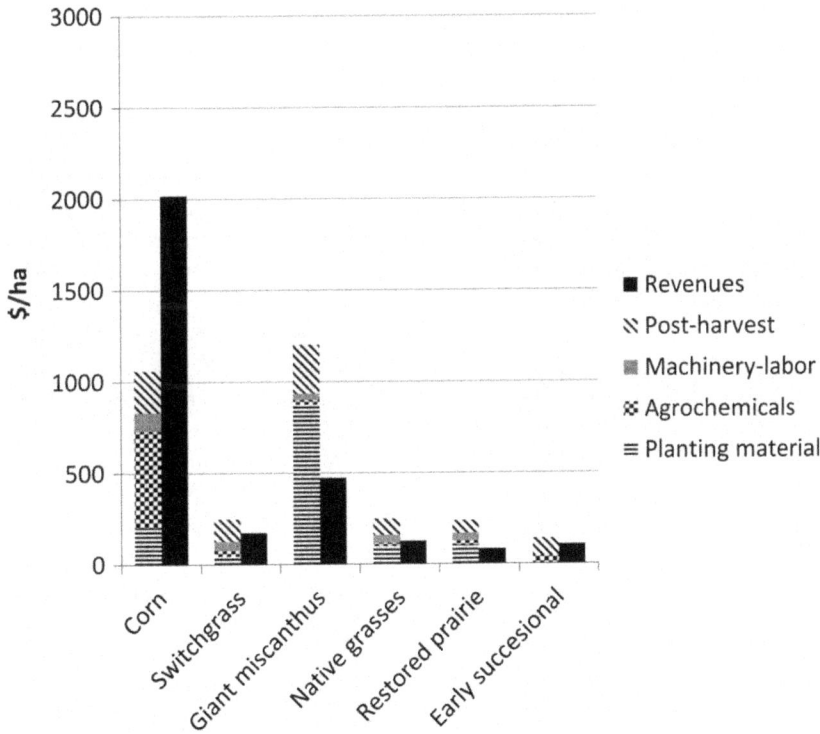

Fig. 4 Revenues and production costs (annualized NPV in $ ha^{-1}) of biomass crops, KBS, MI.

productivity at KBS is less pronounced for bioenergy perennial crops than for corn. The following breakeven analysis examines just how close each site and cropping system comes to matching the profitability of corn.

Comparative breakeven prices

Breakeven prices for cellulosic biomass refer to prices that producers of continuous corn must receive in order to earn equal profit from a cellulosic perennial crop. Table 4 presents comparative breakeven prices for each cropping system assuming a corn grain price of $159 Mg^{-1} ($4 bu^{-1}). The giant miscanthus figures are underestimates, because they ignore the risk of winterkill. Even so, no system can break even at ARL because the mean corn stover yield there exceeds the mean biomass yield of any of the perennial bioenergy crops. At KBS, however, corn stover yields are lower, and three perennial bioenergy crops have the potential to break even at a sufficiently high biomass price. Giant miscanthus, the crop with highest biomass yield, could match the profitability of corn at a biomass price of $203 Mg^{-1}. Switchgrass would require $642 Mg^{-1}, while the native grasses would require the price of a new, small car for each ton of biomass, because their mean yield barely exceeded that of corn stover. Restored prairie and early successional vegetation at KBS produce less biomass than corn stover and so cannot break even at any biomass price.

Comparative breakeven yields

Table 4 also presents comparative breakeven yields for each cropping system at the ARL and KBS sites, assuming a biomass price of $50 Mg^{-1}. Breakeven yield shows the minimum yield required for a producer to earn equal profit to corn. Breakeven yields for all crops are higher at ARL compared to KBS, due to higher yields of the corn system at ARL. The crop with the lowest breakeven yield at ARL is early successional vegetation, which has the lowest costs—just the cost of fertilization and biomass harvest. Next lowest are the native grasses, restored prairie, and switchgrass. At KBS, switchgrass

Table 4 Comparative breakeven prices ($ Mg^{-1}), and yields (Mg ha^{-1}) of biomass feedstocks with respect to a corn grain price of $4.00 bu^{-1} ($159 Mg^{-1}), and biomass price of $50 Mg^{-1} at ARL and KBS sites

Crop	Breakeven prices ($ Mg^{-1})		Breakeven yields (Mg ha^{-1})		Breakeven yield as percent of current yield (%)	
	ARL	KBS	ARL	KBS	ARL	KBS
Switchgrass	N/A*	$642	56	19	1050	362
Giant miscanthus	N/A	$203	104	67	1654	510
Native grasses	N/A	$15 482	52	32	1119	989
Restored prairie	N/A	N/A	55	33	1513	1633
Early successional	N/A	N/A	51	23	1626	882

*N/A denotes that the cropping system cannot break even since it does not produce as much biomass as corn stover.

has the lowest breakeven yield, followed by early successional vegetation, native grasses, and restored prairie. Comparing current yields (Table 3) and breakeven yields (Table 4), at the corn and biomass prices assumed, nearly all of the perennial bioenergy crops would require a tenfold yield boost to break even with corn. However, the magnitude of yield gains needed is much smaller at KBS than at ARL, due to the lower productivity of the corn reference system and the relatively better yields of switchgrass and giant miscanthus at the KBS site.

BCAP and insurance premium results

The USDA Biomass Crop Assistance Program (BCAP) is a current policy designed to enhance the profitability of dedicated bioenergy crops. Figures 5 and 6 compare the profitability of the bioenergy cropping systems at ARL and KBS under no BCAP financial assistance and under four different BCAP scenarios: matching payments for biomass at time of sale, annual rental payments, establishment cost share payments, and all three combined. An important observation is that BCAP payments cannot bridge the profitability gap between corn and bioenergy perennials. However, in the 'all BCAP payments' combined scenario, the profitability of most bioenergy perennials turned from negative to positive. This was the case for all bioenergy perennials except giant miscanthus in both locations. In two cases (early successional vegetation at KBS and switchgrass and early successional vegetation at ARL), individual BCAP payments such as annual rental and matching payments could also reverse the expectation of negative profitability.

Crop revenue insurance offers another potential means to avert negative profitability. Table 5 presents insurance premiums needed to insure against NPV falling below zero. Premiums are very low (i.e., $1–2 ha^{-1}) only in the instances where among the 1000 simulations, the NPV rarely failed to be positive. That occurred only for corn at ARL and corn and switchgrass at KBS. Insurance premiums are higher at ARL for all crops except restored prairie; four bioenergy crops there frequently generated negative annualized NPVs, including giant miscanthus (100% of cases), switchgrass (98%), native grasses (88%), and early successional vegetation (75%). At KBS, only giant miscanthus (98% of cases) and restored prairie (99% of cases) generated negative annualized NPVs most of the time.

Stochastic simulation results

Up to this point, all results have been based on mean values, ignoring production and price risk. Summary statistics from the 1000 stochastic simulations of the six

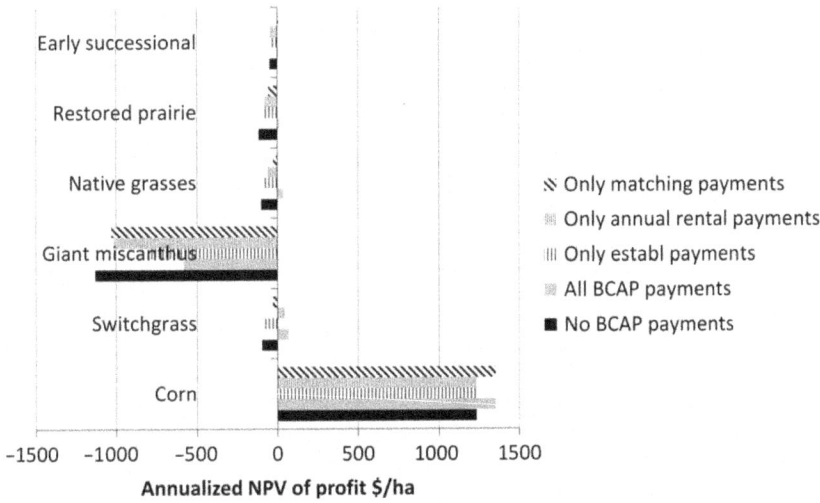

Fig. 5 BCAP scenarios: Annualized NPVs, ARL, WI.

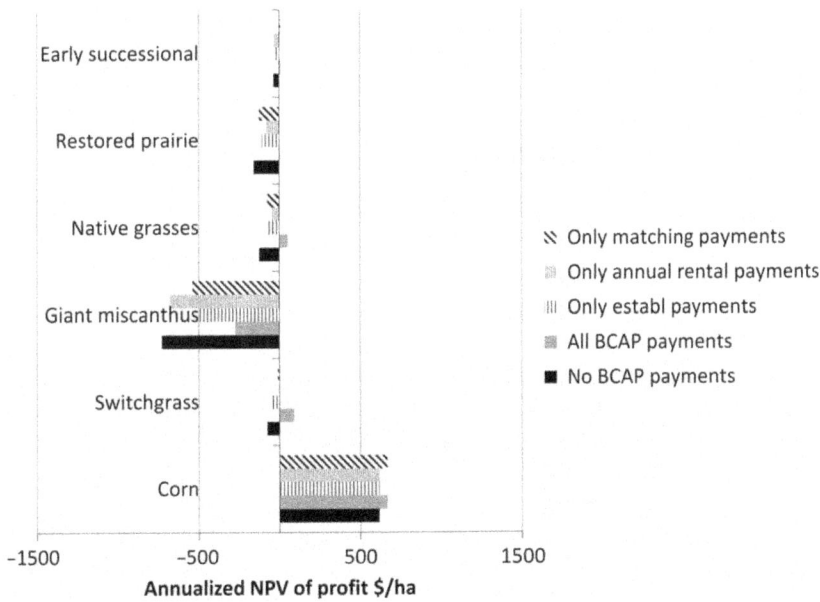

Fig. 6 BCAP scenarios: Annualized NPVs, KBS, MI.

Table 5 Insurance premiums (in $ ha^{-1}) that would support a policy that would pay off whenever the zero threshold (in net returns) is met

	Giant miscanthus	Switchgrass	Native grasses	Restored prairie	Early successional	Corn
ARL	1032	83	45	18	49	2
KBS	626	2	42	136	32	1

bioenergy crops at KBS and ARL are presented in Table 6. Corn stands out as having the highest mean profit, as measured by annualized NPV; it also had the highest maximum at both sites. However, corn presents a high standard deviation, and its minimum values are lower than several perennial bioenergy cropping systems. Giant miscanthus did poorly at both sites because of winter kill. Over the 20-year simulation period, giant miscanthus had a 45% chance of winter kill at ARL and a 10.5% chance at KBS.

First- and second-degree stochastic dominance identified certain systems as relatively efficient in the sense

Table 6 Stochastic annualized NPVs of bioenergy crops at ARL and KBS sites, 1000 simulation iterations (in U.S. dollars)

Crop	ARL (Arlington, WI)					KBS (Hickory Corners, MI)				
	Mean	SD	Median	Min	Max	Mean	SD	Median	Min	Max
Corn	943	439	932	−265	2527	328	168	319	−136	830
Switchgrass	−83	40	−81	−199	34	99	66	98	−113	304
Giant miscanthus	−830	145	−821	−1175	−361	−623	272	−670	−1148	311
Native grasses	−43	39	−40	−153	112	−14	85	−21	−206	276
Restored prairie	94	135	86	−278	503	−136	49	−138	−249	47
Early successional	−39	61	−43	−183	215	−16	55	−23	−130	171

that they were not dominated by any other cropping system at their site. Corn appeared in the efficient set at both sites, joined by native grasses and early successional vegetation at ARL and by switchgrass at KBS. Giant miscanthus was dominated by all other crops under one criterion or the other. At ARL it did so poorly that it lost money even in its best iteration. Consequently, it was strictly dominated under FSD by all of the other crop systems at ARL. At KBS, giant miscanthus was dominated under FSD by switchgrass, and corn and under SSD by restored prairie, native grasses, and early successional. The restored prairie treatment also fared poorly, being dominated at KBS under FSD by switchgrass, native grasses, early successional vegetation and corn, as well as dominated at ARL under SSD by corn. The remaining perennial bioenergy crops differed in their stochastic dominance results between the two sites. Although switchgrass was in the efficient set at KBS, at ARL it was dominated under FSD by native grasses and early successional. The early successional vegetation and native grass treatments that were in the efficient set at ARL were dominated at KBS under FSD by switchgrass (the FSD and SSD results are not reported in full detail in this paper, but can be provided by the authors upon request).

Although corn was accompanied in the FSD and SSD risk efficient sets by switchgrass at KBS and by native grasses and early successional vegetation at ARL, corn was the more profitable system under all but the very worst outcomes simulated. At ARL, corn was more profitable than native grasses and early successional vegetation in over 99.5% of the outcomes. Likewise at KBS, corn was more profitable than switchgrass 95% of the time. Only when the higher cost corn crop failed repeatedly, did it fail to come out ahead of its closest competitors.

Because more than one cropping system remained in the risk efficient sets at each site under FSD and SSD, SERF was used to rank the full set of bioenergy investment projects at each site. Certainty equivalent (CE) values for corn and perennial crops are presented for the range of CARA levels from 0 (risk neutral) to 0.001

(highly risk averse) in Figs 7 and 8. At CARA=0, the CEs equal the mean expected annualized NPV. The CEs decline as risk aversion increases (i.e., as CARA values become larger). In both locations, the locus of CE values for corn is higher everywhere than that for all bioenergy perennials, indicating that producers who are both risk neutral and risk averse over a very wide range of risk aversion would prefer corn to bioenergy perennials. The next best alternative investment is restored prairie in ARL or switchgrass in KBS, but the differences between perennial crops (except giant miscanthus) are very small.

On comparing the capital budgeting (i.e., risk neutral case) and the stochastic budgeting (i.e., risky case) results, we see both similarities and differences in the ranking of risky bioenergy investment projects. Corn is the preferred crop in both the risk neutral and the risky cases and at both locations. The difference between corn and bioenergy perennials is consistently higher at ARL than at KBS, which is attributable to more fertile soils in the former that result in higher corn yields at ARL. The most prominent difference when comparing the results of the risk neutral and the risky case is the change in the ranking of bioenergy perennials (e.g., early successional vegetation ranks second in the risk neutral case in ARL, but when it comes to the risky case, it takes the third place). Small differences in the profitability of most bioenergy perennials (except giant miscanthus) and the fact that stochastic simulation covers a wide range of states of nature may explain ordering changes when moving from the risk neutral to the risky case.

Discussion

This paper supplements standard capital budgeting and comparative breakeven analysis with stochastic simulation to assess the competiveness of bioenergy perennials relative to corn with grain and stover removal. Using data from 2008 to 2013 from two sites in the Great Lakes Region at Arlington, WI, and Kellogg Biological Station (KBS) at Hickory Corners, MI, we simulate four stochastic variables that affect investment returns to bioenergy

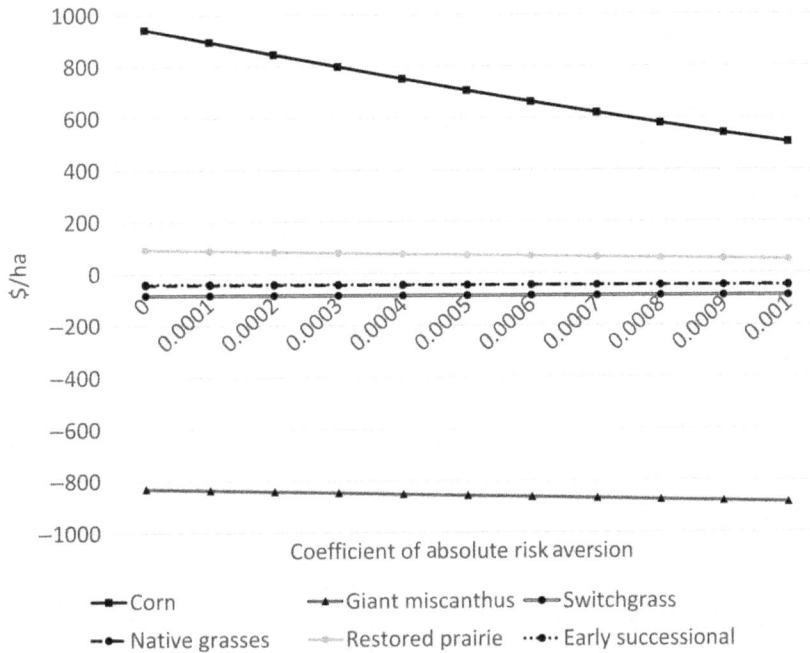

Fig. 7 Certainty equivalents for decision-makers who are risk neutral to highly risk averse with constant absolute risk aversion: stochastic efficiency with respect to a function (SERF) comparison of results from 1000 stochastic simulations of annualized net returns from bioenergy investment projects at Arlington, WI.

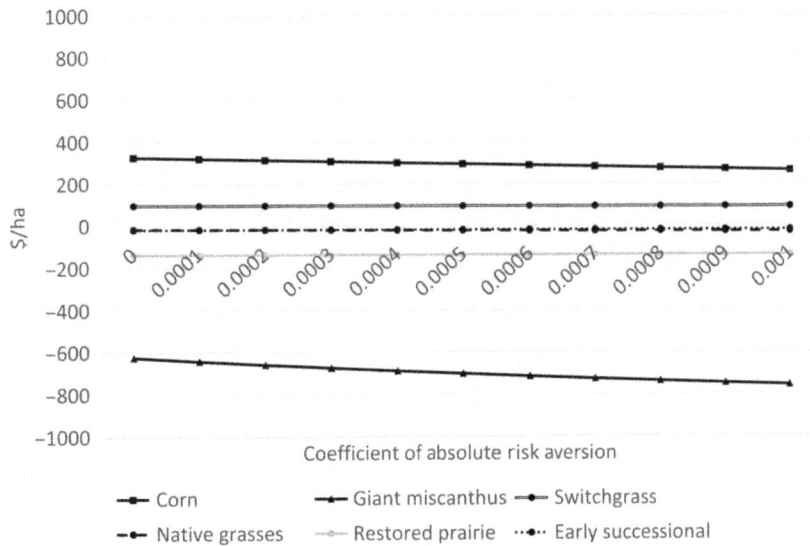

Fig. 8 Certainty equivalents for decision-makers who are risk neutral to highly risk averse with constant absolute risk aversion: stochastic efficiency with respect to a function (SERF) comparison of results from 1000 stochastic simulations of annualized net returns from bioenergy investment projects at Kellogg Biological Station (KBS), MI.

crops: crop failure risk, time to maturity risk, variability in mature yields, and price risk.

The standard, average capital budgeting analyses show that the profitability of corn dominates all other cropping systems at both sites. Corn's dominance comes from 1) providing income from both grain and cellulosic

biomass, and 2) its consistently strong yields as an annual crop (unlike the slow buildup of the perennial crops). Although BCAP payments can reduce profitability losses from adopting perennial bioenergy crops, they are not sufficient to bridge the profitability gap with corn. Future research could seek to assess how much the gap could be narrowed using policies that provide farmers with payments for ecosystem benefits related to perennials.

The comparative breakeven price analysis shows that corn stover yields are so high at ARL that there is no biomass price at which perennial bioenergy crops can match the profitability of corn. Meanwhile, on the poorer soils of KBS, switchgrass, giant miscanthus, and native grasses require very high prices to break even when the price of corn is $4 per bushel. Of the KBS bioenergy crops, giant miscanthus has the lowest break-even price. This result is in line with previous literature that computes breakeven prices for switchgrass and miscanthus in Ontario (DeLaporte et al., 2014), the Midwestern United States (Jain et al., 2010), southern Michigan (James et al., 2010), and Illinois (Khanna et al., 2008). However, the breakeven prices for giant miscant-hus and switchgrass in the current study are higher than those observed elsewhere. The differences may be due to the use of secondary yield and production cost data in those studies vs. primary data in this study. The comparative breakeven yield results find that perennial bioenergy crops require tenfold yield gains at ARL to generate net revenue equal to corn at the prices assumed, with levels at KBS also very high at three- to fivefold gains needed for switchgrass and giant miscant-hus. All values are higher than the significant increases needed that were predicted by James et al. (2010) for southern Michigan. Both the comparative breakeven price and yield analyses demonstrate that although most perennial bioenergy crops are far from achieving average profitability comparable to corn at either site, the potential for bioenergy crops eventually to compete with corn is greater at KBS, where corn productivity is lower.

Results of the investment risk analysis were largely similar. Stochastic efficiency analysis of the investment returns shows annual corn to be an even more resilient benchmark than prior profitability studies that ignored risks of establishment failure and time to maturity of perennial bioenergy crops. Corn was the only crop in the risk efficient set under FSD and SSD at both sites. Under the SERF analysis, corn dominated all other systems over the entire range of risk aversion levels simulated at both locations. No other system came close at ARL in Wisconsin. At KBS in Michigan, switchgrass came second—within competitive range at the $50 Mg^{-1} biomass price if corn grain prices were to fall to by more than half to the $2 per bushel levels of the 1990s and early 2000s. Among the bioenergy perennials, only switchgrass generated positive profits at KBS most of the time (94%). In ARL, apart from corn, only restored prairie generated positive net returns most of the time (73% of cases).

Although earlier studies found that giant miscanthus performs better than other bioenergy perennials (Clancy et al., 2012; Dolginow et al., 2014), we find that in the U.S. Great Lakes region, it has an extremely high proba-bility of generating negative investment returns. Our more negative results were driven by high current rhizome costs and the high probability of winter kill in the establishment year in ARL (45%) and lower but still notable probability of winter kill at KBS (10.5%).

In the absence of changes in agronomic technology or market prices, the pattern of low investment returns from perennial bioenergy crops implies a need for large subsidies to make perennial bioenergy crops equally attractive with corn, with mean differences ranging from $75–385 per acre at KBS to $343–717 per acre at ARL. The bioenergy crops with the lowest subsidy requirements were switchgrass at KBS and restored prairie at ARL. One factor mitigating the cost of poten-tial subsidies required is that the variance of investment returns for bioenergy perennials is lower than for corn (except for giant miscanthus in ARL). Another measure that can increase the attractiveness of bioenergy peren-nials is BCAP payments. Although these payments cannot make bioenergy perennials equally attractive to corn, they can reduce expected losses and (except for giant miscanthus) the probability of a negative invest-ment return.

Overall, the results indicate that these perennial bioenergy crops are currently both less profitable and riskier than corn for farmers in the Great Lakes region. However, the lower corn yields on poorer soils at KBS reduce the revenue gap between corn and most bioen-ergy perennials, compared to the gap at ARL, where soils are highly productive. Like Miao & Khanna (2014), we find that while bioenergy crops remain significantly poorer investments than corn, their lower opportunity cost under more marginal crop production conditions indicates the potential for regional comparative advan-tages at more marginally productive sites if relative prices, technological change, or policy advantages were to favor perennial bioenergy crops. The ranking of biomass investment projects presented here offers infor-mation on the comparative riskiness of bioenergy investment projects in the southern Great Lakes region. Future research could apply this modeling approach to assess the comparative riskiness of bioenergy crops in other regions, where climatic and soil conditions may have different effects on crop establishment risk and the temporal distribution of crop yields.

Acknowledgements

This work was funded by the DOE Great Lakes Bioenergy Research Center (DOE BER Office of Science DE-FC02-

07ER64494). Support for this research was also provided by the NSF Long-term Ecological Research Program (DEB 1027253) at the Kellogg Biological Station and by Michigan State University AgBioResearch. The authors thank J.D. Wesley for meticulous compilation of cost and price data; Joe Simmons for recording field activity data at KBS; Wyatt Thompson and Patrick Westhoff for sharing the 2014 stochastic price simulation data for selected bioenergy crops from the Food and Agricultural Policy Research Institute (FAPRI-MO); and Jeff Andresen for sharing winter soil temperature time series from East Lansing and Hickory Corners, MI. The majority of this research was conducted, while T. Skevas was a research associate with the GLBRC at Michigan State University.

References

Anderson RC, Weersink A (2014) A real options approach for the investment decisions of a farm-based anaerobic digester. *Canadian Journal of Agricultural Economics/Revue canadienne d'agroeconomie*, **62**, 69–87.

Bocquého G, Jacquet F (2010) The adoption of switchgrass and giant miscanthus by farmers: impact of liquidity constraints and risk preferences. *Energy Policy*, **38**, 2598–2607.

Brechbill SC, Tyner WE (2008) *The Economics of Biomass Collection, Transportation, and Supply to Indiana Cellulosic and Electric Utility Facilities.* Department of Agricultural Economics, Purdue University, West Lafayette, IN, Working Paper, (08–03).

Carey JM, Zilberman D (2002) A model of investment under uncertainty: modern irrigation technology and emerging markets in water. *American Journal of Agricultural Economics*, **84**, 171–183.

Chavas JP (1994) Production and investment decisions under sunk cost and temporal uncertainty. *American Journal of Agricultural Economics*, **76**, 114–127.

Chavas JP, Posner JL, Hedtcke JL (2009) Organic and conventional production systems in the Wisconsin Integrated Cropping Systems Trial: II. Economic and risk analysis 1993–2006. *Agronomy Journal*, **101**, 288–295.

Clancy D, Breen JP, Thorne F, Wallace M (2012) A stochastic analysis of the decision to produce biomass crops in Ireland. *Biomass and Bioenergy*, **46**, 353–365.

Clifton-Brown JC, Lewandowski I (2000) Overwintering problems of newly established Giant miscanthus plantations can be overcome by identifying genotypes with improved rhizome cold tolerance. *New Phytologist*, **148**, 287–294.

Cochran MJ, Robison LJ, Lodwick W (1985) Improving the efficiency of stochastic dominance techniques using convex set stochastic dominance. *American Journal of Agricultural Economics*, **67**, 289–295.

DeLaporte AV, Weersink AJ, McKenney DW (2014) A spatial model of climate change effects on yields and break-even prices of switchgrass and miscanthus in Ontario, Canada. *GCB Bioenergy*, **6**, 390–400.

Dillon CR (1993) Advanced breakeven analysis of agricultural enterprise budgets. *Agricultural Economics*, **9**, 127–143.

Dinkins CP, Jones C (2008) Soil Sampling Strategies. Department of Land Resources and Environmental Sciences, Montana State University. Available at: http://store.msuextension.org/publications/agandnaturalresources/mt200803AG.pdf (accessed 14 December 2014).

Dohleman FG, Long SP (2009) More productive than maize in the Midwest: how does Giant miscanthus do it? *Plant Physiology*, **150**, 2104–2115.

Dolginow J, Massey RE, Kitchen NR, Myers DB, Sudduth KA (2014) A stochastic approach for predicting the profitability of bioenergy grasses. *Agronomy Journal*, **106**, 2137–2145.

Erickson KW, Moss CB, Mishra AK (2004) Rates of return in the farm and nonfarm sectors: how do they compare? *Journal of Agricultural and Applied Economics*, **36**, 789–795.

Ghadim AKA, Pannell DJ, Burton MP (2005) Risk, uncertainty, and learning in adoption of a crop innovation. *Agricultural Economics*, **33**, 1–9.

Goodwin BK (1994) Premium rate determination in the Federal crop insurance program: what do averages have to say about risk? *Journal of Agricultural and Resource Economics*, **19**, 382–395.

Griffin RC, Montgomery JM, Rister ME (1987) Selecting functional form in production function analysis. *Western Journal of Agricultural Economics*, **12**, 216–227.

Hardaker BJ (2006) Farm risk management: past, present and prospects. *Journal of Farm Management*, **12**, 593–612.

Hardaker JB, Richardson JW, Lien G, Schumann KD (2004) Stochastic efficiency analysis with risk aversion bounds: a simplified approach. *Australian Journal of Agricultural and Resource Economics*, **48**, 253–270.

Harwood J, Heifner R, Coble K, Perry J, Somwaru A (1999).*Managing Risk in Farming: Concepts, Research and Analysis.* Agricultural Economic Report AER 774. Economic Research Service, U.S. Department of Agriculture. Washington, DC.

Heaton EA, Long SP, Voigt TB, Jones MB, Clifton-Brown J (2004) Giant miscanthus for renewable energy generation: European Union experience and projections for Illinois. *Mitigation and Adaptation Strategies for Global Change*, **9**, 433–451.

Heaton EA, Dohleman FG, Long SP (2008) Meeting US biofuel goals with less land: the potential of Giant miscanthus. *Global Change Biology*, **14**, 2000–2014.

Hernstein IN, Milnor J (1953) An axiomatic approach to expected utility. *Econometrica*, **21**, 291–297.

Isik M, Yang W (2004) An analysis of the effects of uncertainty and irreversibility on farmer participation in the conservation reserve program. *Journal of Agricultural and Resource Economics*, **29**, 242–259.

Jain AK, Khanna M, Erickson M, Huang H (2010) An integrated biogeochemical and economic analysis of bioenergy crops in the Midwestern United States. *GCB Bioenergy*, **2**, 217–234.

James LK, Swinton SM, Thelen KD (2010) Profitability analysis of cellulosic energy crops compared with corn. *Agronomy Journal*, **102**, 675–687.

Just RE, Zilberman D (1983) Stochastic structure, farm size and technology adoption in developing agriculture. *Oxford Economic Papers*, 307–328.

Kells BJ, Swinton SM (2014) Profitability of cellulosic biomass production in the northern Great Lakes Region. *Agronomy Journal*, **106**, 397–406.

Khanna M, Dhungana B, Clifton-Brown J (2008) Costs of producing miscanthus and switchgrass for bioenergy in Illinois. *Biomass and Bioenergy*, **32**, 482–493.

Khanna M, Isik M, Winter-Nelson A (2000) Investment in site-specific crop management under uncertainty: implications for nitrogen pollution control and environmental policy. *Agricultural Economics*, **24**, 9–12.

King RP, Robison LJ (1981) An interval approach to measuring decision maker preferences. *American Journal of Agricultural Economics*, **63**, 510–520.

Koundouri P, Nauges C, Tzouvelekas V (2006) Technology adoption under production uncertainty: theory and application to irrigation technology. *American Journal of Agricultural Economics*, **88**, 657–670.

Kucharik CJ, VanLoocke A, Lenters JD, Motew MM (2013) Miscanthus establishment and overwintering in the Midwest USA: a regional modeling study of crop residue management on critical minimum soil temperatures. *PLoS ONE*, **8**, e68847.

Landers GW, Thompson AL, Kitchen NR, Massey RE (2012) Comparative breakeven analysis of annual grain and perennial switchgrass cropping systems on claypan soil landscapes. *Agronomy Journal*, **104**, 639–648.

Lewandowski I, Scurlock JMO, Lindvall E, Christou M (2003) The development and current status of potential rhizomatous grasses as energy crops in the U.S and Europe. *Biomass and Bioenergy*, **25**, 335–361.

Marra M, Pannell DJ, Ghadim AA (2003) The economics of risk, uncertainty and learning in the adoption of new agricultural technologies: where are we on the learning curve? *Agricultural Systems*, **75**, 215–234.

Miao R, Khanna M (2014) Are bioenergy crops riskier than corn? implications for biomass price. *Choices*, **29**, 1–6.

Mongin P (1997) Expected utility theory. *Handbook of Economic Methodology*, (eds Davis J, Hands W, Maki U), pp. 342–350. Edward Elgar, London.

Mooney DF, Roberts RK, English BC, Tyler DD, Larson JA, Brown BA (2009) Yield and breakeven price of 'Alamo' switchgrass for biofuels in Tennessee. *Agronomy Journal*, **101**, 1234–1242.

Nuñez HM, Trujillo-Barrera A (2014) Impact of US biofuel policy in the presence of uncertain climate conditions. *Biofuels*, **5**, 517–532.

Odening M, Mußhoff O, Balmann A (2005) Investment decisions in hog finishing: an application of the real options approach. *Agricultural Economics*, **32**, 47–60.

Pannell DJ, Malcolm B, Kingwell RS (2000) Are we risking too much? Perspectives on risk in farm modeling. *Agricultural Economics*, **23**, 69–78.

Parry ML, Carter TR (1985) The effect of climatic variations on agricultural risk. *Climatic Change*, **7**, 95–110.

Pietola KS, Myers RJ (2000) Investment under uncertainty and dynamic adjustment in the Finnish pork industry. *American Journal of Agricultural Economics*, **82**, 956–967.

Pope RD, Just RE (1991) On testing the structure of risk preferences in agricultural supply analysis. *American Journal of Agricultural Economics*, **73**, 743–748.

Pratt JW (1964) Risk aversion in the small and in the large. *Econometrica*, **32**, 122–136.

Price TJ, Wetzstein ME (1999) Irreversible investment decisions in perennial crops with yield and price uncertainty. *Journal of Agricultural and Resource Economics*, **24**, 173–185.

Pyter R, Voight T, Heaton E, Dohleman F, Long S (2007). *Growing giant miscanthus in Illinois. Univ. of Illinois, Urbana-Champaign, IL.Growing giant miscanthus in Illinois.* Univ. of Illinois, Urbana-Champaign, IL.

Sanford GR, Oates LG, Jasrotia P, Thelen KD, Robertson GP, Jackson RD (2016) Comparative productivity of alternative cellulosic bioenergy cropping systems in the North Central USA. *Agriculture Ecosystems and Environment*, **216**, 344–355.

Schoengold K, Sunding DL (2014) The impact of water price uncertainty on the adoption of precision irrigation systems. *Agricultural Economics*, **45**, 729–743.

Skevas T, Oude Lansink AGJM, Stefanou SE (2013) Designing the emerging EU pesticide policy: a literature review. *NJAS-Wageningen Journal of Life Sciences*, **64**, 95–103.

Song F, Zhao J, Swinton SM (2011) Switching to perennial energy crops under uncertainty and costly reversibility. *American Journal of Agricultural Economics*, **93**, 768–783.

Thinggaard K (1997) Study of the role of *Fusarium* in the field establishment problem of giant *Miscanthus. Acta Agriculturae Scandinavica, Section B — Soil & Plant Science*, **47**, 238–241.

Tozer PR (2009) Uncertainty and investment in precision agriculture–Is it worth the money? *Agricultural Systems*, **100**, 80–87.

USDA (2014) United States Department of Agriculture. Biomass crop assistance program. Available at: http://www.fsa.usda.gov/FSA/webapp?area=home&subject=ener&topic=bcap (accessed 14 December 2014).

Weston JF, Copeland TE (1986) *Managerial Finance*, 8th edn. Dryden, New York, NY.

9

Food and bioenergy: reviewing the potential of dual-purpose wheat crops

TOBY J. TOWNSEND[1], DEBBIE L. SPARKES[2] and PAUL WILSON[1]

[1]*Division of Agricultural and Environmental Sciences, School of Biosciences, University of Nottingham, Sutton Bonington Campus, Loughborough LE12 5RD, UK*, [2]*Division of Plant and Crop Sciences, School of Biosciences, University of Nottingham, Sutton Bonington Campus, Loughborough LE12 5RD, UK*

Abstract

Within the bioenergy debate, the 'food vs. fuel' controversy quickly replaced enthusiasm for biofuels derived from first-generation feedstocks. Second-generation biofuels offer an opportunity to produce fuels from dedicated energy crops, waste materials or coproducts such as cereal straw. Wheat represents one of the most widely grown arable crops around the world, with wheat straw, a potential source of biofuel feedstock. Wheat straw currently has limited economic value; hence, wheat cultivars have been bred for increased grain yield; however, with the development of second-generation biofuel production, utilization of straw biomass provides the potential for 'food and fuel'. Reviewing the evidence for the development of dual-purpose wheat cultivars optimized for food grain and straw biomass production, we present a holistic assessment of a potential ideotype for a dual-purpose cultivar (DPC). An ideal DPC would be characterized by high grain and straw yields, high straw digestibility (i.e. biofuel yield potential) and good lodging resistance. Considerable variation in these traits exists among current wheat cultivars, facilitating the selection of improved individual traits; however, increasing straw yield and digestibility could potentially have negative trade-off impacts on grain yield and lodging resistance, reducing the feasibility of a single ideotype. Adoption of alternative management practices could potentially increase straw yield and digestibility, albeit these practices are also associated with potential trade-offs among cultivar traits. Benefits from using DPCs include reduced logistics costs along the biofuel feedstock supply chain, but practical barriers to differential pricing for straw digestibility traits are likely to reduce the financial incentive to farmers for growing higher 'biofuel-quality' straw cultivars. Further research is required to explore the relationships among the ideotype traits to quantify potential DPC benefits; this will help to determine whether stakeholders along the bioenergy feedstock supply chain will invest in the development of DPCs that provide food and fuel potential.

Keywords: crop residues, dual-purpose cultivars, lodging, second-generation biofuel, straw digestibility, straw yield, wheat

Introduction

Biofuels, which are liquid or gaseous fuels produced from plant biomass, are being produced for use in the transport sector with the purpose of reducing greenhouse gas (GHG) emissions and increasing energy security (Valentine *et al.*, 2012; Khanna & Chen, 2013). Currently, the majority of biofuel produced is first-generation biofuel (FGB), in particular bioethanol, which is produced from edible plant material such as wheat grain (Simbolotti, 2013). The use of FGB has been criticized due to potential competition with food production (Oladosu & Msangi, 2013) and loss of natural ecosystems through indirect land-use change (Kim & Dale,

2011). These issues have helped drive the development of second-generation biofuel (SGB), which is produced from lignocellulosic biomass such as crop and forestry residues, waste paper and dedicated energy crops (DECs), and is, therefore, considered to have minimal impact on food production (Gnansounou, 2010). European Union legislation has set a mandatory minimum target of a 10% share of energy from renewable sources in transport fuels by 2020, with the majority of this expected to be from biofuels (EU, 2009a).

The biggest potential source of feedstock for SGB production in Europe is straw from wheat (*Triticum aestivum* L.; Scarlat *et al.*, 2010). The area of wheat grown and the potential supply of straw vary with country. In the UK, approximately 2 million hectares of wheat is grown (Anon, 2014a). The straw from wheat production has multiple applications including use as animal

Correspondence: Toby J. Townsend
e-mail: toby.townsend@nottingham.ac.uk

bedding, mushroom production substrate and feedstock for biomass-burning power stations (Copeland & Turley, 2008). However, supply greatly exceeds demand and a large amount of straw is chopped and incorporated into the soil after grain harvest (Glithero et al., 2013a). Some of this incorporated straw could be baled and used as feedstock for bioenergy production without leading to competition with other straw users.

When left in situ, the straw provides benefits such as improved soil structure and water retention, reduced soil erosion, maintained or increased soil organic matter, carbon sequestration and nutrient return (Smil, 1999; Blanco-Canqui & Lal, 2009; Huggins et al., 2011). The impacts of removing residues on soil properties are, however, highly variable and site specific. Consequently, very few consistent effects have been found; however, Blanco-Canqui & Lal (2009) conclude that it is feasible to remove some residues, but this will depend on the specific location and cropping system. Therefore, individual sites need to be assessed to determine whether straw can be sustainably removed (Powlson et al., 2011). Resources are available to estimate the quantities that can be removed without excessive soil erosion (e.g. models such as RUSLE2, WEQ or SCI; see Andrews, 2006), but these may not be appropriate for individual farmers making decisions about residue management and also do not estimate soil impacts other than erosion. It may be more appropriate to use general guidelines such as that by Lafond et al. (2009) who suggested that, for the region there were assessing, it was sustainable to remove the straw provided that it did not happen more than two of every 3 years. In estimating straw availability, Scarlat et al. (2010) found that the literature estimates for sustainable removal ranged from 15% to 60%. In Europe, there are very few resources available for determining how much straw can be sustainably removed; this may be due to soil erosion historically being considered less of a problem in Europe compared to countries such as the USA where guidelines are available. However, the sustainability of straw removal is of direct importance for ethical and environmental acceptability of SGBs; for example, it is important to know whether the fossil fuel GHG emissions displaced through the use of biofuels are outweighed by the reduction in soil carbon sequestration and increase in fertilizer requirements resulting from straw removal.

There is uncertainty regarding the amount of straw chopped and incorporated, and, taken together with the uncertainty regarding the amount of straw that can be sustainably harvested, it is unclear how much straw is available for bioenergy production. In the UK, for example, estimates of straw availability vary widely (c.f. Copeland & Turley, 2008 vs. Glithero et al., 2013a) due to uncertainty about the amount of straw currently used

and straw yield per se, which are often calculated from average straw-to-grain ratios that might not reflect the actual yield relationship during the year the estimate is made. Straw availability may also be overestimated as calculations often assume that all farmers who can sustainably supply straw will supply that straw, whereas, in reality, many farmers are unwilling to do so because of, for example, concerns about negative soil impacts and potential delays in planting subsequent crops (Glithero et al., 2013b).

The amount of straw required for economically sustainable biofuel production at the individual processing plant level is currently unknown. The cost of SGB production depends on the trade-off between economy of scale of biorefinery size and biomass transportation costs (Aden et al., 2002). This suggests that the feasibility of biofuel production will depend on the availability of large amounts of easily accessible biomass. In modelling bioethanol production, Littlewood et al. (2013) considered a biorefinery feedstock demand of approximately 750 000 t yr^{-1}. In contrast, the world's first commercial SGB biorefinery, Beta Renewables' Crescentino biorefinery in Italy, has a maximum feedstock demand of 270 000 t yr^{-1} (comprising a mixture of rice straw, wheat straw and Arundo donax, the common giant reed; Anon, 2013), which suggests feasibility of smaller scale production. Running a biorefinery below capacity is likely to be economically unfeasible, and therefore, estimates of straw availability should be based on feedstock availability from unfavourable years, when straw yields are low, to ensure resilience of the biorefinery (Scarlat et al., 2010). In the UK, competition for straw supply is also increasing with a number of straw-burning power stations planned (e.g. the Sleaford and Brigg plants being constructed by Eco2 UK). Already, this expansion of straw use is causing problems with Eco2 UK failing to gain planning permission for their proposed straw-burning power station in Mendlesham, Suffolk, partly due to concerns about competition with other wheat straw users (Simkins, 2014). This highlights that although straw is available, its high transportation costs, due to its low bulk density, restricts supply to local areas (Valentine et al., 2012) and means that excessive demand in a particular region leads to local competition. As with the Crescentino plant, it is likely that a diverse mix of lignocellulosic material will be required for commercial success so as to not place too much pressure or reliance on a single feedstock resource.

The above issues strongly suggest that for a biorefinery to be able to operate biomass availability will have to increase. One possibility for increasing biomass availability is through growing DECs, such as Miscanthus or short-rotation coppice willow or poplar (Valentine et al., 2012). In the UK in 2013, there was an estimated 7078 ha *Miscanthus* and 2650 ha short-rotation coppice in England (Defra, 2014). There is considerable land suitable for growing DECs, but the majority of this is already being used for agriculture and other uses (Lovett *et al.*, 2014). It has been further proposed that DECs can be grown on marginal land (i.e. poor yielding land) to avoid or reduce competition with food production; however, there is uncertainty as to how much land is available due to the difficulty in defining 'marginal land' (Shortall, 2013). Furthermore, there is uncertainty about the feasibility of using poor quality land for the production of DECs; crucially, the financial benefit of growing DECs on marginal land is dependent on the relative yield penalties of DECs and typical arable crops (Glithero *et al.*, 2015). Regardless of this, both arable and livestock farmers have shown little interest in growing DECs in England (Glithero *et al.*, 2013c; Wilson *et al.*, 2014), implying there is limited scope for increasing DEC production. Allen *et al.* (2014) considered DEC production across Europe and also suggested that there was limited land available for their production and it is unlikely to be economically feasible to grow DECs as much of the available land was poor yielding, fragmented, difficult to access and with limited water supply. Increasing production beyond the limits of the marginal land would, therefore, require expansion of the growing of DECs to non-marginal land, which as previously highlighted is undesirable given issues of competition with food production.

These factors collectively suggest that increases in SGB feedstock availability will, therefore, require an increase in lignocellulosic biomass yields on the land already being used for crops, such as through increased residue yields. Thus, the development of a SGB production industry could lead to a greater importance being placed on crop residues and could encourage farmers to select wheat cultivars with higher straw yields. A focus on increasing yields of the residue component of crops could, in turn, lead to the development of a dual-purpose cultivar (DPC) that is optimized for both grain and straw production.

Contemporary ideas of improving cultivars for dual purposes include de Leon & Coors (2008) and Salas Fernandez *et al.* (2009) who have suggested the breeding of maize cultivars that have characteristics beneficial to both food production and energy production, delivering both high grain and residue yields. Lorenz *et al.* (2010) investigated the possibility of increasing maize stover yield, suggesting that this would be possible without reducing grain yield. Himmel *et al.* (2007) and Harris & DeBolt (2010) have proposed the development of crops with increased potential biofuel yield from the straw (referred to in this study as *digestibility*). Phitsuwan & Ratanakhanokchai (2014) consider the development of

Elite Rice that has improved grain yield, straw yields and digestibility plus improved lodging resistance. Nasidi *et al.* (2015) compared sorghum cultivars for use as feedstock for SGB production by considering biomass yield and digestibility.

DPCs could be chosen from among currently grown cultivars, or new cultivars could be bred, to have improved traits for both food and energy production. Genetic modification techniques could potentially improve traits in a DPC and have been suggested by some authors, although current constraints on the use of genetic modification in Europe currently restrict the selection of a DPC to those that can be produced from conventional crop breeding. There is some work assessing currently grown wheat cultivars for their use as DPCs (e.g. Larsen *et al.*, 2012), but, in general, there is only limited information available on this topic.

This review examines the literature on key traits, and the potential trade-offs among these traits, for a wheat DPC to provide a basis for further research into DPCs. It also considers how management practices could provide additional or alternative strategies with respect to the use of DPCs to improve these traits. The review concludes with a set of recommendations for the further development of DPCs. The review draws on the literature from throughout the world but has a particular focus on the potential development of a DPC for use in the UK and northern Europe in general.

The key DPC traits considered in this review are grain yield, straw yield and straw digestibility. Another trait, lodging resistance, is also considered important due to potential trade-offs with the other traits. Within this review, the DPC concept is considered within the context of SGB production but, with the exception of digestibility, the key traits still apply to a cultivar for bioenergy production via straw combustion.

The article has the following structure: following this introduction, section 'Straw yield' considers straw yields of wheat cultivars. Wheat straw digestibility (section 'Straw digestibility') and lodging in wheat (section 'Lodging susceptibility') are then considered. Grain yield is considered in section 'Trade-offs' via an examination of the potential grain yield trade-offs associated with these traits. Key management practices that influence traits are presented in section 'Crop management' which highlights potential management trade-offs for optimising each trait. Section 'Recommendations' presents recommendations for the use and development of DPCs.

Straw yield

The first key trait of a DPC to be considered is straw yield. Cultivars differ considerably in their straw yields (Donaldson *et al.*, 2001; Engel *et al.*, 2003; Skøtt, 2011;

Larsen *et al.*, 2012). Yield variation is most obvious when comparing modern cultivars with older cultivars as in general straw yields have decreased (both actual straw yields and as a fraction of total biomass) over the past 100 years, in particular with the development of semidwarf cultivars (e.g. Austin *et al.*, 1980; Shearman *et al.*, 2005 – see section 'Grain and straw yields' and Fig. 1a–c).

Straw yields have been assessed for modern wheat cultivars: Larsen *et al.* (2012), in attempting to identify

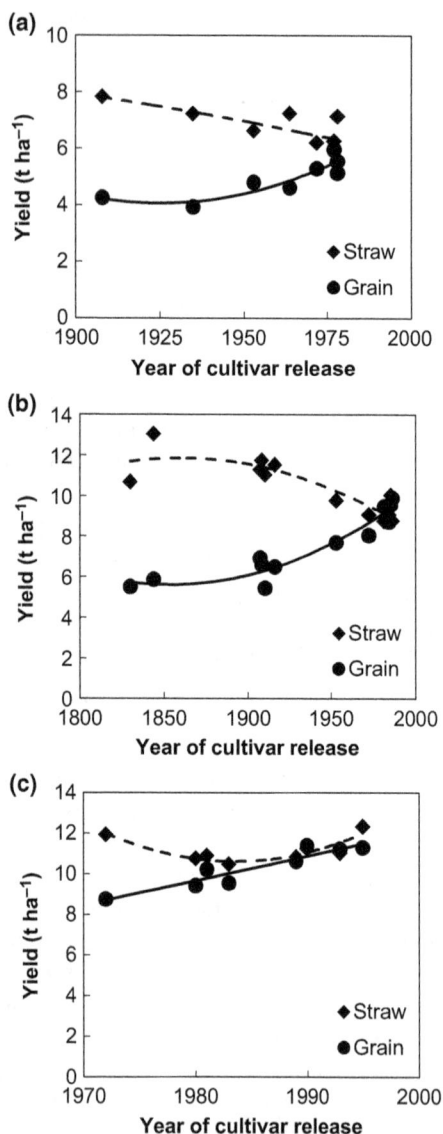

Fig. 1 Grain and straw yields for winter wheat cultivars with different dates of release with data reproduced from: (a) Austin *et al.* (1980); (b) Austin *et al.* (1989); (c) Shearman *et al.* (2005). Straw yields calculated as total AGDM less the grain yield. Dashed line represents general trend for straw yields, and solid line represents general trend for grain yields.

the cultivars with high straw yields for use as feedstock for biofuel production found yields ranged from 2.7 t ha^{-1} to 4.2 t ha^{-1} in one field experiment and 3.4 t ha^{-1} to 4.6 t ha^{-1} in another. [n.b. straw yield here refers to the amount that is baled and removed from the field. Some straw will be left on the field as stubble whilst other straw, in particular leaf and chaff (i.e. the nongrain biomass from the ear), will be lost during combine harvesting and baling; this could potentially account for 60% of total straw (Boyden *et al.*, 2001)]. Agronomists at the University of Kentucky in the US have provided straw yield data in their cultivar trials (e.g. Bruening *et al.*, 2014), with the suggestion that cultivars can be selected based on straw yield to provide a secondary commodity (Lee & Herbek, 2009). In the 2014 variety performance test of US wheat cultivars, straw yields ranged from 1.23 t ha^{-1} to 3.88 t ha^{-1} with an average of 2.67 t ha^{-1}. Straw yields were unrelated to grain yields suggesting that cultivars can be selected for high straw yields from among cultivars with high grain yield. However, the relative rankings of 37 cultivars common to the 2012, 2013 and 2014 field trials demonstrate inconsistencies over time for some cultivars; for example, the cultivar *Pioneer variety 25R32* had the lowest straw yield in 2014, the fourth highest in 2013 and the seventh lowest in 2012. This is in contrast to *Syngenta SY 483* that had the highest straw yields in 2014 and 2013, and the third highest in 2012.

In general, there is limited straw yield data available as straw yields are rarely quantified. There are two main reasons for this: firstly, straw is seen as a by-product to the more important grain, with less incentive for it to be quantified as its economic value is much lower; secondly, straw yields are more difficult to quantify than grain yields, particularly on trial plots, due to straw losses and movement between combining and baling, as well as the need for specialist equipment to take account of topography so as to have an even level of stubble for each plot. In the UK, it is likely that knowledge of cultivar straw yields does exist within the farming community (i.e. anecdotal), but there are currently no published resources available to aid farmers in selecting for wheat straw yields. The cultivar lists produced by the University of Kentucky appear to be unique among recommended lists (RLs) in offering straw yield data for wheat cultivars. Straw yields are not currently given in RLs for UK cultivars, and there are no published records of straw yields for individual cultivars.

One difficulty in identifying cultivars with high straw yields is that there are many environmental and management factors (discussed further in section 'Crop management') that influence straw yield (Engel *et al.*, 2003) such as sowing date and sowing density

(Donaldson *et al.*, 2001), nitrogen and water availability (Engel *et al.*, 2003), and fungal infections and, therefore, fungicide treatment (Jørgensen & Olesen, 2002). Climatic conditions also have a large influence on straw yields; large-scale assessment of wheat straw yields (see Larsen *et al.*, 2012; for references) found that there was considerable temporal variation, with 46% variation in the yearly averages, which was hypothesized to be a result of differences in weather between years. These environmental factors interact with genotypic factors (Engel *et al.*, 2003), further complicating the identification of high straw-yielding cultivars.

Attempts have been made to understand the environmental influences on straw yield. Engel *et al.* (2003) produced an equation linking straw yield to plant height, grain yield and either grain protein content or straw nitrogen (N) concentration (these give an indication of N availability); however, the authors found this relationship varied with water availability, which had an inconsistent influence on yields.

More frequently measured is above ground dry matter (AGDM) and harvest index (HI), which is the ratio of grain to AGDM (see section 'Grain and straw yields'). Using these the non-grain biomass can be calculated, which can be used as a proxy for straw yield (although this will be an overestimate of baled straw yield due to it including stubble, leaf and chaff material that would be left on the field). Hay (1995), in reviewing cereal HI, suggests that HI is reasonably fixed unless there are severe unfavourable conditions. This would suggest that straw yields follow those of grain so in conditions favourable to high grain yields and high straw yields will be achieved. When unfavourable conditions occur, it is likely that straw yields are more heavily impacted than grain yields as the plant increases resource allocation to the grain (Linden *et al.*, 2000), although this will vary with the extent and type of unfavourable conditions as well as the period in which they occur during the crop's life cycle.

Straw digestibility

The biofuel yield of straw depends not only on the total sugars present in the material but also the ease at which these sugars are made accessible to fermentation during processing. *Digestibility*, also referred to as *degradability* and *saccharification potential*, refers to the amount of sugar released from a feedstock under specific processing conditions. This is considered an important trait for the DPC ideotype as using plant material with higher digestibility could reduce SGB production costs (Lindedam *et al.*, 2012; Oakey *et al.*, 2013), for example through requiring lower enzyme amounts and milder pretreatment conditions (Lindedam *et al.*, 2014), or lowering the

amount of feedstock required to produce a set amount of biofuel.

A number of studies have considered the digestibility of wheat cultivars. Early work considered wheat straw digestibility from the perspective of its use as animal feed (reviewed in McCartney *et al.*, 2006) or for mushroom production (e.g. Savoie *et al.*, 1994), which are analogous to its digestibility for biofuel production. A number of these studies showed wheat straw digestibility varied with cultivar (e.g. Knapp *et al.*, 1983; Kernan *et al.*, 1984; Capper, 1988; Habib *et al.*, 1995) and with environmental conditions (Tolera *et al.*, 2008).

Although these studies provide an indication of digestibility, recently, the digestibility of wheat cultivars has been investigated for biofuel production. When considered as feedstock for biofuel production, differences in digestibility among cultivars have been identified (Lindedam *et al.*, 2010a, 2012; Jensen *et al.*, 2011; Wu *et al.*, 2014; Murozuka *et al.*, 2015) although Larsen *et al.* (2012) did not find a significant difference in the cultivars they assessed. The extent of the range of digestibility varied among the studies (Fig. 2). This might be due to the studies using different assays, which prevents direct comparisons, but it could also result from differences in the type and number of cultivars assayed (c.f. five cultivars in Lindedam *et al.*, 2010a; vs. 109 in Jensen *et al.*, 2011).

Developing DPCs with improved digestibility will depend on identifying the genetic and environmental factors determining digestibility. Various factors have been shown to influence the recalcitrance of lignocellulosic material (reviewed in Zhao *et al.*, 2012). These include the following: chemical structural features such as the relative contents of cell-soluble matter, acetate,

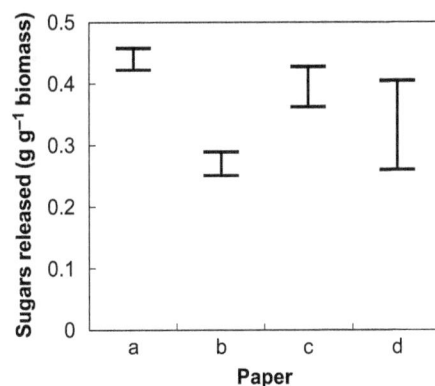

Fig. 2 Minimum and maximum values for the quantity of sugar released per g of biomass for multiple wheat cultivars, from four assessments. References for the four assessments and number of cultivars (n) assessed in each study: (a) Larsen *et al.* (2012), *n* = 10; (b) Lindedam *et al.* (2010a, *n* = 5; (c) Lindedam *et al.* (2012), *n* = 20; (d) Jensen *et al.* (2011), *n* = 109.

ash, cellulose, hemicellulose and lignin; and physical structural features such as accessibility, particle size, surface area, pore volume, degree of polymerization and crystallinity.

Although these factors influence digestibility, it is often unclear why cultivars differ in digestibility. Knowing which features are responsible for this could allow more targeted crop breeding for increased digestibility. The majority of studies comparing digestibility among cultivars considered reasons for any differences seen; however, due to the many factors that could have an influence, these studies were limited in the range of factors they considered. These physical and chemical factors are also differently affected by pretreatment methods (Zhao *et al.*, 2012) and, therefore, the specific assays used for measuring digestibility; this might explain why different studies identified different factors as being important in determining digestibility.

Plant material digestibility can be influenced by variation in the proportions of lignin, cellulose and hemicellulose composing the cell wall (Pauly & Keegstra, 2008). There is considerable variation in the contents of these in wheat (Collins *et al.*, 2014). For example, Wu *et al.* (2014), in assessing 115 wheat ascensions, found that the plant dry matter consisted of 27.5–36.4% cellulose, 28.4–35.1% hemicellulose and 19.3–24.5% lignin.

If during the processing of lignocellulosic material into biofuel, all of the cellulose was converted into biofuel, then biofuel yield would be entirely dependent on the amount of cellulose. However, it is likely that only part of the cellulose will be extracted making the ease at which that cellulose is extracted, rather than the total amount, important. This may explain why cellulose content tends not to determine the differences in digestibility (e.g. Murozuka *et al.*, 2015). Habib *et al.* (1995) actually found a negative relationship between cellulose content and digestibility; Lindedam *et al.* (2012) suggested that this result may be due to the positive correlation seen between cellulose and hemicellulose contents and the negative impact hemicellulose has on sugar release. This may result from differences in the amount of leaf material between samples as leaf material has been shown to have higher digestibility but lower proportions of cellulose and hemicellulose than stem (see below); the negative relationship between cellulose content and digestibility could, therefore, be because as leaf proportion increases relative to stem proportion, digestibility increases whilst the overall proportion of cellulose and hemicellulose decreases.

Lignin and hemicellulose restrict access to the cellulose in the cell wall, so they need to be removed during pretreatment to allow hydrolysis of the cellulose. Therefore, the amounts of lignin and hemicellulose in the lignocellulose material are likely to influence digestibility. Lindedam *et al.* (2012) found digestibility was negatively correlated with lignin content, but Habib *et al.* (1995) did not find a relationship. Wu *et al.* (2014) found a negative correlation between sugar release and hemicellulose content, but Lindedam *et al.* (2012) found a positive relationship, although their assay was considering sugar release from both cellulose and hemicellulose, which is in contrast to most other assays that just considered sugar release from cellulose. Lindedam *et al.* (2010a) found that differences in digestibility were independent of chemical composition. Regarding lignin, an explanation for the inconsistency in these results could be due to the ease at which lignin is removed being determined by the specific location of the lignin and the mechanism in which the lignin is interacting with other components of the cell wall, rather than the total amount of lignin present.

As well as the carbohydrates in the cell wall, soluble carbohydrates may also influence digestibility. Incomplete translocation of soluble carbohydrates from the stem to the grain means that these are available for conversion to biofuel (Capper *et al.*, 1992). Tolera *et al.* (2008) suggested that the differences in digestibility between years that they found in their study could be due to differing amounts of soluble carbohydrates in the stem resulting from variation in rainfall between years; they hypothesized that high rainfall in one year led to higher digestibility either by increasing leaching of soluble carbohydrates from the stem or by increasing the rate of soluble carbohydrate translocation to the grain.

Plant matter also consists of ash, which is the inorganic residue. Negative correlations between ash content and digestibility have been identified (e.g. Lindedam *et al.*, 2012). One possibility for this negative correlation is that additional ash is present at the expense of cellulose, although it is also likely that additional ash is inhibiting sugar release through processes that have yet to be determined. When considering individual elements, Murozuka *et al.* (2015) found that digestibility was not related to silicon content but was negatively correlated to potassium and sulphur contents, which suggests that nutrient availability in the soil could influence digestibility.

Another factor that is important in determining digestibility is the gross morphology of the plant (Capper, 1988). Digestibility varies with cell and tissue types, with overall digestibility varying with the proportions of each (Travis *et al.*, 1996). One of the main explanations for differences in digestibility among cultivars is the relative proportions of the different plant components. Leaf material has been shown to be more digestible than stem resulting in overall digestibility varying with the leaf-to-stem ratio (i.e. the leaf proportion [LP];

Kernan et al., 1984; Capper, 1988; Habib et al., 1995; Tolera et al., 2008). Leaf blade has greater digestibility than the leaf sheath (Ohlde et al., 1992) whilst chaff is also more digestible than the stem (Kernan et al., 1984).

Leaf proportion is both genetically and environmentally determined. Increasing days to heading leads to a greater LP (e.g. in barley and rice, Capper, 1988) and digestibility is positively correlated with days to heading (e.g. Tolera et al., 2008) although this was not seen in Ramanzin et al. (1991). Plant height also influences digestibility through taller plants tending to have lower LP (e.g. Collins et al., 2014). Jensen et al. (2011) found digestibility decreased with increasing plant height; however, this is not always seen with Habib et al. (1995) reporting no relationship whilst Lindedam et al. (2012) found digestibility increased with increasing plant height (with overall digestibility unrelated to LP). Earlier harvesting has been found to lead to higher digestibility through greater leaf retention (McCartney et al., 2006). Weather conditions can also influence LP (Capper, 1988).

In contrast to this, Ramanzin et al. (1991) found that LP was only of minor importance in determining overall digestibility of cultivars as the digestibility of individual components (i.e. leaf and stem) varied among cultivars as well. Ohlde et al. (1992) found that digestibility decreased along the stem with the lower stem the least digestible; this could be due to variation in the proportion of specific tissues comprising these components (e.g. epidermis, mesophyll, parenchyma, sclerenchyma, xylem and phloem) as these are thought to have an influence on digestibility (Capper, 1988; Goto et al., 1991). Travis et al. (1996) found that digestibility was related to the thickness of sclerenchyma and epidermis, and the density of epidermis. Lindedam et al. (2012) suggested that the positive correlation between digestibility and plant height were due to the greater growth of the stem meaning that the tissue was easier to convert, rather than resulting from a difference in LP.

The importance of these differences for the selection of cultivars for use as DPCs and for breeding purposes depends on the stability of this digestibility. Determining this stability requires quantification of the extent that digestibility is influenced by genetic and nongenetic determinants (Oakey et al., 2013). Environmental effects were seen in Jensen et al. (2011), where digestibility differed between the two locations, but not in Larsen et al. (2012). Some studies did not use samples from multiple sites (e.g. Wu et al., 2014) preventing an assessment of environmental impacts. Differences in the relationships between digestibility and other factors, such as conflicting relationships between plant height and digestibility, could be due to environmental and management conditions favouring different plant growth and development; however, the information provided in the literature does not facilitate further consideration of these environmental drivers on digestibility.

Studies have attempted to determine the heritability of digestibility to determine whether it is an appropriate target for breeding programmes. Based on differences among both cultivars and locations, Jensen et al. (2011) and Lindedam et al. (2012) calculated 29% and 57% heritability of digestibility, respectively, suggesting that breeding programmes could increase the digestibility of future cultivars. However, the design of these experiments, along with the digestibility assessments, is likely to only capture some environmental variability as only a limited number of locations were compared, often without intralocation replication, and only a single season of data was collected. The biomass sampling might also not be representative of the straw used for biofuel production as the majority of studies used hand-collected straw samples, which might include more chaff and leaf blades than the baled straw used for biofuel production; only Lindedam et al. (2010a) used baled straw for their assessments.

Oakey et al. (2013) argue that experimental design must consider both the field trial and laboratory work as these both significantly influence digestibility. A robust experimental design can better predict the genetic determinant of digestibility. For future assessments of digestibility, experimental design must, therefore, be carefully considered so as to provide better estimates of the potential increase in digestibility that is achievable.

For the traits identified as being related to digestibility, it is suggested that plants should be bred to have lower ash and lignin contents and higher LP. It is unclear how feasible it is to alter these traits using conventional breeding programmes. Transgenic technologies may provide a more effective means of improving digestibility. They could provide a more targeted method for changes, such as altering the lignin biosynthesis mechanism (Phitsuwan et al., 2013).

Lodging susceptibility

Lodging is defined as the state of permanent displacement of cereal stems from their upright position (Pinthus, 1973; see Berry et al., 2004; for a review of lodging in cereals). Lodging events result from complex interactions between the plant, wind, rain and soil (Baker et al., 1998). The impacts of lodging vary greatly with many lodging events only causing small grain yield reductions whilst others can lead to reductions of up to 80% (Berry et al., 2004). It is estimated that in 1992, severe lodging in the UK cost growers up to £130 million (Sterling et al., 2003). Lodging, particularly towards the end of the growing season, can reduce grain quality, such as by reducing Hagberg falling number, which limits the uses of the grain and likelihood of it achieving a premium price (Berry *et al.*, 2004, 2007) hence further incentivising farmers to reduced lodging susceptibility. Lodging events can also increase farm operation costs, such as through increasing combine harvester costs (ABC, 2014), and can slow harvesting, potentially delaying field preparation for the next crop (Refsgaard *et al.*, 2002). Lodging can also reduce moisture loss from the grain prior to harvest, increasing the need for grain drying postharvest (Baker *et al.*, 1998). It is estimated that, on average, severe lodging occurs in UK wheat crops every 3–4 years when 15–20% of the area lodges (Berry *et al.*, 2004).

There are two distinct types of lodging: stem lodging, which is caused by the breaking of lower culm internodes and occurs when the stem bending moment exceeds the strength of the stem base, and root lodging, which is caused by disturbance to the root-soil interface and occurs when the total bending moment of a plant exceeds the strength of the root-soil interface (Berry *et al.*, 2004). The plant structure influences the likelihood that a plant will lodge; in modelling the failure wind speed of wheat (i.e. the minimum wind speed that is likely to cause lodging in a particular plant at a particular time), Baker *et al.* (1998) and Berry *et al.* (2003a) modelled the bending moment (also known as the leverage force), calculated from the height at the centre of gravity (HCG) of the plant, the natural frequency and the drag of the plant based on the ear area. The strength of the stem base is based on the stem material strength, which is determined by the breaking strength of the stem (tensile failure strength), internode length, and the stem radius and wall width of the lower internodes. For root lodging, the root–soil interface strength is based on the root plate spread and depth.

As cultivars vary in their structural characteristics (e.g. HCG), this leads to cultivars varying in their lodging resistance. Berry *et al.* (2003b) found that, for a selection of 15 cultivars, stem failure wind speed ranged from 9.79 m s^{-1} to 12.71 m s^{-1} and root failure wind speed ranged from 7.15 m s^{-1} to 11.81 m s^{-1}.

Farmers seek to minimize lodging through cultivar selection (Berry *et al.*, 2004). RLs, such as those provided by the Home Grown Cereals Authority (e.g. HGCA, 2012), provide metrics for lodging resistance. As lodging events can cause substantial yield and quality losses, it is unlikely that farmers would to be willing to grow cultivars with higher straw yields or digestibility if they are more susceptible to lodging; however, the relationship between plant structure and lodging susceptibility suggests that there are potential trade-offs between having good lodging resistance and improving other key DPC

traits. The next section examines these aspects as well as other potential trade-offs between the key DPC traits.

Trade-offs

The literature described above has demonstrated cultivar variation in the straw yields, straw digestibility and lodging resistance. Some of these studies have considered the relationship between these traits and grain yield. Relationships among the other traits have only received minor attention, and there are only limited direct measurements comparing these traits; however, correlations between various traits suggest that there may be trade-offs among these key crop traits.

Considering potential trade-offs with grain yield is important because the breeding of cultivars with increased straw yield, at the expense of reduced grain yield, might exclude that straw from use as biofuel feedstock in the EU due to legislative definitions in addition to lowering returns from grain sales. Proposed revisions (EU, 2014) to the Renewable Energy Directive (EU, 2009a) and Fuel Quality Directive (EU, 2009b) suggest that a by-product (e.g. straw) will not gain classification as a 'processing residue' (which gives benefits in double counting of energy contribution and the allocation of zero life cycle GHG emissions prior to its collection; EU, 2009a) if the by-product component of the crop has been increased at the expense of the main product. However, it is unclear whether this definition precludes changing cultivars to increase straw digestibility (rather than biomass *per* se) if the increase in digestibility leads to a reduction in grain yield; it would, however, be expected that this development would be undesirable.

Grain and straw yields

Farmers in the UK are unlikely to sacrifice grain yield for increased straw yield because the value of straw is considerably lower than that of grain due to both less return per tonne and lower yields per unit area. Among high grain-yielding cultivars, there is variation in straw yield (Larsen *et al.*, 2012; Bruening *et al.*, 2014), which suggests that cultivars can be selected for higher straw yields without compromising grain yield. The possibility of increasing straw yields beyond that of modern cultivars without compromising grain yields will depend on the relationship between the yield components.

Breeding progress has led to increases in grain yields (absolute yield as well as a proportion of total biomass) over the past hundred years whilst straw yields have decreased (Austin *et al.*, 1980; Shearman *et al.*, 2005 – see Fig. 1a–c). There are two mechanisms through which the grain yield can increase: through an increase in partitioning of resources to the grain (i.e. increase HI) and through an increase in AGDM. In a comparison of British wheat cultivars released between 1908 and 1978, Austin *et al.* (1980) found nongrain biomass tended to decrease whilst grain yields increased with newer cultivars; this change was attributed to increases in HI without an increase in AGDM. Extending the comparison to cultivars released between 1830 and 1986, Austin *et al.* (1989) found the same pattern although AGDM was slightly higher for the newest cultivars they measured. Shearman *et al.* (2005) considered cultivars released between 1972 and 1995 and found that, whilst improvement in grain yield up to 1983 resulted from increases in HI, after 1983, these were mainly the result of increases in AGDM.

Similar studies have been conducted in other major wheat-producing countries. In a comparison of Argentinian cultivars released between 1912 and 1980, Slafer & Andrade (1989) found that grain yield increased with date of release, but overall, there was no increase in AGDM. Brancourt-Hulmel *et al.* (2003) found the same relationship when comparing French cultivars released between 1946 and 1980. Zhou *et al.* (2014) considered cultivars released from 1995 to 2012 in Henan Province China and found an increase in both AGDM and HI. Waddington *et al.*'s (1986) investigation of Mexican wheat cultivars found higher AGDM in newer cultivars. Donmez *et al.* (2001) considered cultivars grown on the American Great Plains released between 1873 and 1995 and found four cultivars released between 1992 and 1995 as having greater AGDM.

The purpose of these studies has been to use past trends to infer future potential for grain yield increases. As can be seen from the studies discussed above, increases in HI have been mainly responsible for increases in yield, but there is evidence that AGDM has also increased. The potential for increasing wheat grain yields has been extensively reviewed (see Reynolds *et al.*, 2009, 2012; Foulkes *et al.*, 2011; Parry *et al.*, 2011). The feasibility of increasing straw yields in wheat has not been given attention but, interestingly, increases in grain yield might necessitate increases in straw yield. This is because there is a limit to the HI and increasing grain yields beyond, this will require an increase in AGDM (Shearman *et al.*, 2005; Lorenz *et al.*, 2010). The limit to HI is unknown but Austin *et al.* (1980) hypothesized that there is an upper limit to HI of 0.62 based on extrapolating from an average HI and assuming leaf sheath and stem biomass could decrease by 50% (Austin *et al.*, 1980). Foulkes *et al.* (2011) revised this to ~0.64 based on assumptions about additional AGDM production. However, in recent years, there has not been a systematic increase in HI of wheat cultivars (Reynolds *et al.*, 2009). These HI might prove to be infeasible, in particular as they do not take account of the need for adequate stem biomass to prevent lodging (Foulkes *et al.*, 2011), necessitating the need for higher AGDM.

There is the possibility of increasing AGDM through increasing radiation-use efficiency (Long *et al.*, 2006) such as through increasing photosynthetic capacity and efficiency (Parry *et al.*, 2011). With this increase in AGDM, it would be expected that straw yields will also increase; however, there might be a practical limit to AGDM increases due to the interaction between straw yield and lodging, which is explored in the following section.

Straw yields and lodging resistance

In general, there is a strong correlation between straw yield and plant height (Engel *et al.*, 2003; Larsen *et al.*, 2012; Long & McCallum, 2013). This correlation suggests selecting cultivars with high straw yield is likely to lead to the selection of taller cultivars and, because plant height correlates with lodging risk (e.g. Baker *et al.*, 1998; Berry *et al.*, 2003a), to an increased risk of lodging. This is supported by Berry *et al.* (2004) who showed that increasing biomass leads to a greater HCG, hence increasing lodging risk (though distribution of dry matter along the stem was important in the overall influence of biomass on HCG). However, among current cultivars, the relationship between plant height and straw yield is not always observed; for example, Donaldson *et al.* (2001) found that the straw yields of a semidwarf cultivar did not differ significantly from standard height or tall cultivars. It may be possible to find cultivars that have high straw yield whilst also maintaining good lodging resistance through having shorter stems.

As discussed in the previous section, greater straw yields could result from increased AGDM. To avoid reduced lodging resistance, this increase in AGDM will need to be achieved without significantly increasing plant height. Increasing AGDM could, in fact, lead to greater lodging resistance: Berry *et al.* (2007) suggested lodging resistance could be improved by having greater weight per unit length of the lower internodes, which would increase stem material strength but would also necessitate additional AGDM. Further work is needed to explore the relationship between lodging susceptibility and straw yields.

Straw digestibility and grain yields

There appears to be no relationship between grain yield and straw digestibility (e.g. Ramanzin *et al.*, 1991; Habib *et al.*, 1995; Jensen *et al.*, 2011; Lindedam *et al.*, 2012), suggesting that selecting or developing cultivars with higher digestibility will not negatively impact on grain yields. However, attempts to increase digestibility could lower grain yield through compromising plant fitness. Altering the cell wall components could result in weakening of the plant tissues, leading to reduced integrity (Pauly & Keegstra, 2008) potentially leaving the plant more susceptible to pathogens and pests (Li et al., 2008). In studies where Arabidopsis had been genetically modified to have lower recalcitrance to digestion, some, but not all, of these studies found the plants had poor growth due to growth defects or altered susceptibility to pests or pathogens (Zhao & Dixon, 2014). It can be surmised that there is a limit to how much digestibility can be improved without compromising grain yields. Whether the variability in digestibility seen among current cultivars also reflects variability in susceptibility to pests or disease has not been assessed in the available literature.

Straw digestibility and lodging resistance

It has been hypothesized that stem digestibility is negatively correlated with lodging resistance. It has been suggested that breeding to reduce stem lodging through greater straw stiffness (i.e. stem material strength, see section 'Lodging susceptibility') resulted in modified anatomical features of the stem that decrease the digestibility of the straw. Data presented in the literature have been conflicted: Lindedam et al. (2010a) suggested that the low digestibility of one cultivar resulted from it having stiff straw, whereas Travis et al. (1996) found that a stiff-strawed wheat cultivar was more digestible than a soft-strawed wheat cultivar.

It is thought that lignin plays a role in lodging resistance with greater lignin content increasing lodging resistance (Ma, 2009). As some assays have found a negative relationship between lignin content and digestibility (e.g. Lindedam et al., 2012), this would support that there is a trade-off between lodging resistance and digestibility. However, there is little experimental evidence for a correlation between lignin content and digestibility with some studies not finding a correlation (e.g. Kong et al., 2013). It is possible that any impact that the variation in lignin content has on lodging resistance is outweighed by structural characteristics such as leverage force, making it difficult to identify a relationship between lignin and lodging. Also, not all of the lignin may be playing a mechanical role (Köhler & Spatz, 2002), and therefore, the total amount of lignin present is not as important as how much of the total lignin is contributing to mechanical strength. This offers the possibility that nonstructural lignin could be removed without negatively impacting on stem strength; however,

this lignin may be important for other processes unrelated to mechanical strength. Wang et al. (2012) suggested that cellulose might be more important in determining lodging resistance, which could further support the possibility of removing lignin without significantly affecting lodging resistance.

As discussed in the previous section, modifications of the cell wall for improved digestibility could reduce integrity, which could result in greater lodging risk. Interestingly, Li et al. (2015) found that overexpressing the genes GH9B and XAT in rice simultaneously increased both digestibility and lodging resistance.

Among the currently grown cultivars, there might actually be a positive correlation between straw digestibility and lodging resistance. This is because shorter cultivars tend to be associated with higher LP, giving higher digestibility, and lower stem leverage force, giving greater lodging resistance. Complicating this, however, are the contradictory results from Jensen et al. (2011) and Lindedam et al. (2012), so it is unlikely this is a consistent relationship. It is unclear why these studies differed in this relationship as the cultivars were grown at the same locations and followed the same management practices.

Interestingly, depending on the strength of the relationship between plant height and digestibility, this suggests that taller cultivars, which are likely to produce the most straw, are likely to have lower digestibility, potentially indicating a trade-off between digestibility and straw yield.

Crop management

As grain yield has been the priority in crop production, management practices have been optimized to maximize this. It is possible that management practices could be used to maximize the other key traits of a DPC. This section considers how management practices, other than cultivar selection, can be used to increase straw yield and digestibility and considers how these may influence grain yield and lodging resistance.

Plant growth regulators (PGRs) are used to reduce the lodging risk by shortening plant stems through reducing cell elongation and decreasing cell division (Berry et al., 2004) Berry et al., 2007). In the UK, PGRs were applied to 88% of the winter wheat area in 2010 (Garthwaite et al., 2011). With the reduction in height, it would be expected that straw yields would also decrease; however, there are few studies that have compared straw yields between PGR treatments. Bragg et al. (1984) found that although application of the PGR chlormequat reduced plant height, it did not significantly influence straw or grain yields. PGR application shortened plant height but did not influence overall

AGDM in winter wheat (Cox & Otis, 1989) or triticale (Naylor, 1989). In a glasshouse trial of cereals, Rajala & Peltonen-Sainio (2001) found that PGR application reduced main stem growth and weight; however, this was for an early application of PGRs and measurements were taken 14 days after application.

A limited number of studies found PGR application slightly reduced stem strength (Crook & Ennos, 1995; Berry et al., 2000), suggesting that PGR application could affect digestibility. It would also be expected that by reducing plant height, PGRs would lead to an increase in digestibility through an increase in the LP. Taken together, these suggest that PGR application might increase digestibility; however, in the two studies considering this a consistent pattern was not found. Sharma et al. (2000) found that PGRs, when applied with fungicide, increased digestibility though the independent effects of the PGR and fungicides were not determined. Savoie et al. (1994) did not find a consistent PGR effect on digestibility, but PGRs were only applied 62 days before harvest, so it is unlikely that they had a large influence on plant form.

From these studies, it is unclear how much of an impact PGR application has on straw yield or digestibility and further work is needed to quantify these. From a trade-off perspective, PGR application can lead to a reduction in the area lodged by anything up to 70% (Berry et al., 2004), so the benefits of PGR application are likely to outweigh a small decrease in straw yield and digestibility.

Lowering the height of the combine harvester cutter bar decreases stubble height enabling a greater amount of straw to be baled (Allen, 1988); however, farmers might not want to set the cutter bar too low as lowering it increases the straw moving through the combine and this can slow work rate and increase fuel requirements (Hill et al., 1987; Allen, 1988; Kehayov et al., 2004). There is also a risk of damage to the cutters and contamination of grain with soil from cutting too low. Summers et al. (2003) found that the height that rice straw is cut influences not only the straw yield but also the composition of the biomass material. Lowering the cutter bar could decrease overall straw digestibility due to increasing the proportion of straw consisting of lower stem, which is less digestible than the upper stem (Ohlde et al., 1992); this has been shown in barley straw feeding value (Wilson & Brigstocke, 1977). It has been suggested that better quality straw can be achieved by just harvesting the tops of the plants (Kernan et al., 1984) although this would lead to low straw yields. As lowering cutter height is associated with potential costs, the merit of doing so would depend on how much additional straw can be collected. There is also the problem of reducing the amount of residues being returned to the soil, which may not be sustainable (see discussion of sustainable straw removal in section 'Introduction').

Altering N fertilizer application rate influences crop characteristics. The amount of N fertilizer applied influences grain yield (Hay & Walker, 1989). Recommended levels of fertilizer are based on the economic optimum application rate that takes account of the grain yield having a curvilinear response to N application rate. The N applied also depends on the use of the grain, with bread-making quality grain requiring extra N fertilizer for the grain to have higher protein content (ABC, 2014).

Above ground dry matter responds to increasing N in a similar manner to grain (White & Wilson, 2006), but there is limited work investigating how straw yield responds to N fertilizer rates above the economically optimal rates for grain production. Pearman et al. (1978) suggested that straw yield (stem weight and leaf area) was increased relatively greater than grain yield with the application of extra N; it is, therefore, feasible that straw yields could be increased further.

Fertilizer levels have been shown to have an influence on digestibility though the influence is inconsistent: Flachowsky et al. (1993) found that very high applications of N led to higher digestibility, whereas Murozuka et al. (2014) found that higher N fertilizer application rate led to lower straw digestibility. Tolera et al. (2008) found that increasing N and P fertilizer application did not change digestibility; and Kernan et al. (1984) found that increasing N led to higher leaf digestibility, the same or lower digestibility in stem and no difference in chaff; the actual difference in digestibility among fertilizer treatments was very small. The reason for the inconsistency is unclear although these articles do represent different locations and use different timings for N fertilizer application. Murozuka et al. (2014) suggested decreasing digestibility from increasing fertilizer level might be due to an increase in inhibitory factors or a decrease in LP. It is possible that several factors, such as the composition of cell wall components and LP, are interacting, which could explain the different results. In these studies, LP was not measured and it is unclear how LP varies with N fertilizer rate.

If benefits to straw yield and digestibility are shown for high N fertilizer rates, this could shift the economically optimal fertilizer level towards the application of greater amounts of fertilizer. However, any additional N could increase lodging risk as Berry et al. (2000) found that the timing of N application and the amount of N applied, as well as the amount of soil residual N influenced lodging resistance, with low soil residual N or low and delayed spring N application leading to stronger stem bases. There are also limitations to the amount of N fertilizer that can be used due to environmental considerations, such as nitrate leaching (Di &

Cameron, 2002) and the high GHG emissions associated with N fertilizer production and use (Snyder *et al.*, 2009). This suggests that changing fertilizer practices would be unlikely to happen specifically for improved traits although the removal of straw from the field does necessitate increased fertilizer requirements, in particular phosphorus and potassium (Whittaker *et al.*, 2014); however, there is debate about whether straw removal leads to the need for additional N relative to leaving the straw *in situ* as the microbial breakdown of incorporated straw tends to require additional N itself (Powlson *et al.*, 2011).

Earlier sowing can lead to higher straw yields (Donaldson *et al.*, 2001) as well as higher grain yields (Hay & Walker, 1989) although this could come at the expense of increased lodging risk (Berry *et al.*, 2000), as well as disease, weed, pest and drought risks (Hay & Walker, 1989). There are also limitations to how early the crop can be sown due to a need to harvest the previous crop and prepare the land; this is likely to be even more limited if time is required for baling straw from the previous crop. Reducing the tillage operations (e.g. using no-till or reduced tillage) could speed up sowing of the wheat crop (Morris *et al.*, 2010). It does not appear that wheat straw digestibility has been measured under different sowing dates although it would be expected that sowing date would affect LP, which would lead to differences in digestibility. In winter barley, sowing date did not influence LP or the digestibility (Capper *et al.*, 1992); however, this was for a single field experiment and, in general, the influence of sowing date on crops is highly variable as it is dependent on weather conditions (Hay & Walker, 1989).

Sowing density can influence grain and straw yields. Whaley *et al.* (2000) found that plants at lower seed rates compensated by increasing tillering duration, green area per shoot and shoot survival, which resulted in no significant difference in AGDM with sowing density; however, they suggested that at medium sowing density, HI would be higher, which suggests this would have lower straw production. Donaldson *et al.* (2001) found that low straw yields were achieved at low sowing densities regardless of sowing date, whereas grain yield was only reduced from lower sowing density for later sowing dates. Lower sowing density has been shown to reduce lodging risk (Berry *et al.*, 2000). To our knowledge, no work has considered how sowing density influences digestibility; however, Whaley *et al.* (2000) found a general increase in green area per potentially fertile shoot with decreasing plant density suggesting an increase in LP at lower sowing densities. Differences in digestibility could potentially result from different ratios of main stems to tillers, possibly through changing LP. Increasing sowing density has been shown

to reduce stem diameter (Easson *et al.*, 1993), which is associated with lower lodging resistance. It has been suggested that reduced stem diameter could have an influence on digestibility through changing the proportions of structural tissues; however, a correlation between digestibility and stem diameter has not been found (Capper, 1988; Habib *et al.*, 1995).

Recommendations

From the literature presented in this review, there is the potential to develop a wheat DPC although the scope to improve individual traits is limited. Specific cultivars that have high grain yields alongside high straw yields have been identified although there are currently limits on increasing straw yields further without compromising grain yields. Should radiation-use efficiency be improved then this offers the opportunity for higher straw yields. The variation in straw digestibility among cultivars and the fact that digestibility is not correlated with grain yield suggest that cultivars can be bred to have greater digestibility; however, improving digestibility beyond that seen in current cultivars might not be possible without compromising plant fitness.

There does not currently appear to be sufficient incentives for a dedicated breeding programme for developing wheat DPCs. Before embarking on a breeding programme, it needs to be determined whether there would be a sufficient market for these cultivars. Wilson *et al.* (2014) found that if wheat straw were to reach £100 t^{-1}, some farmers in England with both livestock and arable land would be willing to grow cultivars with longer straw (i.e. greater straw yields). Considering the recent straw prices of approximately £45 t^{-1} in England and Wales (Anon, 2014b), it is unlikely farmers would currently choose longer strawed cultivars. Taken together with the lack of market for high digestibility straw (for either bioenergy or animal feed), it appears that there is currently not a market for DPCs.

There are a number of potential benefits to the use of DPCs. Provided that it does not decrease grain yields, increasing straw yields should increase farm revenue. Higher straw yields also lead to increased baling efficiency, which means farmers can supply a set amount of baled straw to a bioenergy plant in a shorter amount of time and at lower cost (Nilsson, 1999; Kühner, 2013). Increasing feedstock supply density (e.g. by increasing the straw yields) can decrease the area of land required to meet a specific feedstock demand, which could decrease transport distances and, therefore, transport costs for large-scale biomass use for biofuel production (Hamelinck *et al.*, 2005). Similarly, increasing digestibility could lower feedstock requirements, and this could also result in reduced logistics costs. Because of this, bioenergy producers might be interested in encouraging biomass suppliers to grow DPCs. Work is required to quantify these benefits. Alongside an increase in straw demand, these benefits could create a demand for DPCs.

In section 'Straw digestibility', it was suggested that digestibility could be increased through having lower ash and lignin contents and higher LP; however, before embarking on a breeding programme to optimize these traits, it is important to determine the strength of the relationships between these traits and digestibility. As there are many factors that can potentially influence digestibility, changing a single trait may only have a minor impact on overall digestibility. Research is also needed to determine the variability in these traits and how this influences overall digestibility; as marketing cultivars as being more digestible will require a certain level of consistency in digestibility, which may not be possible if these traits show high levels of variability. Alongside this it is recommended that cultivars are assessed to identify the potential cultivars for use as DPCs, or to identify the genetic material for use in DPC breeding programmes.

One difficulty in determining whether to breed cultivars for higher digestibility is whether these cultivars will have a premium for the higher digestibility. Uncertainties about which types of pretreatment conditions and enzymes that will be used mean that the assays used in the assessments are unlikely to correspond to those used at the industrial scale; hence, the variation in digestibility seen among cultivars in the current assessments might not be seen in practice.

There is also the issue of whether the results of the assays scale up from laboratory to commercial production scale. Lindedam et al. (2010a) used pilot plant-scale production and found that cultivar variation was still seen; however, differences were only seen under specific enzyme loadings suggesting that differences in digestibility will depend on the system utilized for commercial-scale production. It is recommended that efforts to develop cultivars with higher digestibility are aligned with the development of specific biofuel conversion processes.

Although higher digestibility could reduce costs for biofuel production, it is unclear whether biofuel producers will pay more for higher digestibility straw. This would require being able to determine the digestibility of the feedstock when it arrives at the biorefinery. Current assays are expensive and time-consuming (Collins et al., 2014), which is likely to prohibit them from being used. However, work is considering methods involving the use of spectroscopy to determine the digestibility (e.g. Bruun et al., 2010; Lindedam et al., 2010b), which might allow quick assessment at the biorefinery. But

even with this, biofuel producers might not be willing to offer different prices based on digestibility. For example, bioethanol yield of wheat grain used for FGB production varies with cultivar (Smith et al., 2006) but, for example, at the Ensus plant in Teeside, UK, there is currently a flat rate paid per tonne of grain, regardless of potential bioethanol yield (Nick Oakhill, pers. comm., Glencore). Another example is that of straw used for straw-burning power stations in Denmark; leaving the straw on the field to be exposed to rain produces a better fuel by washing away substances such as chlorine, but farmers are not financially rewarded for doing this as straw is priced based only on weight and water content (Skøtt, 2011). Wheat straw digestibility might also be excluded from differentiated pricing in line with the above examples.

As management practices have been optimized to provide the highest grain yields, it is unlikely that farmers would be willing to change these. Some practices that maximize grain yield also maximize straw yield, whereas others can lead to trade-offs. The use of PGRs clearly benefits grain yield through improved lodging resistance but does not appear to have a significant impact on straw yields. Selecting taller cultivars could lead to higher straw yields though that is not always found, and it would lead to greater risk of lodging. Lowering the cutter bar height can increase straw yield, but there are other considerations, such as sustainable residue removal rates and additional fuel costs for harvesting. Should farmers wish to increase straw yield, earlier sowing with a medium or high sowing density offers one way of increasing straw yield though there is an increased risk of lodging and carry-over of disease and pests. Earlier sowing requires quick field preparation after the previous crop, and this may necessitate changes to tillage practices, such as reducing tillage intensity. Careful consideration of the rotation alongside other management practices for pest and disease control is also needed to reduce the risk of carry-over of disease and pests.

Before a biofuel industry is established, work is required to determine how much feedstock can be made available without compromising food production or the environment and without competing with other straw users. This includes determining sustainable residue removal rates for specific locations; this may necessitate the development of new methodologies for testing soil properties for quick determination of safe levels of straw removal.

Overall, there is considerable uncertainty in the future of the biofuel sector in Europe. This suggests that currently there is little value in developing DPCs. The research presented in this review supports the possibility of improving straw yields and digestibility although cautions that care must be taken to avoid negative impacts on grain yields and lodging resistance. Should a biofuel sector develop then there is scope for developing DPCs.

Acknowledgements

This work is part of TJT's PhD project funded by the Home Grown Cereals Authority (RD-2010-3741) and the University of Nottingham.

References

ABC (2014) *The Agricultural Budgeting and Costing Book, November 2014* (79th edn). Agro Business Consultants Ltd., Melton Mowbray, UK.

Aden A, Ruth M, Ibsen K et al. (2002) *Lignocellulosic Biomass to Ethanol Process Design and Economics Utilizing co-Current Dilute Acid Prehydrolysis and Enzymatic Hydrolysis for Corn Stover*. National Renewable Energy Laboratory, Golden, CO, USA. Available at: http://www.nrel.gov/docs/fy99osti/26157.pdf (accessed 18 August 2015).

Allen RR (1988) Straw recovery as affected by wheat harvest method. *Transactions of the ASAE*, **31**, 1656–1659.

Allen B, Kretschmer B, Baldock D, Menadue H, Nanni S, Tucker G (2014) *Space for Energy Crops – Assessing the Potential Contribution to Europe's Energy Future*. Report produced for the Institute for European Environmental Policy, London. Available at: http://www.eeb.org/EEB/?LinkServID=F6E6DA60-5056-B741-DBD250D05D441B53 (accessed 18 August 2015).

Andrews SS (2006) *Crop residue removal for biomass energy production: effects on soils and recommendations*. White paper, USDA-Natural Resource Conservation Service. Available at: http://www.nrcs.usda.gov/Internet/FSE_DOCUMENTS/nrcs142p2_053255.pdf (accessed 18 August 2015).

Anon (2013) *Crescentino in Figures*. Available at: http://www.betarenewables.com/crescentino/project (accessed 18 August 2015).

Anon (2014a) *Agriculture in the United Kingdom 2013*. Department for Environment, Food and Rural Affairs, London, UK. Available at: https://www.gov.uk/government/uploads/system/uploads/attachment_data/file/315103/auk-2013-29may14.pdf (accessed 18 August 2015).

Anon (2014b) *Hay & Straw, England and Wales Average Prices*. British Hay and Straw Merchants' Association. Available at: https://www.gov.uk/government/statistical-data-sets/commodity-prices (accessed 18 August 2015).

Austin RB, Bingham J, Blackwell RD, Evans LT, Ford MA, Morgan CL, Taylor M (1980) Genetic improvements in winter wheat yields since 1900 and associated physiological changes. *Journal of Agricultural Science*, **94**, 675–689.

Austin BYRB, Ford MA, Morgan CL (1989) Genetic improvement in the yield of winter wheat: a further evaluation. *Journal of Agricultural Science*, **112**, 295–301.

Baker CJ, Berry PM, Spink JH, Sylvester-Bradley R, Griffin JM, Scott RK, Clare RW (1998) A method for the assessment of the risk of wheat lodging. *The Journal of Theoretical Biology*, **194**, 587–603.

Berry PM, Griffin JM, Sylvester-Bradley R, Scott RK, Spink JH, Baker CJ, Clare RW (2000) Controlling plant form through husbandry to minimise lodging in wheat. *Field Crops Research*, **67**, 59–81.

Berry PM, Sterling M, Baker CJ, Spink J, Sparkes DL (2003a) A calibrated model of wheat lodging compared with field measurements. *Agricultural and Forest Meteorology*, **119**, 167–180.

Berry PM, Spink JH, Gay AP, Craigon J (2003b) A comparison of root and stem lodging risks among winter wheat cultivars. *Journal of Agricultural Science*, **141**, 191–202.

Berry PM, Sterling M, Spink JH et al. (2004) Understanding and reducing lodging in cereals. *Advances in Agronomy*, **84**, 217–271.

Berry PM, Sylvester-Bradley R, Berry S (2007) Ideotype design for lodging-resistant wheat. *Euphytica*, **154**, 165–179.

Blanco-Canqui H, Lal R (2009) Crop residue removal impacts on soil productivity and environmental quality. *Critical Reviews in Plant Sciences*, **28**, 139–163.

Boyden A, Hill L, Leduc P, Wassermann J (2001) *Field tests to correlate biomass, combine yield and recoverable straw*. Prairie Agricultural Machinery Institute. Project No. 5000H

Bragg PL, Rubino P, Henderson FKG, Fielding WJ, Cannell RQ (1984) A comparison of the root and shoot growth of winter barley and winter wheat, and the effect of an early application of chlormequat. *Journal of Agricultural Science*, **103**, 257–264.

Brancourt-Hulmel M, Doussinault G, Lecomte C, Bérard P, Le Buanec B, Trottet M (2003) Genetic improvement of agronomic traits of winter wheat cultivars released in France from 1946 to 1992. *Crop Science*, **43**, 37–45.

Bruening B, Curd R, Swanson S, Connelley J, Olson G, Clark A, Van Sanford D (2014) *2014 Kentucky Small-Grain Variety Performance Test*. College of Agriculture, Food and Environment, University of Kentucky. Available at: http://www2.ca.uky.edu/agc/pubs/PR/PR674/PR674.pdf (accessed 18 August 2015).

Bruun S, Jensen JW, Magid J, Lindedam J, Engelsen SB (2010) Prediction of the degradability and ash content of wheat straw from different cultivars using near infrared spectroscopy. *Industrial Crops and Products*, **31**, 321–326.

Capper BS (1988) Genetic variation in the feeding value of cereal. *Animal Feed Science and Technology*, **21**, 127–140.

Capper BS, Sage G, Hanson PR, Adamson AH (1992) Influence of variety, row type and time of sowing on the morphology, chemical composition and *in vitro* digestibility of barley straw. *Journal of Agricultural Science*, **118**, 165–173.

Collins S, Wellner N, Martinez Bordonado I, Harper AL, Miller CN, Bancroft I, Waldron KW (2014) Variation in the chemical composition of wheat straw: the role of tissue ratio and composition. *Biotechnology for Biofuels*, **7**, 121.

Copeland J, Turley D (2008) *National and Regional Supply/Demand Balance for Agricultural Straw in Great Britain*. The National Non-Food Crops Centre, York, UK.

Cox WJ, Otis DJ (1989) Growth and yield of winter wheat as influenced by chlormequat chloride and ethephon. *Agronomy Journal*, **81**, 264–270.

Crook MJ, Ennos AR (1995) The effect of nitrogen and growth regulators on stem and root characteristics associated with lodging in two cultivars of winter wheat. *Journal of Experimental Botany*, **46**, 931–938.

Defra (2014) *Area of Crops Grown For Bioenergy in England and the UK: 2008-2013*. Department for Environment, Food and Rural Affairs, London, UK. Available at: www.gov.uk/government/statistics/area-of-crops-grown-for-bioenergy-in-england-and-the-uk-2008-2013 (accessed 18 August 2015).

Di HJ, Cameron KC (2002) Nitrate leaching in temperate agroecosystems: sources, factors and mitigating strategies. *Nutrient Cycling in Agroecosystems*, **64**, 237–256.

Donaldson E, Schillinger WF, Dofing SM (2001) Straw production and grain yield relationships in winter wheat. *Crop Science*, **41**, 100–106.

Donmez E, Sears RG, Shroyer JP, Paulsen GM (2001) Genetic gain in yield attributes of winter wheat in the Great Plains. *Crop Science*, **41**, 1412–1419.

Easson DL, White EM, Pickles SJ (1993) The effects of weather, seed rate and cultivar on lodging and yield in winter wheat. *Journal of Agricultural Science*, **121**, 145–156.

Engel RE, Long DS, Carlson GR (2003) Predicting straw yield of hard red spring wheat. *Agronomy Journal*, **95**, 1454–1460.

EU (2009a) Directive 2009/28/EC of the European Parliament and of the Council of 23 April 2009 on the promotion of the use of energy from renewable sources and amending and subsequently repealing Directives 2001/77/EC and 2003/30/EC. *Official Journal of the European Union*, 16–62. Available at: http://eur-lex.europa.eu/legal-content/EN/ALL/?uri=CELEX:32009L0028 (accessed 18 August 2015).

EU (2009b) Directive 2009/30/EC of the European Parliament and of the Council of 23 April 2009 amending Directive 98/70/EC as regards the specification of petrol, diesel and gas-oil and introducing a mechanism to monitor and reduce greenhouse gas emissions and amending Council Directive 1999/32/EC as regards the specification of fuel used by inland waterway vessels and repealing Directive 93/12/EEC. *Official Journal of the European Union*, 88–113. Available at: http://eur-lex.europa.eu/legal-content/EN/TXT/?uri=celex:32009L0030 (accessed 18 August 2015).

EU (2014) Fuel quality directive and renewable energy directive ***II European Parliament legislative resolution of 28 April 2015 on the Council position at first reading with a view to the adoption of a directive of the European Parliament and of the Council amending Directive 98/70/EC relating to the quality of petrol and diesel fuels and amending Directive 2009/28/EC on the promotion of the use of energy from renewable sources (10710/2/2014 – C8-0004/2015 – 2012/0288(COD)) (Ordinary legislative procedure: second reading). Available at: http://www.europarl.europa.eu/sides/getDoc.do?pubRef=-//EP//NONSGML+TA+P8-TA-2015-0100+0+DOC+PDF+V0//EN (accessed 18 August 2015).

Flachowsky G, Vanselow G, Schneider A (1993) Influence of nitrogen fertilization of wheat on *in sacco* dry matter degradability of wheat straw. *Journal of Applied Animal Research*, **3**, 91–96.

Foulkes MJ, Slafer GA, Davies WJ et al. (2011) Raising yield potential of wheat. III. Optimizing partitioning to grain while maintaining lodging resistance. *Journal of Experimental Botany*, **62**, 469–486.

Garthwaite DG, Barker I, Parrish G, Smith L, Chippindale C, Pietravalle S (2011) *Pesticide Usage Survey Report 235: Arable Crops in the United Kingdom 2010 (Including*

Aerial Applications). Agricultural Survey Team Food & Environment Research Agency, York, UK, and Department for Environment, Food and Rural Affairs, London, UK.

Glithero NJ, Wilson P, Ramsden SJ (2013a) Straw use and availability for second generation biofuels in England. *Biomass and Bioenergy*, 55, 311–321.

Glithero NJ, Ramsden SJ, Wilson P (2013b) Barriers and incentives to the production of bioethanol from cereal straw: a farm business perspective. *Energy Policy*, 59, 161–171.

Glithero NJ, Wilson P, Ramsden SJ (2013c) Prospects for arable farm uptake of Short Rotation Coppice willow and miscanthus in England. *Applied Energy*, 107, 209–218.

Glithero NJ, Wilson P, Ramsden SJ (2015) Optimal combinable and dedicated energy crop scenarios for marginal land. *Applied Energy*, 147, 82–91.

Gnansounou E (2010) Production and use of lignocellulosic bioethanol in Europe: current situation and perspectives. *Bioresource Technology*, 101, 4842–4850.

Goto M, Morita O, Chesson A (1991) Morphological and anatomical variations among barley cultivars influence straw degradability. *Crop Science*, 31, 1536–1541.

Habib G, Shah SBA, Inayat K (1995) Genetic variation in morphological characteristics, chemical composition and *in vitro* digestibility of straw from different wheat cultivars. *Animal Feed Science and Technology*, 55, 263–274.

Hamelinck CN, Suurs RAA, Faaij APC (2005) International bioenergy transport costs and energy balance. *Biomass and Bioenergy*, 29, 114–134.

Harris D, DeBolt S (2010) Synthesis, regulation and utilization of lignocellulosic biomass. *Plant Biotechnology Journal*, 8, 244–262.

Hay RKM (1995) Harvest Index: a review of its use in plant breeding and crop physiology. *Annals of Applied Biology*, 126, 197–216.

Hay RKM, Walker AJ (1989) *An Introduction to the Physiology of Crop Yield*. Longman Scientific and Technical, Essex, UK.

HGCA (2012) *HGCA Recommended List Winter Wheat 2012/13*. Agriculture and Horticulture Development Board, Stoneleigh Park, Kenilworth, Warwickshire.

Hill LG, Frehlich GF, Wasserman JD (1987) *Effect of MOG/G Ratio and Grain Moisture Content on Combine Performance*. Agriculture Canada, Ottawa, ON, Canada.

Himmel ME, Ding S-Y, Johnson DK, Adney WS, Nimlos MR, Brady JW, Foust TD (2007) Biomass recalcitrance: engineering plants and enzymes for biofuels production. *Science*, 315, 804–807.

Huggins DR, Karow RS, Collins HP, Ransom JK (2011) Introduction: evaluating long-term impacts of harvesting crop residues on soil quality. *Agronomy Journal*, 103, 230–233.

Jensen JW, Magid J, Hansen-Møller J, Andersen SB, Bruun S (2011) Genetic variation in degradability of wheat straw and potential for improvement through plant breeding. *Biomass and Bioenergy*, 35, 1114–1120.

Jørgensen LN, Olesen JE (2002) Fungicide treatments affect yield and moisture content of grain and straw in winter wheat. *Crop Protection*, 21, 1023–1032.

Kehayov D, Vezirov C, Atanasov A (2004) Some technical aspects of cut height in wheat harvest. *Agronomy Research*, 2, 181–186.

Kernan JA, Coxworth EC, Crowle WL, Spurr DT (1984) The nutritional value of crop residue components from several wheat cultivars grown at different fertilizer levels. *Animal Feed Science and Technology*, 11, 301–311.

Khanna M, Chen X (2013) Economic, energy security, and greenhouse gas effects of biofuels: implications for policy. *American Journal of Agricultural Economics*, 95, 1325–1331.

Kim S, Dale BE (2011) Indirect land use change for biofuels: testing predictions and improving analytical methodologies. *Biomass and Bioenergy*, 35, 3235–3240.

Knapp JS, Parton JH, Walton NI (1983) Enzymic saccharification of wheat straw – differences in the degradability of straw derived from different cultivars of winter wheat. *Journal of the Science of Food and Agriculture*, 34, 433–439.

Köhler L, Spatz H-C (2002) Micromechanics of plant tissues beyond the linear-elastic range. *Planta*, 215, 33–40.

Kong E, Liu D, Guo X et al. (2013) Anatomical and chemical characteristics associated with lodging resistance in wheat. *The Crop Journal*, 1, 43–49.

Kühner S (2013) *Deliverable feedstock costs*. Project co-funded by the European Commission FP7 Directorate-General for Transport and Energy Grant No. 282873. Available at: http://bioboost.eu/uploads/files/bioboost_d1.1-syncom_feedstock_cost-vers_1.0-final.pdf (accessed 18 August 2015).

Lafond GP, Stumborg M, Lemke R, May WE, Holzapfel CB, Campbell CA (2009) Quantifying straw removal through baling and measuring the long-term impact on soil quality and wheat production. *Agronomy Journal*, 101, 529.

Larsen SU, Bruun S, Lindedam J (2012) Straw yield and saccharification potential for ethanol in cereal species and wheat cultivars. *Biomass and Bioenergy*, 45, 239–250.

Lee C, Herbek J (2009) *A Comprehensive Guide to Wheat Management in Kentucky*. University of Kentucky College of Agriculture. Available at: http://www2.ca.uky.edu/agc/pubs/id/id125/id125.pdf (accessed 18 August 2015).

de Leon N, Coors JG (2008) Genetic improvement of corn for lignocellulosic feedstock production. In: *Genetic Improvement of Bioenergy Crops* (ed. Vermerris W), pp. 185–210. Springer, New York, USA.

Li X, Weng J-K, Chapple C (2008) Improvement of biomass through lignin modification. *The Plant Journal*, 54, 569–581.

Li F, Zhang M, Guo K et al. (2015) High-level hemicellulosic arabinose predominately affects lignocellulose crystallinity for genetically enhancing both plant lodging resistance and biomass enzymatic digestibility in rice mutants. *Plant Biotechnology Journal*, 13, 514–525.

Lindedam J, Bruun S, Jørgensen H, Felby C, Magid J (2010a) Cellulosic ethanol: interactions between cultivar and enzyme loading in wheat straw processing. *Biotechnology for Biofuels*, 3, 25.

Lindedam J, Bruun S, DeMartini J et al. (2010b) Near infrared spectroscopy as a screening tool for sugar release and chemical composition of wheat straw. *Journal of Biobased Materials and Bioenergy*, 4, 378–383.

Lindedam J, Andersen SB, DeMartini J et al. (2012) Cultivar variation and selection potential relevant to the production of cellulosic ethanol from wheat straw. *Biomass and Bioenergy*, 37, 221–228.

Lindedam J, Bruun S, Jørgensen H et al. (2014) Evaluation of high throughput screening methods in picking up differences between cultivars of lignocellulosic biomass for ethanol production. *Biomass and Bioenergy*, 66, 261–267.

Linden DR, Clapp CE, Dowdy RH (2000) Long-term corn grain and stover yields as a function of tillage and residue removal in east central Minnesota. *Soil and Tillage Research*, 56, 167–174.

Littlewood J, Murphy RJ, Wang L (2013) Importance of policy support and feedstock prices on economic feasibility of bioethanol production from wheat straw in the UK. *Renewable and Sustainable Energy Reviews*, 17, 291–300.

Long DS, McCallum JD (2013) Mapping straw yield using on-combine light detection and ranging (lidar). *International Journal of Remote Sensing*, 34, 6121–6134.

Long SP, Zhu XG, Naidu SL, Ort DR (2006) Can improvement in photosynthesis increase crop yields? *Plant, Cell and Environment*, 29, 315–330.

Lorenz AJ, Gustafson TJ, Coors JG, de Leon N (2010) Breeding maize for a bioeconomy: a literature survey examining harvest index and stover yield and their relationship to grain yield. *Crop Science*, 50, 1–12.

Lovett A, Sünnenberg G, Dockerty T (2014) The availability of land for perennial energy crops in Great Britain. *Global Change Biology Bioenergy*, 6, 99–107.

Ma QH (2009) The expression of caffeic acid 3-O-methyltransferase in two wheat genotypes differing in lodging resistance. *Journal of Experimental Botany*, 60, 2763–2771.

McCartney DH, Block HC, Dubeski PL, Ohama AJ (2006) Review: the composition and availability of straw and chaff from small grain cereals for beef cattle in western Canada. *Canadian Journal of Animal Science*, 86, 443–455.

Morris NL, Miller PCH, Froud-Williams RJ (2010) The adoption of non-inversion tillage systems in the United Kingdom and the agronomic impact on soil, crops and the environment—a review. *Soil and Tillage Research*, 108, 1–15.

Murozuka E, Laursen KH, Lindedam J et al. (2014) Nitrogen fertilization affects silicon concentration, cell wall composition and biofuel potential of wheat straw. *Biomass and Bioenergy*, 64, 291–298.

Murozuka E, de Bang TC, Frydenvang J et al. (2015) Concentration of mineral elements in wheat (Triticum aestivum L.) straw: genotypic differences and consequences for enzymatic saccharification. *Biomass and Bioenergy*, 75, 134–141.

Nasidi M, Agu R, Deeni Y, Walker G (2015) Improved production of ethanol using bagasse from different sorghum cultivars. *Biomass and Bioenergy*, 72, 288–299.

Naylor REL (1989) Effects of the plant growth regulator chlormequat on plant form and yield of triticale. *Annals of Applied Biology*, 114, 533–544.

Nilsson D (1999) SHAM – a simulation model for designing straw fuel delivery systems. Part 1: model description. *Biomass and Bioenergy*, 16, 25–38.

Oakey H, Shafiei R, Comadran J et al. (2013) Identification of crop cultivars with consistently high lignocellulosic sugar release requires the use of appropriate statistical design and modelling. *Biotechnology for Biofuels*, 6, 185.

Ohlde GW, Beck K, Akin DE, Rigsby LL, Lyon CE (1992) Differences in rumen bacterial degradation of morphological fractions in eight cereal straws and the effect of digestion on different types of tissues and mechanical properties of straw stalks. *Animal Feed Science and Technology*, 36, 173–186.

Oladosu G, Msangi S (2013) Biofuel-food market interactions: a review of modeling approaches and findings. *Agriculture*, 3, 53–71.

Parry MAJ, Reynolds M, Salvucci ME et al. (2011) Raising yield potential of wheat. II. Increasing photosynthetic capacity and efficiency. *Journal of Experimental Botany*, 62, 453–467.

Pauly M, Keegstra K (2008) Cell-wall carbohydrates and their modification as a resource for biofuels. *The Plant Journal*, 54, 559–568.

Pearman I, Thomas SM, Thorne GN (1978) Effect of nitrogen fertilizer on growth and yield of semi-dwarf and tall varieties of winter wheat. *The Journal of Agricultural Science*, **91**, 31–45.

Phitsuwan P, Ratanakhanokchai K (2014) Can we create "Elite Rice"—a multifunctional crop for food, feed, and bioenergy production? *Sustainable Chemical Processes*, **2**, 10.

Phitsuwan P, Sakka K, Ratanakhanokchai K (2013) Improvement of lignocellulosic biomass in planta: a review of feedstocks, biomass recalcitrance, and strategic manipulation of ideal plants designed for ethanol production and processability. *Biomass and Bioenergy*, **58**, 390–405.

Pinthus MJ (1973) Lodging in wheat, barley and oats; the phenomenon, its causes and preventative measures. *Advances in Agronomy*, **14**, 209–263.

Powlson DS, Glendining MJ, Coleman K, Whitmore AP (2011) Implications for soil properties of removing cereal straw: results from long-term studies. *Agronomy Journal*, **103**, 279–287.

Rajala A, Peltonen-Sainio P (2001) Plant growth regulator effects on spring cereal root and shoot growth. *Agronomy Journal*, **943**, 936–943.

Ramanzin M, Bailoni L, Beni G (1991) Varietal differences in rumen degradation of barley, wheat and hard wheat straws. *Animal Production*, **53**, 143–150.

Refsgaard K, Flaten O, Gudem R, Lien G (2002) A multicriteria evaluation by the public approval of pesticides – a case with the plant-growth regulators in grain. In: *13th International Farm Management Congress*. 2002, Wageningen, The Netherlands.

Reynolds M, Foulkes MJ, Slafer GA, Berry P, Parry MAJ, Snape JW, Angus WJ (2009) Raising yield potential in wheat. *Journal of Experimental Botany*, **60**, 1899–1918.

Reynolds M, Foulkes J, Furbank R *et al.* (2012) Achieving yield gains in wheat. *Plant, Cell and Environment*, **35**, 1799–1823.

Salas Fernandez MG, Becraft PW, Yin Y, Lübberstedt T (2009) From dwarves to giants? Plant height manipulation for biomass yield. *Trends in Pant Science*, **14**, 454–461.

Savoie J-M, Minvielle N, Chalaux N (1994) Estimation of wheat straw quality for edible mushroom production and effects of a growth regulator. *Bioresource Technology*, **48**, 149–153.

Scarlat N, Martinov M, Dallemand J-F (2010) Assessment of the availability of agricultural crop residues in the European Union: potential and limitations for bioenergy use. *Waste Management*, **30**, 1889–1897.

Sharma HSS, Faughey G, Chambers J, Lyons G, Sturgeon S (2000) Assessment of winter wheat cultivars for changes in straw composition and digestibility in response to fungicide and growth regulator treatments. *Annals of Applied Biology*, **137**, 297–303.

Shearman VJ, Sylvester-Bradley R, Scott RK, Foulkes MJ (2005) Physiological processes associated with wheat yield progress in the UK. *Crop Science*, **45**, 175–185.

Shortall OK (2013) 'Marginal land' for energy crops: exploring definitions and embedded assumptions. *Energy Policy*, **62**, 19–27.

Simbolotti G (2013) *Production of Liquid Biofuels: Technology Brief*. The International Renewable Energy Agency (IRENA) and The Energy Technology Systems Analysis Programme (ETSAP). Available at: https://www.irena.org/DocumentDownloads/Publications/IRENA-ETSAP%20Tech%20Brief%20P10%20Production_of_Liquid%20Biofuels.pdf (accessed 18 August 2015).

Simkins G (2014) *40 MW Suffolk biomass project abandoned*. Available at: http://www.endswasteandbioenergy.com/article/1305899/40mw-suffolk-biomass-project-abandoned (accessed 18 August 2015).

Skøtt T (2011) *Straw to Energy: Status, Technologies and Innovation in Denmark 2011*. Innovation Network for Biomass, Agro Business Park A/S, Tjele, Denmark.

Slafer G, Andrade FH (1989) Genetic improvement in bread wheat (*Triticum aestivum*) yield in Argentina. *Field Crops Research*, **21**, 289–296.

Smil V (1999) Crop residues: agriculture's largest harvest. *BioScience*, **49**, 299–308.

Smith TC, Kindred DR, Brosnan JM, Weightman RM, Shepherd M, Sylvester-Bradley R (2006) *Wheat as a Feedstock for Alcohol Production*. Research review no. 61. HGCA, Agriculture and Horticulture Development Board, Stoneleigh Park, Kenilworth, Warwickshire.

Snyder CS, Bruulsema TW, Jensen TL, Fixen PE (2009) Review of greenhouse gas emissions from crop production systems and fertilizer management effects. *Agriculture, Ecosystems and Environment*, **133**, 247–266.

Sterling M, Baker CJ, Berry PM, Wade A (2003) An experimental investigation of the lodging of wheat. *Agricultural and Forest Meteorology*, **119**, 149–165.

Summers MD, Jenkins BM, Hyde PR, Williams JF, Mutters RG, Scardacci SC, Hair MW (2003) Biomass production and allocation in rice with implications for straw harvesting and utilization. *Biomass and Bioenergy*, **24**, 163–173.

Tolera A, Tsegaye B, Berg T (2008) Effects of variety, cropping year, location and fertilizer application on nutritive value of durum wheat straw. *Journal of Animal Physiology and Animal Nutrition*, **92**, 121–130.

Travis AJ, Murison SD, Hirst DJ, Walker KC, Chesson A (1996) Comparison of the anatomy and degradability of straw from varieties of wheat and barley that differ in susceptibility to lodging. *Journal of Agricultural Science*, **127**, 1–10.

Valentine J, Clifton-Brown J, Hastings A, Robson P, Allison G, Smith P (2012) Food vs. fuel: the use of land for lignocellulosic 'next generation' energy crops that minimize competition with primary food production. *Global Change Biology Bioenergy*, **4**, 1–19.

Waddington SR, Ransom JK, Osmanzai M, Saunders DA (1986) Improvement in the yield potential of bread wheat adapted to Northwest Mexico. *Crop Science*, **26**, 698–703.

Wang J, Zhu J, Huang R, Yang Y (2012) Investigation of cell wall composition related to stem lodging resistance in wheat (Triticum aestivum L.) by FTIR spectroscopy. *Plant Signaling & Behavior*, **7**, 856–863.

Whaley JM, Sparkes DL, Foulkes MJ, Spink JH, Semere T, Scott RK (2000) The physiological response of winter wheat to reductions in plant density. *Annals of Applied Biology*, **137**, 165–177.

White EM, Wilson FEA (2006) Responses of grain yield, biomass and harvest index and their rates of genetic progress to nitrogen availability in ten winter wheat varieties. *Irish Journal of Agricultural and Food Research*, **45**, 85–101.

Whittaker C, Borrion AL, Newnes L, McManus M (2014) The renewable energy directive and cereal residues. *Applied Energy*, **122**, 207–215.

Wilson PN, Brigstocke T (1977) The commercial straw process. *Process Biochemistry*, **12**, 17–20.

Wilson P, Glithero NJ, Ramsden SJ (2014) Prospects for dedicated energy crop production and attitudes towards agricultural straw use: the case of livestock farmers. *Energy Policy*, **74**, 101–110.

Wu Z, Hao H, Tu Y *et al.* (2014) Diverse cell wall composition and varied biomass digestibility in wheat straw for bioenergy feedstock. *Biomass and Bioenergy*, **70**, 347–355.

Zhao Q, Dixon RA (2014) Altering the cell wall and its impact on plant disease: from forage to bioenergy. *Annual Review of Phytopathology*, **52**, 69–91.

Zhao X, Zhang L, Liu D (2012) Biomass recalcitrance. Part 1: the chemical compositions and physical structures affecting the enzymatic hydrolysis of lignocellulose. *Biofuels, Bioproducts and Biorefining*, **6**, 465–482.

Zhou B, Sanz-Sáez Á, Elazab A *et al.* (2014) Physiological traits contributed to the recent increase in yield potential of winter wheat from Henan Province, China. *Journal of Integrative Plant Biology*, **56**, 492–504.

10

More plant growth but less plant defence? First global gene expression data for plants grown in soil amended with biochar

MAUD VIGER[1], ROBERT D. HANCOCK[2], FRANCO MIGLIETTA[3,4] and GAIL TAYLOR[1]

[1]Centre for Biological Sciences, University of Southampton, Southampton SO17 1BJ, UK, [2]Cell &Molecular Sciences, The James Hutton Institute, Invergowrie, Dundee DD2 5DA, UK, [3]Institute of Biometeorology (IBIMET), National Research Council (CNR), Via Caproni 8, Firenze 50145, Italy, [4]FoxLab, Forest & Wood Science, San Michele All'Adige 38010, Italy

Abstract

Biochar is a carbon (C)-rich solid formed when biomass is used to produce bioenergy. This 'black carbon' has been suggested as a solution to climate change, potentially reducing global anthropogenic emissions of greenhouse gases by 12%, as well as promoting increased crop growth. How biochar application to soil leads to better crop yields remains open to speculation. Using the model plant *Arabidopsis* and the crop plant lettuce (*Lactuca sativa* L.), we found increased plant growth in both species following biochar application. Statistically significant increases for *Arabidopsis* in leaf area (130%), rosette diameter (61%) and root length (100%) were observed with similar findings in lettuce, where biochar application also increased leaf cell expansion. For the first time, global gene expression arrays were used on biochar-treated plants, enabling us to identify the growth-promoting plant hormones, brassinosteroid and auxin, and their signalling molecules, as key to this growth stimulation, with limited impacts on genes controlling photosynthesis. In addition, genes for cell wall loosening were promoted as were those for increased activity in membrane transporters for sugar, nutrients and aquaporins for better water and nutrient uptake and movement of sugars for metabolism in the plant. Positive growth effects were accompanied by down-regulation of a large suite of plant defence genes, including the jasmonic acid biosynthetic pathway, defensins and most categories of secondary metabolites. Such genes are critical for plant protection against insect and pathogen attack, as well as defence against stresses including drought. We propose a conceptual model to explain these effects in this biochar type, hypothesizing a role for additional K^+ supply in biochar amended soils, leading to Ca^{2+} and Reactive Oxygen Species (ROS) –mediated signalling underpinning growth and defence signalling responses.

Keywords: Arabidopsis thaliana, biochar stimulated growth, carbon sequestration, geoengineering, global gene expression, microarrays, plant immunity

Introduction

Biochar is formed as a byproduct when biomass, from bioenergy crops, organic wastes or crop residues is used to produce liquid or gaseous bioenergy with no or a limited supply of oxygen, at temperatures between 300 and 1200 °C. Biochar application to soil has been widely suggested as an option to mitigate climate change through carbon sequestration in long-term soil pools (Lehmann *et al.*, 2006). A consequential reduction in global anthropogenic emissions of greenhouse gases by up to 12% has been proposed, (Woolf *et al.*, 2010), but the wide scale use of biochar application is controversial, since much remains unknown on the long-term impacts in managed and natural ecosystems. For example, many reports show that biochar application has a positive effect in enhancing crop yields. A meta-analysis for plant productivity after biochar application revealed a significant mean increase of 10% (Jeffery *et al.*, 2011), with some studies reporting growth stimulations of more than 50%, although the response was highly variable with both negative and positive effects observed. There remains limited understanding of how enhanced crop growth is achieved, but improved water and nutrient retention, increased soil pH, effects on soil microbes and ethylene production have all been proposed to have a role (Lehmann & Joseph, 2009; Sohi *et al.*, 2010; Spokas *et al.*, 2010). Properties of biochar can also vary depending on different feedstocks (Sohi, 2012) and results on crop production are influenced by experimental design, soil and biochar properties (Jeffery *et al.*, 2011).

Correspondence: Gail Taylor
e-mail: g.taylor@soton.ac.uk

Interest in biochar has increased considerably in the past decade with research focussed on the potential for durable carbon sequestration and increased plant productivity. Despite this, there are relatively few studies on 'modern biochar'—made from pyrolysis or gasification processes in a bioenergy system and the mechanisms of action for altered soil and plant functioning remain largely unproven. There has been no reported information on the underlying plant genetic processes following biochar application that lead to stimulated plant growth. Global gene expression arrays are now used routinely for gene discovery in plants, animal and microbes and have enabled increased understanding of how these systems work, particularly in relation to the environment. For example, we now understand how plants respond to drought (Street et al., 2006), to attack by pests and pathogens (Barah et al., 2013) and variation between difference ecotypes (Kusnierczyk et al., 2007). Microarrays are a particularly useful technology to discover genes expressed during a process that were not predicted a priori (Richmond & Somerville, 2000), revealing novel insight into a problem that may then be followed up in 'hypothesis driven' experimentation. No global gene expression profiling has as yet been reported for plants grown in soil amended with biochar and our understanding of gene regulation following this treatment is limited to just an handful of genes. For example, the expression of five defence-related genes (Meller Harel et al., 2012) were assessed in plants growing in soil mixed with biochar.

Here, we report the effects of biochar application on the model plant Arabidopsis as well as for the crop, lettuce. We provide the first global gene expression data for plants exposed to biochar and reveal important insights into the growth-promoting effects of biochar and unpredicted and highly novel effects on plant defence signalling.

Material and methods

Biochar, soil and plant material and growth conditions

Biochar was obtained through gasification of poplar wood chips (where 80% of the chips were sized 16–30 mm diameter), from a 5 year old short rotation forest grown in Northern Italy, where carbonaceous material is partially oxidized by heating at

1200 °C. The gasifier was a fixed bed, down draft, open core design of AGT (Advanced Gasification Technology, Italy), with biochar characterized as having a compacted bulk density of 252 g l^{-1}: pH, 10.6; ash, 15% Dry Mass (DM); total N, P, K by DM, 1.6, 0.2, 1.8% respectively. Depending on the experiment, biochar was applied at rates equivalent to field application: 0, 20, 30, 50 and 100 t ha^{-1}, assuming a soil bulk density of 0.4 g cm^{-3} to a depth of 30 cm, achieved by mixing 0%, 1.6%, 2.5%, 4.2% and 8.3% (kg of biochar per kg of soil), following the methods in Baronti et al. (2010). Soil and biochar were precisely weighed for each condition and mixed in large boxes to homogenize the mixture before filling the pots. We note that soil bulk density of this soil was low, but within the observed range, particularly for organic soils. Even in organic soils, samples derived from soil cores are likely to be more closely compacted with less porosity than the soil mixture that was used in this laboratory experiment and should be considered in interpreting our results more widely.

In a series of experiments, seeds of Arabidopsis thaliana (ecotype Columbia-0) and lettuce (Lactuca sativa L.) were sown into commercially available top soil (Table 1). The texture of the top soil was loamy. When fertilizer was used, the rate of application was 45 g m^{-2}, corresponding to 108, 22.5, 36 and 9 kg ha^{-1} of nutrients applied of nitrogen, phosphorus, potassium and magnesium oxide, respectively. Temperatures were maintained at 23 °C daytime and 18 °C night-time, humidity at 60% and 55% (day and night, respectively) with a photoperiod of 16 h per day at a PAR of 130 μmol m^{-2} s^{-1}. Soil, with and without biochar, was analysed for pH, P, K, Mg, total nitrogen and Soil Organic Matter (Table 1).

Growth measurements and global gene expression with Arabidopsis thaliana

Twenty replicates of Arabidopsis thaliana were planted for each of five biochar conditions (0, 20, 30, 50 and 100 t ha^{-1}), 10 with and 10 without fertilizer, in a fully replicated and randomized design, where pots were rerandomized within treatments every few days throughout the experiment. Images of each pot were taken from 13 Days After Planting (DAP) every 3 days until 28 DAP and processed using ImageJ for leaf area, rosette diameter and leaf number. Measurements of height, stem diameter and root length were recorded at 32 DAP.

Leaves were sampled for microarrays 33 DAP and snap frozen in liquid nitrogen. RNA was extracted (Chang et al., 1993) for three replicates of control and five of biochar (50 t ha^{-1}) with fertilizer, and were hybridized to GeneChip Arabidopsis Genome Arrays (Affymetrix, Santa Clara, CA, USA) by the European Arabidopsis Stock Centre (NASC, Nottingham, UK). Data were

Table 1 Soil analysis, pH, P, K, Mg, nitrogen contents and Soil Organic Matter (SOM)

Soil samples	Soil pH	P (kg ha^{-1})	K (kg ha^{-1})	Mg (kg ha^{-1})	Total Nitrogen (% w/w)	SOM (%)
Soil	6.6	21	108	516	0.15	15.6
Soil + Biochar (50 t ha^{-1})	7.2	30.6	795	513	0.21	15.9

analysed in GeneSpring (Agilent technologies, Santa Clara, CA, USA). Chips were normalized by MAS5. Differentially expressed genes were identified through t-test between samples of control and biochar ($P < 0.05$, 2-fold change difference) with a multiple testing correction (Benjamini-Hochberg). Genes were annotated using GeneSpring (Agilent technologies, Santa Clara, CA, USA) and the website Tair (www.arabidopsis.org). Pathways were analysed using the software Mapman (http://mapman.gabipd.org/web/guest/mapman). A gene ontology analysis was also performed on AGRIGO (Du *et al.*, 2010) using the tool PAGE (Parametric Analysis of Gene Set Enrichment). A hierarchical clustering was performed focusing on five Gene Ontology groups using MeV MultiExperiment Viewer (http://www.tm4.org/mev/). Candidate genes were selected for real-time PCR. Forward and Reverse primers were designed using Primer-BLAST (www.ncbi.nlm.nih.gov/tools/primer-blast). Amplification efficiency was measured following Liu & Saint (2002).

Secondary metabolites

A second *Arabidopsis* experiment was conducted after microarray analysis using 20 replicates at 0 and 50 t ha^{-1} with fertilizer to confirm the role of biochar in altering the biosynthesis of secondary metabolites. Lyophilized leaf samples were extracted and partitioned according to the method of Foito *et al.* (2013) with the exception that internal standards were omitted. Following partitioning, morin was added to the polar fraction as internal standard to a final concentration of 0.5 mM. Glucosinolates were separated on a Synergi Hydro RP18 column (3 μm, 150 × 2 mm, Phenomenex) using a gradient consisting of 0.2% formic acid in water (A) and 0.2% formic acid in 90% acetonitrile (B). The flow rate was a constant 200 μl min^{-1} and the flow gradient was 0–50% B over 30 min. Glucosinolates, anthocyanins and flavonols were identified and quantified as previously described (Tohge *et al.*, 2005; Rochfort *et al.*, 2008; Yonekura-Sakakibara *et al.*, 2008) and are presented as peak area of the identified compound relative to peak area of the internal standard.

Growth measurements with lettuce

Following initial analysis with *Arabidopsis*, a model plant, we wished to confirm our findings with a second crop species and chose lettuce. Ten replicates for three biochar conditions were planted for lettuce, selected from across the range used in *Arabidopsis* experiments: 0, 50 and 100 t ha^{-1} which all had fertilizer added to the soil. As with the *Arabidopsis* experiment, images of each pot were taken from 15 DAP every 3 days until 25 DAP and processed using ImageJ for leaf area, rosette diameter and leaf number. Aboveground fresh and dry weight was measured at the end (35 DAP). Leaf cell expansion was further investigated following results with microarrays that identified cell walls as key targets for gene expression change, where leaf cell size was assessed at 0 and 50 t ha^{-1} biochar application rates. Cell measurements were taken on the adaxial side of a mature leaf (number five from the newly emerged leaf) using nail vanish (Gardner *et al.*, 1995). Imprints were collected with a Zeiss microscope at × 40 magnification attached with a camera capturing the images. Cell area (μm^2) was calculated using ImageJ from an average of 10 cells in the field of view (40 000 μm^2).

Statistical analysis

Statistical analysis was undertaken using the SPSS software (SPSS, Chicago, IL, USA). Normality was tested (Kolmogorov–Smirnov test) and data transformed as necessary (square root or log$_{10}$). For the growth experiment using *Arabidopsis* plants, biochar and fertilizer effects on leaf area, rosette diameter, leaf number, height, stem diameter and root length were analysed using a two-way ANOVA test to quantify the effects of each treatment and their interaction (biochar × fertilizer):

$$Y_{ij} = \mu + \alpha_i + \beta j + \varepsilon$$

where Y_{ij} is the phenotype in the ith biochar condition and in the jth fertilizer condition, α_i is the biochar effect, β_j is the fertilizer effect, and ε is the residual error. A Student Newman Keuls *post-hoc* test was also performed. Secondary metabolites measured in *Arabidopsis* plants were analysed using a t-test between the two biochar rate applications (0 and 50 t ha^{-1}). For the growth experiment and the cell measurements using lettuce plants, only the biochar effect was tested with a one-way ANOVA test as fertilizer was applied on all the pots:

$$Y_i = \mu + \alpha_i + \varepsilon$$

where Y_j is the phenotype in the ith biochar condition, α_i is the biochar effect, β_j is the fertilizer effect, and ε is the residual error. A Student Newman Keuls *post-hoc* test was also performed.

Leaf area for *Arabidopsis* and lettuce experiments was also analysed over time with a repeated measurement test with time, biochar and fertilizer as effects.

Results

Plant growth is enhanced by biochar

Analysis of soil with and without biochar for pH, phosphorus (P), potassium (K), magnesium (Mg) and total nitrogen content is given (Table 1). An increase in pH, P and K contents was observed in soil mixed with biochar. We used the model plant *Arabidopsis* grown in a soil closer to *realistic* soil conditions (commercial top soil) in comparison to the typical *Arabidopsis* soil mixture of compost and vermiculite, to study growth response following biochar application (Figs 1, 2; Table 2). Leaf area of the *Arabidopsis* plants was observed over time for all biochar application rates and fertilizer treatments from images of the pots (Fig. 1a, 2a, b). Using a repeated measurement test the whole dataset showed significant biochar ($F_{4,90} = 5.74$, $P < 0.001$), fertilizer ($F_{1,90} = 37.71$, $P < 0.001$) and time effects ($F_{1,90} = 182.77$, $P < 0001$), but no interactions.

Fig. 1 *Arabidopsis* rosette growth (a) at 0, 50 and 100 t ha^{-1} equivalent biochar application with fertilizer over time and percentage change of biomass (b) between control and 50 t ha^{-1} biochar rate application. Asterisks indicate a significant biochar effect (*** $P \le 0.001$, ** $P \le 0.01$, * $P \le 0.05$, n.s. nonsignificant) by ANOVA test.

When studying the traits towards the end of the experiment (28 DAP), leaf area, leaf number and rosette diameter (Fig. 2c, d; Table 2) showed a significant increase due to biochar application (Leaf area: $F_{4,90} = 7.16$, $P < 0.001$; Rosette diameter: $F_{4,90} = 11.25$, $P < 0.001$; Leaf number: $F_{4,90} = 10.07$, $P < 0.001$). Fertilizer also significantly increased leaf area ($F_{1,90} = 87.06$, $P < 0.001$), rosette diameter ($F_{1,90} = 57.05$, $P < 0.001$) and leaf number ($F_{1,90} = 51.60$, $P < 0.001$). Indeed plants were larger when fertilizer was applied regardless of the biochar application. No significant interaction of biochar x fertilizer was observed. A post-hoc test, revealed plants grown with biochar at rates of application of 20, 30, 50 and 100 t ha^{-1} were significantly bigger than those grown without biochar for leaf area, rosette diameter and leaf number. At 28 DAP; leaf area was largest when biochar was applied at 50 t ha^{-1} in combination with fertilizer (Table 2).

Height, stem diameter and root length were also measured (Table 2; Fig. 2e, f) at the end of the experiment (32 DAP), all showing a significant increase in response to biochar application (Height: $F_{4,40} = 10.12$, $P < 0.001$; Stem diameter: $F_{4,40} = 6.14$, $P < 0.001$; Root length: $F_{4,40} = 4.95$, $P < 0.01$). With the exception of root length ($F_{1,40} = 3.38$, $P = 0.074$), fertilizer also had a positive effect on height and stem diameter ($F_{1,40} = 6.08$, $P < 0.05$ and $F_{1,40} = 19.47$, $P < 0.001$, respectively). No interaction of biochar x fertilizer was observed for these traits.

Growth measurements were also made on lettuce (Figs 3, 4; Table 3), with similar results. Fertilizer was applied to all lettuce plants. Like *Arabidopsis*, lettuce produced larger leaves over time (Figs 3, 4a) and the repeated measurement test showed a biochar ($F_{2,27} = 29.09$, $P < 0.001$) and time effect ($F_{1,27} = 2042.59$, $P < 0.001$). Towards the end of the experiment, leaf area, rosette diameter and leaf number were increased when biochar was applied (Leaf area: $F_{2,27} = 39.91$, $P < 0.001$; Rosette diameter: $F_{2,27} = 33.23$, $P < 0.001$; Leaf number: $F_{2,27} = 20.87, P < 0.001$). As the *post-hoc* test revealed (Table 3), values were similar between the biochar rate at 50 and 100 t ha^{-1} (Fig. 4a–c) and significantly different to plants grown without biochar. Leaf area increased from 3360 mm^2 without biochar to 5195 mm^2 on average with biochar, showing a 55% increase with biochar application. Rosette diameter increased by 31% from 159 mm without biochar to 209 mm with biochar. Finally, lettuce plants had on average 6 leaves per plant without biochar compared to 7.5 leaves with biochar, showing a percentage change of 25% in response to biochar application.

Fresh and dry weight were also measured for the lettuce plants 35 DAP (Fig. 4d) showing a significant increase in response to biochar application (Fresh: $F_{2,12} = 28.86$, $P < 0.001$; Dry: $F_{2,12} = 42.87$, $P < 0.001$). Dry weight increased by 111% from 0.58 g at 0 t ha^{-1} biochar to an average of 1.24 g at 50 and 100 t ha^{-1}.

Gene expression in response to biochar application

The microarray analysis revealed a total of 1076 genes differently expressed when comparing control plants with plants grown in biochar at 50 t ha^{-1}, both with fertilizer (Table S1). Leaf area was the largest at the biochar application rate of 50 t ha^{-1} with fertilizer and was thus selected for the global gene expression analysis and compared to plants growth without biochar but with fertilizer to focus on the biochar effect. We identified 571 genes that were down-regulated and 505 that were up-regulated in response to biochar application. Using

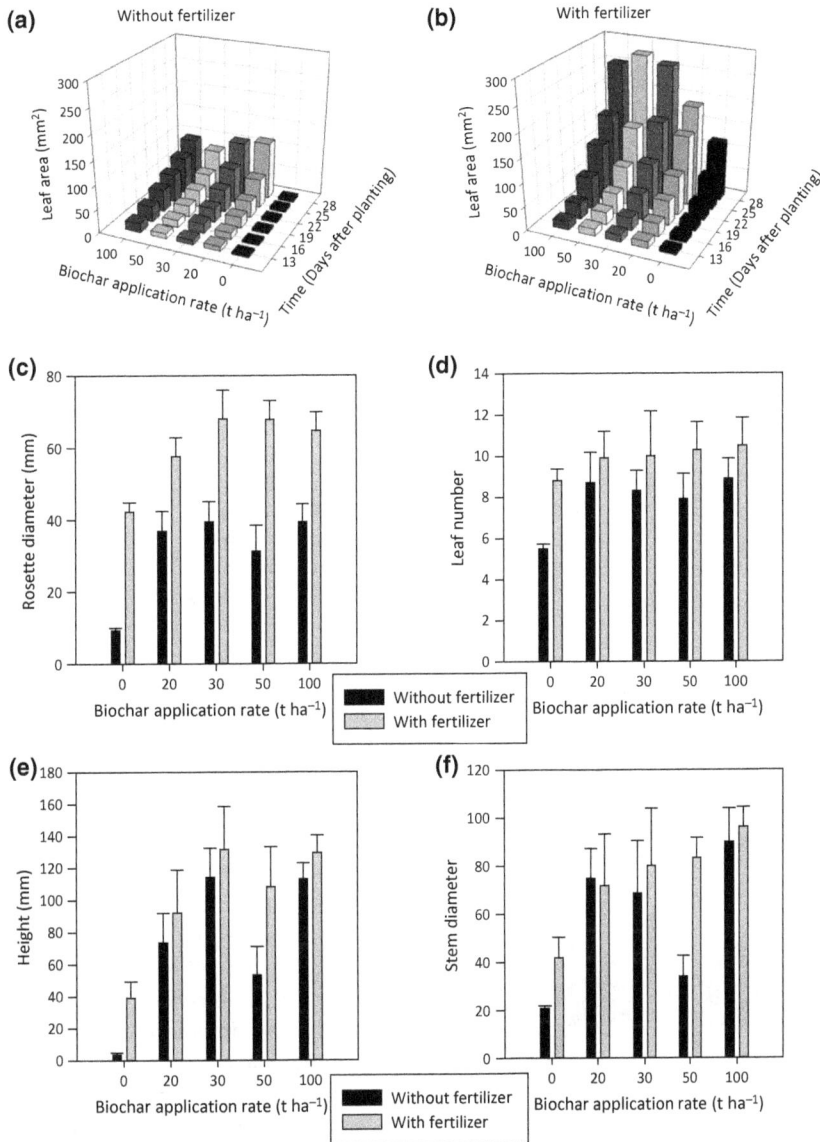

Fig. 2 Growth in response to biochar application in *Arabidopsis thaliana*: leaf area evolution over time related to biochar rate application without (a) and with fertilizer (b), rosette diameter (c) and leaf number (d) at 28 DAP, height (e) and stem diameter (f) at 32 DAP.

the AGRIGO software (Du *et al.*, 2010), a complete view of the gene ontology (GO) interaction was completed using the gene expression data (Figure S3; Table S2). Recurrent groups were observed, showing an effect of biochar application, included growth (Figure S3a), cell morphogenesis (Figure S3a), response to stimulus or stress (Figure S3b), hormones, such as jasmonic acid, auxin and cytokinin (Figure S3b), and secondary metabolism (Figure S3c). The three GO categories significantly represented with the highest z-score were xyloglucan: xyloglucosyl transferase activity, intrinsic to membrane and plant-type cell wall, whilst the groups with the lowest z-score were secondary metabolic process, response

to wounding and response to jasmonic acid stimulus (Table S2). Genes within those groups, except response to wounding, were extracted and used to construct a hierarchical clustering (Figure S1). The latter showed a consistent up-regulation for all the biochar samples for xyloglucan:xyloglucosyl transferase activity, intrinsic to membrane and plant-type cell wall and a down-regulation for secondary metabolic process and response to jasmonic acid stimulus.

Few effects were found for genes controlling photosynthetic pathways (Table S1). In contrast, signalling, transport and biosynthesis of two plant hormones central to growth stimulation – auxin and brassinosteroid – were

Table 2 Average values and standard errors of growth in *Arabidopsis* at different rates of biochar, with and without fertilizer, and statistical results presenting the p-value for each trait using a GLM test for the biochar and fertilizer effects and the interaction biochar x fertilizer effect. Bold values are significant ($P < 0.05$). The letters correspond to the *post-hoc* results

Biochar application (t ha^{-1})	Fertilizer	Leaf area 28 DAP (mm^2)	Rosette diameter 28 DAP (mm)	Leaf number 28 DAP	Height 32 DAP (mm)	Stem diameter 32 DAP (mm)	Root length 32 DAP (mm)
0	No	7.7 ± 0.6^a	9.3 ± 0.4^a	5.5 ± 0.2^a	3.7 ± 1.1^a	0 ± 0^a	20.8 ± 1.1^a
20	No	118.0 ± 37.4^{a,b}	36.9 ± 4.9^{b,c}	8.7 ± 0.3^{b,c,d}	73.5 ± 18.5^{b,c}	0.5 ± 0.09^{a,b}	75.1 ± 12.3^{a,b}
30	No	109.2 ± 33.6^{a,b}	39.5 ± 6.4^{b,c}	8.3 ± 0.3^{b,c}	114.3 ± 18.3^{b,c}	0.5 ± 0.11^{a,b}	68.6 ± 21.9^{a,b}
50	No	81.8 ± 39.4^{a,b}	31.3 ± 8.9^b	7.9 ± 0.8^b	53.3 ± 17.8^{a,b,c}	0.2 ± 0.08^{a,b}	34.0 ± 8.6^{a,b}
100	No	96.7 ± 21.3^{a,b}	39.4 ± 4.9^{b,c}	8.9 ± 0.5^{b,c,d}	113.1 ± 10.2^{b,c}	0.4 ± 0.05^{a,b}	90.0 ± 13.9^b
0	Yes	122.0 ± 22.7^{a,b}	42.2 ± 3.6^{b,c}	8.8 ± 0.6^{b,c,d}	39.1 ± 10.1^{a,b}	0.5 ± 0.09^{a,b}	41.8 ± 8.4^{a,b}
20	Yes	190.1 ± 31.4^{b,c}	57.6 ± 5.8^{c,d}	9.9 ± 1.4^{c,d}	92.1 ± 26.8^{b,c}	0.7 ± 0.08^{b,c}	71.9 ± 21.5^{a,b}
30	Yes	265.9 ± 53.8^{b,c}	68.1 ± 8.8^d	10.0 ± 1.6^{c,d}	132.0 ± 26.7^c	0.6 ± 0.15^{b,c}	80.2 ± 23.9^{a,b}
50	Yes	280.6 ± 34.8^{b,c}	67.8 ± 5.0^d	10.3 ± 1.5^{c,d}	108.4 ± 25.1^{b,c}	0.7 ± 0.12^{b,c}	83.4 ± 8.4^{a,b}
100	Yes	255.1 ± 41.1^{b,c}	64.8 ± 6.7^d	10.5 ± 1.2^d	129.7 ± 10.9^c	0.9 ± 0.06^c	96.3 ± 8.2^b
Statistics P-value	Biochar	**0.003**	**<0.001**	**<0.001**	**<0.001**	**0.001**	**0.002**
	Fertilizer	**<0.001**	**<0.001**	**<0.001**	**0.018**	**<0.001**	0.074
	Biochar × Fertilizer	0.418	0.717	0.060	0.799	0.409	0.450

Time

13 days   19 days   25 days

Biochar rate application (t ha^{-1})

0 t ha^{-1}

50 t ha^{-1}

100 t ha^{-1}

Fig. 3 Lettuce growth over time (13 to 25 days after planting) at 0, 50 and 100 t ha^{-1} equivalent biochar application with fertilizer.

up-regulated for plants grown with biochar (Fig. 5a) Here, auxin conjugation, for example, controlled by *IAR3*, *ILL5* and *ILL6* was reduced by −4.13, −2.36 and −3 fold change, respectively (Table S3), leading to the production of more free auxin (Leclere *et al.*, 2002) and auxin responsive proteins were stimulated (Table S3). *IAR3* was also identified by Leclere *et al.* (2002) as a possible link between wound responses and auxin with *IAR3* proposed to control hydrolysis of amino acid

conjugates of jasmonic acid. Similarly for brassinosteroids, genes promoting hormone biosynthesis were up-regulated (*DWF4*, *CYP72C1*, *SMT2*, *SMT3*: 2.19, 6.15, 2.18 and 4.01 fold change, respectively). *DWF4*, for example, encodes an important steroid for a rate limiting step in BR biosynthesis (Choe *et al.*, 2001). Increased plant biomass in biochar in this study was therefore characterized by large stimulations in genes controlling brassinosteroid and auxin biosynthesis and signalling, with limited impacts on genes controlling the photosynthetic machinery (Fig. 5a). In addition, genes controlling cell wall loosening, including numerous xyloglucan endotransglucosylases and expansins (Fig. 5a; Table S3) were up-regulated in biochar suggesting enhanced growth was underpinned by increased cell expansion. For example, *XTH9*, *XTH33*, *XTR3*, *XTR4*, *XTR6*, *XTR7*, *XTH17*, *TCH4*, *EXGT-A3* and *EXGT-A1* were xyloglucan endotransglucosylases and xyloglucan:xyloglucosyl transferases that were up-regulated in response to biochar with a fold change between 2.3 to 16.57. *ATEXLA1*, *ATEXLA2* and *ATEXPA11* were expansins that were up-regulated in plants growing with biochar by 15.20, 4.42 and 2.44 fold change respectively while only two expansins *ATEXPA4* and *ATEXLB1* were down regulated with a fold change of −2.44 and −2.66, respectively (Table S3). Cell wall proteins were also promoted, such as *AGP24*, *AGP7*, *AGP20*, *AGP17*, *AGP22*, *FLA13* and *FLA9*, with a fold change varying from 2.22 to 11.76.

For lettuce, cell expansion was assessed on a mature leaf growing with and without biochar (50 and 0 t ha^{-1}), with larger cells observed for plants grown with added biochar, although this was only apparent as a

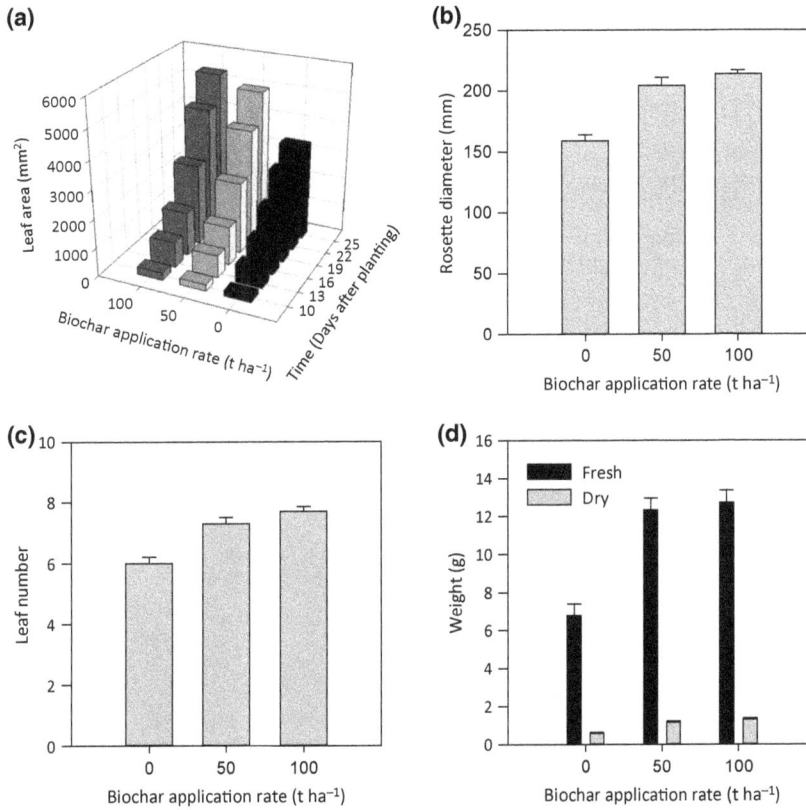

Fig. 4 Growth in response to biochar application in lettuce: leaf area evolution over time depending on biochar rate application with fertilizer (a), rosette diameter (b) and leaf number (c) at 25 DAP, fresh and dry weight (d) at 35 DAP.

Table 3 Average values and standard errors of growth in lettuce at different rates of biochar, with fertilizer and statistical results presenting the *P*-value for each trait using a GLM test for biochar effect. Bold values are significant (*P* < 0.05). The letters correspond to the *post-hoc* results

Biochar application (t ha^{-1})	Fertilizer	Leaf area 25 DAP (mm2)	Rosette diameter 25 DAP (mm)	Leaf number 25 DAP	Fresh weight 35 DAP (g)	Dry weight 35 DAP (g)
0	Yes	3359.6 ± 155.3[a]	159.0 ± 4.9[a]	6 ± 0.2[a]	6.79 ± 0.6[a]	0.59 ± 0.05[a]
50	Yes	5001.3 ± 218.0[b]	204.4 ± 6.5[b]	7.3 ± 0.2[b]	12.33 ± 0.6[b]	1.16 ± 0.07[b]
100	Yes	5389.6 ± 125.1[b]	213.8 ± 3.3[b]	7.7 ± 0.1[b]	12.71 ± 0.6[b]	1.32 ± 0.06[b]
Statistics *P*-value	Biochar	**<0.001**	**<0.001**	**<0.001**	**<0.001**	**<0.001**

trend and significant at the 10% level of probability (Fig. 6, $F_{1,98}$ = 2.95, *P* = 0.089 898 μm^2 compared to 805 μm^2 in control. Cell number (per field of view), stomatal density and stomatal index (not reported) did not differ between treatments ($F_{1,98}$ = 0.12, *P* = 0.73; $F_{1,98}$ = 0.01, *P* = 0.91; $F_{1,98}$ = 0.12, *P* = 0.73, respectively).

Increased growth was also associated with enhanced activity in membrane transporter genes for sugar, nutrients and aquaporins for better water and nutrient uptake and movement of sugars for metabolism in the plant. *ERD6*, *STP1* and *STP4* were up-regulated sugar and sucrose transporter (fold change: 3.15, 7.98 and 2.37,

respectively). *PIP1C*, *PIP1,5* and *DELTA-TIP* were water channels up-regulated in response to biochar application with fold change from 2.21 to 3.19 (Table S3). One of the largest up-regulations was observed in two genes related to signalling in sugar and nutrient physiology, *EXO* – a phosphate-responsive protein, which also responds to brassinosteroid stimulus – and *PHI-1* – a phosphate-induced protein (Table S3). Signalling G-proteins which are involved in the regulation of signal transduction and cell proliferation (Vernoud *et al.*, 2003) were mainly up-regulated in response to biochar application, such as *ROP2*, *RAB7*, *RABE1d*, *RAB7A* and *RABA1g*. Expression

BIOCHAR APPLICATION

(a) Plant growth stimulated

BR increased ← Brassinosteroid biosynthesis and signalling EXL5, DWF4, SMT2, CYP72C1, BIM1

Plant growth, cell elongation and division promoted (Choe et al., 1998; Shröder et al., 2011)

Auxin biosynthesis IAR3, ILL6

Auxin freely released (LeClere et al., 2002)

Auxin transport and signalling CYP83B1, CYP711A1, GRH1, SAUR-like proteins, IAA19, IAA6, SHY2, MYB proteins

Cell wall modification XTH9, XTH33, ATEXLA1, ATEXLA2, XTR4, XTR7, TCH4

Cell wall proteins AGP7, AGP20, AGP22, FLA9

Stimulated tissue morphogenesis, cell growth and division (Du et al., 1996)

Cell elongation promoted (McQueen-Mason, 1995)

Improved transmembrane sugar transport in sink tissues (Szenthe et al., 2007)

Sugar and water transport SPT1, ERD6, STP4, PIP1D, PIP1C, TIP2

Aquaporins improved water transport (Postaire et al., 2010)

Log2: >3, 2 to 3, 1 to 2, 0, −1 to −2, −2 to −3, >−3

Signalling in sugar and nutrient physiology EXO, PHI-1

Signalling G-proteins ARAC4, RAB7, RAB7A, RAB8C

Involved in signalling process coordination BR responses and growth promoted (Shröder et al., 2011)

Regulated signal transduction, cell proliferation, cytoskeletal organisation and intracellular membrane trafficking (Vernoud et al., 2003)

(b) Plant defence reduced

Salicylic acid biosynthesis ↓ FAMT ↔ Jasmonic acid biosynthesis LOX2, AOS, AOC4, AOC1 → JA reduced

SA reduced

Jasmonic acid and Jasmonates signalling JAS1, JAZ2, JAZ5, JAZ9, JR1, JAL31, MYC2

Ethylene signalling ERF2, ERF6, ERF15, MES9

Seed germination increased (Wasternack & Hause, 2013)

Defensins PDF1.2, ATHCHIB, THI2.1

Decreased signal transduction and gene activation for defence (Wasternack & Hause, 2013)

Flower and pollen development reduced (Wasternack & Hause, 2013)

Secondary metabolism Glucosinolates, Anthocyanins, Dihydroflanols, Phenylpropanoids...

Plant defences against insects, fungi and pathogens reduced (Farmer et al, 2003)

Root elongation increased (Wasternack & Hause, 2013)

Response to abiotic stress reduced (Farmer et al., 2003)

Glutathione-S-transferases GSTU7, GSTU5, GSTU6, GSTF5, GSTF12

Redox – Ascorbate and glutathione VTC2, VTC5, DHAR2

Leaf senescence decreased (Wasternack & Hause, 2013)

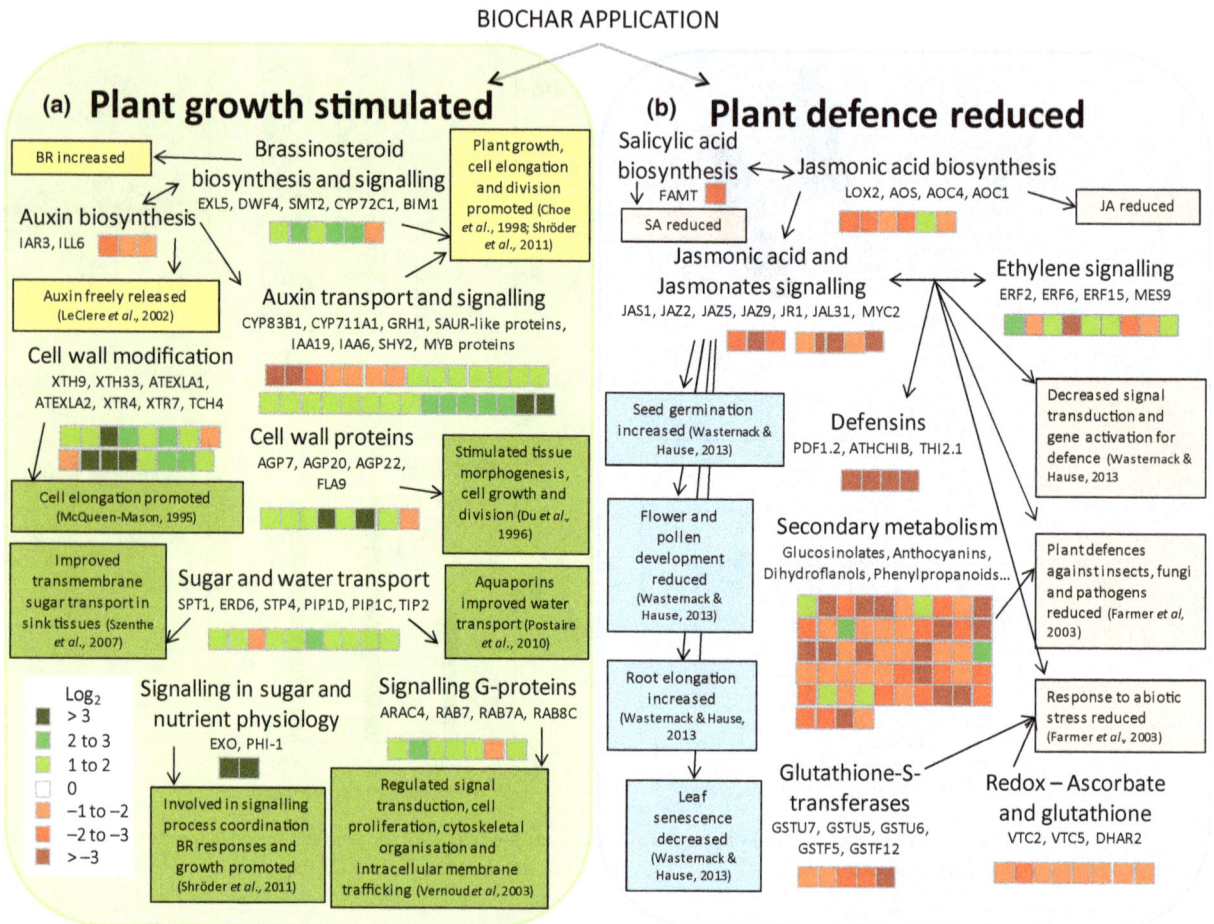

Fig. 5 A summary of gene expression analysis in response to biochar application (50 t ha^{-1}), 33 days after planting, focusing on gene expression related to (a) plant growth response and (b) plant defence. Each square represents a single gene probe, with green up-regulated and red down-regulated in biochar treated soil (biochar) compared to soil without addition (control).

values from the microarrays were confirmed by real-time PCR on a selection of genes (Figure S2), including an ethylene responsive binding element (ATERF15), a sugar transporter (STP14), a phosphate responsive (EXO) and a raffinose metabolism related gene (DIN10). STP14, EXO and DIN10 were all up regulated in plants grown in biochar.

However, these positive effects on growth were linked to large and consistent down-regulation of genes controlling plant defence mechanisms, including the jasmonic and salicylic acid biosynthetic pathways, defensins and most categories of secondary metabolites (Fig. 5b; Table S3), which all have roles in plant defence against pathogens and pest attacks (Abe et al., 2008). Genes linked to the JA biosynthesis pathway included LOX2, AOS, AOC4 and AOC1, were all down-regulated in biochar grown plants (Fold change −5.11, −4.55, −2.81 and −4.01, respectively). LOX2 was a candidate gene used for microarray confirmation by real-time PCR and showed a down regulation in response to biochar

both in the microarray and real-time PCR analysis (Figure S2). Numerous genes related to jasmonic acid and jasmonate signalling were also down-regulated in plants grown with biochar (Fig. 5b; Table S3) such as JAL31, JR1, JAZ2, JAZ5, JAZ9 and JAZ10 with fold change from −3.07 to −10.42. The largest decreases in gene expression in biochar were observed for ethylene- and jasmonate-responsive plant defensins, for example CHIB, THI2.1, PDF1.2b and PDF1.2 with observed fold changes from −16.96 to −49.47 (Table S3). Decreased expression in genes related to secondary metabolites in biochar was detected including the largest negative fold change of the microarray analysis for TPS04 of −155.33 (Table S3). This observation was confirmed with the analysis of glucosinolates and flavonoids in leaf samples collected 19 DAP from a separate replicated experiment using Arabidopsis. A significant reduction of anthocyanins, flavonols and glucosinolates was observed for the plants growing in biochar (50 t ha^{-1}) compared with the control (Table 4).

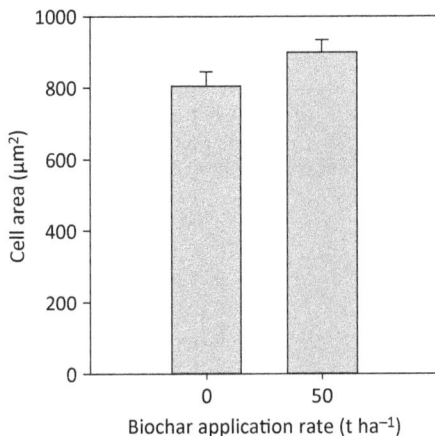

Fig. 6 Leaf epidermal cell area (μm^2) at 0 and 50 t ha^{-1} biochar application rate. in lettuce, with fertilizer addition.

Discussion

We report the first global gene expression study, to our knowledge, following plant growth in soil amended with biochar made from poplar wood chips, providing a clear insight into how this type of biochar results in growth promotion. These data give the first glimpse of an important mechanistic understanding and may help to elucidate why some but not all biochars are effective in promoting crop yield. We now have a model system where many biochars could be tested since it has been observed that plants react differently to different biochar types, but the reasons for different responses remain unclear (Jeffery et al., 2011). Our data show overwhelming evidence that auxin is central to biochar stimulated growth, occurring largely through enhanced plant cell expansion. Leaf expansion was stimulated in both the model plant *Arabidopsis* and the crop plant lettuce (130% and 49% respectively), and when investigated in lettuce, we observed that increased leaf size was attributed to increased cell expansion rather than production (12%), but only significant at the 10% level of probability. Combined with these observations, a high number of cell wall related genes such as xyloglucan endotransglucosylases, expansins and arabinogalactan-proteins were up-regulated in plants grown in biochar (Fig. 6). Those have a role in cell wall loosening and growth (Mcqueen-Mason, 1995; Du et al., 1996) and our data suggest that this is an important part of the mechanism explaining enhanced growth in biochar, with several GO categories in the functional analysis showing an up-regulation following exposure to biochar (Table S2; Figure S3a). Several strands of evidence point to auxin as the hormone controlling this response. Firstly, the auxin receptor *GRH1*, identified as an F-box protein belonging to the *TIR1* subfamily which mediates transcriptional res-

ponses to auxin (Kepinski & Leyser, 2005), was up-regulated in plants grown with biochar. This suggests that more down-stream effects of auxin action are likely. Similarly several auxin response factors – coding for ARF proteins - were up-regulated in plants grown with biochar, including *ARF7*, which has a known role in promoting leaf expansion (Wilmoth et al., 2005), as well as inducing lateral root formation. Several components of the auxin biosynthesis pathway were also stimulated in biochar (*IAR3, ILL5, ILL6*), providing further evidence that biochar results in an increase in auxin biosynthesis perhaps in both young leaves and roots. Auxin is known to be important in promoting a variety of plant growth processes including increased shoot elongation, leaf growth, meristematic activity and root and shoot branching, with the expression of many genes regulated by auxin including those identified here for cell growth. Why auxin biosynthesis is stimulated remains open to speculation but both altered pH and soluble sugar content are thought to impact on auxin biosynthesis and action (Lager et al., 2010), and it is possible that biochar exposure had a primary effect on ion uptake, (through altered soil pH and increased nutrient availability, Table 1), leading to improved transport of water and nutrients and altered cell signalling. Soil analysis revealed an increase in pH, phosphate, potassium and total nitrogen when biochar was applied to soil (Table 1). Soil pH increased from slightly acidic (6.6) to neutral (7.2). In Jeffery et al. (2011), the meta-analysis showed that increase of soil pH had a positive effect on crop growth. Other factors also influenced biochar effects including soil structure, crop species or biochar feedstock (Jeffery et al., 2011). A second meta-analysis confirmed this with significant biochar effects on soil pH, soil P, K content, total N and C, crop yield and aboveground biomass (Biederman & Harpole, 2012). K concentration in plant tissue was also significantly increased after biochar application in their study. Our findings add significantly to these meta-analyses and also revealed no significant interaction between biochar addition and nutrient status of the plant (Table 2). For plants grown on biochar, there was an up-regulation of genes related to water transport such as *TIP* and *PIP* and to sugar transport and signalling (Szenthe et al., 2007) (Fig. 5a). *PIP* genes are known to enable improved root hydraulic conductivity (Postaire et al., 2010). *EXO* (a phosphate-responsive protein) and *PHI-1* (a phosphate-induced protein) showed the highest gene expression change in response to biochar and are considered essential for leaf cell expansion and growth promotion (Schröder et al., 2011), achieved by mediating brassinosteroid (BR)- induced growth promotion. BR is also known to be primarily involved in growth-promotion through its action on plant cell expansion, and several

Table 4 Quantification of secondary metabolites (glucosinolates and flavonoids) by LC/MS in control (**0 t ha^{-1}**) and biochar (**50 t ha^{-1}**) with fertilizer in *Arabidopsis thaliana*. Metabolite concentrations are presented as values relative to internal standards as described in materials and methods. Bold values are significant ($P < 0.05$). $n = 20$

Class	Name	Control (0 t ha^{-1})	Biochar (50 t ha^{-1})	t-test
Glucosinolates				
Methionine derived	Glucoerucin	2.536	3.475	0.067
	Glucoberteroin	0.290	0.270	0.656
	Glucolesquerellin	0.102	0.096	0.656
	Heptyl GLS	0.101	0.077	**0.035**
	Glucoarabishin	2.552	2.240	0.385
	Glucoibarin	0.442	0.423	0.722
	Glucoraphanin 1	2.515	1.913	**0.030**
	Glucoraphanin 2	2.426	1.794	**0.005**
	Glucohesperin	0.050	0.035	**0.005**
	Glucoibarin	0.373	0.261	**0.006**
	Glucoalyssin	0.225	0.127	**<0.001**
	Glucohirsutin	5.776	4.394	**0.030**
	Pro/Epi-goitrin	0.037	0.028	**0.035**
Tryptophan/ Phenylalanine derived	Glucobrassicin	10.321	6.190	**0.001**
	4-Methoxyglucobrassicin	2.351	2.235	0.591
	Neoglucobracissin	3.355	0.808	**<0.001**
	Gluconasturtiin	0.143	0.105	**0.031**
Unknown	Indole 4/5-OH	0.100	0.069	**0.007**
	488.033	0.195	0.306	**0.009**
	321.0096	0.039	0.051	**0.034**
	233.061	0.928	0.990	0.088
Flavonoids				
Anthocyanins	Cyanidin 3-O-[2'-O-(xylosyl)-6'-O-(p-O-(glucosyl)-p-coumaroyl)glucosude] 5-O-glucoside	0.017	0.007	0.085
	Cyanidin 3-O-[2'-O-(xylosyl)6'-O-(p-O-(glucosyl) p-coumaroyl) glucoside] 5-O-[6'-O-(malonyl)glucoside]	0.184	0.060	**0.006**
	Cyanidin 3-O-(2'-O-(2' '-O-(sinapoyl)xylosyl) 6'-O-(p-O-(glucosyl) p-coumaroyl) glucoside] 5-O-glucoside	0.077	0.021	**0.004**
	Cyanidin 3-O-[2'-O-(6' '-O-(sinapoyl)xylosyl) 6'-O-(p-O-(glucosyl)-p-coumaroyl) glucoside] 5-O-(6' '-O-malonyl) glucoside	0.687	0.189	**0.001**
	Cyanidin 3-O-[2' '-O-(sinapoyl)xylosyl) 6'-O-(p-O-coumaroyl) glucoside] 5-O-[6' '-O-(malonyl) glucoside]	0.044	0.015	**0.001**
	Cyanidin 3-O-[2'-O-(xylosyl) 6'-O-(p-coumaroyl) glucoside] 5-O-malonyl glucoside	0.176	0.060	**0.016**
	Cyanidin 3-O-[2' '-O-(sinapoyl)xylosyl) 6'-O-(p-O-coumaroyl) glucoside] 5-O-[6' '-O-(malonyl) glucoside]	0.522	0.119	**0.004**
Flavonols	Quercetin 3-O-[6''-O-(rhamnosyl)glucoside] 7-O-rhamnoside	0.694	0.241	**<0.001**
	Kaempferol 3-O-[6'-O-(rhamnosyl)glucoside] 7-O-rhamnoside	3.963	2.831	**<0.001**
	Quercetin 3-O-glucosyl hexose	0.000	0.004	0.065
	Quercetin 3-O-glucoside 7-O-rhamnoside	0.935	0.399	**<0.001**
	Quercetin 3-O-arbinoside 7-O-rhamnoside	0.011	0.007	**0.027**
	Kaempferol 3-O-[6'-O-(glucosyl) glucoside] 7-O-rhamnoside	0.053	0.042	**0.033**
	Kaempferol 3-O-glucoside 7-O-rhamnoside	6.100	4.399	**<0.001**
	Quercetin 3-O-rhamnoside 7-O-rhamnoside	5.819	4.206	**<0.001**
	Kaempferol 3-O-arabinoside 7-O-rhamnoside	0.008	0.007	0.221
	Isorhamnetin 3-O-glucoside 7-O-rhamnoside	0.098	0.072	0.103

Table 4 (continued)

Class	Name	Control (0 t ha^{-1})	Biochar (50 t ha^{-1})	t-test
	Kaempferol 3-O-arabinoside 7-O-rhamnoside	0.040	0.024	**<0.001**
	Quercetin 3-O-glucoside	0.017	0.012	**0.003**
	Kaempferol 3-O-rhamnoside 7-O-rhamnoside	7.124	5.547	**<0.001**
	Isorhamnetin 3-O-rhamnoside 7-O-rhamnoside	0.056	0.047	0.413
	Quercetin 3-O-rhamnoside	0.015	0.009	0.193
	Kaempferol 3-rhamnoside	0.102	0.057	0.183

genes involved in BR biosynthesis, signalling and action were also up-regulated in plants grown with biochar, including *STM2* and *STM3* and *CYP72C1*, while *BIM1* was down-regulated (Vert *et al.*, 2005). BR has previously been shown to be closely involved in auxin signalling (Nakamoto *et al.*, 2006). It is also linked to delayed leaf senescence and to promoted plant growth, cell elongation and cell division. For example, *DWF4* was significantly up-regulated in biochar grown plants. The study of its mutant revealed a dwarfed phenotype due to the reduction of cell elongation and that it had a defective step which was rate-limiting in BR biosynthesis pathway (Choe *et al.*, 1998). The magnitude of the growth effect found here was similar to that observed previously for a range of crop plants (Jeffery *et al.*, 2011), with height growth increased by 177%, root length by 100% and leaf area (size) by 130% when comparing control with 50 t ha^{-1} biochar application with fertilizer (Fig. 1b). Similar observations on lettuce revealed a significant increase in leaf area, leaf number, rosette diameter and fresh and dry weight following biochar application with fertilizer (Table 3; Fig. 4). It was also observed in the *Arabidopsis* experiment that leaf area, rosette diameter, leaf number, height and stem diameter were larger when biochar and fertilizer were combined, compared with plants where fertilizer was applied without biochar. Similarly, Steiner *et al.* (2007) grain production was doubled when charcoal and biochar were combined than fertilizer without biochar.

Although no obvious toxicity was observed even at high rates of biochar (100 t ha^{-1}) in plants of *Arabidopsis* and lettuce, a novel and unpredicted finding, revealed by the microarray, was the consistent and large down-regulation of a suite of genes known to control plant defence and response to both biotic and abiotic stress. There was a clear down-regulation in genes related to jasmonates (defined to include biologically active intermediates in the pathway for jasmonic acid biosynthesis), as well as the biologically active derivatives of jasmonic acid, JA (Turner *et al.*, 2002) but also defence and secondary metabolism (Fig. 5b). Response to jasmonic acid was a GO category that was significantly and highly down-regulated in plants grown in biochar treatment (Table S2). This phytohormone is involved in plant immunity and resistance to abiotic stresses (Farmer *et al.*, 2003) and influences the expression of defence genes. It is also more generally involved in the hypersensitive response, including plant exposure to acute concentrations of ozone (Rao *et al.*, 2000). A number of genes involved in JA biosynthesis and action were down-regulated following biochar application, but more importantly, the *JAZ* proteins, that are known to be targets of the SCF^{COI1} complex, the JA receptor, were also down-regulated. This complex and its interaction with the *JAZ* proteins, is central to JA biosynthesis perception and signalling (Wasternack & Hause, 2013), enabling JA induced gene expression to be initiated. *JAZ* acts as a negative regulator of JA action, whilst *MYC2* is a transcription factor that promotes JA-responsive gene expression, but both were down-regulated here. Other examples of highly reduced gene expression after biochar application included the defensin genes such as *CHIB*, *THI2.1*, *PDF1.2*, which when promoted enhance resistance to different biotic stress (Penninckx *et al.*, 2003; Chan *et al.*, 2005). Coupled to decreased JA biosynthesis and action, there was also evidence for biochar impacts on associated ethylene and salicylic acid perception and signalling. For example, JA and ethylene are known to act together to signal the expression of *PDF1.2* (Pré *et al.*, 2008), through the action of the *AP2/ERF* domain transcription factor. We provide evidence that this type of cross-talk exists for response to biochar since several ERFs from the superfamily were affected, although both up- and down-regulation were observed (Table S2), confirming earlier findings that ERFs can both negatively and positively regulate the expression of *PDF1.2* (Pré *et al.*, 2008). Evidence for reduced salicylic acid (SA) biosynthesis was also apparent with *FAMT* reduced following biochar application, although no regulatory genes involved in SA-JA cross-talk responded to biochar application. Taken together, these data provide powerful evidence that gene expression related to plant immunity and defence was reduced following biochar application to soil. Very few data on

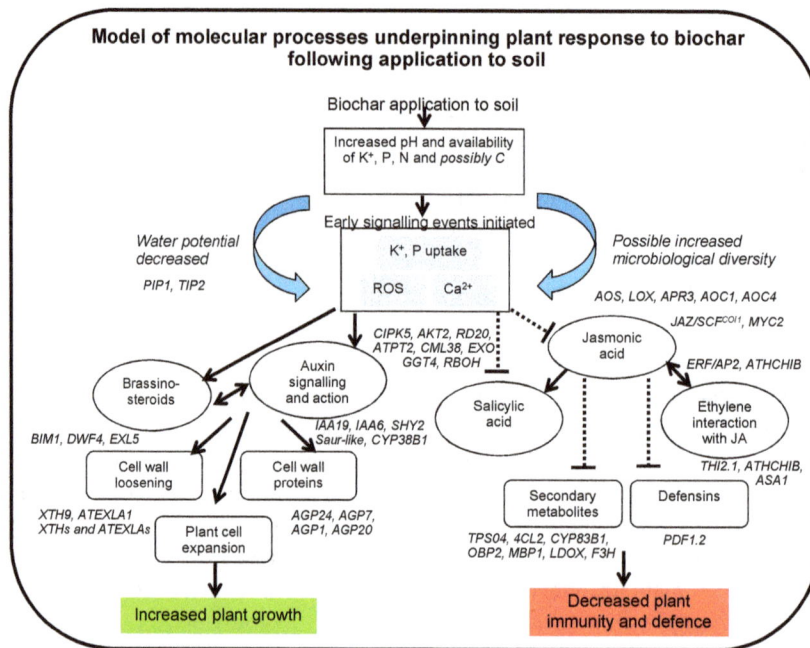

Fig. 7 A conceptual model to explain the early responses, signalling and altered gene expression following exposure of *Arabidopssis thaliana* to biochar, derived from pyrolysis and proposed consequences for plant growth, and immunity and defence. Text in italics represents changes likely from the literature on biochar but not measured here, whilst genes are those identified from analysis that were significantly differentially expressed and are key to the proposed mechanisms.

gene expression in plants following exposure to biochar have been published to date and none on global gene expression, to our knowledge, but the few data that are available are for defence genes. In contrast with the data provided here, a previous report suggests a positive role for biochar in the Systemic Acquired Resistance (*SAR*) of strawberry following fungal pathogen attacks (Elad et al., 2011; Meller Harel et al., 2012). The expression of five defence genes including *LOX* was followed and showed an up-regulation in genes related to defence in response to biochar, conflicting with our results. However, the growth medium for the strawberries in this prior work was coconut fibre and peat and the biochar made from different feedstock to that reported here, highlighting a need for future research to unravel these different responses. In particular, *Arabidopsis* grown in soil amended with biochar should be challenged with pests and pathogens to test the validity of the idea that their defense responses are impaired, including the development of a dose-response relationship for the efficacy of this effect.

A pressing question from this current study is how application of biochar to the soil, results in the altered expression of approximately 1000 genes in our model system, leads to stimulated plant growth and possibly, reduced plant immunity and defence. What are the key signalling mechanisms? We can propose some potential mechanisms given the extensive literature of the

impacts of altered gene expression and links to function in the model *Arabidopsis*. It seems likely that the primary effects of biochar on soil pH, potassium (K^+), phosphorous (P) and nitrogen (N, Table 1), result in conditions that enable the plant to take up more nutrients, perhaps K^+ in particular, but also P and N (Fig. 7). The consequences of this would be an increased osmoticum in plant cells with reduced water potential triggering for alter gene expression. When plants are exposed to K^+ starvation, Armengaud et al. (2004) intriguingly, have revealed that the most prominent response found was for genes linked to jasmonic acid biosynthesis and action, including many of those reported here and was proposed as a novel signalling molecule to regulate plant response to this stress. Following K^+ resupply, analogous to our 'high K^+ biochar', gene expression categories found to be most sensitive and up-regulated were the aquaporins including *PIP1*, and a *TIP* which were also up-regulated here following biochar application. Alongside this category, cell wall proteins and calcium signalling molecules were most sensitive to K^+ supply, again categories also highly sensitive to biochar. It seems likely that these categories represent some of the earliest responses to biochar. Similarly the most prominent categories of genes down-regulated on resupply of K^+ were those related to JA biosynthesis. There is therefore a striking similarity between our results in biochar and those related to gene expression following changes in K$^+$ supply (Armengaud *et al.*, 2004), suggesting that similar signalling and response pathways occur that explain our findings (Fig. 7). Further evidence for this comes from a proposed role for redox status and calcium signalling, with redox homeostasis a highly up-regulated functional category in response to biochar (Fig. 5) and several calmodulin related proteins and Ca^{2+} sensors including *CIPK5* and *CIPK15*, both down-regulated in biochar. Interaction between protein kinases and calcium sensing proteins (*CBLs*) has previously been shown to be central for regulating K$^+$ uptake may be increased when plants are exposed to K$^+$ stress (Li *et al.*, 2006; Liu *et al.*, 2013). Here, the K$^+$ transporter *AKT2*, known to have a role in K$^+$ transport for source to sink (Lacombe *et al.*, 2000) was also down-regulated. Similar to our finding for K$^+$ sensing and signalling, we also found a large suite of genes previously associated with P starvation to be down-regulated in our experiment. This suggests that as with potassium, the plant is sensing available phosphate and moderating gene expression accordingly and that this regulation involving transporters (*atPT2*), transcription factors (*PAP1*), and a suite of other genes (*MGD2, SEN1,*) may represent the up-stream sensing and initiation of signalling that leads to the downstream impacts on growth and defence. The availability of extensive mutant collections in *Arabidopsis* should enable these ideas to be tested as single gene knockouts, for example for gene related to auxin biosynthesis and Ca^{2+} signalling can be grown in biochar and their response tested.

Taken together, our study showed that biochar, at an application rate equivalent to 50–100 t ha^{-1}, resulted in increased plant growth for both the model plant *Arabidopsis* (leaf area, plant height and root growth) and for a leafy crop, lettuce (leaf area, fresh and dry weight). For the first time, we have quantified global gene expression for biochar treated plants, with 507 genes up- and 571 genes down-regulated in response to biochar application. From these gene expression data, we identified auxin and brassinosteriod signalling as central for the control of enhanced growth following biochar application. Our data are limited in that they cannot conclusively unravel the mechanistic understanding that links increased soil pH, availability of soil K$^+$ and P to biochar impacts but nevertheless, the literature provides strong evidence that increased pH and K$^+$ in particular, could trigger a series of signalling and functional changes in the plant that lead to enhanced cell expansion regulated through auxin and brassinosteroid action. Many of the genes that were down regulated were related to plant immunity and defence and this is a novel finding, contrary to a previous study (Meller Harel *et al.*, 2012). This highlights the complex interaction between different plant, soil and biochar types and suggests that future investigations are required, at a range of biochar application rates and in both model and crop plants, to determine if these changes in gene expression related to defence, result in reduced plant immunity and defence when plants are subjected to pathogen and pest attack. Our measurements of secondary metabolites, particularly glucosinolates, flavonoids and flavonols, provide further support for the idea that defence mechanisms may be impaired. Future research should consider a much lower concentration of biochar, below 10 t ha^{-1}, as likely to be commercially relevant. They should consider gene expression studies that target signalling and metabolic pathways involved in auxin, brassinosteriod and jasmonic acid action, where cell expansion and cell wall biophysical properties are used as 'indicator traits' for impact of biochar on growth and where metabolites linked to plant defence are quantified in a wide range of tissues when plants are subjected to abiotic and biotic stresses.

Acknowledgements

This research was supported by the Seventh Framework For Research of the European Commission within the project EuroChar (Contract No 265179). We thank E. Miranda, C. Vidal, C. Zavalloni for assistance and A. Pozzi (Advanced Gasification Technology, AGT, Cremona, Italy) for providing the biochar. Research in the laboratory of GT on bioenergy is also supported by NERC as part of the Carbo-BioCrop project (Grant reference number: NE/H010742/1).

References

Abe H, Ohnishi J, Narusaka M, Seo S, Narusaka Y, Tsuda S, Kobayashi M (2008) Function of Jasmonate in Response and Tolerance of *Arabidopsis* to Thrip Feeding. *Plant Cell Physiology*, **49**, 68–80.

Armengaud P, Breitling R, Amtmann A (2004) The potassium-dependent transcriptome of *arabidopsis* reveals a prominent role of jasmonic acid in nutrient signaling. *Plant Physiology*, **136**, 2256–2576.

Barah P, Winge P, Kusnierczyk A, Tran DH, Bones AM (2013) Molecular Signatures in *Arabidopsis thaliana* in Response to Insect Attack and Bacterial Infection. *PLoS ONE*, **8**, e58987.

Baronti S, Alberti G, Delle Vedove G et al. (2010) The biochar option to improve plant yields: first results from some field and pot experiments in italy. *Italian Journal of Agronomy*, **5**, 3–11.

Biederman LA, Harpole WS (2012) Biochar and its effects on plant productivity and nutrient cycling: a meta-analysis. *Global Change Biology Bioenergy*, **5**, 202–214.

Chan Y-L, Prasad V, Sanjaya, Chen K, Liu PC, Chan M-T, Cheng C-P (2005) Transgenic tomato plants expressing an *Arabidopsis* thionin (*Thi2.1*) driven by fruit-inactive promoter battle against phytopathogenic attack. *Planta*, **221**, 386–393.

Chang S, Puryear J, Cairney J (1993) A simple and efficient method for isolating RNA from pine trees. *Plant Molecular Biology Reporter*, **11**, 113–116.

Choe S, Dilkes BP, Fujioka S, Takatsuto S, Sakurai A, Feldmann KA (1998) The DWF4 gene of arabidopsis encodes a cytochrome P450 that mediates multiple 22a-hydroxylation steps in brassinosteroid biosynthesis. *The Plant Cell*, **10**, 231–243.

Choe S, Fujioka S, Noguchi T, Takatsuto S, Yoshida S, Feldmann KA (2001) Overexpression of DWARF4 in thebrassinosteroid biosynthetic pathway results in increased vegetative growth and seed yield in Arabidopsis. *The Plant Journal*, **26**, 573–582.

Du H, Clarke AE, Bacic A (1996) Arabinogalactan-proteins: a class of extracellular matrix proteoglycans involved in plant growth and development. *Trends in Cell Biology*, **6**, 411–414.

Du Z, Zhou X, Ling Y, Zhang Z, Su Z (2010) AgriGO: a GO analysis toolkit for the agricultural community. *Nucleic Acids Research*, **38**, W64–W70.

Elad Y, Cytryn E, Meller Harel Y, Lew B, Graber ER (2011) The biochar effect: plant resistance to biotic stresses. *Phytopathol. Mediterr.*, **50**, 335–349.

Farmer EE, Almeras E, Krishnamurthy V (2003) Jasmonates and related oxylipins in plant responses to pathogenesis and herbivory. *Current Opinion in Plant Biology*, **6**, 372–378.

Foito A, Byrne SL, Hackett CA, Hancock RD, Stewart D, Barth S (2013) Short-term response in leaf metabolism of perennial ryegrass (*Lolium perenne*) to alterations in nitrogen supply. *Metabolomics*, **9**, 145–156.

Gardner S, Taylor G, Bosac C (1995) Leaf growth of hybrid poplar following exposure to elevated CO_2. *New Phytologist*, **131**, 81–90.

Jeffery S, Verheijen FGA, van der Velde M, Bastos AC (2011) A quantitative review of the effects of biochar application to soils on crop productivity using meta-analysis. *Agriculture, Ecosystems & Environment*, **144**, 175–187.

Kepinski S, Leyser O (2005) The *Arabidopsis* F-box protein TIR1 is an auxin receptor. *Nature*, **435**, 446–451.

Kusnierczyk A, Winge P, Midelfart H, Armbruster WS, Rossiter JT, Bones AM (2007) Transcriptional responses of *Arabidopsis thaliana* ecotypes with different glucosinolate profiles after attack by polyphagous *Myzus persicae* and oligophagous *Brevicoryne brassicae*. *Journal of Experimental Botany*, **58**, 2537–2552.

Lacombe B, Pilot G, Michard E, Gaymard F, Sentenac H, Thibaud J-B (2000) A shaker-like k⁺ channel with weak rectification is expressed in both source and sink phloem tissues of *Arabidopsis*. *American Society of Plant Physiologists*, **12**, 837–851.

Lager I, Andreasson O, Dunbar TL, Andreasson E, Escobar MA, Rasmusson AG (2010) Changes in external pH rapidly alter plant gene expression and modulate auxin and elicitor responses. *Plant, Cell and Environment*, **33**, 1513–1528.

Leclere S, Tellez R, Rampey RA, Matsuda SPT, Bartel B (2002) Characterization of a family of IAA-amino acid conjugate hydrolases from Arabidopsis. *The Journal of Biological Chemistry*, **277**, 20446–20452.

Lehmann J, Joseph S (2009) Biochar for environmental management: an introduction. In: *Biochar for Environmental Management*. (eds Lehmann J, Joseph S), pp. 1–12. Earthscan Publishers Ltd, London, UK.

Lehmann J, Gaunt J, Rondon M (2006) Bio-char sequestration in terrestrial ecosystems - a review. *Mitigation and Adaptation Strategies, for Global Change*, **11**, 395–419.

Li L, Kim B-G, Cheong YH, Pandey GK, Luan S (2006) A Ca^{2+} signaling pathway regulates a K⁺ channel for low-K response in *Arabidopsis*. *PNAS*, **103**, 12625–12630.

Liu W, Saint DA (2002) A new quantitative method of real time reverse transcription polymerase chain reaction assay based on simulation of polymerase chain reaction kinetics. *Analytical Biochemistry*, **302**, 52–59.

Liu L-L, Ren H-M, Chen L-Q, Wang Y, Wu W-H (2013) A protein kinase, calcineurin b-like protein-interacting protein kinase9, interacts with calcium sensor calcineurin b-like protein3 and regulates potassium homeostasis under low-potassium stress in *Arabidopsis*. *Plant Physiology*, **161**, 266–277.

Mcqueen-Mason SJ (1995) Expansins and cell wall expansion. *Journal of Experimental Botany*, **46**, 1639–1650.

Meller Harel Y, Elad Y, Rav-David D, Borenshtein M, Shulchani R, Lew B, Graber ER (2012) Biochar mediates systemic response to strawberry to fungal pathogens. *Plant and Soil*, **357**, 245–257.

Nakamoto D, Ikeura A, Asami T, Yamamoto KT (2006) Inhibition of brassinosteroid biosynthesis by either a *dwarf4* mutation or a brassinosteroid biosynthesis inhibitor rescues defects in tropic responses of hypocotyls in the arabidopsis mutant *nonphototropic hypocotyl 4*. *Plant Physiology*, **141**, 456–464.

Penninckx MA, Eggermont K, Schenk PM, van den Ackerveken G, Cammue BPA, Thomma BPHJ (2003) The *Arabidopsis* mutant *iop1* exhibits induced over-expression of the plant defensin gene *PDF1.2* and enhanced pathogen resistance. *Molecular Plant Pathology*, **4**, 479–486.

Postaire O, Tournaire-Roux C, Grondin A, Boursiac Y, Morillon R, Shaffner AR, Maurel C (2010) A PIP1 aquaporin contributes to hydrostatic pressure-induced water transport in both the root and rosette of Arabidopsis. *Plant Physiology*, **152**, 1418–1430.

Pré M, Atallah M, Champion A, De Vos M, Pieterse CMJ, Memelink J (2008) The ap2/erf domain transcription factor ora59 integrates jasmonic acid and ethylene signals in plant defense. *Plant Physiology*, **147**, 1347–1357.

Rao MV, Lee H-I, Creelman RA, Mullet JE, Davis KR (2000) Jasmonic acid signaling modulates ozone-induced hypersensitive cell death. *American Society of Plant Physiologists*, **12**, 1633–1646.

Richmond T, Somerville S (2000) Chasing the dream: plant EST microarrays. *Current Opinion in Plant Biology*, **3**, 108–116.

Rochfort SJ, Trenerry VC, Imsic M, Panozzo J, Jones R (2008) Class targeted metabolomics: ESI ion trap screening methods for glucosinolates based on MSn fragmentation. *Phytochemistry*, **69**, 1671–1679.

Schröder F, Lisso J, Muessig C (2011) EXORDIUM-LIKE1 (EXL1) promotes growth during low carbon availability in Arabidopsis thaliana. *Plant Physiology*, **156**, 1620–1630.

Sohi S (2012) Carbon storage with benefits. *Science*, **338**, 1034–1035.

Sohi SP, Krull E, Lopez-Capel E, Bol R (2010) A review of biochar and its use and function in soil. *Advances in Agronomy*, **105**, 47–82.

Spokas KA, Baker JM, Reicosky DC (2010) Ethylene: potential key for biochar amendment impacts. *Plant and Soil*, **333**, 443–452.

Steiner C, Teixeira WG, Lehmann J, Nehls T, de Macedo JLV, Blum WEH, Zech W (2007) Long term effects of manure, charcoal and mineral fertilization on crop production and fertility on a highly weathered central Amazonian upland soil. *Plant and Soil*, **291**, 275.

Street NR, Skogstrom O, Sjodin A *et al.* (2006) The genetics and genomics of the drought response in Populus. *Plant J.*, **48**, 321–341.

Szenthe A, Schäfer H, Hauf J, Schwend T, Wink M (2007) Characterisation and expression of monosaccharide transporters in lupins, Lupinus polyphyllus and L. albus. *Journal of Plant Research*, **120**, 697–705.

Tohge T, Nishiyama Y, Yokota Hirai M *et al.* (2005) Functional genomics by integrated analysis of metabolome and transcriptome of *Arabidopsis*plants overexpressing an MYB transcription factor. *The Plant Journal*, **42**, 218–235.

Turner JG, Ellis C, Devoto A (2002) The Jasmonate signal pathway. *American Society of Plant Biologists*, **14**, S153–S164.

Vernoud V, Horton AC, Yang Z, Nielsen E (2003) Analysis of the small GTPase gene superfamily of Arabidopsis. *Plant Physiology*, **131**, 1191–1208.

Vert G, Nemhauser JL, Geldner N, Hong FX, Chory J (2005) Molecular mechanisms of steroid hormone signaling in plants. *Annual Review of Cel and Developmental Biology*, **21**, 177–201.

Wasternack C, Hause B (2013) Jasmonates: biosynthesis, perception, signal transduction and action in plant stress response, growth and development. An update to the 2007 review in Annals of Botany. *Annals of Botany*, **111**, 1021–1058.

Wilmoth JC, Wang S, Tiwari SB *et al.* (2005) NPH4/ARF7 and ARF19 promote leaf expansion and auxin-induced lateral root formation. *The Plant Journal*, **43**, 118–130.

Woolf D, Amonette JE, Street-Perrott A, Lehmann J, Joseph S (2010) Sustainable biochar to mitigate global climate change. *Nature Communications*, **1**, 56.

Yonekura-Sakakibara K, Tohge T, Matsuda F *et al.* (2008) Comprehensive flavonol profiling and transcriptome coexpression analysis leading to decoding gene-metabolite correlations in Arabidopsis. *The Plant Cell*, **20**, 2160–2176.

11

Comparing predicted yield and yield stability of willow and Miscanthus across Denmark

SØREN LARSEN[1], DEEPAK JAISWAL[2], NICLAS S. BENTSEN[1], DAN WANG[3] and STEPHEN P. LONG[2,4]

[1]Department of Geosciences and Natural Resource Management, University of Copenhagen, Rolighedsvej 23, 1958 Frederiksberg C, Denmark, [2]Energy Bioscience Institute, University of Illinois, Urbana, IL 61801, USA, [3]International Center for Ecology, Meteorology and Environment, School of Applied Meteorology, Nanjing University of Information Science and Technology, Nanjing 210044, China, [4]Departments of Crop Sciences and of Plant Biology, University of Illinois, Urbana, IL 61801, USA

Abstract

To achieve the goals of energy security and climate change mitigation in Denmark and the EU, an expansion of national production of bioenergy crops is needed. Temporal and spatial variation of yields of willow and Miscanthus is not known for Denmark because of a limited number of field trial data. The semi-mechanistic crop model BioCro was used to simulate the production of both short-rotation coppice (SRC) willow and Miscanthus across Denmark. Predictions were made from high spatial resolution soil data and weather records across this area for 1990–2010. The potential average, rain-fed mean yield was 12.1 Mg DM ha^{-1} yr^{-1} for willow and 10.2 Mg DM ha^{-1} yr^{-1} for Miscanthus. Coefficient of variation as a measure for yield stability was poorest on the sandy soils of northern and western Jutland, and the year-to-year variation in yield was greatest on these soils. Willow was predicted to outyield Miscanthus on poor, sandy soils, whereas Miscanthus was higher yielding on clay-rich soils. The major driver of yield in both crops was variation in soil moisture, with radiation and precipitation exerting less influence. This is the first time these two major feedstocks for northern Europe have been compared within a single modeling framework and providing an important new tool for decision-making in selection of feedstocks for emerging bioenergy systems.

Keywords: BioCro, bioenergy, C4 photosynthesis, crop model, geospatial modeling, mechanistic model, Miscanthus, perennial grasses, short-rotation coppice, Willow, Wimovac

Introduction

The European Union has agreed upon ambitious policies on energy supply, climate change mitigation and environmental sustainability. To meet the targets, EU countries have issued so-called National Renewable Energy Action Plans (NREAP) specifying the development of renewable energy generation till 2020 (Beurskens & Hekkenberg, 2011). Biomass is a cornerstone of the NREAPs and is stipulated to account for 56% of renewable energy generation by 2020 (Beurskens & Hekkenberg, 2011), corresponding to an increase in bioenergy generation from 2.4 EJ in 2005 to 5.7 EJ in 2020. It has been estimated that the biomass consumption will increase from 3.8 EJ in 2005 to 10.0 EJ in 2020 due to the increase in bioenergy generation during this period (Bentsen & Felby, 2012).

Correspondence: Søren Larsen
e-mail: slar@ign. ku.dk

Bioenergy is also expected to play a significant role in the Danish efforts to secure supply and mitigate climate change. To comply with EU policy, Denmark's target for the share of renewable energy is at least 30% of the gross final energy consumption by 2020 (European Parliament and the Council, 2009).

Willow and Miscanthus cultivation in Denmark

Willow (*salix spp.*) and Miscanthus (*Miscanthus × giganteus*, (Greef et. Deu.)) have not yet gained momentum as energy crops in Denmark, and only a very small area is used for cultivation of these. Both are considered key opportunities for achieving an increase in sustainable national biomass production and are used more extensively the neighboring countries (Alexander *et al.*, 2014; Sevel *et al.*, 2012; The Danish AgriFish Agency, 2013). Perennials are favored because of their long growing seasons, efficient recycling of nutrients, stabilization of soil and ability to accumulate soil carbon (Heaton *et al.*, 2010; Jørgensen *et al.*, 2013; Voigt, 2015).

Achieving the 2020 bioenergy supply, goal of Denmark might require planting of large additional areas of these feedstocks. Many factors will determine the appropriate feedstock for a given location. However, major considerations are yield and stability of yield at each location. Without widespread trials, it is difficult to know which would have the higher yield at a given location. Mechanistically rich models provide the means to predict beyond experience. Such models have been developed for Miscanthus (Clifton-Brown *et al.*, 2000, 2004; Richter *et al.*, 2008; Hastings *et al.*, 2009a,b; Bauen *et al.*, 2010; Pogson, 2011) and for willow (Lindroth & Båth, 1999; Aylott *et al.*, 2008; Mola-Yudego & Aronsson, 2008; Mola-Yudego, 2010; Tallis *et al.*, 2013), but each within its own unique modeling framework.

We use the mechanistic model BioCro, which is a generic crop model based on the WIMOWAC model, Humphries & Long (1995), adapted for Miscanthus by Miguez *et al.* (2012, 2009) and for willow by Wang *et al.* (2015). BioCro was designed to provide a single framework for predicting growth and yield of perennial bioenergy crops to avoid confounding species differences with differences in modeling assumptions and structure. It has been successfully applied previously to compare switchgrass and Miscanthus in the USA (Miguez *et al.*, 2012). Here, it is applied to compare Miscanthus and willow in Denmark, so providing a further key tool in decision making on the choice of feedstock for different locations.

This is the first time the model has been used to model both Miscanthus and willow in Europe, and this approach allows us to model potential yields for both crops within the same modeling framework. When comparing yields simulated by different models, one often risks comparing model structures and assumptions instead of comparing model results and biological differences between crops (Nair *et al.*, 2012; Wang *et al.*, 2015). This risk is avoided using one model for the two different crops.

This study (a) maps potential yield and yield stability of Miscanthus and willow in Denmark, using weather data for 1990–2010 to quantify the effects of year-to-year variation in weather, combined with high resolution soil maps, (b) compares the potential yields of the two crops across the country and (c) determines which factors appear most important in determining yield and yield stability of these crops.

Materials and methods

Model description

The BioCro model is extensively described by (Humphries & Long, 1995; Miguez *et al.*, 2009, 2012; Wang *et al.*, 2015), there-

fore the following only provides a short overview, focusing on the set up for this study.

Miscanthus and willow in BioCro are simulated through its detailed mechanistic biochemical and biophysical multilayer canopy model that partitions assimilate between different plant organs (stem, leaf, root and storage) according to phenological development stages as determined by thermal time. Using hourly weather data, the model calculates direct and diffuse light for dynamically changing sunlit and shaded portions of the canopy layers and computes carbon and water exchange with the atmosphere by interface with leaf biochemical and biophysical submodels for each hour of the day and each day of the growing season. The canopy module is dynamically linked to a multilayer soil/hydrology module. Soil water status coupled with canopy properties is used to calculate leaf water potential which modulates stomatal conductance and which together with temperature and assimilate supply determines rates of leaf expansion and senescence.

Soil data

BioCro requires soil rooting depth, wilting point and field capacity for each location simulated. In Denmark, there is no national database that includes these properties, but instead a database has been established with soil textural properties in three layers: 0–30 cm, 30–70 cm and 70–120 cm, bulk density and rooting depth. This database is based on all available soil data (around 54 000 soil samples in total). The two topmost layers are constructed by kriging interpolation, and for the bottommost layer, median georegionalized values are used. This allows for a national map with soil textural properties in three layers with a resolution of 250 m × 250 m for the top layer and 500 m × 500 m for the two bottommost layers. All soils are ascribed to one of the 9–10 soil types most prevalent in each of Denmark's 5 georegions or one of two different wetland (which are generated separately from the minerogenic soil types) soil types (Børgesen *et al.*, 2009).

To simplify the calculations and to use the same method as previously used for BioCro, a weighted average rooting depth was calculated for each soil type and used as input to the model. The soil water content at the beginning of the growing season was set to field capacity each year which in most years is reasonable because of a precipitation surplus during the winter making the soils saturated when the growing season starts (Madsen *et al.*, 1992). The rooting depth for each soil type is taken from Børgesen *et al.* (2009) and has previously been used for crop modeling. Rooting depth varies between 50 cm and 150 cm depending on soil type. Soil hydrological parameters, field capacity and wilting point, are determined on the basis of textural properties using the equations shown in Supporting information, eq. 1 and 2.

Weather data

Daily weather data for the simulations were obtained from the Danish Meteorological Institute, Scharling (2012), for 1990–2010 for total precipitation, average temperature, accumulated potential evaporation, average wind speed and accumulated global radiation. Daily precipitation is the only data from the 10 km × 10 km grid, and the other data are from the 20 km × 20 km grid. From the Danish 40 km × 40 km climate grid, we got daily mean relative humidity and daily minimum and maximum (Plauborg & Olesen, 1991; Scharling, 1999). Daily minimum and maximum relative humidity were calculated from the recorded temperature and absolute humidity (Allen et al., 1998). Day of the year, hour and latitude were used to determine the hourly solar declination and solar zenith angle. Hourly weather data were estimated from the daily data by the interpolation methods included in BioCro and described in (Humphries & Long, 1995).

Regional simulations

BioCro was parameterized as described and validated previously (Miguez et al., 2009, 2012; Wang et al., 2015). The full equation set and parameter tables are given in these prior publications. Simulations were performed to predict the course of growth and final yield for each year from 1990 to 2010 at the high resolution provided by the geospatial soil data available for the country (250 m × 250 m).

To perform the simulations, a climate grid was generated in ArcGIS, ESRI (2010), so that each 10 × 10 km climate cell was also filled with data from the 20 × 20 and 40 × 40 km climate data. This gives 609 unique climate cells covering Denmark and each soil cell is given climate values from the climate cell that it lies within. A very limited part of the land area was not covered by the climate grid, that is small tongues of land and small forelands. These small areas were assigned the values from the adjacent climate cell and covers <1% of the simulated area.

The highest resolution of climate data available was 10 km × 10 km. As several soil cells (250 × 250 m) within one climate cell (10 × 10 km) often would be of the same type, to avoid repeating calculations, the result from one soil cell would be applied to all other cells with the same soil type within the climate cell. This reduced the number of cells simulated from potentially about 80 000 to 4852. For each cell, BioCro calculates net carbon exchange, canopy microclimate and evapotranspiration on an hourly basis, and growth, biomass partitioning, canopy structure and soil moisture dynamics on a daily basis. As such, it is computationally intensive. To complete calculations, it was necessary to parallelize the code to allow computation on a cluster (at time of computation the cluster consisted of Dell Poweredge 1950 servers with 24 nodes each with 8 cores of 2.8 GHz CPUs).

In the simulation, willow was assumed to be grown on a 3 year coppice cycle, but annual yields are given by averaging across the 3 years. After the first year, the willow is cut back to induce coppicing. Miscanthus was simulated for an annual harvest. It was assumed that both crops would be harvested in the late winter or early spring as often done in Denmark (Larsen et al., 2013, 2014a).

To determine the harvestable yield of willow, it was assumed that there was a 10% loss during harvest, and for Miscanthus, it was assumed that 67% of the peak biomass could be harvested due to losses during senescence and harvest (Beale

& Long, 1995; Venendaal et al., 1997; Hastings et al., 2009b; Miguez et al., 2012). Winter losses in willow are not well documented and leaf biomass lost due to frost is the same as in Wang et al. (2015). The assumption regarding harvest loss used here is based on practical experience in experimental and commercial plantations in Denmark, personal communication with L. Sevel and S. U. Larsen. The results were summarized by calculating mean annual yield for each location across the 21 years together with the coefficient of variation as a measure of yield stability, that is, year-to-year variation driven by weather conditions relative to averaged yields. Yield maps were generated at 250 m × 250 m resolution equal to that of the soil data.

Climatic and soil variable sensitivity

To determine which soil and climatic variables were most important in determining yield, we calculated a number of parameters to test with a generalized linear model (GLM). These were precipitation and radiation sum during the growing season (April–October), the available water content (AWC – difference between field capacity and wilting point for the soil profile from surface to rooting depth), and lastly, we included the Danish georegion because the soil data are generated in such a way where only the 10 most abundant soil types of each georegion are present in each (Børgesen et al., 2009). The GLM procedure was performed in R (R Core Team, 2013) with the above mentioned parameters. The procedure is performed for both willow and Miscanthus.

Results

Yield predictions

Large spatial variation of harvestable yields and yield stability were found. In general, the sandy soils of western and northwestern Denmark show much lower harvestable yields than the more clay-rich soils of central and eastern Denmark (Fig. 1a,b). This holds true for both crop species. The area-weighted mean yield was 12.1 Mg DM ha^{-1} yr^{-1} for willow and 10.2 Mg DM ha^{-1} yr^{-1} for Miscanthus. The lowest annual willow yields were much higher than the lowest Miscanthus yields. This is in part because the willow yields are a mean of 3 years of production so years with weather conducive for high yield offset those causing poorer yields and vice versa.

Stability of yields

The coefficient of variation (CV) for annual biomass yield was calculated for Miscanthus. For willow, the results were calculated on the basis of the yield of a 3 year period corresponding to a cutting cycle. These results show that the largest coefficient of variation, and therefore lowest yield stability, was found in western

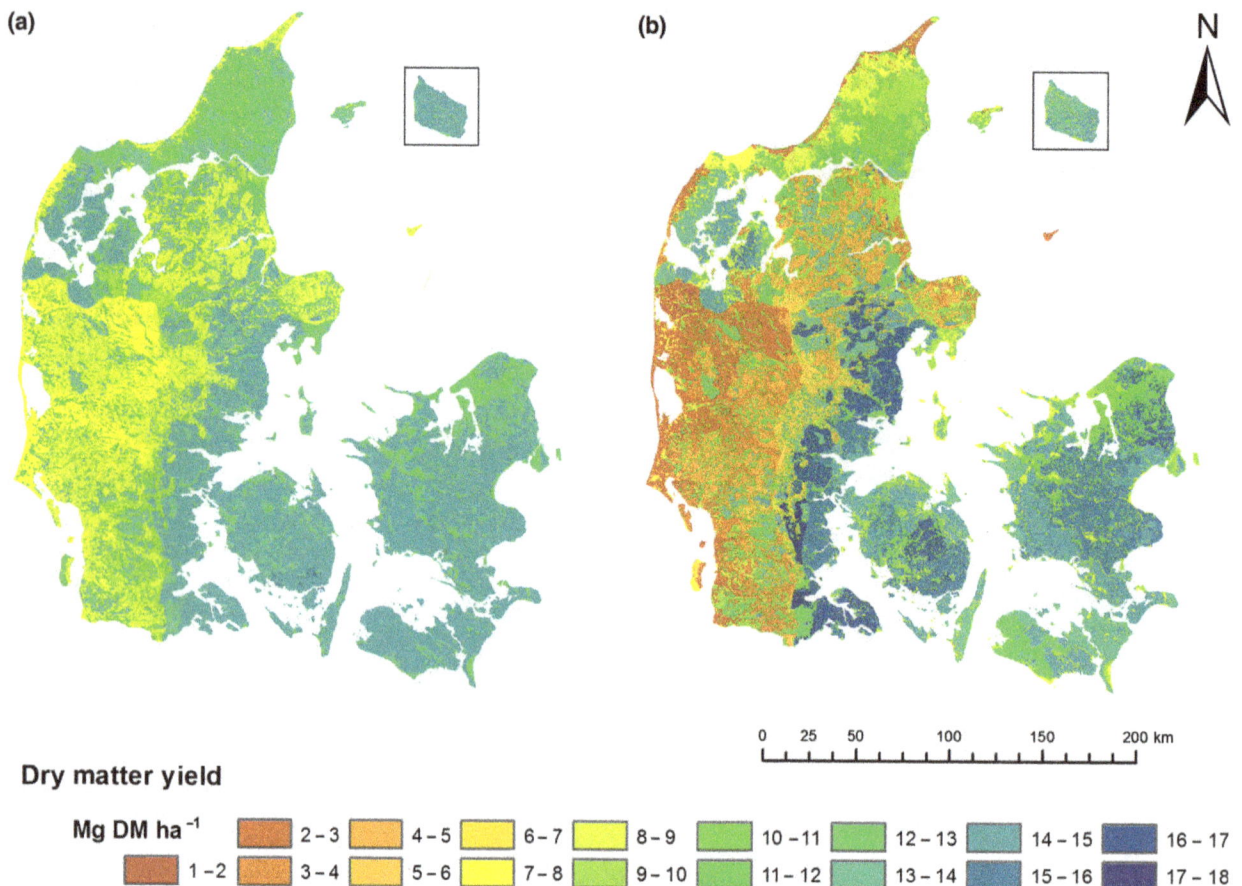

Dry matter yield

Mg DM ha^{-1}

2–3	4–5	6–7	8–9	10–11	12–13	14–15	16–17
1–2	3–4	5–6	7–8	9–10	11–12	13–14	15–16	17–18

Fig. 1 Simulated mean annual harvested biomass (Mg DM ha^{-1} yr^{-1}), as dry weight, for (a) SRC willow and (b) Miscanthus over the period 1990–2010.

Denmark for both crop species, (Fig. 2a,b). However, stability was lower at all locations for Miscanthus. The poor, sandy soils are primarily found in western and northwestern Denmark (Fig. S1b).

Difference in harvestable yields

The difference in yield was calculated as a difference between the mean harvestable yields for 1990–2010 for the two species, that is, the difference between the yields illustrated in Fig. 1(a,b).

The results show that on the poor soils in western and northwestern Denmark, willow has an advantage over Miscanthus (blue shading), but on better, clay-rich soils of central and eastern Denmark, Miscanthus has a higher productivity than willow (red to green shading), Fig. 3.

Relationship between crop yield and biophysical factors

The results of the GLM procedure show that AWC is the most important factor for yield in both willow and Miscanthus. The higher the AWC, the higher the simulated yields. Precipitation, radiation sum and georegion

are also significant, but exert less influence. See Fig. S1 (a) for an AWC map of Denmark.

Discussion

Model performance

The yields predicted by the model are potential yields in the sense that they are only water limited. The model assumes good agronomy with adequate fertilization and no pests, diseases or damage from extreme climatic events (Miguez *et al.*, 2009). This leads to a discussion of how realistic the yields we report for the two crops are, when there is only very limited yield data available, especially for Miscanthus. Karp & Shield (2008) and Lobell *et al.* (2009) discuss the difference between theoretical, potential and actual yield. The yields simulated here are theoretical water-limited yields, and consequently, they are higher than both potential and practically achieved yields. However, predicted growth and final yield predicted with BioCro were very close to those observed in research trials, at separate sites, for both Miscanthus (Miguez *et al.*, 2009) and willow (Wang

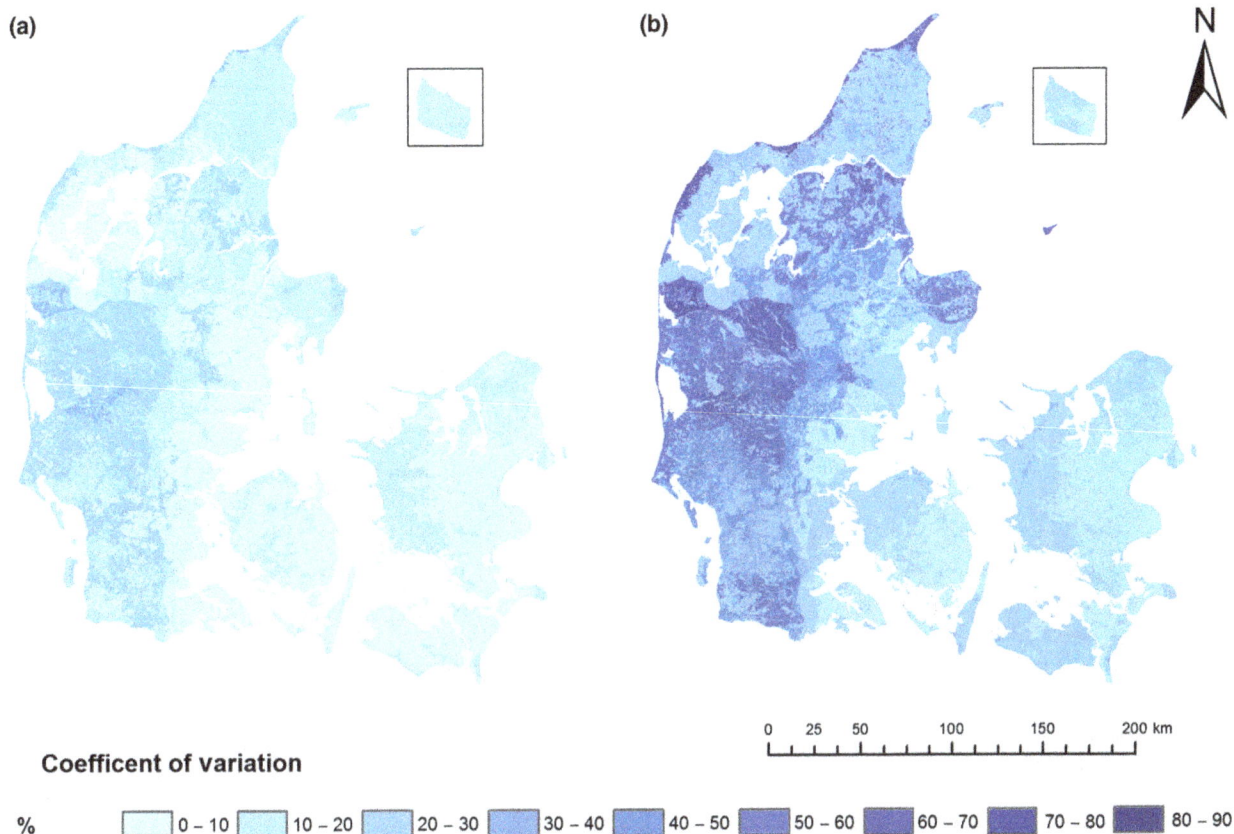

Fig. 2 Coefficient of variation in % of annual biomass productivity for the years 1990–2010 for (a) SRC willow on a 3 year coppice cycle, and (b) Miscanthus on an annual harvest cycle.

et al., 2015). Yields in research trials are commonly found to exceed those experienced in practice, but are a good representative of what may be achieved with good agronomy.

Comparison with yields in Denmark

In Denmark, a small number of experiments and trials have looked into willow productivity. Sevel *et al.* (2012) report average productivities between 5.2 and 8.8 Mg DM ha^{-1} yr^{-1} in a commercial plantation over a two-year rotation. Other willow trials in commercial plantations in Denmark have found average yields of 2–8 Mg DM ha^{-1} yr^{-1}, but with a large variation in yields indicating that the potential yield is much higher than the reported averages (Morsing & Nielsen, 1995, Venendaal *et al.*, 1997; Landbrug og Fødevarer, 2010, 2012). Other studies have found higher average yields of around 10–12 Mg DM ha^{-1} yr^{-1} for the best yielding clones and treatments (Sevel *et al.*, 2013) (Larsen *et al.*, 2014b). These trials are in line with the yields modeled with BioCro and show the potential for the best yielding clones in Denmark under close to optimal management regimes. In a general sense, the trial results show higher

yields on clay-rich soil, exactly as BioCro predicts hereby showing that BioCro is well suited to take the spatial variability of Danish soils into account (Mortensen *et al.*, 1998, Landbrug Og Fødevarer, 2012).

For willow, we have compared measured and modeled yields at one location in Denmark, Fig. S2. This shows that BioCro overestimates willow yields at this location, but also shows that the best yielding treatments and years are able to produce at a level similar to that predicted by BioCro.

The only other modeling study covering Denmark predicts an average productivity of 9.5 Mg DM ha^{-1} yr^{-1} if the production is only water limited, but higher yields can be achieved when considering the best growers or 2010 production (Mola-Yudego, 2010). This model uses a completely different method to achieve its results and uses much larger spatial units, but still achieves results comparable to both the ones of BioCro and trials; especially if you compare optimally managed trials and models where optimal management is an assumption such as BioCro.

There have only been a few studies of Miscanthus cultivation in Denmark. Larsen *et al.* (2013) studied the long term (1993–2012) yield of Miscanthus (*M. giganteus*

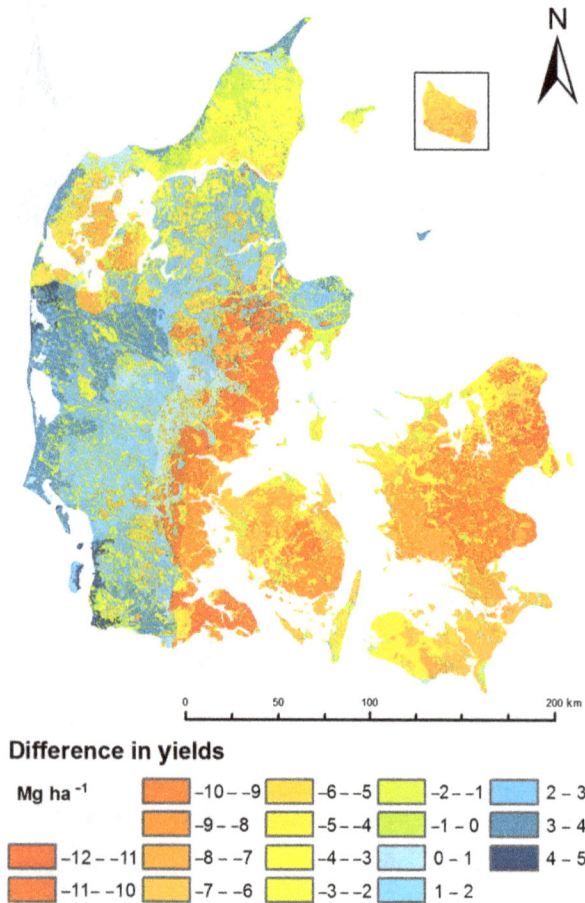

Difference in yields

Mg ha^{-1}			
	−10 − −9	−6 − −5	−2 − −1	2 − 3
	−9 − −8	−5 − −4	−1 − 0	3 − 4
−12 − −11	−8 − −7	−4 − −3	0 − 1	4 − 5
−11 − −10	−7 − − 6	−3 − −2	1 − 2	

Fig. 3 Difference in mean productivity of willow and Miscanthus 1990–2010, using the data of Fig. 1. Numbers are relative to the predicted yield of willow at any one location. Therefore, a negative value is where Miscanthus is more productive than willow and *vice versa*.

and *M. goliath*) at two locations in Denmark and found that the highest yielding *M.* x *giganteus* treatment had a mean yield of 13.1 Mg DM ha^{-1} yr^{-1} with late autumn harvest. Spring harvest is shown to reduce the yield by 34–42%, which is a little higher compared to the assumption of 33% used here, but the fraction lost depends on the exact harvest dates.

Venendaal *et al.* (1997) report mean yields of 7–8 (sandy soil) and 8–9 (clay soil) Mg DM ha^{-1} yr^{-1} for spring harvested Miscanthus in Denmark under commercial conditions.

Again, we have compared measured and modeled yield for one location in Denmark, Fig. S3. BioCro overestimates the yields, except for one year. There can be a number of reasons for this, for instance nonoptimal management of the experiments, poor BioCro performance for this location and soil or a yield decline as discussed below. One should exercise great caution to conclude anything from this comparison, but it is

evident that for this location BioCro vastly overestimates productivities of Miscanthus.

Larsen *et al.* (2013) also report a yield decline after 5–8 years and Arundale *et al.* (2014) similarly reports a decline. As a relatively new crop, these are the only studies to report beyond 5 years of experience and so it is difficult at this point to understand whether this should be expected wherever the crop is grown or if this is specific to given climates, soils or agronomy. Given the limited information, this effect cannot be simulated in BioCro so it would be more appropriate to compare BioCro simulations with the yields achieved in the maturity phase in Larsen *et al.* (2013), which are 8–12 Mg DM ha^{-1} yr^{-1} for spring harvested *M.* x *giganteus* in a location in the central western part of Denmark (Foulum) and thus more comparable to the yields simulated by BioCro.

Crop yield and biophysical factors

As shown in other studies, climate parameters are important for determining yield (Hastings *et al.*, 2009b; Miguez *et al.*, 2012; Wang *et al.*, 2015).

The GLM procedure shows that precipitation has a negative influence on yields. This might seem strange, but the reason for this should be that the regions in Denmark with the highest precipitation (the western and central parts of the peninsula Jutland) are also regions where sandy soils dominate. So even if there is high precipitation, the sandy soils dictate that the plant available water storage capacity is low.

Miscanthus and willow harvest losses

We assume that 10% of the stem biomass is lost for willow and 33% for Miscanthus between the time of peak biomass and harvest, due to stubble and translocation during senescence and shoot fragmentation in the case of Miscanthus. For Miscanthus, the assumption is corroborated by experimental trials in Denmark and abroad (Lewandowski & Heinz, 2003; Heaton *et al.*, 2009; Larsen *et al.*, 2013). Our assumption of 10% harvest loss is based on practical experience as mentioned above. However, it is reasonable to anticipate smaller losses in willow. The stem serves as the key perennation organ, so less material is translocated below ground in the autumn while the woody and living stems will be far less vulnerable to fragmentation losses in high winds.

The reason for reporting the harvestable yield instead of total aboveground biomass is to make it easier to compare the amounts of biomass that would be available for bioenergy processing for the two crops. In particular, for Miscanthus, there is a difference concerning mass and quality of the biomass depending on harvest time. Autumn harvest results in higher yields of wetter biomass, whereas delaying harvest until late winter or spring results in a smaller but drier biomass yield (Heaton et al., 2010). Winter harvest is better for thermal conversion of the biomass, whereas autumn harvest can be better suited for fermentation of sugars in the biomass (Lewandowski et al., 2003; Le Ngoc Huyen et al., 2010; Hodgson et al., 2011).

Difference in yields

In the case of willow, Sevel et al. (2012) showed a higher production on organic soil compared to sandy soil in southern Sweden. These results support the findings of this study that willow biomass production is higher on clayey soils compared to sandy soils and that willow productivity is positively correlated with available water content. Miscanthus is considered more water use efficient, because of its use of C4 photosynthesis. These biochemical differences and their effects on leaf level water use efficiency are described fully in BioCro (Miguez et al., 2009; Wang et al., 2015). On the other hand with a longer growing season, willow can take advantage of a longer period of precipitation, which will have particular benefit in the early spring when potential evapotranspiration is low. This may explain the superior yields predicted for willow on the lighter soils of western Denmark (Fig. 3). Average growing season temperatures are also lower on the western part of Denmark, and this would also favor willow over Miscanthus (cf. (Miguez et al., 2009; Wang et al., 2015)).

Water availability is important to the yields of both crops. Although Denmark may be considered an area of high precipitation relative to potential evapotranspiration, the stochastic nature of precipitation events means that transient periods of water shortage occurs. These are ameliorated on deep and clay or organic matter rich soils by better water storage capacity. This is offset on the most clay-rich soils, by the fact that clay particles bind water generating a low matric potential and causing less of the water present to be available to the plant. Water availability is therefore a combination of soil type, precipitation and evapotranspiration. These transient effects are captured by BioCro, which dynamically simulates water transfer between ten soil layers in the rooting zone. Effects of soil composition on the availability of water are accounted for by calculating water potential from volumetric soil water content in each layer from first principles (Miguez et al., 2009).

Yield stability

The coefficient of variation (CV) in annual yields is a measure of yield stability, or the year-to-year variation

in yield. This is an important property with respect to biomass facilities, because it affects the security of supply. For both crops, yield stability was lowest on the poor, sandy soils. In this situation, willow has a major advantage, since on a 3-year cycle, it will tend to average poor with good years. This is an artifact of how yields are calculated. In addition, willow biomass can in effect be stored live until sufficient yield is obtained. However, Miscanthus has to be harvested each year. The higher variability is driven by the poorer ability of these soils to store water, making them more vulnerable to transient droughts. Arundale et al., 2014 showed larger year-to-year variation in yields in Illinois on the sandy soil of Havana compared to the deep loam soil of Urbana over a 7-year study.

In previous applications, BioCro has shown the lowest CV on the soils giving the highest yields of both willow and Miscanthus within a region (Miguez et al., 2012; Wang et al., 2015).

Limitations of BioCro

If there had been a large body of field data for these crops across Denmark, an empirical model interpolating between this data may have been more appropriate. Inevitably, it does leave the question of what faith can be placed in largely untested predictions. However, parameterization of the model based on data from one site in south England allowed a remarkably close prediction of the measured growth and production of Miscanthus across sites from Portugal and Greece to Ireland and south Sweden, capturing the experienced year-to-year variation at individual sites (Miguez et al., 2009). As in the present study, the model was run with soil and weather data for the individual sites. The BioCro model has not been validated for Denmark as a part of this analysis because of limitations in field trial data, but data from temperate regions all over the world have been used to develop and validate the model as described in (Miguez et al., 2009; Wang et al., 2015).

Another limitation of BioCro is that it does not take frost kills of Miscanthus into account when simulating yields and establishment. Several studies and reviews indicate that Miscanthus has problems with frost during establishment in Europe and Denmark (Venendaal et al., 1997; Heaton et al., 2004; Larsen et al., 2013). Miscanthus has, however, been able to survive very low temperatures and there should be breeding resources available to improve the cold tolerance of several Miscanthus species by different techniques (Heaton et al., 2008, 2010; Głowacka et al., 2014). So although the cold tolerance aspect is a limitation of the model, there is scope for improvement of the cold tolerance of Miscanthus. Other modeling studies show that frost kill is taking place in

Denmark and Europe, but new hybrids and a changing climate may limit these impacts in the future (Hastings *et al.*, 2009a,b).

Willow does not have the same problems with frost because cold tolerant hybrids have been developed and willow has also been grown for many years in climates far colder than Denmark (Ledin, 1996; Larsson, 1998). Some Danish experiments have, however, shown problems with frost damage in Denmark (Sevel *et al.*, 2012).

Model uncertainties

The BioCro model has some uncertainties on top of those limitations reported above. Some of these uncertainties are mentioned in (Miguez *et al.*, 2009, 2012; Wang *et al.*, 2015).

There is a specific uncertainty connected with the low-lying, organic soil types. The hydrological properties of these soils are not well simulated because they are groundwater fed and available water is very important for yield. This leads to added uncertainty for the 16.2% of the area occupied by these soil types (Madsen *et al.*, 1992; Børgesen *et al.*, 2009). But, low-lying, organic soils with high ground water tables can be productive in Denmark, at least for willow (Sevel *et al.*, 2012).

Similarly, other aspects of soil properties are uncertain: Soil hydrological parameters are established using equations based on a limited dataset and the rooting depth is established as a general value for crops, not specifically for perennial bioenergy crops (Madsen *et al.*, 1992; Børgesen *et al.*, 2009). We have, however, used the same data for both crops, so any uncertainties are the same for both crops.

Yield improvements and scope of Miscanthus and willow cultivation in Denmark

As discussed above, there is a gap between the model simulations and achieved yields for both crops. Agronomy of both crops is in its infancy and yields will increase from increased experience. Further breeding for improved yield and climatic tolerance has only just begun for Miscanthus. Therefore, there is considerable potential for closing the yield gap. The mechanistic basis of BioCro allows reparameterization to include new developments in genetics and agronomy, and allow recasting of the projected yields presented as innovations emerge.

For willow, there is a clear trend of increasing yields in Sweden caused by both improved genetic material and management. The historic yield increase has been shown to be 0.34 Mg DM ha^{-1} yr^{-1} for Swedish growers from 1986 to 2000 (Mola-Yudego, 2011). Similar results are seen in the UK where breeding efforts have improved the yield with 2 Mg DM ha^{-1} yr^{-1} from 1974 to 2005 (Karp *et al.*, 2011).

There is much less experience with growing Miscanthus in Denmark and Europe. But it is often stated that there is a large potential for Miscanthus to improve its productivity (Heaton *et al.*, 2008, 2010). This is partly due to Miscanthus being genetically unimproved, so a breeding and selection effort is likely to improve its productivity or other key traits (Heaton *et al.*, 2010). For example, germplasm with greater freezing and chilling tolerance has recently been identified in tests within Denmark (Głowacka *et al.*, 2014).

In 2013, there was only a small area in Denmark grown with willow (5633 ha) and Miscanthus (66 ha), but it is expected that perennial biomass crops can play a vital role in the future agriculture of Denmark where biomass crops are used in a biorefinery concept and can be used for both feed and fuels (Alexander *et al.*, 2014; Gylling *et al.*, 2013; Jørgensen *et al.*, 2013). This study shows what yields can be expected if willow and Miscanthus areas are expanded to areas where there currently is no production. Furthermore, Denmark has a high proportion of CHP and district heating plants that are able to use wood chips and straw as a feedstock for energy production and even more is expected in the future (Danish Energy Agency, 2012, Energistyrelsen, 2012). These aspects make it very important to be able to accurately estimate the feedstock production of biomass crops. A crop model is very useful in this regard because it offers opportunities to investigate how much feedstock that can be produced, but also offers information on the yield variation and spatial patterns exhibited by these crops. This aspect will be very important for making decisions on where and which feedstock to grow in Denmark. It is obvious that perennial biomass crops such as willow and Miscanthus can help to achieve the ambitious climate change mitigation policies of Denmark. The most recent analysis of bioenergy in Denmark suggests increasing use of biomass in the Danish energy system in both near- and medium-term future. Similarly, there will be an increase in area available for biomass production, so there are ample opportunities to increase production (Dalgaard *et al.*, 2011; The Danish Energy Agency, 2014).

Acknowledgements

This study was funded by the BIORESOURCE project funded by the Danish Council for Strategic Research. DJ, DW and SPL were also supported by Energy Biosciences Institute award OO1G20. We thank Finn Plauborg and Christen Duus Børgesen, Aarhus University, for their help in obtaining the climate and soil data used here. We thank researchers and support staff at the Long Lab and Institute for Genomic Biology at the University of Illinois for their support and help throughout this research. At last, we thank two anonymous reviewers who gave us valuable feedback on an earlier version of this manuscript.

References

Alexander P, Moran D, Smith P et al. (2014) Estimating UK perennial energy crop supply using farm-scale models with spatially disaggregated data. Global Change Biology Bioenergy, 6, 142–155.

Allen RG, Pereira LS, Raes D, Smith M (1998) Crop evapotranspiration - Guidelines for computing crop water requirements. FAO Irrigation and Drainage Paper 56, pp. 1–15. FAO, Rome, Italy.

Arundale RA, Dohleman FG, Heaton EA, Mcgrath JM, Voigt TB, Long SP (2014) Yields of Miscanthus × giganteus and Panicum virgatum decline with stand age in the Midwestern USA. Global Change Biology Bioenergy, 6, 1–13.

Aylott MJ, Casella E, Tubby I, Street NR, Smith P, Taylor G (2008) Yield and spatial supply of bioenergy poplar and willow short-rotation coppice in the UK. New Phytologist, 178, 358–370.

Bauen AW, Dunnett AJ, Richter GM, Dailey AG, Aylott M, Casella E, Taylor G (2010) Modelling supply and demand of bioenergy from short rotation coppice and Miscanthus in the UK. Bioresource Technology, 101, 8132–8143.

Beale CV, Long SP (1995) Can perennial C4 grasses attain high efficiencies of radiant energy conversion in cool climates? Plant, Cell & Environment, 18, 641–650.

Bentsen N, Felby C (2012) Biomass for energy in the European Union - a review of bioenergy resource assessments. Biotechnology for Biofuels, 5, 25.

Beurskens LWM, Hekkenberg M (2011) Renewable energy projections as published in the national renewable energy action plans of the European member states. pp 244, Petten, NL and Copenhagen, DK, Energy Research Centre of the Netherlands and European Environment Agency.

Børgesen CD, Waagepetersen J, Iversen TM, Grant R, Jacobsen B, Emlholt S (2009) Midtvejsevaluering af Vandmiljøplan III. Hoved- og Baggrundsnotater [in Danish]. In: DJF Rapport Markbrug. pp. 238, Det Jordbrugsvidenskabelige Fakultet, Aarhus Universitet, Tjele.

Clifton-Brown JC, Neilson B, Lewandowski I, Jones MB (2000) The modelled productivity of Miscanthus x giganteus (GREEF et DEU) in Ireland. Industrial Crops and Products, 12, 97–109.

Clifton-Brown JC, Stampfl PF, Jones MB (2004) Miscanthus biomass production for energy in Europe and its potential contribution to decreasing fossil fuel carbon emissions. Global Change Biology, 10, 509–518.

Dalgaard T, Olesen JE, Petersen SO et al. (2011) Developments in greenhouse gas emissions and net energy use in Danish agriculture - how to achieve substantial CO(2) reductions? Environmental Pollution, 159, 3193–3203.

Danish Energy Agency (2012) Energy Policy in Denmark. Danish Energy Agency, Copenhagen.

Energistyrelsen (2012) Danmarks Energifremskrivning 2012 [in Danish]. pp. 60, Energistyrelsen, Copenhagen.

ESRI (2010) ArcGIS ver. 10.1. ESRI, Redlands, CA.

European Parliament and the Council (2009) Directive 2009/28/EC of the European Parliament and of the Council of 23 April 2009 on the promotion of the use of energy from renewable sources and amending and subsequently repealing Directives 2001/77/EC and 2003/30/EC. 2009/28/EC.

Głowacka K, Adhikari S, Peng J, Gifford J, Juvik JA, Long SP, Sacks EJ (2014) Variation in chilling tolerance for photosynthesis and leaf extension growth among genotypes related to the C(4) grass Miscanthus ×giganteus. Journal of Experimental Botany, 65, 5267–5278.

Gylling M, Jørgensen U, Bentsen NS, Kristensen IT, Dalgaard T, Felby C, Johansen VK (2013) The + 10 million tonnes study. Increasing the sustainable production of biomass for biorefineries. pp 32, Frederiksberg, Department of Food and Resource Economics, Faculty of Science, University of Copenhagen.

Hastings A, Clifton-Brown J, Wattenbach M, Mitchell CP, Stampfl P, Smith P (2009a) Future energy potential of Miscanthus in Europe. Global Change Biology Bioenergy, 1, 180–196.

Hastings A, Clifton-Brown J, Wattenbach M, Mitchell P, Smith P (2009b) The development of MISCANFOR, a new Miscanthus crop growth model: towards more robust yield predictions under different climatic and soil conditions. Global Change Biology Bioenergy, 1, 154–170.

Heaton E, Long S, Voigt T, Jones M, Clifton-Brown J (2004) Miscanthus for renewable energy generation: European Union experience and projections for Illinois. Mitigation and Adaptation Strategies for Global Change, 9, 433–451.

Heaton EA, Flavell RB, Mascia PN, Thomas SR, Dohleman FG, Long SP (2008) Herbaceous energy crop development: recent progress and future prospects. Current Opinion in Biotechnology, 19, 202–209.

Heaton EA, Dohleman FG, Long SP (2009) Seasonal nitrogen dynamics of Miscanthus× giganteus and Panicum virgatum. Global Change Biology Bioenergy, 1, 297–307.

Heaton EA, Dohleman FG, Fernando Miguez A et al. (2010) Miscanthus: a promising biomass crop. Advances in Botanical Research, 56, 76.

Hodgson EM, Nowakowski DJ, Shield I, Riche A, Bridgwater AV, Clifton-Brown JC, Donnison IS (2011) Variation in Miscanthus chemical composition and implications for conversion by pyrolysis and thermo-chemical bio-refining for fuels and chemicals. Bioresource Technology, 102, 3411–3418.

Humphries SW, Long SP (1995) WIMOVAC: a software package for modelling the dynamics of plant leaf and canopy photosynthesis. Computer Applications in the Biosciences: CABIOS, 11, 361–371.

Jørgensen U, Elsgaard L, Sørensen P et al. (2013) Biomasseudnyttelse i Danmark - Potentielle ressourcer og bæredygtighed [in Danish]. In: DCA Rapport. DCA-Nationalt Center for Fødevarer og Jordbrug, Tjele.

Karp A, Shield I (2008) Bioenergy from plants and the sustainable yield challenge. New Phytologist, 179, 15–32.

Karp A, Hanley SJ, Trybush SO, Macalpine W, Pei M, Shield I (2011) Genetic Improvement of Willow for Bioenergy and Biofuels. Journal of Integrative Plant Biology, 53, 151–165.

Landbrug Og Fødevarer (2010) Oversigt over Landsforsøgene 2010 [in Danish]. Videncentret for Landbrug.

Landbrug Og Fødevarer (2012) Oversigt over landsforsøgene 2012 [In Danish]. pp. 488, Videncentret for landbrug.

Larsen S, Jørgensen U, Kjeldsen J, Lærke P (2013) Long-term Miscanthus yields influenced by location, genotype, row distance, fertilization and harvest season. Bioenergy Research, 7, 620–635.

Larsen SU, Jørgensen U, Kjeldsen JB, Lærke PE (2014a) Long-term yield effects of establishment method and weed control in willow for short rotation coppice (SRC). Biomass and Bioenergy, 71, 266–274.

Larsen SU, Jørgensen U, Lærke PE (2014b) Willow yield is highly dependent on clone and site. Bioenergy Research, 7, 1280–1292.

Larsson S (1998) Genetic improvement of willow for short-rotation coppice. Biomass and Bioenergy, 15, 23–26.

Le Ngoc Huyen T, Rémond C, Dheilly RM, Chabbert B (2010) Effect of harvesting date on the composition and saccharification of Miscanthus x giganteus. Bioresource Technology, 101, 8224–8231.

Ledin S (1996) Willow wood properties, production and economy. Biomass and Bioenergy, 11, 75–83.

Lewandowski I, Heinz A (2003) Delayed harvest of miscanthus—influences on biomass quantity and quality and environmental impacts of energy production. European Journal of Agronomy, 19, 45–63.

Lewandowski I, Clifton-Brown JC, Andersson B et al. (2003) Environment and harvest time affects the combustion qualities of miscanthus genotypes. Agronomy Journal, 95, 1274–1280.

Lindroth A, Båth A (1999) Assessment of regional willow coppice yield in Sweden on basis of water availability. Forest Ecology and Management, 121, 57–65.

Lobell DB, Cassman KG, Field CB (2009) Crop yield gaps: their importance, magnitudes, and causes. Annual Review of Environment and Resources, 34, 179–204.

Madsen HB, Nørr AH, Holst KA (1992) Den Danske Jordklassificering [in Danish]. Det Kongelige Geografiske Selskab, København.

Miguez FE, Zhu XG, Humphries S, Bollero GA, Long SP (2009) A semimechanistic model predicting the growth and production of the bioenergy crop Miscanthus x giganteus: description, parameterization and validation. Global Change Biology Bioenergy, 1, 282–296.

Miguez FE, Maughan M, Bollero GA, Long SP (2012) Modeling spatial and dynamic variation in growth, yield, and yield stability of the bioenergy crops Miscanthus x giganteus and Panicum virgatum across the conterminous United States. Global Change Biology Bioenergy, 4, 509–520.

Mola-Yudego B (2010) Regional potential yields of short rotation willow plantations on agricultural land in northern Europe. Silva Fennica, 44, 63–76.

Mola-Yudego B (2011) Trends and productivity improvements from commercial willow plantations in Sweden during the period 1986-2000. Biomass and Bioenergy, 35, 446–453.

Mola-Yudego B, Aronsson P (2008) Yield models for commercial willow biomass plantations in Sweden. Biomass and Bioenergy, 32, 829–837.

Morsing M, Nielsen KH (1995) Tørstofproduktionen i Danske Pilekulturer 1989–1994 [In Danish]. (ed Koch NE), pp. 35. Forskningscentret for Skov og Landskab, Hørsholm.

Mortensen J, Hauge Nielsen K, Jørgensen U (1998) Nitrate leaching during establishment of willow (Salix viminalis) on two soil types and at two fertilization levels. *Biomass and Bioenergy*, **15**, 457–466.

Nair SS, Kang SJ, Zhang XS *et al.* (2012) Bioenergy crop models: descriptions, data requirements, and future challenges. *Global Change Biology Bioenergy*, **4**, 620–633.

Plauborg F, Olesen JE (1991) Development and validation of the model MARK-VAND for irrigation scheduling in agriculture [in Danish]. In: *Tidsskrift for Planteavls Specialserie*. pp. 103. Landbrugsministeriet, Statens Planteavlsforsøg, Tjele.

Pogson M (2011) Modelling Miscanthus yields with low resolution input data. *Ecological Modelling*, **222**, 3849–3853.

R Core Team (2013) *R: A Language and Environment for Statistical Computing*. R Foundation for Statistical Computing, Vienna, Austria.

Richter GM, Riche AB, Dailey AG, Gezan SA, Powlson DS (2008) Is UK biofuel supply from Miscanthus water-limited? *Soil Use and Management*, **24**, 235–245.

Scharling M (1999) KLIMAGRID DANMARK - Nedbør, lufttemperatur og potentiel fordampning 20*20 & 40*40 km [in Danish]. *Technical Report*, pp. 1–48. Danish Meteorological Institute, Copenhagen.

Scharling M (2012) Climate Grid Denmark. *Technical Report*, pp. 1–12. Danish Meteorological Institute, Copenhagen.

Sevel L, Nord-Larsen T, Raulund-Rasmussen K (2012) Biomass production of four willow clones grown as short rotation coppice on two soil types in Denmark. *Biomass and Bioenergy*, **46**, 664–672.

Sevel L, Nord-Larsen T, Ingerslev M, Jørgensen U, Raulund-Rasmussen K (2013) Fertilization of SRC willow, I: biomass production response. *Bioenergy Research*, **7**, 319–328.

Tallis MJ, Casella E, Henshall PA, Aylott MJ, Randle TJ, Morison JIL, Taylor G (2013) Development and evaluation of ForestGrowth-SRC a process-based model for short rotation coppice yield and spatial supply reveals poplar uses water more efficiently than willow. *Global Change Biology Bioenergy*, **5**, 53–66.

The Danish Agrifish Agency (2013) Markblokkort [in Danish]. Copenhagen.

The Danish Energy Agency (2014) *Analyse af Bioenergi i Danmark (eng: Analysis of Bioenergy in Denmark)*. The Danish Energy Agency, Copenhagen.

Venendaal R, Jørgensen U, Foster C (1997) European energy crops: a synthesis. *Biomass and Bioenergy*, **13**, 147–185.

Voigt TB (2015) Are the environmental benefits of Miscanthus × giganteus suggested by early studies of this crop supported by the broader and longer-term contemporary studies? *Global Change Biology Bioenergy*, **7**, 567–569.

Wang D, Jaiswal D, Lebauer DS, Wertin TM, Bollero GA, Leakey AD, Long SP (2015) A physiological and biophysical model of coppice willow (Salix spp.) production yields for the contiguous USA in current and future climate scenarios. *Plant, Cell and Environment*, **38**, 1850–1865.

12

Dedicated biomass crops can enhance biodiversity in the arable landscape

ALISON J. HAUGHTON[1], DAVID A. BOHAN[2], SUZANNE J. CLARK[1], MARK D. MALLOTT[1], VICTORIA MALLOTT[3], RUFUS SAGE[4] and ANGELA KARP[1]

[1]*Rothamsted Research, West Common, Harpenden, Hertfordshire AL5 2JQ, UK,* [2]*INRA, UMR 1347 Agroécologie, Pôle ECOLDUR, 17 rue Sully, Dijon CEDEX 21065, France,* [3]*168 Putteridge Road, Luton LU2 8HJ, UK,* [4]*Game and Wildlife Conservation Trust, Burgate Manor, Fordingbridge, Hampshire SP6 1EF, UK*

Abstract

Suggestions that novel, non-food, dedicated biomass crops used to produce bioenergy may provide opportunities to diversify and reinstate biodiversity in intensively managed farmland have not yet been fully tested at the landscape scale. Using two of the largest, currently available landscape-scale biodiversity data sets from arable and biomass bioenergy crops, we take a taxonomic and functional trait approach to quantify and contrast the consequences for biodiversity indicators of adopting dedicated biomass crops on land previously cultivated under annual, rotational arable cropping. The abundance and community compositions of biodiversity indicators in fields of break and cereal crops changed when planted with the dedicated biomass crops, miscanthus and short rotation coppiced (SRC) willow. Weed biomass was consistently greater in the two dedicated biomass crops than in cereals, and invertebrate abundance was similarly consistently higher than in break crops. Using canonical variates analysis, we identified distinct plant and invertebrate taxa and trait-based communities in miscanthus and SRC willows, whereas break and cereal crops tended to form a single, composite community. Seedbanks were shown to reflect the longer term effects of crop management. Our study suggests that miscanthus and SRC willows, and the management associated with perennial cropping, would support significant amounts of biodiversity when compared with annual arable crops. We recommend the strategic planting of these perennial, dedicated biomass crops in arable farmland to increase landscape heterogeneity and enhance ecosystem function, and simultaneously work towards striking a balance between energy and food security.

Keywords: biodiversity indicators, bioenergy, biomass crops, canonical variates analysis, functional traits, invertebrates, miscanthus, seedbank, short rotation coppiced willow, weed biomass

Introduction

Anthropogenic-induced climate change continues to be the single, overriding challenge to the future of humans and ecosystems, and reductions in emissions of CO_2 are essential to limit the risks of climate change (IPCC, 2014). Balancing the food and fuel security demands of a growing human population, in the context of climate change, has led to a global drive to increase production from land that has resulted in unforeseen land use conflicts, particularly for crops traditionally grown for food being diverted for use in the transport biofuel industry (Searchinger et al., 2015). These conflicts compound genuine concerns that a shift in focus on to cheaper sources of gas, including the recent developments in the shale gas industry, could disrupt progress in the development

Correspondence: Alison Haughton
e-mail: alison.haughton@rothamsted.ac.uk

and adoption of sustainable renewable technologies and significantly delay efforts to further reduce global emissions of CO_2 (Davis & Shearer, 2014; Jackson et al., 2014; McJeon et al., 2014).

Non-food, perennial, dedicated biomass crops, such as trees grown as short rotation coppice and grasses, are potentially integral to reducing CO_2 emissions and many studies have documented positive benefits of growing perennial biomass crops, including for ecosystem services (Berndes et al., 2008; Baum et al., 2013a; Meehan et al., 2013) and biodiversity (Haughton et al., 2009; Rowe et al., 2009; Dauber et al., 2010; Baum et al., 2012; Stanley & Stout, 2013; Bourke et al., 2014). However, much of the ecological evidence is directly (e.g. Rowe et al., 2009) or indirectly (e.g. Holland et al., 2015) based on studies conducted on small, temporal (single samples within a single season), spatial (localized, experimental plots) scales, whilst sustainability concerns relate to longer term, landscape-scale expansion (Fargione, 2010; Dauber & Bolte,

2014). Furthermore, many studies assess biodiversity taxa of one type of biomass crop, without drawing comparisons with the land uses they may replace (see review by Dauber *et al.*, 2010) and use coarse levels of identification (e.g. Rowe *et al.*, 2011) resulting in misleading interpretation of responses for ecosystem service provision (e.g. Holland *et al.*, 2015). Nevertheless, models using data derived from small-scale experiments have predicted that perennial, dedicated biomass crops could have beneficial environmental impacts if integrated into agricultural landscapes (e.g. (Meehan *et al.*, 2010; Tilman *et al.*, 2009). There have been well documented declines in farmland biodiversity and ecosystem service provision in the latter half of the 20th Century (e.g. Donald *et al.*, 2001; Bianchi *et al.*, 2006; Storkey *et al.*, 2012; Woodcock *et al.*, 2014; Senapathi *et al.*, 2015). These reductions in biodiversity and ecosystem function have been attributed to an homogenization of the farmed landscape, in terms of reduction in the area and diversity of semi-natural habitats and diversity of on-farm cropping and management systems (Benton *et al.*, 2003; Bianchi *et al.*, 2006). It is therefore important to test whether cultivating perennial, dedicated biomass crops in annual arable crop-dominated landscapes could be used to enhance and conserve farmland biodiversity and ecosystem function.

The agronomic management and growth characteristics of perennial, dedicated biomass crops, such as willows (*Salix* spp), poplars (*Populus* spp) and miscanthus (*Miscanthus* spp.), contrast with those of food crops typically grown for biofuel (e.g. wheat, maize, soy). Once established, these crops can reach 3–4 m in height and have the potential to produce large yields from very low fertilizer and pesticide inputs and provide structure in the landscape right through the winter, as they are normally harvested after senescence (miscanthus) and leaf drop (usually between December and April). As they are perennials, remaining *in situ* for ca. 20 years, the soil is not cultivated annually and they provide more stable habitats punctuated only by annual (for energy grasses) or triennial (for trees grown as short rotation coppice) harvesting. For trees like poplar, that are also grown as short rotation forestry, harvesting cycles are even longer (>15 years).

Planting perennial biomass crops in farmland should, therefore, result in contrasting abundances and compositions of plants and invertebrates compared with annual arable crops, reflecting differences in both crop growth and management. To test this, we carried out extensive sampling of established fields of two perennial, dedicated biomass crops [miscanthus and short rotation coppiced (SRC) willows] and used taxonomic, functional trait and phylogenetic groupings to compare the abundance and community compositions of key biodiversity indicators with those of arable crops. Thus, we test the

null hypothesis that there is no change in biodiversity in perennial, dedicated biomass crops planted on land previously under annual arable crop management.

Materials and methods

Experimental design

We undertook the most intensive temporal- and spatial-scale sampling of biodiversity in perennial biomass crops reported to date (Karp *et al.*, 2009) and compared these data with the most complete study of biodiversity previously carried out in arable crops (Coghlan, 2003; Perry *et al.*, 2003) that is currently available. Although these large-scale experiments were carried out independently of each other in different years, they were designed using the same methodologies, such that indicators of weed and invertebrate biodiversity were intensively sampled across entire, commercial fields over a single growing season and represent the most comprehensive, standardized assessment of regional- and national-scale patterns of biodiversity in the farmed landscape of Great Britain. One concern with comparing data collected at different times is that populations of biodiversity indicators in farmland, per se, may have changed, making such a comparison problematic. Butterfly Lepidoptera are used as an indicator of environmental change (Defra, 2014), in part because they exhibit rapid (between-year) response to environmental stresses. Butterfly populations were found to be stable during the period 2000–2006, when these studies were carried out (Defra, 2014) and have been used previously (Haughton *et al.*, 2009) to provide confidence that any differences in abundance and community composition in the crop types are crop management-mediated.

Biomass crops

Using questionnaires similar to those used for study site selection in the Farm Scale Evaluation (FSE) (Champion *et al.*, 2003), we selected 17 established fields of miscanthus and 15 of SRC willows that were distributed in the East Midlands, South-west and Southern regions of England and reflected the geographical locations of commercial dedicated biomass crops (Table S1). All study fields were in standard commercial production on land that had previously been used for arable crop production and represented a range of inherent weediness from farms of varying cropping intensities that yielded between 7 and 11 tonnes winter wheat ha^{-1}. The fields were planted between 1999 and 2004 and the fields of miscanthus had been harvested annually in the winter, while the SRC willows had passed through at least two coppice rotations and were due to be harvested during the winter following data collection. The biomass crops were thus representative of established, and for SRC willows, mature phase crops.

Arable crops

The FSEs of genetically modified, herbicide-tolerant break crops (Firbank *et al.*, 2003) have previously been used to compare butterfly abundance in field margins of arable and dedicated biomass crops (Haughton et al., 2009). Thus, the data for the arable crops came from 255 fields sampled as part of the FSEs (Champion et al., 2003; Firbank et al., 2003; Bohan et al., 2005), made up of 65 fields of spring-sown beet (*Beta vulgaris* L.), 58 fields of spring-sown maize (*Zea mays* L.), 67 fields of spring-sown oilseed rape (*Brassica napus* L.) and 65 fields of winter-sown oilseed rape (*B. napus* L.). The fields represented the range of agricultural and environmental conditions found in commercial practice with regard to geographical distribution, agronomy, soil type and field size (Champion et al., 2003; Bohan et al., 2005) and treatment effects on the abundance of plant and invertebrate indicators were shown not to co-vary with year, study site or geographical location (Haughton et al., 2003; Heard et al., 2003; Bohan et al., 2005). The FSEs used a split-field design, where the effect on biodiversity indicators of 'conventional' arable practice was compared with that of a modified herbicide management regime associated with genetically modified, herbicide-tolerant break crops. Here, only data from the conventional half of the split field (herein after termed 'field') are used. The crops were established from 2000 to 2002 and sampled throughout the growing seasons from 2000 to 2003. In the years subsequent to growing contrasting GMHT and conventional varieties of break crops, farmers followed their usual rotation and the fields were sown with a non-GMHT crop and plant biodiversity indicators were assessed for the first 2 years following the FSEs. Biodiversity data for the cereal crops came from these follow-up assessments of the conventional half of the split field (Heard et al., 2005) in fields of inter-sown barley (*n* = 19) and winter-sown wheat (*n* = 72).

Weeds

Methods for sampling biodiversity indicators were standardized for all crops, using the approach taken in the FSEs, described in detail in (Firbank et al., 2003; Haughton et al., 2003; Heard et al., 2003; Bohan et al., 2005). A total of twelve, evenly spaced transects, extending 32 m into the crops, were placed around and perpendicular to the field edges (Firbank et al., 2003) and biodiversity indicators were sampled as follows. Soil core samples of the seedbank were taken from five loci at 2 and 32 m along four transects, prior to the break crops being sown (year *t*), and at 1 (*t* + 1) and 2 (*t* + 2) years after drilling, and in April in the biomass crops. The seeds contained within the cores were germinated and identified following the methods in Heard et al. (2003). Abundance is reported here as the density (numbers m^{-2}) to a depth of 0.15 m, where one seed per field sample was equivalent to 18.75 m^{-2} (Heard et al., 2005). Seedbank data representing the effect of the four main conventional break crops of the FSE were taken a year after drilling, year *t* + 1 (Heard et al., 2003) and data that reflect the effect of growing cereals in the year subsequent to the break crops were taken from year *t* + 2 samples of those fields sown to cereals in year *t* + 1 (Heard et al., 2005).

Noncrop plant (weed) biomass, representative of a single crop growing season, was sampled in 1 m × 1 m quadrats at 2 and 32 m along all 12 transects in the month before harvest for the arable crops, and in August for the biomass crops. All plants rooted within the quadrat were cut at ground level,

identified and sorted into species and dried for 24 h at 80 °C before weighing (Heard et al., 2003). Biomass data reported here are g m^{-2}. Plant species were assigned to monocot or eudicot phylogenetic group (APG, 2003) following (Stace, 2010) and allocated to primary growth strategy following (Grime et al., 2007) prior to analysis (see Table S2).

Invertebrates

Within-crop invertebrates from the soil and weeds were sampled using a Vortis suction sampler (Arnold 1994), where five, 10-second 'sucks' were taken 1 m apart at 2 and 32 m along three transects in June (Haughton et al., 2003) and identified to various taxonomic levels and assigned to appropriate trophic (functional) group for analysis (Table 1) (Hawes et al., 2003). An area of 0.6 m^2 per field was sampled and abundance is reported here as density of invertebrates m^{-2}.

Statistical analyses

To determine whether the densities of phylogenetic and growth strategy groups of weeds and trophic groups of invertebrates differed between biomass and arable crops, field totals were transformed to common logarithms, after adding an offset of one to seedbank and invertebrate counts and 0.005 to biomass measurements. Sites for which the total count was zero or one were excluded (c.f. Heard et al., 2003). The number of fields included in each analysis is reported as N. For each biodiversity indicator group and biomass-arable crop comparison of interest, the null hypothesis of no difference between means (H$_0$: $\delta = 0$, H$_1$: $\delta \neq 0$, where $\hat{\delta} = d$) was tested using a *t*-test, with degrees of freedom adjusted using Satterthwaite's formula when crop variances were unequal (based on an *F*-test, $P > 0.05$). Crop means are presented back-transformed to the original scales. Relative crop effects for each biodiversity indicator group are reported as *R*, the multiplicative ratio (biomass crops : arable crops), calculated as $R = 10^d$, where *d* is the difference between the crop means on the logarithmic scale. Upper and lower 95% confidence limits for δ were back-transformed similarly to give confidence limits for the true value of *R*.

Canonical variates analysis (CVA) (Gardner et al., 2006) was used to detect differences in the weed and invertebrate communities of miscanthus, SRC willows and break crops; and for weeds only, between miscanthus, SRC willows and cereal crops. To avoid the effects of dominance of a few, highly abundant species (Smith et al., 2008), field abundance of individual weed and invertebrate taxa or grouping was transformed to proportions of the total abundance per field. Significant differences reported for the compositions of weed and invertebrate communities therefore indicate differences in proportions rather than abundance. Taxa that were considered to have occurred by chance were excluded from the analysis, such that data for taxa that were present either in only a single crop with an abundance of <1%, or in two or more crops at <0.1% abundance were removed. Where removal of these low-abundance taxa resulted in the remaining proportion of abundance at individual sites being less than 80% of the original site total, these

Table 1 Levels of identification and assignment to trophic group of invertebrates

Taxa	Level of identification	Trophic group			
		Detritivore	Herbivore	Predator	Mix
Collembola	Family	y			
Orthoptera	Order		y		
Hemiptera					
Heteroptera	Species		y	y	y
Auchenorrhyncha	Sub-order		y		
Aphidoidea	Superfamily		y		
Neuroptera	Order			y	
Lepidoptera					
Larvae	Order		y		
Diptera	Order				y
Hymenoptera					
Symphyta larvae	Sub-order		y		
Parasitica	Superfamily			y	
Coleoptera					
Coccinellidae	Family	y		y	
Curculionidae	Family		y		
Staphylinidae	Family				y
Carabidae	Species		y	y	
Others	Order				y
Araneae					
Linyphiidae	Family			y	
Tenuiphantes tenuis	Species			y	
Erigone	Genus			y	
Oedothorax	Genus			y	
Others	Order			y	

sites were removed from the analysis. Proportion data were arcsine square-root transformed (Sokal & Rohlf, 2012) prior to analysis, with crop type (miscanthus, SRC willows, break, cereal crops) as the grouping factor. All analyses were done using GenStat 17th Edition (VSNI, 2014).

Results

Seedbank

Seedbank densities tended not to differ between biomass and arable crops for all plant groupings (Fig. 1, Table 2). Total seedbank densities did not differ in either of the biomass crops compared with break crops or cereals, but there was a trend for seedbank densities of the plant groups to be greater in miscanthus than in break or cereal crops. There was no consistent direction of differences in seedbank densities between SRC willows and the arable crops (Fig. 1, Table 2); however, there were lower densities of ruderals in SRC willows than in break crops ($R = 0.37$) and cereals ($R = 0.38$). *Poa annua* L. was the most dominant species in break and cereals crops, while *Matricaria* spp. and *Epilobium* spp. dominated in miscanthus and SRC willows respectively.

Canonical variates analysis of the proportion of the densities of 64 taxa recorded from the seedbanks of the four crop types identified distinct communities in miscanthus and SRC willows; however, seedbank communities of cereal and break crops were not distinct from each other (Fig. 2a). The first two axes explained 97.9% of the variation accounting for 94.1% ($\chi^2_{192} = 897.97$, $P < 0.001$) and 3.8% ($\chi^2_{126} = 166.55$, $P < 0.009$) of the variation, respectively, for axes 1 and 2.

Canonical variates analysis of the proportion of the densities of ten plant strategies recorded from the seedbanks of the four crop types identified distinct strategy-based communities in some of the crop types (Fig. 2b). The communities in miscanthus and SRC willows were distinct from each other and from the composite community of break crops and cereals. The first two axes represented 97.7% of the variation, with separation between the biomass crops and the arable crops along axis 1, that accounted for 92.0% ($\chi^2_{30} = 246.17$, $P < 0.001$) of the variation. Separation along axis 2 was not statistically significant, representing only 5.7% of the variation ($\chi^2_{18} = 26.59$, $P < 0.087$).

Weed biomass

Weed biomass varied between biomass and arable crops, where differences were of many orders of magnitude, ranging from 0.01 to 12.33-fold (Fig. 3, Table 3).

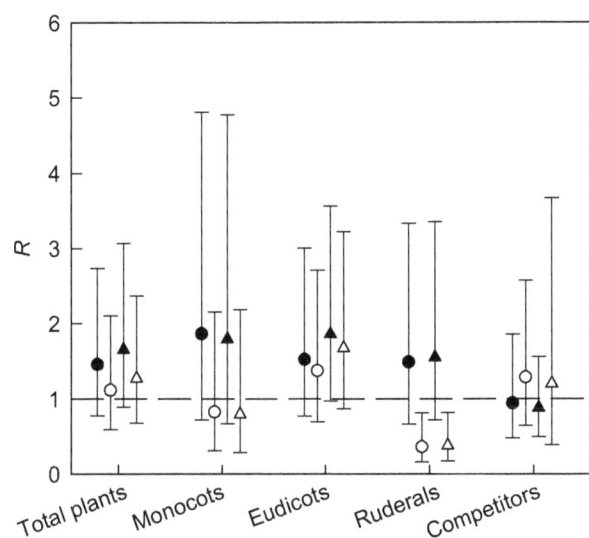

Fig. 1 Ratio (R) of seed density in the seedbanks of miscanthus and SRC willows to break and cereal crops. Solid symbols: miscanthus; open symbols: SRC willows; circles: break crops; triangles: cereals. R is computed as 10^d, where d is the difference between the means (over sites) of the logarithmically transformed seed density m^{-2} per field to a depth of 0.15 m. Dashed line is line of equality ($d = 0$ or $R = 1$). Error bars are 95% confidence limits for R, also back-transformed to the ratio scale (hence asymmetry).

The total amount of weed biomass in both miscanthus and SRC willows was lower than that of break crops ($R = 0.04$; $R = 0.10$, respectively), but higher than that in cereals ($R = 1.92$; $R = 4.76$, respectively). The difference in total weed biomass in miscanthus compared with break crops was reflected in amounts of biomass of monocots ($R = 0.18$), eudicots ($R = 0.01$) and ruderals ($R = 0.01$). There were no differences in amount of competitor biomass in miscanthus compared with break crops. The greater total weed biomass in miscanthus crops compared with cereals was not reflected consistently across the other groupings, with greater

competitor biomass ($R = 3.31$) and lower ruderal biomass ($R = 0.39$) in miscanthus. Monocot and eudicot biomasses did not differ between miscanthus and cereals. *Poa annua* L. was the most dominant species in break and cereals crops, while *Cirsium arvense* (L.) and *Elytrigia repens* (L.) dominated miscanthus and SRC willows respectively.

The lower total weed biomass in SRC willows compared with break crops was not reflected across all groupings, where competitor biomass was greater ($R = 12.33$), and biomasses of eudicots and ruderals were lower ($R = 0.05$; $R = 0.01$, respectively). Monocot biomass did not differ between SRC willows and break crops. The greater total weed biomass in SRC willows compared with cereals was reflected in many, but not all, plant groupings. Biomasses of monocot, eudicot and competitor were greater in SRC willows compared with cereals ($R = 3.82$; $R = 5.07$; $R = 5.75$, respectively), but that of ruderal plants was lower ($R = 0.32$).

Canonical variates analysis of the proportion of biomass of 92 taxa recorded from all four crop types revealed clearly defined species compositions in three of the crop types (Fig. 4a). Just as for seedbanks, communities in miscanthus and SRC willows were distinct both from each other and from those in arable crops, but those in break and cereal crops were indistinguishable. The first two axes accounted for 86.4% of the variation in plant species composition, with clear separation along axis 1 that represented 69.5% of the variation ($\chi^2_{276} = 1333.53$, $P < 0.001$) and axis 2 that represented 16.85% of the variation ($\chi^2_{182} = 653.05$, $P < 0.001$).

Canonical variates analysis of the proportion of biomass of nine plant strategies recorded from the four crop types identified distinct strategy-based communities in all crop types (Fig. 4b). The first two axes explained 96.9% of the variation, accounting for 75.5% ($\chi^2_{27} = 357.26$, $P < 0.001$) and 21.4% ($\chi^2_{16} = 106.88$, $P < 0.001$) of the variation, respectively, for axes 1 and 2. The compositions of the plant-strategy communities

Table 2 Back-transformed mean of densities of seeds (counts m^{-2}) in the top 0.15 m of soil per field in break crops (break), cereals (cereal), miscanthus (misc) and SRC willows (SRC) and t-statistics for comparisons between biomass and arable crop means, with observed significance levels

| | Mean seed density | | | | Comparisons with Miscanthus | | | | | | Comparisons with SRC | | | | | |
| | | | | | Break crops | | | Cereals | | | Break crops | | | Cereals | | |
	Break	Cereal	Misc	SRC	t	df	P	t	df	P	t	df	P	t	df	P
Total plants	141.2	124.0	206.0	157.9	1.18	249.0	0.240	1.62	97.0	0.109	0.35	249.0	0.728	0.76	97.0	0.450
Monocots	32.5	33.8	61.4	26.7	1.30	249.0	0.196	1.19	97.0	0.239	−0.39	249.0	0.696	−0.45	97.0	0.654
Eudicots	84.7	69.3	129.9	116.8	1.23	249.0	0.221	1.90	97.0	0.060	0.92	249.0	0.359	1.57	97.0	0.120
Ruderals	71.3	68.2	106.6	25.5	0.97	249.0	0.332	1.14	97.0	0.259	−2.46	249.0	0.015	−2.49	97.0	0.014
Competitors	1.7	1.9	1.6	2.5	−0.17	249.0	0.865	−0.45	97.0	0.656	0.72	249.0	0.473	0.38	7.4	0.714

Number of study fields (N): break crops = 243; cereals = 91; miscanthus = 8; SRC willows = 8.

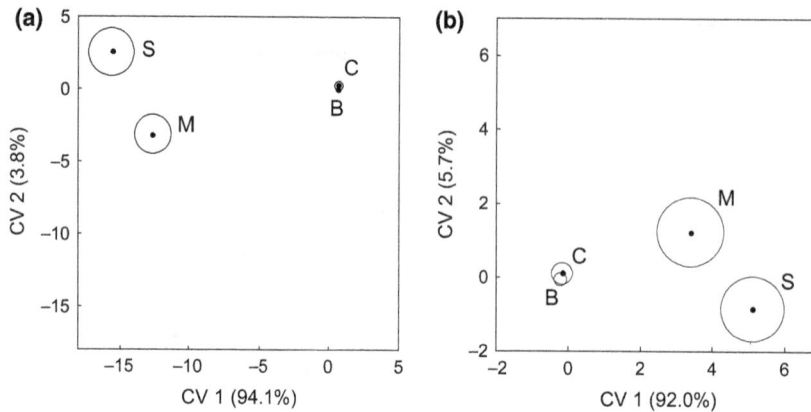

Fig. 2 Canonical variates for seedbanks in biomass and arable crops. CVA group mean scores (•) with 95% confidence regions for proportions recorded in the seedbank of (a) taxa and (b) plant growth strategies (after Grime 2001) in break crops (B), cereals (C), miscanthus (M) and SRC willows (S). The percentage variation explained by each canonical variate is given in parentheses.

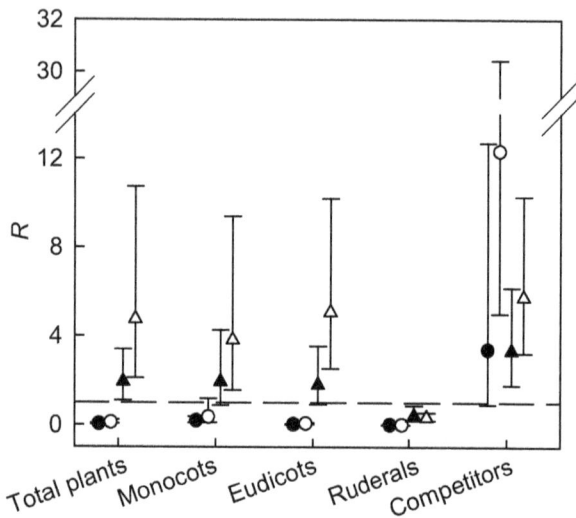

Fig. 3 Ratio (R) of weed biomass m^{-2} in miscanthus and SRC willows to break and cereal crops. Solid symbols: miscanthus; open symbols: SRC willows; circles: break crops; triangles: cereals. R is computed as 10^d, where d is the difference between the means (over sites) of the logarithmically transformed densities m^{-2} per field. Dashed line is line of equality ($d = 0$ or $R = 1$). Error bars are 95% confidence limits for R, also back-transformed to the ratio scale (hence asymmetry).

in each of the four crop types were distinct from each other, with separation between cereals, miscanthus and SRC willows occurring on axis 1, and separation between the two arable crops types occurring on axis 2.

Invertebrates

The densities of all invertebrate groupings were many times greater in the two biomass crops compared with break crops (Fig. 5, Table 4), ranging from 4.64 to 38.37-

fold differences. Isotomid Collembola was the most dominant taxon in both break crops and SRC willows, while entomobryid Collembola were most dominant in miscanthus.

The defined trophic groups comprised 47% of total invertebrates in break crops, and 92% and 84% in miscanthus and SRC willows, respectively, with detritivores consistently the dominant trophic group, representing 35%, 84% and 72% of total invertebrates in break crops, miscanthus and SRC willows respectively. CVA of the compositions of 41 taxa recorded from the three crop types identified distinct communities (Fig. 6a). The communities were separated on axis 1 ($\chi^2_{82} = 357.07$, $P < 0.001$) representing 94.2% of the variation, but not on axis 2 ($\chi^2_{40} = 38.43$, $P < 0.541$). Trophic communities of the two biomass crops were different from those of the break crops, but not from each other (Fig. 6b), with separation along axis 1 ($\chi^2_8 = 43.40$, $P < 0.001$) accounting for 94.9% of the variation, but not axis 2 ($\chi^2_3 = 2.39$, $P < 0.495$).

Discussion

Our findings corroborate our, and previous authors' (e.g. (Meehan *et al.*, 2013), expectations that replacing annual arable crops with perennial, dedicated biomass crops results in significant, large-scale changes to the abundance and composition of plant and invertebrate biodiversity indicators. We suggested that such changes would be a result of differences in crop management. Apart from the differences in physical structure of biomass crops, the between-year timing and frequency of inputs, harvesting and other disturbance in perennial crops are both consistent and reduced in comparison with intensively farmed arable crops that constitute the annual rotation-based farming system. We believe these

Table 3 Back-transformed mean of weed biomass (g m^{-2}) per field in break crops (break), cereals (cereal), miscanthus (misc) and SRC willows (SRC), and t-statistics for comparisons between biomass and arable crop means, with observed significance levels

	Mean biomass				Comparisons with Miscanthus						Comparisons with SRC					
					Break crops			Cereals			Break crops			Cereals		
	Break	Cereal	Misc	SRC	t	df	P	t	df	P	t	df	P	t	df	P
Total plants	613.8	13.3	25.6	63.4	−12.42	21.7	<0.001	2.33	34.0	0.026	−5.50	259.0	<0.001	3.80	97.0	<0.001
Monocots	61.4	6.8	13.2	25.1	−4.82	23.2	<0.001	1.64	100.0	0.104	−1.70	259.0	0.091	2.96	97.0	0.004
Eudicots	404.6	4.6	8.2	23.2	−9.36	262.0	<0.001	1.73	100.0	0.086	−11.02	19.7	<0.001	4.60	97.0	<0.001
Ruderals	150.3	5.2	2.1	1.7	−10.84	262.0	<0.001	−2.35	100.0	0.021	−10.71	259.0	<0.001	−4.43	37.0	<0.001
Competitors	0.4	1.1	3.7	6.5	1.90	21.8	0.070	4.04	17.1	<0.001	5.71	26.7	<0.001	6.45	14.1	<0.001

Number of study fields (N): break crops = 247; cereals = 85; miscanthus = 17; SRC willows = 14.

characteristics of perennial cropping led to increases in abundance of competitors and decreases in ruderals and we expect these differences to persist through the lifetime of the crop. Similar responses by plant traits to changes in frequency and intensity of crop management have been reported in arable crops (Froud-Williams *et al.*, 1983); field margins (Critchley *et al.*, 2006) and set-aside (Boatman *et al.*, 2011). Rowe *et al.* (2011) assessed responses to biomass and cereal crops by plant traits and found a statistically, nonsignificant trend towards greater numbers of perennial plants in biomass crops. It is likely that high variability between the low number of study sites (three) in the study by Rowe *et al.* (2011) contributed to the lack of a statistically significant result and highlights the value of large-scale studies such as these we report here. Storkey *et al.* (2013) assessed the effects of arable cropping on plant assemblages in greater detail, by measuring the response by plant traits to a disturbance gradient that ranged from annual cultivations and inputs to perennial noncrop habitat and demonstrated that frequency of disturbance was an important driver of trait-based community assembly in the arable systems tested.

Functional approaches have been argued to provide a more parsimonious explanation and understanding of management effects on biodiversity and ecosystem functioning in comparison to species-based, taxonomic approaches (Hooper *et al.*, 2005; Cadotte *et al.*, 2011; Gagic *et al.*, 2015). Analysis of functional groups has allowed us greater insight into ecosystem function responses to levels of disturbance, as we note the amount of variation accounted for by the first two axes in the weed CVA was greater for the functional group analyses than in the taxonomic groups (96.9% vs. 86.4% for weed biomass; 97.7% vs. 86.4% for seedbank). Although not measured here, functional indices have been shown to be positively related to ecosystem function (Hoehn *et al.*, 2008) and a next step would be to assess the functional diversity of the communities.

We used two contrasting indicators of weed biodiversity: seedbanks and biomass. Seedbanks are a repository of the effect of previous management (Bohan *et al.*, 2011), reflecting the longer term effects of field management and cropping system (Hawes *et al.*, 2010) and our results appear to be consistent with this, because crop effect ratios (R) of seedbank densities in the biomass-break crops and biomass-cereal comparisons are similar in terms of magnitude and direction of difference. Measures of weed biomass, however, reflect within-season effects of growing a particular crop type (Heard *et al.*, 2003; Hawes *et al.*, 2009), and this has been inferred by Baum *et al.* (2013b) and in our results, as the crop effect ratios (R) of weed biomass densities are very different both in their order of magnitude and direction of

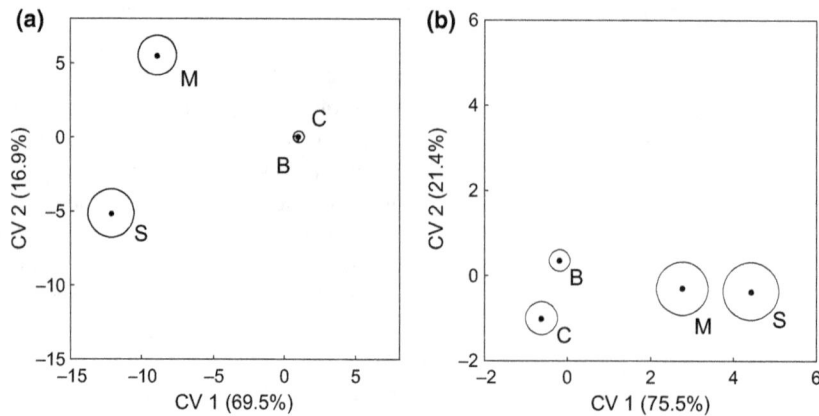

Fig. 4 Canonical variates for weed biomass in biomass and arable crops. CVA group mean scores (•) with 95% confidence regions for proportions of biomass of (a) taxa and (b) plant growth strategies (after Grime 2001) in break crops (B), cereals (C), miscanthus (M) and SRC willows (S). The percentage variation explained by each canonical variate is given in parentheses.

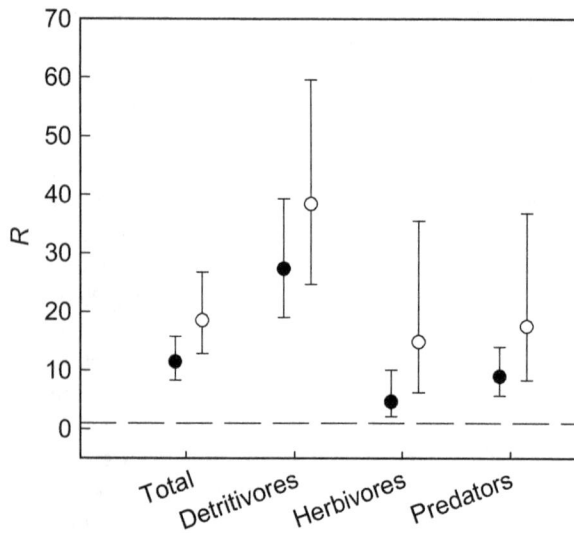

Fig. 5 Ratio (R) of the density of invertebrates m^{-2} in miscanthus and SRC willows to break crops. Solid symbols: miscanthus; open symbols: SRC willows. R is computed as 10^d, where d is the difference between the means (over sites) of the logarithmically transformed density of invertebrates m^{-2} per field. Dashed line is line of equality ($d = 0$ or $R = 1$). Error bars are 95% confidence limits for R, also back-transformed to the ratio scale (hence asymmetry).

difference in the biomass-break crop and biomass-cereal comparisons. Thus, for a given field, the choice of crop in an annual cropping system results in different effects on abundance of weed taxa and growth strategies between cropping seasons. However, the trend of direction of crop effect on indicators observed in biomass is not seen in the results for seedbanks. Our work suggests that weed biomass could be a predictor of the development of seedbanks in established, perennial crops. Where we recorded significant and consistent crop effects on weed

biomass for biomass-break crops and biomass-cereal comparisons, we expect the same effects to develop in the seedbank of miscanthus and SRC willows, such that there would be a longer term shift towards a flora dominated by perennials and competitors.

This work has shown responses by biodiversity indicators to crop planting vary according to the type and longevity of the crop. The similarity and consistency in direction and magnitude of crop effect on the seedbank suggest that, when assessed across rotations of an arable system, cereals and break crops are components of a single cropping system. Indeed, the community analyses of weed taxa recorded as biomass and in the seedbank identify a unified arable community. While our findings support the theory of spatially structured arable weed communities (Freckleton & Watkinson, 2002), the regional- and national-scale data used here suggest that these communities operate at greater scales than suggested. Previous analyses of weed and invertebrate communities in arable crops have also identified taxonomic community response to farm management at different temporal and spatial scales. Hawes *et al.* (2010) found longer term, farm-scale cropping system-mediated responses by weed seedbank communities in conventional, organic and integrated fields, while within-year, field-scale effects of crop were identified by Smith *et al.* (2008), who found that weed and invertebrate communities were associated with individual break crops. Our work demonstrates an intermediate level of response to cropping system, as arable and biomass cropping is possible within a single farm unit.

We have previously reported results for butterfly data collected in this multi-site, regional-scale experiment (Haughton *et al.*, 2009), where the abundance of nonpest butterfly species was significantly higher, and that of pest butterflies was significantly lower in the field mar-

Table 4 Back-transformed mean of densities of invertebrates (counts m^{-2}) per field in break crops, miscanthus and SRC willows (SRC), and t-statistics for comparisons between biomass and arable crop means, with observed significance levels

	Mean invertebrate density			Comparisons with Miscanthus			Comparisons with SRC		
	Break crops	Miscanthus	SRC	t	df	P	t	df	P
Total	160.8	1852.5	2998.2	1.18	249.0	0.240	0.35	249.0	0.728
Detritivores	55.6	1547.8	2171.7	1.30	249.0	0.196	−0.39	249.0	0.696
Herbivores	9.7	48.7	157.9	1.23	249.0	0.221	0.92	249.0	0.359
Predators	10.5	101.6	199.9	0.97	249.0	0.332	−2.46	249.0	0.015

Number of study fields (N): break crops = 233; miscanthus = 14; SRC willows = 11.

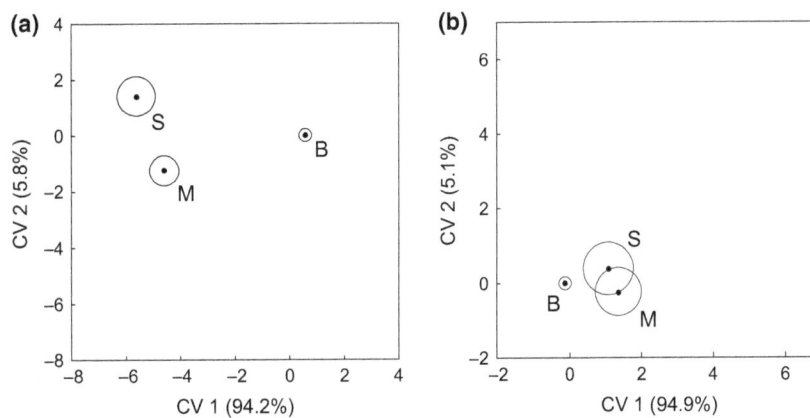

Fig. 6 Canonical variates for invertebrates in biomass and break crops. CVA group mean scores (•) with 95% confidence regions for proportions of invertebrate (a) taxa and (b) trophic functional group recorded in break crops (B), miscanthus (M) and SRC willows (S). The percentage variation explained by each canonical variate is given in parentheses.

gins of both miscanthus and SRC willows than arable break crops. The results presented here, based on large numbers of entire, commercial fields and study fields distributed regionally and nationally, show greater abundances of biodiversity indicators in biomass crops at the landscape scale. This concurs with the limited number of previous, predominantly small-scale, local studies of comparative impacts on biodiversity of cultivating miscanthus, SRC willows and arable crops. Bellamy et al. (2009) found greater weed cover and abundance of canopy invertebrates in miscanthus than in cereals led to greater numbers and diversity of bird species and Rowe et al. (2011) and Baum et al. (2012) reported greater weed biomass and species richness, respectively, in SRC willows than in cereals. Stanley & Stout (2013) found significant benefits of miscanthus to solitary, nesting bees compared with cereal crops and suggest this could be a result of the enhanced floral resource in miscanthus. In a similar, but larger, experiment carried out in a different geographical region to Bellamy et al. (2009), Sage et al. (2010) found fewer bird species and individuals in miscanthus, and suggested that location and differences in the levels of weediness

contributed to this disparity in their results, thus indirectly suggesting that biodiversity studies should be widely spatially distributed if they are to account for regional variability. Miscanthus is known to be patchy in its early establishment, which in turn leads to patchy distributions of weeds (Zimmermann et al., 2014) and Dauber et al. (2015) caution against long-term expectations of biodiversity benefits of miscanthus, as they suggest farmers would eliminate such patches to maximize crop yield. The fields of miscanthus studied in this experiment were the oldest commercially managed crops available at the time of the experiment and it is possible that they were in the late establishment phase (Karp & Shield, 2008). Nevertheless, the study fields were managed for yield, and we would expect patches of weeds in these late establishment phase crops to remain a feature in older crops.

An unexpected and surprising outcome of this work was the contrast in the magnitudes of the crop effect on weed biomass and the invertebrates, suggesting that, unlike in arable crops, there is a significant positive crop–resource relationship in perennial biomass crops. Previous studies in annual arable crops (Hawes et al.,

2003, 2009; Bohan *et al.*, 2007) have demonstrated positive relationships between weed resource and invertebrate functional groups, and we suggest that in addition to the effect of the biomass crop itself, the competitor plant community in biomass crops could, in part, drive the detritivore-dominated invertebrate fauna in biomass crops, as the competitors were the only group of plants to show greater densities in biomass crops than break crops. Competitor plant types typically exhibit a perennial reproductive strategy (Grime *et al.*, 2007) and positive benefits of perennial vegetation on species richness and abundance of parasitic tachinid Diptera have been reported by Letourneau *et al.* (2012). Storkey *et al.* (2013) found that plant herbivores respond positively to ruderal plants in arable systems; however, we found no evidence of this in perennial biomass crops, as the density of plant herbivores was greater in biomass crops, despite statistically significant lower densities of ruderal plant biomass. It is unfortunate that similar data for invertebrates in cereals were not collected in the FSEs, and to our knowledge, are not available elsewhere at scales equivalent to those analysed here; however, if the same pattern we have found in break crops were followed in cereals, we would predict that invertebrate abundance could be somewhat greater in cereals, due to the marginally greater densities of competitors than in break crops.

In conclusion, our analyses of regional- and national-scale data have shown that indicators of biodiversity are more abundant in perennial biomass cropping systems than annual cropping systems and we identified divergent functional compositions of plant and invertebrate communities in the arable and biomass crops. Our analyses also confirm the value of break crops for biodiversity indicators in arable rotations. These findings support the view that strategic planting of dedicated biomass crops, in intensively managed, arable-dominated farmland, can be used as a powerful tool for increasing landscape heterogeneity in the bid to create resilient, multifunctional landscapes (Rader *et al.*, 2014).

Acknowledgements

We thank the farmers for allowing us access to their crops, and the field ecologists who collected the data. The FSEs were funded through the Scottish Executive and Defra. RELU-Biomass (RES-227-25-0020) was funded under the Rural Economy and Land Use programme of the ESRC, BBSRC and NERC. Rothamsted Research is a national institute of bioscience strategically funded by the BBSRC.

References

APG (2003) An update of the Angiosperm Phylogeny Group classification for the orders and families of flowering plants: APG II. *Botanical Journal of the Linnean Society*, **141**, 399–436.

Arnold AJ (1994) Insect suction sampling without nets, bags or filters. *Crop Protection*, **13**, 73–76.

Baum S, Bolte A, Weih M (2012) High value of short rotation coppice plantations for phytodiversity in rural landscapes. *Global Change Biology Bioenergy*, **4**, 728–738.

Baum C, Eckhardt KU, Hahn J, Weih M, Dimitriou I, Leinweber P (2013a) Impact of poplar on soil organic matter quality and microbial communities in arable soils. *Plant Soil and Environment*, **59**, 95–100.

Baum S, Weih M, Bolte A (2013b) Floristic diversity in Short Rotation Coppice (SRC) plantations: comparison between soil seed bank and recent vegetation. *Landbauforschung*, **63**, 221–228.

Bellamy PE, Croxton PJ, Heard MS *et al.* (2009) The impact of growing miscanthus for biomass on farmland bird populations. *Biomass and Bioenergy*, **33**, 191–199.

Benton TG, Vickery JA, Wilson JD (2003) Farmland biodiversity: is habitat heterogeneity the key? *Trends in Ecology & Evolution*, **18**, 182–188.

Berndes G, Borjesson P, Ostwald M, Palm M (2008) Multifunctional biomass production systems - an overview with presentation of specific applications in India and Sweden. *Biofuels Bioproducts & Biorefining*, **2**, 16–25.

Bianchi F, Booij CJH, Tscharntke T (2006) Sustainable pest regulation in agricultural landscapes: a review on landscape composition, biodiversity and natural pest control. *Proceedings of the Royal Society B-Biological Sciences*, **273**, 1715–1727.

Boatman ND, Jones NE, Conyers ST, Pietravalle S (2011) Development of plant communities on set-aside in England. *Agriculture, Ecosystems and Environment*, **143**, 8–19.

Bohan DA, Boffey CWH, Brooks DR *et al.* (2005) Effects on weed and invertebrate abundance and diversity of herbicide management in genetically modified herbicide-tolerant winter-sown oilseed rape. *Proceedings of the Royal Society B*, **272**, 463–474.

Bohan DA, Hawes C, Haughton AJ, Denholm I, Champion GT, Perry JN, Clark SJ (2007) Statistical models to evaluate invertebrate-plant trophic interactions in arable systems. *Bulletin of Entomological Research*, **97**, 265–280.

Bohan DA, Powers SJ, Champion G *et al.* (2011) Modelling rotations: can crop sequence explain arable weed seedbank abundance? *Weed Research*, **51**, 422–432.

Bourke D, Stanley D, O'rourke E *et al.* (2014) Response of farmland biodiversity to the introduction of bioenergy crops: effects of local factors and surrounding landscape context. *Global Change Biology Bioenergy*, **6**, 275–289.

Cadotte MW, Carscadden K, Mirotchnick N (2011) Beyond species: functional diversity and the maintenance of ecological processes and services. *Journal of Applied Ecology*, **48**, 1079–1087.

Champion GT, May MJ, Bennett S *et al.* (2003) Crop management and agronomic context of the Farm Scale Evaluations of genetically modified herbicide-tolerant crops. *Philosophical Transactions of the Royal Society*, **358**, 1801–1818.

Coghlan A (2003) Farming 1, wildlife 0. *New Scientist*, **2418**, 21.

Critchley CNR, Fowbert JA, Sherwood AJ (2006) The effects of annual cultivation on plant community composition of uncropped arable field boundary strips. *Agriculture, Ecosystems and Environment*, **113**, 196–205.

Dauber J, Bolte A (2014) Bioenergy: challenge or support for the conservation of biodiversity? *Global Change Biology Bioenergy*, **6**, 180–182.

Dauber J, Jones MB, Stout JC (2010) The impact of biomass crop cultivation on temperate biodiversity. *Global Change Biology Bioenergy*, **2**, 289–309.

Dauber J, Cass S, Gabriel D, Harte K, Astrom S, O'rourke E, Stout JC (2015) Yield-biodiversity trade-off in patchy fields of *Miscanthus x giganteus*. *Global Change Biology Bioenergy*, **7**, 455–467.

Davis SJ, Shearer C (2014) A crack in the natural-gas bridge. *Nature*, **514**, 436–437.

Defra (2014) *UK Biodiversity Indicators 2014*. Defra, London, UK.

Donald PF, Green RE, Heath MF (2001) Agricultural intensification and the collapse of Europe's farmland bird populations. *Proceedings of the Royal Society B*, **268**, 25–29.

Fargione J (2010) Is bioenergy for the birds? An evaluation of alternative future bioenergy landscapes. *Proceedings of the National Academy of Sciences of the United States of America*, **107**, 18745–18746.

Firbank LG, Heard MS, Woiwod IP *et al.* (2003) An introduction to the Farm-Scale Evaluations of genetically modified herbicide-tolerant crops. *Journal of Applied Ecology*, **40**, 2–16.

Freckleton RP, Watkinson AR (2002) Large-scale spatial dynamics of plants: metapopulations, regional ensembles and patchy populations. *Journal of Ecology*, **90**, 419–434.

Froud-Williams RJ, Chancellor RJ, Drennan DSH (1983) Influence of cultivation regime on weed floras of arable cropping systems. *Journal of Applied Ecology*, **20**, 187–197.

Gagic V, Bartomeus I, Jonsson T *et al.* (2015) Functional identity and diversity of animals predict ecosystem functioning better than species-based indices. *Proceedings of the Royal Society B*, **282**, 20142620.

Gardner S, Gower JC, Le Roux NJ (2006) A synthesis of canonical variate analysis, generalised canonical correlation and Procrustes analysis. *Computational Statistics & Data Analysis*, **50**, 107–134.

Grime JP (2001) *Plant Strategies, Vegetation Processes, and Ecosystem Processes*, 2nd edn. John Wiley and Sons, Chichester.

Grime JP, Hodgson JC, Hunt R (2007) *Comparative Plant Ecology: A Functional Approach to Common British Species*. Castlepoint, Dalbeattie.

Haughton AJ, Champion GT, Hawes C *et al.* (2003) Invertebrate responses to the management of genetically modified herbicide-tolerant and conventional spring crops. II. Within-field epigeal and aerial arthropods. *Philosophical Transactions of the Royal Society*, **358**, 1863–1877.

Haughton AJ, Bond AJ, Lovett AA *et al.* (2009) A novel, integrated approach to assesseing social, economic and environmental implications of changing rural land-use: a case study of perennial biomass crops. *Journal of Applied Ecology*, **46**, 315–322.

Hawes C, Haughton AJ, Osborne JL *et al.* (2003) Responses of plants and invertebrate trophic groups to contrasting herbicide regimes in the Farm Scale Evaluations of genetically modified herbicide-tolerant corops. *Philosophical Transactions of the Royal Society*, **358**, 1899–1913.

Hawes C, Haughton AJ, Bohan DA, Squire GR (2009) Functional approaches for assessing plant and invertebrate abundance patterns in arable systems. *Basic and Applied Ecology*, **10**, 34–42.

Hawes C, Squire GR, Hallett PD, Watson CA, Young M (2010) Arable plant communities as indicators of farming practice. *Agriculture, Ecosystems and Environment*, **138**, 17–26.

Heard MS, Hawes C, Champion GT *et al.* (2003) Weeds in fields with contrasting conventional and genetically modified herbicide-tolerant crops. I. Effects on abundance and diversity. *Philosophical Transactions of the Royal Society*, **358**, 1818–1822.

Heard MS, Rothery P, Perry JN, Firbank LG (2005) Predicting longer-term changes in weed populations under GMHT crop management. *Weed Research*, **45**, 331–338.

Hoehn P, Tscharntke T, Tylianakis JM, Steffan-Dewenter I (2008) Functional group diversity of bee pollinators increases crop yield. *Proceedings of the Royal Society B*, **275**, 2283–2291.

Holland RA, Eigenbrod F, Muggeridge A, Brown G, Clarke D, Taylor G (2015) A synthesis of the ecosystem services impact of second generation bioenergy crop production. *Renewable and Sustainable Energy Reviews*, **46**, 30–40.

Hooper DU, Chapin FS, Ewel JJ *et al.* (2005) Effects of biodiversity on ecosystem functioning: a consensus of current knowledge. *Ecological Monographs*, **75**, 3–35.

IPCC (2014) *Climate Change 2014: Synthesis Report*. Contribution of Working Groups I, II and III to the Fifth Assessment Report of the Intergovernmental Panel on Climate Change (eds Core Writing Team, Pachauri RK, Meyer LA). IPCC, Geneva, Switzerland.

Jackson RB, Vengosh A, Carey JW, Davies RJ, Darrah TH, O'sullivan F, Petron G (2014) The environment costs and benefits of fracking. *Annual Reviews of Environment and Resources*, **39**, 327–362.

Karp A, Shield I (2008) Bioenergy from plants and the sustainable yield challenge. *New Phytologist*, **179**, 15–32.

Karp A, Haughton AJ, Bohan DA et al. (2009) Perennial energy crops: implications and potential. In: *What is Land For? The Food, Fuel and Climate Change Debate* (eds Winter M, Lobley M), pp. 47–72. Earthscan, London.

Letourneau DK, Bothwell Allen SG, Stireman JO (2012) Perennial habitat fragments, parasitoid diversity and parasitism in ephemeral crops. *Journal of Applied Ecology*, **49**, 1405–1416.

McJeon H, Edmonds J, Bauer N *et al.* (2014) Limited impact on decadal-scale climate change from increased use of natural gas. *Nature*, **514**, 482–485.

Meehan TD, Hurlbert AH, Gratton C (2010) Bird communities in future bioenergy landscapes of the Upper Midwest. *Proceedings of the National Academy of Sciences of the United States of America*, **107**, 18533–18538.

Meehan TD, Gratton C, Diehl E *et al.* (2013) Ecosystem-service tradeoffs associated with switching from annual to perennial energy crops in riparian zones of the US Midwest. *PLoS ONE*, **8**. doi: 10.1371/journal.pone.0080093.

Perry JN, Rothery P, Clark SJ, Heard MS, Hawes C (2003) Design, analysis and statistical power of the Farm-Scale Evaluations of genetically modified herbicide-tolerant crops. *Journal of Applied Ecology*, **40**, 17–31.

Rader R, Birkhofer K, Schmucki R, Smith HG, Stjernman M, Lindborg R (2014) Organic farming and heterogeneous landscapes positively affect different measures of plant diversity. *Journal of Applied Ecology*, **51**, 1544–1553.

Rowe RL, Goulson D, Street NR, Taylor G (2009) Identifying potential environmental impacts of large-scale deployment of dedicated bioenergy crops in the UK. *Renewable & Sustainable Energy Reviews*, **13**, 271–290.

Rowe RL, Hanley ME, Goulson D, Clarke DJ, Doncaster CP, Taylor G (2011) Potential benefits of commercial willow Short Rotation Coppice (SRC) for farm-scale plant and invertebrate communities in the agri-environment. *Biomass and Bioenergy*, **35**, 325–336.

Sage R, Cunningham M, Haughton AJ, Mallott MD, Bohan DA, Riche A, Karp A (2010) The environmental impacts of biomass crops: use by birds of miscanthus in summer and winter in southwestern England. *IBIS*, **152**, 487–499.

Searchinger T, Edwards R, Mulligan D, Heimlich R, Plevin R (2015) Do biofuel policies seek to cut emissions by cutting food? *Science*, **347**, 1420–1422.

Senapathi D, Carvalheiro LG, Biesmeijer JC *et al.* (2015) The impact of over 80 years of land cover changes on bee and wasp pollinator communities in England. *Proceedings of the Royal Society B*, **282**. doi: 10.1098/rspb.2015.0294.

Smith V, Bohan DA, Clark SJ, Haughton AJ, Bell JR, Heard MS (2008) Weed and invertebrate community compositions in arable farmland. *Arthropod-Plant Interactions*, **2**, 21–30.

Sokal RR, Rohlf FJ (2012) *Biometry: The Principles and Practice of Statistics in Biological Research*. Freeman and Co, New York.

Stace C (2010) *New Flora of the British Isles*. Cambridge University Press, Cambridge.

Stanley DA, Stout JC (2013) Quantifying the impacts of bioenergy crops on pollinating insect abundance and diversity: a field-scale evaluation reveals taxon-specific responses. *Journal of Applied Ecology*, **50**, 335–344.

Storkey J, Meyer S, Still KS, Leuschner C (2012) The impact of agricultural intensification and land-use change on the European arable flora. *Proceedings of the Royal Society B*, **279**, 1421–1429.

Storkey J, Brooks D, Haughton A, Hawes C, Smith BM, Holland JM (2013) Using functional traits to quantify the value of plant communities to invertebrate ecosystem service providers in arable landscapes. *Journal of Ecology*, **101**, 38–46.

Tilman D, Socolow R, Foley JA *et al.* (2009) Beneficial biofuels-the food, energy, and environment trilemma. *Science*, **325**, 270–271.

VSNI (2014) *GenStat for Windows*, 17th edn. VSN International, Hemel Hempstead.

Woodcock BA, Harrower C, Redhead J *et al.* (2014) National patterns of functional diversity and redundancy in predatory ground beetles and bees associated with key UK arable crops. *Journal of Applied Ecology*, **51**, 142–151.

Zimmermann J, Styles D, Hastings A, Dauber J, Jones MB (2014) Assessing the impact of within crop heterogeneity ('patchiness') in young *Miscanthus × giganteus* fields on economic feasibility and soil carbon sequestration. *Global Change Biology Bioenergy*, **6**, 566–576.

13

Reconciling food security and bioenergy: priorities for action

KEITH L. KLINE[1,†], SIWA MSANGI[2], VIRGINIA H. DALE[3], JEREMY WOODS[4], GLAUCIA M. SOUZA[5], PATRICIA OSSEWEIJER[6], JOY S. CLANCY[7], JORGE A. HILBERT[8], FRANCIS X. JOHNSON[9], PATRICK C. MCDONNELL[10] and HARRIET K. MUGERA[11]

[1]Environmental Science Division, Climate Change Science Institute, Oak Ridge National Laboratory, TN 37831, USA, [2]International Food Policy Research Institute, 2033 K St NW, Washington, DC 20006, USA, [3]Center for Bioenergy Sustainability, Environmental Science Division, ORNL, Oak Ridge, TN 37831, USA, [4]Centre for Environmental Policy, Imperial College London, Exhibition Road, London SW7 1NA, UK, [5]Instituto de Química, Universidade de São Paulo, Av. Prof. Lineu Prestes 748, São Paulo, Brazil, [6]Department of Biotechnology, Delft University of Technology, 2628 BC Delft, The Netherlands, [7]CSTM, University of Twente, 7500AE Enschede, The Netherlands, [8]Instituto de Ingeniería Rural INTA, cc 25 1712 Castelar, Buenos Aires, Argentina, [9]Stockholm Environment Institute Africa Centre, World Agroforestry Centre (ICRAF), United Nations Avenue, Gigiri PO Box 30677, Nairobi, Kenya, [10]BEE Energy, 2000 Nicasio Valley Rd., Nicasio, CA 94946, USA, [11]World Bank, 1818 H Street NW, Washington, DC 20433, USA

Abstract

Understanding the complex interactions among food security, bioenergy sustainability, and resource management requires a focus on specific contextual problems and opportunities. The United Nations' 2030 Sustainable Development Goals place a high priority on food and energy security; bioenergy plays an important role in achieving both goals. Effective food security programs begin by clearly defining the problem and asking, 'What can be done to assist people at high risk?' Simplistic global analyses, headlines, and cartoons that blame biofuels for food insecurity may reflect good intentions but mislead the public and policymakers because they obscure the main drivers of local food insecurity and ignore opportunities for bioenergy to contribute to solutions. Applying sustainability guidelines to bioenergy will help achieve near- and long-term goals to eradicate hunger. Priorities for achieving successful synergies between bioenergy and food security include the following: (1) clarifying communications with clear and consistent terms, (2) recognizing that food and bioenergy need not compete for land and, instead, should be integrated to improve resource management, (3) investing in technology, rural extension, and innovations to build capacity and infrastructure, (4) promoting stable prices that incentivize local production, (5) adopting flex crops that can provide food along with other products and services to society, and (6) engaging stakeholders to identify and assess specific opportunities for biofuels to improve food security. Systematic monitoring and analysis to support adaptive management and continual improvement are essential elements to build synergies and help society equitably meet growing demands for both food and energy.

Keywords: bioenergy, biofuels, energy, flex crops, food insecurity, food security and nutrition, natural resource management, poverty reduction, sustainable development goals

†This manuscript was coauthored by UT-Battelle, LLC, under Contract No. DE-AC05-00OR22725 with the U.S. Department of Energy. The United States Government retains and the publisher, by accepting the article for publication, acknowledges that the United States Government retains a non-exclusive, paid-up, irrevocable, world-wide license to publish or reproduce the published form of this manuscript, or allow others to do so, for United States Government purposes.

Correspondence: Keith L. Kline
e-mail: klinekl@ornl.gov

The most serious mistakes are not being made as a result of wrong answers. The truly dangerous thing is asking the wrong questions.   —Peter Drucker (1971)

Introduction

Understanding the nexus of food security, bioenergy sustainability, and resource management facilitates achievement of the 2030 Sustainable Development Goals (SDGs) to end hunger and ensure access to modern energy for all (United Nations (UN) 2015), as well as the Paris Agreement under the UN Convention on

Climate Change. Contextual conditions determine costs, benefits, and strategic opportunities that foster food and energy security for all (DeRose *et al.*, 1998; FAO, 2015b; FAO, IFAD and WFP 2014). However, it is important to acknowledge that public perception about the interaction of bioenergy, in particular biofuels, and food security is mostly negative. Popular media reinforce beliefs reflected in the assumption used in economic models that biofuels produced from crops or on cropland compete with food production and increase food prices. Cartoons of hungry children juxtaposed to corn being 'fed' to cars have generated an emotional response to biofuel policies that is difficult to overcome (Osseweijer *et al.*, 2015; The Economist, 2015). Sensational news garners attention while subsequent corrections are overlooked (Flipse & Osseweijer, 2013). In this report, we review the underlying evidential and theoretical basis concerning the impacts of bioenergy, in general, and biofuels, in particular, on food security and offer steps that can help society achieve SDGs for food and energy security.

A science-based examination of evidence linking food security and bioenergy illuminates practical solutions when problems are well defined. Good science is essential to inform decisions in a world of strong beliefs (Hecht *et al.*, 2009). An initial step must be to understand relationships between biomass production, food production, and hunger. Food security is recognized as a fundamental human right (UN General Assembly, 2015) with modern energy services being an essential component of food production, supply, and preparation (Woods *et al.*, 2010).

This study describes the complexities in assessing sustainability as related to energy and food security in four parts: (1) food security, (2) interactions among food security, biofuels, and resource management, (3) priorities and conditions for achieving positive synergies, and (4) conclusions and recommendations. We begin by recognizing that food insecurity is typically the indicator, so linkages among resource management, biofuels, and strategies to reduce food insecurity are relevant. We highlight where conventional wisdom could be misleading and identify areas where further research should be a priority. The paper concludes with recommendations for enhancing food and energy security as complementary goals for sustainable development.

An international workshop (IFPRI, 2015) helped frame the key issues evaluated here and underscored the importance of clear definitions and consistent use of terminology. The workshop focused on liquid biofuels, but the discussion and conclusions in this paper aim to be broadly applicable to food security interactions with an expanding bio-based economy. Polarization in the food-vs.-fuel debate begins with differing definitions

and assumptions about relationships among biofuels, prices, food, and land security. It is important to analyze the reasons for divergence and to find common ground (Rosillo-Calle & Johnson, 2010).

Food security

Definitions and measures of food security

The definitions used for food and food security are important determinants of the scope and outcomes of analyses. The oft-cited definition from the Food and Agriculture Organization of the United Nations (FAO) reflects broad aspirational goals (FAO 1996, Table 1). Four dimensions of food security emerge from this definition, namely, availability, accessibility, stability, and utilization (Table 2). Thus, one approach to assessing impacts of biofuels on food security examines interactions across these four dimensions. However, many other factors including distributional and contextual issues affect vulnerability and hunger (von Grebmer *et al.*, 2014).

Measuring food insecurity. While the concept of food security is intuitive, underlying data are fraught with uncertainties due to large variations in diets and biophysical conditions, making food security difficult to measure and monitor. Therefore, *manifestations of food insecurity that can be observed and verified* are often used as proxy indicators of hunger and are monitored, rather than monitoring food security itself. For example, three international organizations collaborate to produce annual reports on the 'State of Food Insecurity in the World' (SOFI) (e.g., FAO, IFAD and WFP 2015a, 2014, 2013, FAO, WFP, IFAD, 2012, and previous years).

The terms food security and food insecurity are often used loosely or interchangeably; however, the definitions and approaches for their measurement vary considerably (DeRose *et al.*, 1998). Anthropometric measures of food insecurity are complemented by qualitative surveys of behavior from census data on household income and expenditures. Undernourishment, a common measure of food insecurity, is the probability that an *individual* in the population is undernourished (FAO, 2015a), while other measures focus on household food purchases (USDA, 2015; Coleman-Jensen *et al.*, 2015). A global hunger index combines three equally weighted indicators: (1) undernourishment, defined as people with insufficient caloric intake (percentage of population); (2) children under the age of five with low weight for their age; and (3) mortality rate for children under age five (von Grebmer *et al.*, 2014, Gautam, 2014). The effects of biofuels or a given policy on 'food insecurity' thus depend on the measures used to define who is 'food insecure.'

Table 1 Definitions relating to food security (based on IPC Global Partners 2012 and other sources as noted)

Term	Definition/Examples
Anthropometry	Study of the measurements and proportions of the human body; used as an indicator of malnutrition. Examples include child underweight (weight for age), stunting (height for age), and wasting (weight for height), compared with reference standards (United Nations World Food Program (WFP) Hunger Glossary, 2015)
Commodity	Traded item, especially unprocessed materials. Relevant examples include crude palm oil, raw sugar, #2 yellow corn, wheat, soybeans
Commodity price index	Mathematical value used to measure commodity price movements over a defined time period; typically based on prices registered between suppliers or nations
Consumer food price index	Mathematical measure of price movements over a defined time period for a fixed basket of food items in a given nation, state, region, or group
Famine	Food insecurity causing or potentially causing death in the near term
Food	Source of nutrients required for energy and growth
FAO food price index	Monthly change in international prices of a basket of five food commodity groups (cereals, oils, dairy, meats, sugar), weighted per average export share values of each group for a given period, for example, 2002–2004 (FAO, 2013a)
Flex crop	Cultivated plant grown for both food and nonfood markets.
Food security	Condition that exists when all people, at all times, have physical and economic access to sufficient safe and nutritious food that meets their dietary needs and food preferences for an active and healthy life (FAO, 1996)
Food insecurity (chronic or transitory)	Absence of food security; condition exists when people suffer or are at risk of suffering from inadequate consumption to meet nutritional requirements; may be classified as chronic (long term), acute (transitory), cyclical, or critical (see famine); typically measured via multiple indicators of malnutrition
Hunger (or 'food deprivation')	Degree of discomfort or unpleasant physical sensation associated with insufficient food consumption. World Food Program defines hunger as 'Not having enough to eat to meet energy requirements.' The World Hunger Education Service (2015) refers to hunger as 'aggregated food scarcity exemplified by malnutrition.'
Malnutrition	Condition arising from deficiencies, excesses, or imbalances in the consumption of important macro- and micronutrients. Malnutrition can arise directly from food insecurity or be a result of (1) inadequate childcare practices, (2) inadequate health services, (3) a harmful environment, or (4) excessive intake of unhealthy food
Poverty	State of being that encompasses multiple dimensions of deprivation relating to human capacity and capability, including consumption and food security, health, education, rights, security, dignity, and decent work
Social safety nets	Public programs that provide assistance, often as income transfers, to families or individuals who are unable to work or are temporarily affected by natural disasters, political crises, or other adverse conditions. Programs may involve (1) direct and targeted feeding (school meals, soup kitchens, or food distribution centers), (2) food-for-work programs, (3) cash or in-kind transfers (e.g., food vouchers), (4) subsidized rations, or (5) other support to targeted households
Staple food	Principal or recurring food ingredient in a regional diet

Price indices alone are not indicators of food security. Given the high cost and complexity of field measurements, broad indicators related to prices and regional balances of commodity supplies and utilization are often used for food market assessments. Price, supply, and trade data are readily available from existing sources and do not require primary fieldwork to gather. Further, because these data can be easily plugged into existing market equilibrium models, they have been widely used to estimate the effects of biofuels on food security. Yet, as discussed below, there is little evidence that price indices can tell us much about who actually suffers from malnutrition due to food insecurity or its primary causes. Despite correlations, changes in global commodity prices are distinct from changes in consumer food price indices (Fig. 1).

Table 2 Questions and trade-offs to consider when assessing effects of bioenergy across four dimensions of food security (food security dimensions based on FAO, 1996, 2008)

Dimensions of food security	Key questions: Does the proposed project increase or decrease…	Assessment considerations	Trade-offs
Availability: quantity available for consumption in markets or within households	The quantity of food, especially staples, available for household consumption? coping mechanisms and institutional capacity to respond in times of crisis? quantity of food required for traditional cultural practices and identity?	Which dimensions of food security are the primary causes for food insecurity or risk of insecurity in this area? Which households/subgroups of the local population are most food insecure at present and why?	Can improvement in one dimension offset reductions in another? Will critical aspects for local food security or insecurity be affected?
Accessibility: affordability or other aspects of securing available food	Affordability of food, particularly for low-income households or other at-risk groups? investment in roads, bridges, public transport, or other features that facilitate access to markets and services (particularly in times of crisis)? factors that have caused disruptions in access to food in the past for this area?	Which households and subgroups are at highest risk of becoming food insecure, given current local trends and the context of the proposed project?	Will the project make clean energy services more affordable or widely available? Who gains and who loses in each dimension?
Stability: volatility in prices, availability, or affordability	Market 'floors' or 'ceilings' that reduce price fluctuation in staple foods? diversity of markets for producers (e.g., higher or lower dependence on single buyer or use)? diversity of food sources? diversity of sources of income? the base area of production of staple foods (e.g., changing susceptibility to localized extreme weather events)? other price and supply volatility impacts?	How does local energy use interact with food production, transport, preparation, and processing?	How are project impacts distributed among groups, particularly food-insecure and at-risk groups?
Utilization: retention and use of the nutrients in consumed food to sustain health and well-being	Nutritional value of diet for at-risk population? health and sanitation services? education for at-risk populations? micronutrient deficiencies? food safety, general health, and other factors influential in utilization?		

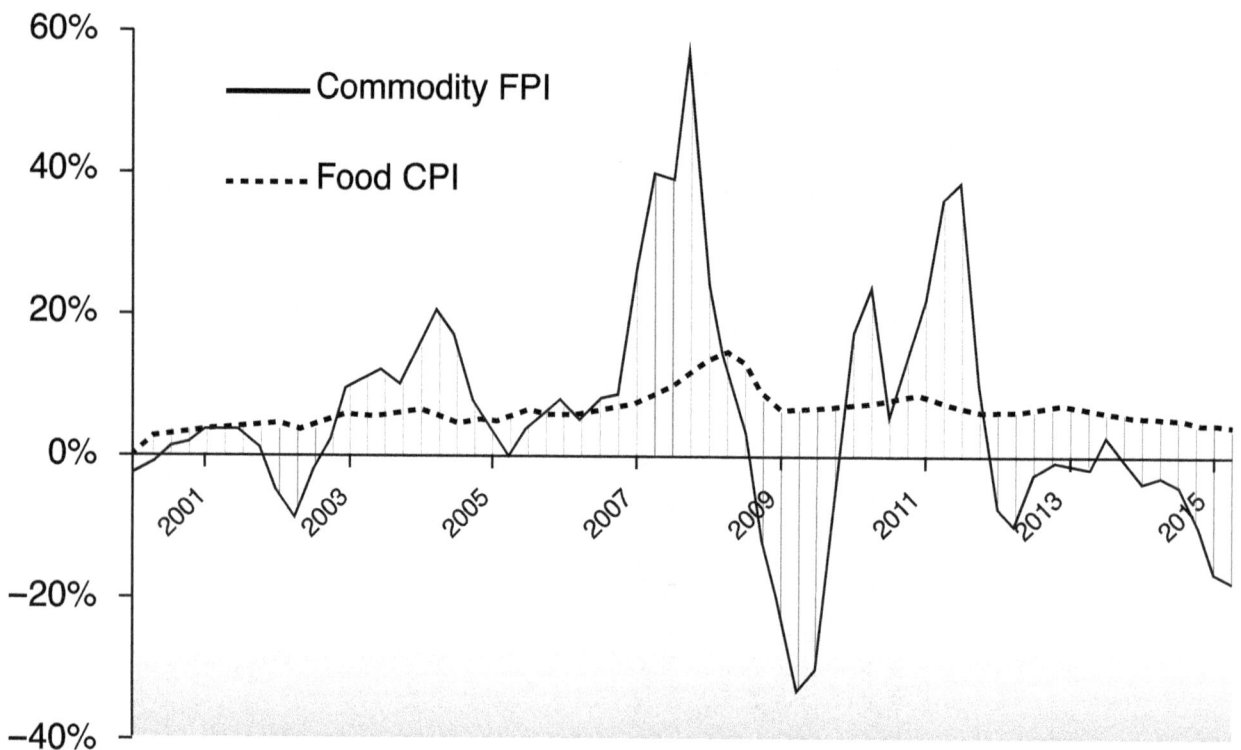

Fig. 1 The FAO global Food Price Index (FPI) based on commodities vs. the FAO global food Consumer Price Index (CPI), 2000–2015 (FAOStat, 2015). See Table 1 for definitions. Percentage change is relative to the 2002–2004 average for FPI and year 2000 for CPI (FAO, 2015c). The food CPI increased each year at an average annual rate of 6% (2000–2015), while the annual average global FPI varied sharply and was negative in 7 of the 15 years.

FAO notes that its food price index (FPI) is not an indicator of food insecurity. Rather, the FPI is based on weighted indices of trade data (Table 1) which may not reflect: (1) foods needed by food-insecure countries; (2) price changes relevant to food security; and (3) the actual prices for households which 'may be quite different from the border prices' (FAO, 2013a). Furthermore, in nations where high numbers of people are food insecure, staples such as rice are managed or regulated explicitly to protect local consumers from external price fluctuations (FAO, 2014, 2015c). Finally, FPI weighting creates bias favoring expensive commodities that are less important for populations at risk; for example, meat has the highest weight, 0.35, while sugar has a weight of 0.07.

National and regional 'consumer food price indices' (CFPIs) provide a higher resolution than the FPI but are still insufficient indicators of food insecurity due to similar dollar-value weighting bias and reliance on formal market prices. The people most susceptible to severe food insecurity typically live in isolated areas and rely on informal markets or subsistence production (Rose, 1999; FAO, 2015a; FAO, IFAD and WFP, 2015b). Rice, wheat, millet, white maize, and yams are staples in Asia and Africa, where 94% of the world's

hungry reside (FAO, 2015a), but their local prices have minimal influence on CFPI values. When these staples are grown and consumed locally, they are omitted by both the aggregate trade models and CFPIs, despite being crucial sources of nutrition for vulnerable households.

The annual SOFI reports highlight dozens of context-specific factors, other than CFPI changes, that determine who goes hungry in times of crisis (e.g., FAO, WFP, 2010). Malnutrition is associated with many factors other than food intake (e.g., Smith & Haddad, 2000; Gautam, 2014; Lombard, 2014). Thus, biofuel effects on food security could be determined by a project's influence on physical infrastructure, asset accrual, institutional capacity, training, technologies that enhance food safety or resilience, ecosystem stability, cultural well-being, or other drivers and coping mechanisms omitted from food price indices (Rose, 1999; RTI, 2014; Coleman-Jensen et al., 2015; Gustafson et al., 2016).

Finally, analyses that rely on FPIs tend to focus on price spikes while ignoring long periods of depressed prices. This can mislead policymakers and the public because depressed prices discourage agricultural investment and can be more detrimental to long-term food

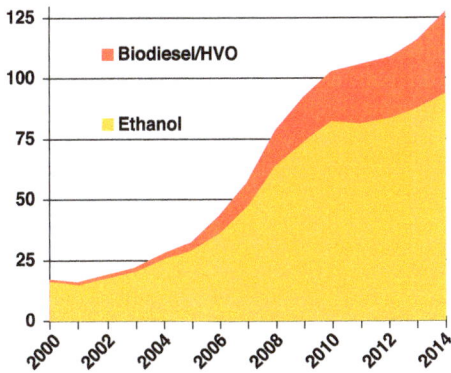

Fig. 2 Global biofuel consumption (billion liters) 2000–2014 grew steadily, although fuel ethanol production dipped slightly in 2010–2012 due to global recession and poor weather in Brazil (in 2011) and the USA (in 2012). Still, average annual growth in global production over 2009–2014 remained robust, at 5.2% and 11% for fuel ethanol and biodiesel, respectively (REN21, 2015). Chart based on U.S. Energy Information Administration (2015) and REN21, 2015.

security than price spikes (see, e.g., the SOFI reports and Roser, 2015). Projects that contribute to price stability at a level high enough to motivate local investment in food production and its associated infrastructure will improve resilience and food security over the long term (FAO, IFAD, WFP, 2002).

Effective food security strategies address relevant risk factors

To assess how a policy or project affects food security, an understanding of risk factors that lead to food insecurity is needed. As described above, analysis of aggregate commodity data may generate conflicting conclusions, because correlations with biofuels are often extraneous to the causes of local food insecurity. Understanding why and how people become food insecure is prerequisite to developing effective responses. Food insecurity may involve distinct risk factors depending on whether effects are long term (chronic) or short term (acute or transitory).

The type and cause of food insecurity in a particular context determine appropriate responses (IPC Global Partners, 2012) and how the effects of a bioenergy project on food security should be assessed (Table 2). Addressing chronic food insecurity requires coordinated commitments to long-term strategies that reduce household vulnerabilities. Transitory food insecurity requires investments that mitigate or prevent sudden events that can limit access to adequate food for short periods. Transitory food insecurity may be caused by events that impede distribution from areas of food surplus to areas of need (e.g., loss of critical bridge or

road). Thus, the degree to which biofuel production and processing may influence food security depends on the interaction of many variables within a local context including, among others: what feedstocks are grown and where and how feedstocks are distributed, what investments are made, management practices, who benefits, and who loses (Table 2).

Biofuels and food security: short-term correlations vs. long-term trends. The high-profile expansion of ethanol production in the United States and Brazil, in tandem with a global price spike in food and commodities in 2007–2008, led many to contend that a causal relationship exists between biofuels expansion and food insecurity (e.g., Mitchell, 2008; Tenenbaum, 2008; Wenzlau, 2013). The apparent short-term correlations are often cited as evidence of negative impacts of biofuels on food security (e.g., EPI, 2014; Searchinger & Heimlich, 2015). There are several problems with such assertions (Zilberman *et al.*, 2013). First, many studies attribute the food price spikes in 2008 primarily to other factors such as oil prices, economic growth, currency exchange rates, and trade policies (e.g., Baffes & Dennis, 2013; Konandreas, 2012; HLPE, 2011; Foresight, 2011; Trostle *et al.*, 2011; DEFRA, 2010). Speculation in food commodities also contributed to price spikes in 2008 and 2011 (Lagi *et al.*, 2011; Hajkowicz *et al.*, 2012). Second, the correlations did not persist as global biofuel consumption continued to grow (Fig. 2) and cereal prices fell or showed distinct patterns over the last 6 years driven by oil price, national agricultural policies, and exchange rates (FAO, 2015a,c, The Economist Intelligence Unit, 2015). Causation cannot be assumed based on correlation, but the divergence in recent trends is notable, and models using the same data can reach opposing conclusions (Table 3).

A majority of papers and reports that assert that biofuels harm food security rely on assumed relationships between biofuels, rising global 'food' commodity prices, and food insecurity over relatively short time spans (e.g., on the order of months) (Boddiger, 2007; Rajagopal *et al.*, 2007; Tenenbaum, 2008; Wenzlau, 2013). Interestingly, organizations wishing to show that biofuels do not raise food prices often cite the same FAO 'food commodity' data over similar time spans (e.g., see Zhang *et al.*, 2010; Mueller *et al.*, 2011; and GRFA, 2015). The assumptions underlying both sides of this food-vs.-fuel debate are questionable and subjective (Table 3). Long-term trends (over years and decades) for food insecurity and food commodity prices illustrate that the world's most severe famines (Roser, 2015) occur during extended periods of depressed global food prices (Sumner, 2009). The emphasis on biofuels and food commodity price spikes has diverted attention from more

Table 3 Identical data can support contradicting hypotheses about nutritional effects of biofuel-food interactions

Observations: Despite population growth, 167 million fewer people suffered from hunger and undernourishment in 2015 than a decade earlier (FAO, 2015a). Over the same decade, biofuel production expanded rapidly along with the number of people suffering early mortality and disease from consuming too much of the wrong foods. Today, more people are malnourished from overconsumption than are undernourished due to insufficient food. Over the coming decade, the global population suffering from hunger is projected to decline, while the number suffering from diseases caused by overconsumption is projected to steadily rise (WHO, 2015)

Hypothesis 1: The effect of biofuel production on the price of food is most pronounced for commodities that compete directly with bioenergy feedstock. Sugarcane and yellow maize are the two most important biofuel feedstocks. The primary foods derived from sugarcane and yellow maize are sugar and other sweeteners (such as high-fructose corn syrup used globally), and red meat (most yellow maize is fed to cattle). These foods are among the primary sources of malnutrition from overeating (WHO, 2015). If biofuels cause higher prices and higher prices marginally reduce overconsumption, then the expected impacts on health would be beneficial	Hypothesis 2: The effect of biofuel production on the price of food is most pronounced for commodities that compete directly with bioenergy feedstock. Sugarcane and yellow maize are the two most important biofuel feedstocks. Bioenergy markets bolster investment and innovation, reducing long-term costs and increasing global supplies of said commodities. The primary foods derived from sugarcane and yellow maize (sugar, sweeteners, red meat) are more widely available at lower prices than would occur without biofuels. Thus, the impacts would be detrimental to health if biofuels drove sugar and yellow maize prices down so as to marginally increase overconsumption of red meat and sweeteners	Hypothesis 3 (conventional wisdom): The effect of biofuel production on the price of food is most pronounced for commodities such as maize that compete directly with bioenergy feedstock. Biofuels also compete for land, reducing production of other crops. This reduces food supply or increases food prices, thereby contributing to increased hunger. Evidence cited in this paper refutes most assumptions underlying this hypothesis. Whether the issue is hunger or overconsumption, who is impacted depends on who is at risk of malnutrition and other contextual conditions that determine causal relationships. Specific nutrition problems must be clearly defined to identify effective solutions

Conclusion: None of the hypotheses above can be endorsed because they are not supported by evidence of price transmissions to the specific populations at risk. Despite a rapid increase in biofuel production, there is no evidence of biofuel impacts on food-related health, either beneficial or detrimental. Models that simulate demand shocks from biofuels necessarily show price transmission and reduced consumption, but evidence is lacking to support either the assumed 'shock' or the assumed impacts on people at risk. To test a hypothesis, the problem must be clearly defined and the linkages between biofuels and impacts on behavior verified

constructive efforts to improve data (Gibson, 2013) and to identify effective mechanisms to address the food security issues that matter most, namely those having an impact on human health and morbidity.

Priority actions to reduce risks of food insecurity. Biofuel projects can address food security concerns by applying best practices that reduce exposure to risks of food insecurity (Table 4). Many recommendations for investments in biofuels tailored for developing nations have been published (UNCTAD, 2014; FAO 2010, 2011a, 2015b; FAO, IFAD, WFP, 2002, FAO, IFAD, WFP 2013).

Lifting people out of poverty is essential to reduce hunger (von Braun *et al.*, 2009, FAO, IFAD, WFP, 2014, 2015b; Coleman-Jensen *et al.*, 2015). The creation of stable, gainful, rural employment is a high-priority, poverty-reduction strategy (Conway and Wilson, 2012; FAO, IFAD, WFP, 2015b). Improvement in rural household incomes is proposed as a proxy indicator for

improvement in food security when assessing the sustainability of biofuels projects (Dale *et al.*, 2013).

Bioenergy projects that improve resilience can reduce vulnerabilities that lead to food insecurity (Gustafson *et al.*, 2016). Resilience refers to the ability of the system to recover following disturbance, and vulnerability refers to inability to withstand a hostile situation. Reducing risk exposure might take the form of facilitating the transition of households from livelihoods that are subject to high levels of variability – such as low-level subsistence farming dependent on a single crop – toward more stable sources of revenue and income.

Exposure to risk can also be reduced by programs that help build rural assets and diversify income sources. If the exposure of households to environmental or socioeconomic shocks cannot be reduced, then a bioenergy project might aim to increase the capacity of vulnerable households to cope with shocks when they arise. Resilience is achieved by 'strengthening sustainable local food systems, and fostering access to

Table 4 Examples of convergence among recommended practices to enhance food security and to produce sustainable biomass for bioenergy (based on FAO, IFAD, WFP 2002; FAO, 2010, 2011a, 2013b, 2014b, 2015b, 2015e; FAO, IFAD, WFP 2013, 2015b; IMF, 2013; UNCTAD, 2013, 2014; World Bank, 2015)

Dimension	Recommended practices
Access to land, water, and markets	Consultation with stakeholders including smallholders
	Mapping of customary rights and communal environmental services
	Fair compensation to owners and traditional users
	Rule of law and fair mechanisms for conflict resolution
	Infrastructure to access inputs and markets
Employment	Adherence to international conventions (e.g., International Labour Organization guidelines)
	Reliable local jobs and healthy working conditions
	Access to education, vocational skills, and safety
	Incentives to expand local production
	Removal of barriers to trade and market information
Income generation	Contracts with local goods and service providers (e.g., profit-sharing options)
	Freedom of association and collective bargaining
	Access to credit and business management training
	Fair and transparent pricing
	Stable regulatory environment
Local food security	Integrated food and energy systems
	Improved output and nutritional value from urban gardens and small farms
	Provision of agricultural inputs, technologies, and equipment
	Training that is relevant for developing coping strategies (asset building, etc.)
	Distribution and storage systems
Community development	Improved local infrastructure (transportation, water, schools, etc.)
	Women in leadership positions
	Health and safety services and emergency assistance
	Microlending and financial support mechanisms
	Social welfare organizations
Energy security	Improved energy infrastructure and maintenance
	Energy for agricultural technology: cultivation, marketing, irrigation, etc.
	Bio-based fuels and improved stoves for healthy food preparation
	Clean, affordable, and reliable energy for value-added processing
	Equitable and open energy markets
Cross-cutting aspects	Recognition that problems and solutions are context specific
	Focus on women, the poor, and small producers
	Transparency
	Access to financial, technical, 'safety nets' and other social services
	Environmental sustainability

productive resources and to markets that are remunerative and beneficial to smallholders' (FAO, 2015d).

Interactions among bioenergy, food security, and resource management, focusing on more sustainable systems

Making progress toward sustainable development goals requires attention to provision of social and ecosystem services as well as economics across integrated production systems. Sustainability involves assessing trade-offs among multiple dynamic goals and striving for continual improvement, rather than achieving a specific state.

Assessments should compare the relative merits of alternative trajectories in meeting goals. The trade-offs depend on historical developments and prevailing local economic, social, environmental, political, and cultural conditions (Efroymson et al., 2013). Because sustainability is context specific, local stakeholders should help set priorities, define the purposes of the assessment, and establish the temporal and spatial boundaries for consideration (Tarka-Sanchez et al., 2012; Dale et al., 2015). For example, dimensions of sustainability for bioenergy include soil quality, water quality and quantity, greenhouse gases, biodiversity, air quality, productivity, social well-being, energy security, trade, profitability, resource conservation, and social acceptability (McBride et al., 2011; Dale et al., 2013).

Choices inevitably involve trade-offs. Improving one aspect of sustainability may compromise another, and benefits for one group may involve costs for another (Table 2). Complete transformation chains rather than single bioenergy products should be analyzed to understand the interactions across sectors and industries that may influence system efficiencies for bioenergy and food security (Hilbert, 2014). A key goal is to identify opportunities where collective progress can be achieved – sometimes referred to as the triple bottom line of social, economic, and environmental benefits.

Resource management practices are more important in determining many environmental impacts than crop type (Davis et al., 2013). Wise management of available resources supports both bioenergy sustainability and food security (Manning et al., 2014). Hence, interactions among resource management, bioenergy sustainability, and food security are discussed with paired interactions considered first, followed by the three-way nexus (Fig. 3).

Two-way linkages

Bioenergy effects on food security. Bioenergy can foster social development, which is a precondition for food security and sustainability. Bioenergy provides energy security not only for transport (and hence broader access to food, selling markets, employment, and services) but also for food processing, business development, and drying and storage of surplus production (Durham et al., 2012; Lynd et al., 2015). The latter, providing an outlet for surplus, diversifies sources of income and improves supply resilience in the event of market shocks or shortages. Innovation is stimulated as new institutions and actors are empowered to engage in expanding biomass production. The early investments made by developed, developing, and emerging economies alike in biofuels illustrate the universal nature of the linkages between energy security and development (Johnson & Silveira, 2014).

The capacity for biofuels to help balance another commodity market has been demonstrated by the Brazilian sugar–ethanol industries. Similarly, U.S. ethanol

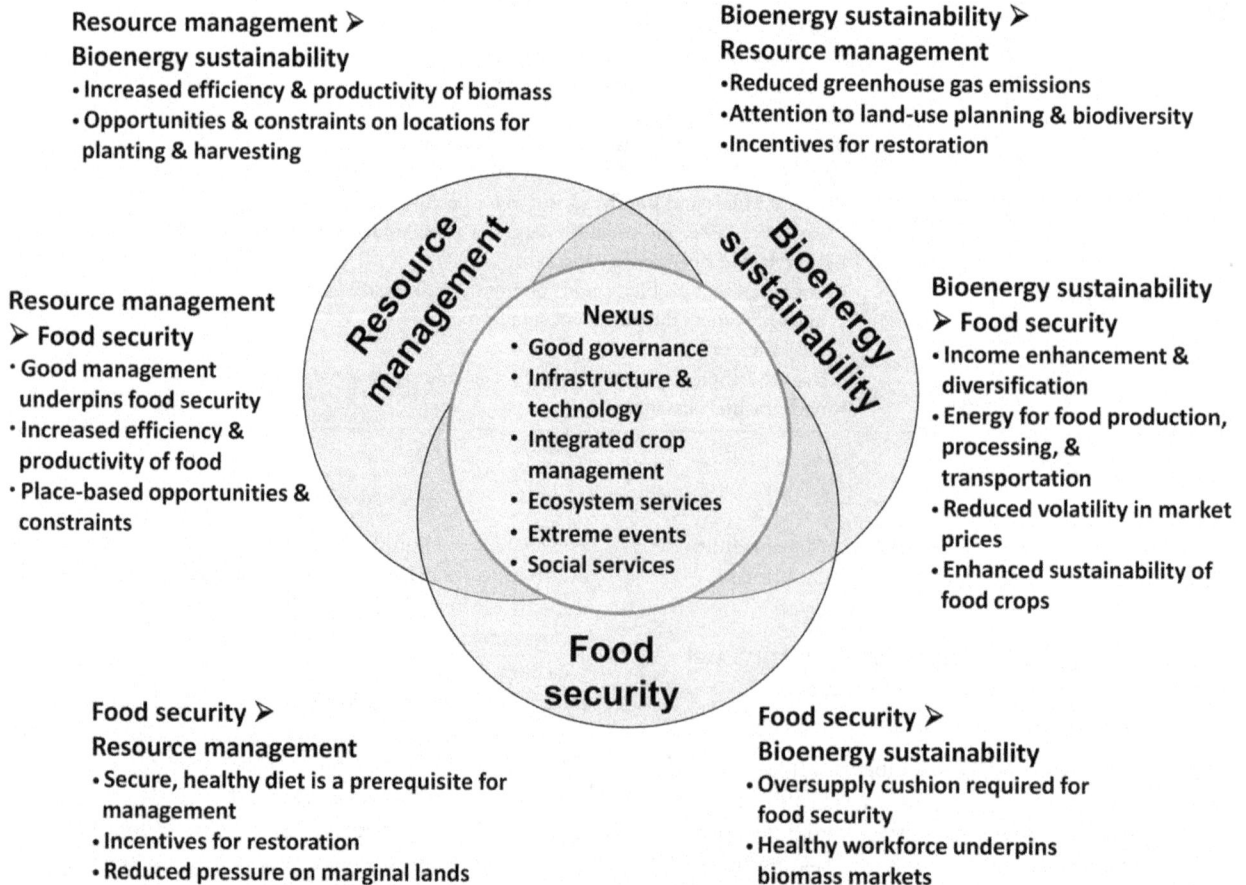

Resource management ➤
Bioenergy sustainability
• Increased efficiency & productivity of biomass
• Opportunities & constraints on locations for planting & harvesting

Bioenergy sustainability ➤
Resource management
• Reduced greenhouse gas emissions
• Attention to land-use planning & biodiversity
• Incentives for restoration

Resource management ➤ Food security
• Good management underpins food security
• Increased efficiency & productivity of food
• Place-based opportunities & constraints

Bioenergy sustainability ➤ Food security
• Income enhancement & diversification
• Energy for food production, processing, & transportation
• Reduced volatility in market prices
• Enhanced sustainability of food crops

Nexus
• Good governance
• Infrastructure & technology
• Integrated crop management
• Ecosystem services
• Extreme events
• Social services

Food security ➤ Resource management
• Secure, healthy diet is a prerequisite for management
• Incentives for restoration
• Reduced pressure on marginal lands

Food security ➤ Bioenergy sustainability
• Oversupply cushion required for food security
• Healthy workforce underpins biomass markets

Fig. 3 The nexus of resource management, bioenergy sustainability, and food security. Key aspects of the six two-way interactions frame the nexus at the center.

legislation passed in part due to recognition of latent productive capacity for maize. In the decade leading up to 2012, U.S. maize production increased steadily and exceeded targets for fuel blending under national legislation. In 2012, the U.S. experienced the most extensive drought recorded since the 1950s (IMF, 2012; USDA, 2013). As impacts of the drought became evident, markets responded; some ethanol plants reduced output; others shut down temporarily. Thanks, in part, to the ethanol 'supply cushion' and market flexibility, there was not a notable jump in commodity prices as the 2012–2013 crop was harvested, despite a drought affecting 80% of U.S. agricultural land.

While several studies discuss potential negative effects of biofuels, few examine the ways that biofuels can positively influence food security. First, adequately planned biofuel production can add value, stabilize, and diversify rural production systems (Kline *et al.*, 2009). Additionally, energy is required throughout the food supply chain; therefore, to the degree that biofuels enhance sustainability and accessibility of energy supplies, particularly energy for households most at risk from poverty, they enhance food security. Furthermore, as long as farmers and agro-industry are free to respond, diversified markets for products can spread risk and reduce price volatility compared with more narrow markets. Adding bioenergy markets to existing uses of local produce can thereby increase price stability. Finally, efforts to enhance sustainability of biofuels have generated spin-off effects in other sectors and placed greater scrutiny on resource management associated with conventional production (Woods & Kalas, 2014). The result is improved sustainability for many nonbiofuel products that constitute the majority of final uses for palm oil, sugar cane, soybean, and maize.

Bioenergy effects on resource management. Bioenergy has spurred well-known efforts to develop best practices that reduce greenhouse gas emissions and negative impacts on soil and water. However, bioenergy sustainability has also called attention to land-use planning and biodiversity protection and provided increased incentives for land restoration (Souza *et al.*, 2015). Specifically, bioenergy sustainability calls for consideration of a diverse set of potential effects on water, soils, air, and biodiversity, with emphasis on understanding baseline conditions and setting targets for continual improvement. These are key steps toward implementation of resource management systems that are resilient and adaptable to climate change.

Resource management effects on bioenergy. In turn, improving resource management influences bioenergy

sustainability by increasing the efficiency and productivity of supply chains. Improved management of soils and water permits higher output of bioenergy, food, and other products coupled to enhanced nutrient and water use efficiencies (FAO-UNEP, 2011). Past and future resource management goals help define both opportunities and constraints for cultivating more sustainable feedstock crops.

Resource management effects on food security. Good resource management underpins food security. Increased efficiency and productivity of crops enhance resilience and are essential for secure food availability. Similar to biofuel sustainability, good resource management allows identification of place-based opportunities and constraints and enhances the efficiency of resource use.

Food security effects on bioenergy. Food security can affect biomass resource management in many ways. A secure, healthy diet provides the biophysical and socioeconomic basis for managing soil, water, nutrients, and related resources. Excess production, desirable to enhance food security as a precautionary measure, can be absorbed by bioenergy markets and expand income opportunities for farmers when that supply cushion is not needed for sustenance.

Food security effects on resource management. Improving food security can reduce pressures on forests and marginal lands, thereby avoiding erosion and other negative consequences for soils, water, and ecosystem functions. Food-secure families are less inclined to risk health and livelihood to set off to distant frontiers and clear new land, whereas migration is often a last resort for food-insecure families. Food-secure families are also less likely to feel a need to cultivate on steep slopes and other fragile areas that involve physical and legal risks (parks and reserves). Desperate actions required to address food crises or famine can lead to displaced populations and emergency actions that have environmental consequences. Finally, food security provides the foundation required for effective outreach and learning about systematic approaches to improving natural resource management.

The three-way nexus between resource management, bioenergy sustainability, and food security

The interactions between these three factors form the central region of the Venn diagram in Fig. 3. Good governance incorporates both political commitment and the institutional capacity to provide effective services and security under the rule of law. Good governance is essential for effective resource management, food security, and bioenergy sustainability. Government institutions provide 'social safety nets,' or create conditions that allow nongovernment organizations to fill this role, to help vulnerable populations cope in times of food crisis. These coping mechanism become unavailable or inoperable when governance fails or is undermined by corruption. Several initiatives promoting sustainable bioenergy (e.g., GBEP, 2011; RSB, 2011; FAO, 2011a) acknowledge this nexus by considering governance, participation of civil society, and development of institutional capacity.

Respect for peoples' rights to land and resources is interwoven with good governance and prerequisite for any project promoting more sustainable production (FAO, 2011a; Dale et al., 2013). The 'Global Commercial Pressures on Land Project' found that failures of governance were causal factors leading to 'land grabs' (Anseeuw et al., 2011). Traditional uses of land and other natural resources by the poor are of special concern when designing policies and projects to enhance food security. Guidelines are available to ensure that biofuels development takes traditional land rights into consideration (e.g., FAO, 2011a, 2013b). Properly applying these guidelines would avoid problems such as the displacement of smallholder farmers by agro-industrial developments as transpired in Colombia (Clancy et al., 2013).

Investments in infrastructure and advances in technology are necessary for all parts of the system. Food security requires the means to produce, package, and distribute high-quality food. Biofuel sustainability relies on efficient systems for production, transport, and processing. As documented in Brazil, investments in bioethanol industries can support spin-off benefits for neighboring productive sectors and local economies. In rural areas where biomass and labor are abundant, but infrastructure is limited by lack of funding, bioenergy investments help fill gaps and facilitate economic development (Batidzirai & Johnson, 2012; Moraes & Zilberman, 2014). In Malawi and Tanzania, contracting with smallholders was found to effectively improve household incomes and community welfare (Sulle & Nelson, 2009; Hermann & Grote, 2015).

Integrated crop management and production systems are necessary for efficient provision of food, feed, fiber, and energy feedstocks. Integration helps minimize use of inputs such as fertilizer or pesticides and helps optimize use of assets such as natural, social, physical, and financial capital (e.g., Pretty, 2008; and Mueller et al., 2012). Combining the goals of food security and biofuel sustainability with other local priorities contributes to increases in total factor productivity that are responsible for the majority of growth in output from global agricultural systems over the last decade (Fuglie & Rada, 2013). Integrated system design can also help to identify opportunities to utilize what might otherwise be considered waste from one part of the system, as input for other parts (Berndes et al., 2015). Reduction in and reallocation of waste offer significant benefits, particularly if the waste would otherwise be burned or require costly removal.

Diverse ecosystem services are influenced by the interactions among resource management, food, and biofuel feedstock production (Gasparatos et al., 2011). For example, enhanced water and air quality, improved soil conditions, stable jobs, and economic benefits can all accrue if the agricultural system is designed and deployed in a way that efficiently meets the demand for food, fiber, and feedstocks (Berndes et al., 2015; Souza et al., 2015).

The occurrence of extreme weather events is unpredictable, but their intensity and frequency are expected to increase because of climate change (IPCC, 2014). Resilience to extreme events is enhanced through diversified production systems and multiple suppliers with flexibility to adjust based on the linkages between resource management, food security, and sustainable bioenergy production systems. This buoyancy can occur whether the disturbance is due to natural events (e.g., hurricanes, droughts, fires), market forces (e.g., sudden sharp decline or rise in prices), or human-induced disasters (social or political conflicts). More diversified production systems have also been shown to be more adaptable to change than traditional monoculture production systems (Woods et al., 2015).

By understanding the nexus and intentionally designing systems to promote beneficial linkages among resource management, bioenergy sustainability, and food security, we can enhance the resilience and adaptability of biofuel and food production systems and the coping mechanisms required in times of crisis. Such integrated systems should be designed to apply best practices and support critical local priorities including food security (Tables 4 and 5).

Priorities and conditions to achieve positive synergies

Many challenges in reconciling bioenergy and food security also present opportunities. Achieving positive synergies between bioenergy and food production requires science-based clarifications about context-specific problems. This also demands science-based validation of assumptions and clear definitions. Therefore, in addition to techno-economic challenges of multiproduct agricultural systems, we also should resolve barriers to social acceptance, clarify terminology, and verify that

Table 5 Benefits arising from the systemic integration of food production and bioenergy (based on Dale *et al.*, 2014, and case studies cited in this paper)

Main principle	Land-efficient food production and consumption	Integrated bioenergy production and use	Comments
Food security	Increased food supply, decreased pressure on land	Direct provision of energy services and income from existing land	Enhanced coping capacity also essential; requires planning for optimized use of limited resources (capital, water, inputs, time)
Climate security	Reduced land clearing and land-use change	Supply of low-carbon energy to agriculture and rural communities	Enhanced soil and above-ground carbon stocks; increased resilience
Energy security and supply	Increased land for bioenergy and ecological habitats	Increased provision of local energy services	Bioenergy providing low-cost drying, processing energy, and transport energy
Preservation of habitat, wild places	Reduced expansion of managed lands	Enhanced vegetation cover, species diversity, and wildlife corridors	Introduction of perennial cropping in riparian zones, on steeper slopes and in vulnerable zones in water catchment
Enhanced soil quality	Increased resiliency and crop yields	Benefits of perennial bioenergy crops in landscape and cropping strategies with greater diversity of options	Novel crop rotations, increased use of perennials to enhance soil organic matter and reduce soil disturbance
Enhanced environmental quality	Increased intensity of production with reduced environmental impacts	Novel landscape planning and cropping strategies to reduce erosion, enhanced nutrient and water availability, and decreased leaching	Benefits not a default outcome but will require careful planning and implementation combined with improved extension, knowledge transfer, and IT-based decision tools
Poverty alleviation	Greater and more resilient yields, reduced storage losses, and improved tillage and transport logistics raise income and reduce economic losses	Enhanced reliability and resilience of local energy supply; hedging strategies provided in case of damaged, condemned, or contaminated crops; improved use of residues to raise income	Direct benefits to rural farmers, processors, and traders. Care required with emerging economies of scale and marginalization of the most vulnerable/poor
Rural economic development	Increased competitiveness, enhanced knowledge and innovation capacity	Increased local economic activity and critical mass	Benefits to urban poor and rural poor

scientifically sound approaches are applied to address real problems. Focusing on positive synergies urges us to ask the right questions and to identify mechanisms for energy investments that improve food security.

Use accurate and consistent terms for analysis and communications

Robust scientific analysis should be grounded in a clear definition of the problem to be assessed and a systemic approach to resolving it. The results of many studies rely on faulty assumptions such as: Global land area is the limiting factor for food production; producing more commodities in the United States will alleviate global hunger; or any increase in commodity prices will cause food insecurity. Furthermore, policymakers and the public are misled by terms used in reporting research about food security. For example, #2 yellow corn, the subject of many reports about U.S. biofuel impacts on 'food security,' is a feed grain unfit for direct use as food. U.S. maize grown for human consumption (sweet corn, white corn, popcorn) represents about 3% of total U.S. corn production (Hansen & Brester, 2012), and from 2010 to 2014, represented only 2% of total U.S. maize exports (USDA-GATS, 2015). Simplified models confuse #2 yellow industrial feed with food. Resulting communications promulgate misconceptions, for example, that food insecurity increases with increasing commodity prices of corn or sugar (Table 3). Authentic communication requires that appropriate terminology is defined clearly and used consistently.

Recognize that food and bioenergy need not compete for land

The idea that bioenergy competes with food for land is predicated on several correlations and assumptions, beginning with land being a limiting factor for global food production. The land scarcity concept is based, in part, on conventional wisdom ('Buy land, they aren't making more of it!') and on an oversimplified interpretation of historical land clearing. Many analyses assume incorrectly that a land-cover class indicates the cause of clearing. In such analyses, forest cover typically change to agricultural cover classified as crops or pastures, and deforestation is attributed to agricultural demand. Yet, when viewed from social and historical perspectives, the actual causes of deforestation can be attributed to many other drivers such as colonization and tenure policies, market-distorting subsidies, speculation based on intrinsic value, new infrastructure, customary practices for claiming frontier land, migration, and extractive enterprises (Scouvart *et al.*, 2007; Kline & Dale, 2008).

Sorting out complex causal relationships for deforestation is difficult (Pacheco *et al.*, 2012). Quantitative models are facilitated by the convenience of remote sensing data and the simplicity of the conventional assumption that causation can be determined by the apparent land cover following deforestation; however, oversimplifications in such models often lead to faulty conclusions (Dale & Kline, 2013). Correlations between deforestation and increasing 'agricultural area' are assumed to reflect agricultural land scarcity. Several studies that use models to support the hypothesis that biofuels compete globally for land with food (Boddiger, 2007; Tenenbaum, 2008; Searchinger & Heimlich, 2015) rely on assumptions that contradict empirical evidence (Kline *et al.*, 2011; Souza *et al.*, 2015).

Indeed, policymakers in major food-producing nations have been challenged by waste, overproduction, and depressed farm-commodity prices for decades. As a result of excess production, policies were developed in the 1980s and 1990s to reduce spoilage, waste, and financial losses associated with excessive stocks of major food commodities. Those policies emphasized land set-asides and environmental protection rather than increased production. Furthermore, food security in some less developed nations was impaired by food 'aid' and subsidized export of surplus production (Thurow & Kilman, 2009; FAO, WFP, 2010). Since the 1990s, innovations in technology, system integration, and logistics have allowed producers to meet the growing global demands for food without requiring additional land (Alexandratos & Bruinsma, 2012; Conway & Wilson, 2012). Yet the belief that biofuel production directly competes with food production and increases food prices remains widely held (e.g., Hajkowicz *et al.*, 2012).

It becomes clear that global land area is not the limiting factor for food and bioenergy production when consistent data on land cover, land use, and productive potential are applied to the analysis (Babcock, 2011; Woods *et al.*, 2015). Despite ongoing population growth and deforestation, the total land area *used* to feed the world has remained steady since 1990 (Ausubel *et al.*, 2013; FAOStat, 2015). The average area of cropland used to feed one person has fallen from 0.45 ha in 1961 to 0.22 ha in 2006 (FAO, 2011b) and is projected to be close to 0.19 ha at present, based on FAOStat 2015. At 0.19 ha per capita, 1.7 billion hectares, or about a third of all arable land available today, could feed the population of 9 billion projected for 2050.

Output from most agricultural land is far below potential yields (Mueller *et al.*, 2012). Thus, the land *required* to feed humanity is a fraction of that currently classified as agricultural (Woods *et al.*, 2015). Most U.S. cities could be fed from a 50-mile-radius 'foodshed'

(Zumkehr & Campbell, 2015). Rooftops and other small urban gardens illustrate that far higher yields per hectare are possible, potentially reducing land *requirements* to as little as 0.01 ha per capita (Orsini *et al.*, 2014; Rockwell, 2015). Still less land would be required for intensive, closed-loop agricultural systems that recycle water and nutrients. Given current trends, some researchers expect that the agricultural area required to support global food needs will decline over coming decades (Roser, 2015).

When considering land, context is critical. Local competition for land reflects historic inertia and can be politically and socially sensitive. Even though no further deforestation is required to feed humanity well into the future, deforestation continues due in part to poor understanding of the local causes. Effective policies to conserve natural areas do not require reducing food or biomass production but may involve incentives for efficient resource management and recycling of water and nutrients.

Invest in technological innovation to build capacity and infrastructure

One of the most persistent recommendations for improving food security is to invest in rural agricultural technology, as discussed in the SOFI reports and reflected in multiple recent initiatives to 'feed the future' (Godfray *et al.*, 2010; IMF, 2013; USG, 2015; World Bank, 2015). However, during periods of historically low real prices for food producers, there is limited motivation for investments in technology or yield improvement. Declining support for agricultural research around the globe since the 1970s is a concern, and the 'significant decline in annual investment in high-income countries between 1991 and 2000 is especially troubling' (Beachy, 2014). Case studies in Brazil have illustrated the potential for investments in bioenergy technology and infrastructure to simultaneously reduce hunger, expand food commodity exports, and promote socioeconomic development (Souza *et al.*, 2015).

Investments in innovation and local infrastructure are promoted at the nexus of sustainable bioenergy, food security, and resource management. Innovations in technology and integrated production systems characterized recent biofuels expansion in the United States and Brazil (Gee & McMeekin, 2011). Bio-based industries that can entice new investments are a prominent part of many rural development strategies (UNCTAD, 2014). Investment is required to complement the land and labor that tend to be plentiful in rural areas at risk to food insecurity (FAO, 2015a). Key constraints, capital and technology, can be alleviated by investments in strong, growing markets.

Promote stable prices high enough to incentivize local food production

Price volatility in a food security context is defined as large, sudden changes in the prices of staples on which at-risk populations depend. Sudden price increases make staples less accessible to urban at-risk groups, while sudden decreases undermine smallholder producers' livelihoods and household incomes in rural areas. More predictable staple prices that create incentives for local investment in food production are important to improving food security (IFPRI, 2015). Declines in prices are more detrimental to food security than temporary price spikes because (1) capacity and investment in local food production supply chains are undermined, (2) over 70% of the global population living with hunger is in rural areas (FAO, 2014, 2015b), and (3) price crashes catalyze rural-to-urban migration, which can further undermine existing productive capacity. Rural areas and uncharted neighborhoods created by recent migrants are more difficult and costly to reach with food assistance than well-established, urban populations. Farmers and agro-industries have demonstrated capacity to respond to local market signals for products that can be grown profitably.

Adopt flex crops that can provide food and other products

Extreme weather events such as drought and flood are inevitable and cause unpredictable supply shocks in affected areas. Trade combined with surplus production from diverse regions can help alleviate such vulnerabilities to extreme events. Remote sensing tools and communication platforms that share crop progress and projected harvest data are increasingly allowing far-flung regions to respond quickly to supply shocks. Producers with competitive technologies and access to markets can boost yields or plant a second crop on existing fields. The supply shock caused by the 2012 drought in North America was offset in part by planting second crops on existing fields in the Southern Hemisphere (USDA, 2013). The increasingly interconnected world is better informed and responsive to arising crises, helping to reduce casualties from famine over the last two decades (Roser, 2015).

Biofuel markets have been proposed as one mechanism that can absorb the surplus production in normal years and provide a cushion in years of unexpected supply disruptions. The opportunities offered and problems created by 'flex crops' that can serve food and other markets merit further study. International organizations concerned with food security (e.g., FAO, IFAD, IMF, OECD, UNCTAD, WFP, the World Bank, WTO, IFPRI) support policies or market mechanisms that allow feedstocks to be diverted from biofuel production to uses that could dampen volatility of food commodity prices (see for example the Committee on Food Security report, HLPE 2013; Locke *et al.*, 2013). This ability to shift end use of available supply as a 'safety valve' to reduce price volatility (Wright, 2011) has been a cornerstone of Brazilian strategies for maintaining strong biofuels and sugar industries (Osseweijer *et al.*, 2015).

Similarly, U.S. maize production capacity expanded from 2002 to 2011 in part as a response to federal biofuel mandates. Investments made during this period in technologies such as precision agriculture, irrigation, and grain storage would have been impossible without favorable profit margins. Federal support to expand biofuel markets increased confidence in the ability to sell crops at a profit. The investments increased efficiency and reduced long-term production costs. Investments in irrigation and storage between 2002 and 2012 also helped to moderate price volatility in the face of the worst drought to hit U.S. farms in more than 50 years (USDA, 2013). A drought of this magnitude represents a 'supply shock' that could have triggered a global food price crisis, but market responses helped avoid a major price spike. Moreover, the drought and its effects were monitored and communicated widely, which allowed Southern Hemisphere nations to respond with second crops of maize. There is growing recognition of the value of flex crops combined with good market intelligence to support predictable and relatively stable commodity prices, as this information influences decisions of buyers and sellers in futures markets (FAO-UNEP, 2011; UNCTAD, 2014).

Identify conditions under which bioenergy improves food security

Integration of land- and resource-efficient food and bioenergy production will increase the sustainability of the system and extend benefits across multiple value-added product chains (Table 5).

Conclusions and recommendations

Relationships among food security and biofuel policies are complex and context specific. Such nuanced local relationships cannot be captured in global-scale analyses, and the validity of simple models for useful policy guidance is questionable. Assessing impacts requires an understanding of the interactions among factors relevant to food security within a specific place and time. The debate needs to transition from irreconcilable generalizations about whether biofuels are 'good or bad' for food security, to constructive understandings of where and how biofuels can help achieve sustainable

development goals including the eradication of hunger. The following recommendations aim to facilitate synergies between food security and energy security through careful planning and development of bioenergy projects and policies.

Ask the right questions

Analysis must consider local contextual conditions to understand the drivers of food insecurity. Multiple causal factors should be addressed using a holistic approach. Developing a bioenergy policy or project designed to improve food security requires that answers to the following questions be applied to a well-defined, local context.

1. Who is most at risk from food insecurity?
2. What factors are causing or increasing the risk of specific food security problems? How do these factors relate to energy and fuels?
3. What actions are feasible and likely to effectively address the causal risk factors?
4. What can be done to mitigate hunger problems in the near term while also building resilience to reduce future risk of food insecurity? And how do these actions and those identified in response to question 3 relate to potential (bio)energy/fuels?
5. How can a bioenergy policy or project be designed to address the local causal risk factors and contribute to reduced food insecurity?
6. Is a regional development plan that integrates sustainable bioenergy more effective and efficient in achieving food security goals than one without bioenergy?

Engage stakeholders to address needs for food and energy security

Consensus-based principles of sustainable global food security underscore the importance of developing projects with local ownership that consider the needs of the most vulnerable populations (FAO, 2015a) (Table 4). Stakeholders can help identify ways in which bioenergy investments can reinforce efficient local food production and other services. Stakeholder engagement also supports adaptive decision-making to enhance goal achievement (Dale *et al.*, 2015).

Encourage coproduct complementarity, diversity, and stable markets

Relatively stable and predictable prices for food and energy are essential for food security. Access to affordable energy supports food security goals, while energy price volatility can exacerbate food crises. Building confidence with long-term policies allows markets to work effectively. For example, to the degree that biofuel policies support a more stable and profitable market-driven price floor, local production can be incentivized by markets that can absorb increasing output. If price caps are used to protect consumers, mechanisms to support local producers may be needed lest food security be undermined. As price crashes are often more detrimental to long-term food security than price spikes, sudden shifts in policies that reduce investment in agricultural production should be avoided.

Diversifying sources of production and end uses of agricultural products enhances local food security. More efficient production of nutritious staples can be promoted through integrated production systems that offer a diversity of coproducts for bioenergy and other markets. Crops that can serve multiple markets reduce risks for producers and possibly enhance food safety by providing noon-food outlets for contaminated or damaged food. It may be beneficial to promote strategic supply chains in order to facilitate access to multiple markets for such 'flex crops.' Investments in better technology and more efficient production (e.g., precision agriculture and efficient irrigation) can help producers respond to market signals for different crops as well as adapt to disturbances such as those caused by weather. Diversity in the geospatial distribution of production and types of production can reduce price sensitivities caused by disruptive events (e.g., political upheaval, flood, or drought).

Support planning and implementation of landscapes designed for multiple uses and waste minimization

Apply landscape design to help stakeholders assess trade-offs when making choices about locations, types, and management of crops, as well as transport, refining, and distribution of products and services. Landscape design refers to a spatially explicit, collaborative plan for management of landscapes and supply chains for food, energy, and other services (Dale *et al.*, 2016), which respects traditional landholdings and farming practices. Proactive resource-use planning can support improvements in management and provision of services based on a set of defined goals (Dale *et al.*, 2014). Such planning should consider shared infrastructure to meet the needs for food, energy, and other markets in a way that reduces costs and waste. Reduction in agricultural wastes provides a means for more efficient crop production. Agro-ecological zoning developed in response to biofuel sustainability concerns in Brazil has influenced other agricultural sectors and helped protect biodiversity and forests, which are important sources for sustained food production in rural areas (Sunderland *et al.*, 2013). The sugarcane–ethanol industry in Brazil

supports 4.5 million jobs, improves livelihoods, and promotes rural infrastructure and development (Moraes & Zilberman, 2014).

Apply adaptive management and promote continual improvement

Adaptive management involves learning from ongoing monitoring so that decisions can be adjusted to changing conditions and needs. Timely information about environmental, social, and economic conditions, local crops, and market intelligence can support more sustainable food and energy production. It is important to collect data and monitor indicators of food and energy security that are most relevant to local context and stakeholders. Local monitoring helps to verify progress, flag problems, and signal requirements for corrective actions. The information gained needs to inform adjustments in management practices and plans that support adaptation to changing conditions. Accurate and timely data on prices, stocks, futures markets, and weather are essential to support monitoring and adaptive management. Crop monitoring and timely information sharing can also help address unplanned supply shortfalls and reduce price volatility, as observed when Southern Hemisphere nations such as Brazil and Argentina planted second crops in response to early reports of the 2012 U.S. drought.

Communicate clearly about barriers and opportunities to address local needs

How food and food security are discussed shapes public opinion. Clear definitions, consistent use of terminology, science-based problem identification, and validation of assumptions help reduce confusing and conflicting messages. Data need to be relevant; communications focusing on global commodity prices may have little bearing on the factors that determine when and where local food insecurity becomes a problem. Reliance on readily available aggregate data distracts attention from aspects of food insecurity that matter most for peoples' health and well-being. Timely information on the status of indicators for environmental, social, and economic effects of development projects needs to be publicly accessible. Long-term commitments to food security, energy security, and environmental quality need to be broadly communicated, and defined goals should be shared widely.

Collaborate with local development programs on common goals

Bioenergy policies can support progress toward the 2030 Sustainable Development Goals of doubling of agricultural productivity, improving incomes of small-scale food producers, and providing clean energy for all (UN, 2015). Research should provide relevant lessons drawn from bioenergy–food interactions over the last decade to guide efforts to provide food and energy while reducing greenhouse gas emissions (Dale et al., 2011). The 2015 assessment of progress toward Millennium Development Goals (MDGs) found that several countries with domestic biofuel production policies, such as Brazil, China, Indonesia, Malawi, Malaysia, and Peru, also achieved or exceeded challenging hunger-reduction goals (FAO, IFAD, WFP, 2015a). Other countries with notable bioenergy potential, but where biofuel policies were not effectively implemented, such as Zambia, Senegal, and Guatemala, fell short on MDG hunger-reduction targets (Tay, 2013; Mukanga, 2014; UNCTAD, 2014; World Bank, 2015). Biofuel projects responsive to site-specific needs in developing nations offer opportunities to support food and energy security goals (Kline et al., 2009; Gasparatos et al., 2011; Mitchell, 2011).

Build on and improve existing systems

Bioenergy is already an integral part of global food production, processing, and consumption systems. Experience indicates that investments in bioenergy can help expand local food supplies, infrastructure, and productive capacity and thereby reduce risks of hunger for specific groups and situations (FAO, 2011a; Durham et al., 2012; Moraes & Zilberman, 2014). The nexus of bioenergy, food security, and resource management is especially significant for the rural poor. Dependence on subsistence agriculture and inefficient traditional biomass use leaves rural populations vulnerable and deepens impoverishment through resource degradation. Current practices can transition and transform through continual improvements to meet the needs of society in a changing world. Institutional capacity for learning and sharing experiences should be developed across the supply chain. Applying science to support continual improvement will help feed more people and provide them with more sustainable energy resources for the future.

Prioritize research investments

Future research priorities include better monitoring and analysis to determine cause-and-effect relationships among factors that determine vulnerabilities to food insecurity. Research should support design and planning so that negative effects are minimized or avoided and persistent improvements in energy and food security are achieved. Better resource management can address both food and energy needs and lift people out of poverty, but this requires governance and policies that create the right

incentives. Case studies that document actual conditions before and after project implementation can support more integrated project designs and adaptive management (FAO, 2011a; Elbehri et al., 2013). Transparent documentation of the problem, hypotheses, research methods, input data sources, and assumptions is essential to avoid potential misrepresentation of analytical results (Dale & Kline, 2013).

Conclusions

Effectively addressing food security and bioenergy sustainability requires a renewed focus on populations at risk. Understanding the local causes of food insecurity is a prerequisite step for designing bioenergy projects that improve food security in a specific place and time. This approach requires multidisciplinary analysis and program design to consider and address key constraints and opportunities. Projects should target rural poor with opportunities to engage in more sustainable, diversified, and integrated systems that provide clean, affordable fuels and nutritious food. Bioenergy can contribute to improved food security through production systems designed to increase adaptability and resilience of human populations at risk and to reduce context-specific vulnerabilities that could limit access to local staples and required nutrients in times of crisis.

Acknowledgements

Work by Keith L. Kline and Virginia H. Dale was supported by the U.S. Department of Energy (DOE) under the Bioenergy Technologies Office; ORNL is managed by UT-Battelle, LLC, for DOE under contract DE-AC05-00OR22725. Harriet K. Mugera was supported by the World Bank. Jeremy Woods was supported by Climate-KIC and Imperial College London. Glaucia M. Souza was supported by a grant from the Sao Paulo State Research Foundation (FAPESP 2012/23765-0) and a Productivity Fellowship from the Brazilian National Council for Scientific and Technological Development (CNPq). Erica Atkin and Gina Busby are thanked for editorial assistance.

References

Alexandratos N, Bruinsma J (2012) World agriculture towards 2030/2050: the 2012 revision. ESA Working paper No. 12-03. Rome, FAO.

Anseeuw W, Wily LA, Cotula L, Taylor M (2011) Land Rights and the Rush for Land. Findings of the Global Commercial Pressures on Land Research Project. International Land Coalition, Rome. www.landcoalition.org.

Ausubel JH, Wernick IK, Waggoner P (2013) Peak farmland and the prospect for land sparing. Population and Development Review, 38, 221–242.

Babcock BA (2011) The impact of U.S. biofuel policies on agricultural price levels and volatility. ICTSD Programme on Agricultural Trade and Sustainable Development Issue Paper 35, International Centre for Trade and Sustainable Development, Geneva, Switzerland.

Baffes J, Dennis A (2013) Long-term drivers of food prices. In: The World Bank Development Prospects Group & Poverty Reduction and Economic Management Network. World Bank Policy Research Working Paper 6455, Washington, DC.

Batidzirai B, Johnson FX (2012) Energy security, agro-industrial development and international trade: the case of sugarcane in southern Africa. In: *Socio-Economic and Environmental Impacts of Biofuels: Evidence from Developing Nations* (eds Gasparatos A, Stromberg P), Chapter 12, PP. 254–277. Cambridge University Press, London.

Beachy RN (2014) Building political and financial support for science and technology for agriculture. *Philosophical Transactions of the Royal Society of London. Series B, Biological Sciences*, **369**, 20120272.

Berndes G, Youngs H, Ballester MVR *et al.* (2015) Land and bioenergy. In: *Soils and water. In: Bioenergy & Sustainability: Bridging the Gaps* (eds Souza GM, Victoria R, Joly C, Verdade L) Chapter 18. pp. 618–659) SCOPE 72, Paris.

Boddiger D (2007) Boosting biofuel crops could threaten food security. *Lancet*, **370**, 923–924.

von Braun J, Hill RV, Pandya-Lorch R (2009) The poorest and the hungry: a synthesis of analyses and actions. In: *The Poorest and Hungry: Assessments, Analyses and Actions: An IFPRI 2020 Book* (eds von Braun J, Hill RV, Pandya-Lorch R), pp. 1–62. IFPRI, Washington, DC.

Clancy J, Marin-Burgos V, Narayanaswamy A (2013) Addressing poverty through inclusion in global production chains: who wants it?. University of Birmingham, UKDSA Annual Meeting.

Coleman-Jensen A, Rabbitt MP, Gregory C, Singh A (2015) Household Food Security in the United States in 2014, ERR-194, U.S. Department of Agriculture, Economic Research Service, USA.

Conway G, Wilson K (2012) *One Billion Hungry. Can we Feed the World?* Cornell University Press, Ithica, NY, USA.

Dale VH, Kline KL (2013) Modeling for integrating science and management. In: *Land Use and the Carbon Cycle: Advances in Integrated Science, Management, and Policy* (eds Brown DG, Robinson DT, French NHF, Reed BC), pp. 209–237. Cambridge University Press, New York, NY, USA.

Dale VH, Efroymson RA, Kline KL (2011) The land use–climate change–energy nexus. *Landscape Ecology*, **26**, 755–773.

Dale VH, Efroymson RA, Kline KL *et al.* (2013) Indicators for assessing socioeconomic sustainability of bioenergy systems: a short list of practical measures. *Ecological Indicators*, **26**, 87–102.

Dale BE, Anderson J, Brown R *et al.* (2014) Take a closer look: biofuels can support environmental, economic and social goals. *Environmental Science & Technology*, **48**, 7200–7203.

Dale VH, Efroymson RA, Kline KL, Davitt M (2015) A framework for selecting indicators of bioenergy sustainability. *Biofuels, Bioproducts & Biorefining*, **9**, 435–446.

Dale VH, Kline KL, Buford MA *et al.* (2016) Incorporating bioenergy into sustainable landscape designs. *Renewable & Sustainable Energy Reviews*, **56**, 1158–1171.

Davis SC, Boddey RM, Alves BJR *et al.* (2013) Management swing potential for bioenergy crops. *GCB-Bioenergy*, **5**, 623–638.

Department for Environment, Food, and Rural Affairs (DEFRA) (2010) *The 2007/08 Agricultural Price Spikes: Causes and Policy Implications.* HM Government, London, UK.

DeRose L, Messer E, Millman S (1998) *Who's Hungry? And How Do We Know? Food Shortage.* United Nations University Press, Tokyo, Japan.

Drucker P (1971) *Men, Ideas & Politics.* Harper & Row, New York, NY.

Durham C, Davies G, Bhattacharyya T (2012) Can biofuels policy work for food security? In: *Food and Farming Industry Policy.* Department for Environment Food and Rural Affairs, London, UK.

Efroymson RA, Dale VH, Kline KL *et al.* (2013) Environmental indicators of biofuel sustainability: what about context? *Environmental Management*, **51**, 291–306.

Elbehri A, Segerstedt A, Liu P (2013) *Biofuels and the Sustainability Challenge: a Global Assessment of Sustainability Issues, Trends and Policies for Biofuels and Related Feedstocks.* FAO, Rome.

EPI (2014) Attacks on renewable energy standards and net metering policies by fossil fuel interests & front groups 2013-2014. Energy and Policy Institute of the University of Michigan. Available at: http://www.energyandpolicy.org/renewable-energy-state-policy-attacks-report (accessed 2 October 2015).

FAO (2008) An introduction to the basic concepts of food security. FAO, Rome. Available at: http://www.fao.org/docrep/013/al936e/al936e00.pdf (accessed 2 October 2015).

FAO (2010) Bioenergy and food security: the BEFS Analytical Framework. Environment and Natural Resources Management Series No. 16, FAO, Rome.

FAO (2011a) BEFSCI brief: good socio-economic practices in modern bioenergy production – minimizing risks and increasing opportunities for food security. FAO, Rome. Available at: http://www.fao.org/bioenergy/31478-0860de0873f5 ca89c49c2d43fbd9cb1f7.pdf (accessed 2 October 2015).

FAO (2011b) *The State of the World's Land and Water Resources for Food and Agriculture (SOLAW) – Managing Systems at Risk.* FAO, Rome, and Earthscan, London.

FAO (2013a) Food Outlook – bi-annual report on global food markets. FAO, special issue expansion. Available at: http://www.fao.org/fileadmin/templates/world food/Reports_and_docs/FO-Expanded-SF.pdf (accessed 2 October 2015).

FAO (2013b) Fortieth Session Report, Committee on World Food Security, Rome, Italy, 7-11 October 2013. Available at: http://www.fao.org/docrep/meeting/029/mi744e.pdf (accessed 6 October 2015).

FAO (2014) Global and regional food consumer price inflation monitoring. Issue 3. Available at: http://www.fao.org/fileadmin/templates/ess/documents/consumer/CPI_Jan_2014.pdf (accessed 2 October 2015).

FAO (2015a) Hunger map 2015. FAO Statistics Division, Rome. Available at: http://www.fao.org/hunger/en/ (accessed 6 October 2015).

FAO (2015b) Forty-second Session Report (Draft), Committee on World Food Security, Rome, Italy, 12-15 October 2015. Global Strategic Framework for Food Security & Nutrition (GSF) Fourth Version. Available at: http://www.fao.org/3/a-mo187e.pdf (accessed 10 October 2015).

FAO (2015c) FAO consumer price index. Available at: http://www.fao.org/filead min/templates/ess/documents/consumer/CPINewsReleaseApril24_EN_v8.pdf (accessed 2 October 2015).

FAO (2015d) Forty-second Session Report #42/4, Committee on World Food Security, Rome, Italy, 12-15 October 2015. Framework for action for food security and nutrition in protracted crises (Sept 2015; 11 pp). Available at: http://www.fao.org/fileadmin/templates/cfs/Docs1415/FFA/CFS_FFA_Final_Draft_Ver2_EN.pdf (accessed 11 December 2015).

FAO (2015e) The state of food and agriculture – social protection and agriculture: breaking the cycle of rural poverty. FAO, Rome. Available at: http://www.fao.org/3/a-i4910e.pdf (accessed 13 May 2016).

FAO, IFAD, WFP (2013) *The State of Food Insecurity in the World 2013 – the Multiple Dimensions of Food Security.* FAO, Rome.

FAO, IFAD, WFP (2014) *The State of Food Insecurity in the World 2014 (Strengthening the Enabling Environment for Food Security and Nutrition.* FAO, Rome.

FAO, IFAD, WFP (2015a) *The State of Food Insecurity in the World 2015 (Meeting the 2015 International Hunger Targets: Taking Stock of Uneven Progress.* FAO, Rome.

FAO, IFAD, WFP (2015b) *Achieving Zero Hunger (The Critical Role of Investments in Social Protection and Agriculture.* FAO, Rome.

FAO, Un World Food Program (WFP) (2010) *The state of food insecurity in the world 2010 – addressing food insecurity in protracted crises.* FAO, Rome.

FAO, United Nations International Fund for Agricultural Development (IFAD), WFP (2002) Reducing poverty and hunger: the critical role of financing for food, agricultural and rural development. FAO, Rome. Available at: ftp://ftp.fao.org/docrep/fao/003/Y6265E/Y6265E.pdf (accessed 2 October 2015).

FAO, WFP, IFAD (2012) *The State of Food Insecurity in the World 2012 – Economic Growth is Necessary but not Sufficient to Accelerate Reduction of Hunger and Malnutrition.* FAO, Rome.

FAOStat (2015) Available at: http://faostat3.fao.org/download/P/CP/E (accessed 2 October 2015).

FAO-UNEP (2011) Bioenergy Decision Support Tool. FAO, Rome. Available at: http://www.bioenergydecisiontool.org/ (accessed 6 October 2015).

Flipse SM, Osseweijer P (2013) Media attention on GM food cases – an innovation perspective. *Public Understanding of Science*, **22**, 185–202.

Food and Agriculture Organization of the United Nations (FAO) (1996) Rome declaration on world food security and world food summit plan of action. World Food Summit 13-17 November 1996. FAO, Rome.

Foresight (2011) *The Future of Food and Farming Final Project Report.* The Government Office for Science, London, UK.

Fuglie K, Rada N (2013) Growth in global agricultural productivity: an update. U.S. Department of Agriculture, Economic Research Service, Amber Waves, November issue. Available at: http://www.ers.usda.gov/amber-waves/2013-November/growth-in-global-agricultural-productivity-an-update.aspx# (accessed 13 May 2016).

Gasparatos A, Stromberg P, Takeuchi K (2011) Biofuels, ecosystem services and human wellbeing: putting biofuels in the ecosystem services narrative. *Agriculture, Ecosystems and Environment*, **142**, 111–128.

Gautam KC (2014) Addressing the challenge of hidden hunger. In: *Global Hunger Index 2014 (Chapter 3), International Food Policy Research Institute (IFPRI).* Available at: http://www.ifpri.org/sites/default/files/ghi/2014/feature_1818.html (accessed 2 October 2015).

GBEP (2011) *Global Bioenergy Partnership Sustainability Indicators for Bioenergy*, 1st edn. FAO, Rome. Available at: http://www.globalbioenergy.org/fileadmin/user_upload/gbep/docs/Indicators/The_GBEP_Sustainability_Indicators_for_Bio energy_FINAL.pdf (accessed 13 May 2016).

Gee S, McMeekin A (2011) Eco-innovation systems and problem sequences: the contrasting cases of US and Brazilian biofuels. *Industry and Innovation*, **18**, 301–315.

Gibson J. (2013) The crisis in food price data. *Global Food Security*, **2**, 97–103.

Godfray HCJ, Beddington JR, Crute IR *et al.* (2010) Food security: the challenge of feeding 9 billion people. *Science*, **5967**, 812–818.

GRFA (2015) Global Renewable Fuels Alliance. Available at: http://globalrfa.org/news-media/un-data-shows-that-ethanol-is-not-causing-food-price-rises (accessed 11 December 2015).

von Grebmer K, Saltzman A, Birol E *et al.* (2014) *2014 Global Hunger Index: the Challenge of Hidden Hunger*. Welthungerhilfe, International Food Policy Research Institute, and Concern Worldwide, Bonn, Washington, DC, and Dublin.

Gustafson D, Gutman A, Leet W *et al.* (2016) Seven food system metrics of sustainable nutrition security. *Sustainability*, **8**, 196.

Hajkowicz S, Negra C, Barnett P *et al.* (2012) Food price volatility and hunger alleviation – can Cannes work? *Agriculture & Food Security*, **1**, 8.

Hansen R, Brester G (2012) White corn profile. Ag Marketing Resource Center, Iowa State University. Available at: http://www.agmrc.org/commodities__products/grains__oilseeds/corn_grain/white-corn-profile/ (accessed 6 December 2015).

Hecht AD, Shaw D, Bruins R *et al.* (2009) Good policy follows good science – using criteria and indicators for assessing sustainable biofuel production. *Ecotoxicology*, **18**, 1–4.

Hermann R, Grote U (2015) Large-scale agro-industrial investments and rural poverty: evidence from sugarcane in Malawi. *Journal of African Economies*, **24**, 645–676.

Hilbert JA (2014) A systemic study of biofuels in complex agriculture markets. In: *Proceedings of the 22nd European Biomass Conference and Exhibition (EUBCE) Hamburg*, pp. 158–164.

HLPE (2013) *Biofuels and Food Security, Report 5*. The High Level Panel of Experts on Food Security and Nutrition of the Committee on World Food Security, Rome.

HLPE (High Level Panel of Experts) (2011) *Price Volatility and Food Security*. The High Level Panel of Experts on Food Security and Nutrition of the Committee on World Food Security, Rome.

IFPRI (2015) Biofuels and food security interactions workshop final report. Available at: http://www.ifpri.org/event/workshop-biofuels-and-food-security-interactions (accessed 11 December 2015).

IMF (2012) Food and fuel prices – policy options for riding out food, fuel price spikes. International Monetary Fund (IMF) Survey Magazine, October 8. Available at: http://www.imf.org/external/pubs/ft/survey/so/2012/INT100712A.htm (accessed 16 December 2015).

IMF (2013) Impact of high food and fuel prices on developing countries—frequently asked questions. International Monetary Fund External Relations Department, Washington, DC. Available at: http://www.imf.org/external/np/exr/faq/ffpfaqs.htm (accessed 18 December 2015).

IPC Global Partners (2012) *Integrated Food Security Phase Classification Technical Manual Version 2.0. Evidence and Standards for Better Food Security Decisions*. FAO, Rome.

IPCC (2014) Summary for policy makers. In: *Climate Change 2014: Impacts, Adaptation, and Vulnerability*. International Panel on Climate Change, Work Group II. Geneva, Switzerland.

Johnson FX, Silveira S (2014) Pioneer countries in the transition to alternative transport fuels: comparison of ethanol programmes and policies in Brazil, Malawi and Sweden. *Environmental Innovation & Societal Transitions*, **11**, 1–24.

Kline KL, Dale VH (2008) Biofuels, causes of land-use change, and the role of fire in greenhouse gas emissions. *Science*, **321**, 199.

Kline KL, Dale VH, Lee R, Leiby P (2009) In defense of biofuels, done right. *Issues in Science and Technology*, **25**, 75–84.

Kline KL, Oladosu GA, Dale VH, McBride AC (2011) Scientific analysis is essential to assess biofuel policy effects. *Biomass and Bioenergy*, **35**, 4488–4491.

Konandreas P (2012) Trade policy responses to food price volatility in poor net food-importing countries. ICTSD Programme on Agricultural Trade and Sustainable Development; Issue Paper No. 42. International Centre for Trade and Sustainable Development, Geneva, Switzerland.

Lagi M, Bar-Yam Y, Bertrand KZ, Bar-Yam Y (2011) The food crises: a quantitative model of food prices including speculators and ethanol conversion. *New England Complex System Institute*, **4859**, 1–56.

Locke A, Wiggins S, Henley G, Keats S (2013) *Diverting Grain from Animal Feed and Biofuels*. Overseas Development Institute, London, UK.

Lombard MJ (2014) Mycotoxin exposure and infant and young child growth in Africa: what do we know? *Annals of Nutrition and Metabolism*, **64** (Suppl2), 42–52.

Lynd LR, Sow M, Chimphango AFA *et al.* (2015) Bioenergy and African transformation. *Biotechnology for Biofuels*, **8**, 18.

Manning P, Taylor G, Hanley ME (2014) Bioenergy, food production and biodiversity – an unlikely alliance? *GCB-Bioenergy*, **7**, 570–576.

McBride A, Dale VH, Baskaran L *et al.* (2011) Indicators to support environmental sustainability of bioenergy systems. *Ecological Indicators*, **11**, 1277–1289.

Mitchell D (2008) *A Note on Rising Food Prices*. World Bank, Washington, DC.

Mitchell D (2011) *Biofuels in Africa: Opportunities, Prospects and Challenges*. World Bank Publications, Washington, D.C..

Moraes MAFD, Zilberman D (2014) *Production of Ethanol from Sugarcane in Brazil: from State Intervention to a Free Market*. Springer International Publishing, Cham, Switzerland.

Mueller SA, Anderson JE, Wallington TJ (2011) Impact of biofuel production and other supply and demand factors on food price increases in 2008. *Biomass and Bioenergy*, **35**, 1623–1632.

Mueller ND, Gerber JS, Johnston M *et al.* (2012) Closing yield gaps through nutrient and water management. *Nature*, **490**, 254–257.

Mukanga C (2014) The economics of biofuels. Zambian Economist. Avaialbe at: http://www.zambian-economist.com/2014/02/economics-of-biofuels.html (accessed 4 December 2015).

Orsini F, Gasperi D, Marchetti L *et al.* (2014) Exploring the production capacity of rooftop gardens in urban agriculture: the potential impact on food and nutrition security, biodiversity and other ecosystem services in the city of Bologna. *Food Security*, **6**, 781–792.

Osseweijer P, Watson HK, Johnson FX *et al.* (2015) Bioenergy and food security. In: *Bioenergy & Sustainability: Bridging the Gaps* (eds Souza GM, Victoria R, Joly C, Verdade L) (Chapter 4) 72, pp. 779. Paris. SCOPE. ISBN 978-2-9545557-0-6. Available at: http://www.bioenfapesp.org/scopebioenergy/index.php (accessed 13 May 2016).

Pacheco P, Wardell A, German L *et al.* (2012) *Bioenergy, Sustainability and Trade-Offs: can we Avoid Deforestation While Promoting Biofuels?* CIFOR Infobrief 54. Center for International Forestry Research, Bogor, Indonesia.

Pretty JN (2008) Agricultural sustainability: concepts, principles and evidence. *The Royal Society Publishing; Philosophical Transactions B: Biological. Science*, **363**, 447–465.

Rajagopal D, Sexton SE, Roland-Holst D, Zilberman D (2007) Challenge of biofuel: filling the tank without emptying the stomach? *Environmental Research Letters*, **2**, 1–9.

REN21 (2015) *Renewables 2015 Global Status Report*. Paris, REN21 Secretariat. ISBN 978-3-9815934-6-4.

Rockwell L (2015) Six thousand pounds of food … 1/10th of an acre. The Burning Platform. Available at: http://www.theburningplatform.com/2015/05/02/six-thousand-pounds-of-food-110th-of-an-acre/ (accessed 2 October 2015).

Rose D (1999) Economic determinants and dietary consequences of food insecurity in the United States. *Journal of Nutrition*, **129**, 517S–520S.

Roser M (2015) Our World in Data. Available at: www.OurWorldinData.org (accessed 6 October 2015).

Rosillo-Calle F, Johnson FX (2010) *Food Versus Fuel: an Informed Introduction to Biofuels*. ZED Books, London.

Roundtable on Sustainable Biomaterials (RSB) (2011) Standard RSB-STD-04-001; and RSB Low iLUC Risk Biomass Criteria and Compliance Indicators. Available at: http://rsb.org/pdfs/standards/RSB-STD-04-001-ver0.3RSBLowiLUCCriteriaIndicators.pdf. (accessed 6 October 2015).

RTI (2014) Current and prospective scope of hunger and food security in america: a review of current research. Research Triangle Institute International Center for Health and Environmental Modeling. Available at: www.rti.org/pubs/full_hunger_report_final_07-24-14.pdf (accessed 8 October 2015).

Scouvart M, Adams RT, Caldas M *et al.* (2007) Causes of deforestation in the Brazilian Amazon: a qualitative comparative analysis. *Journal Land Use Science*, **2**, 257–282.

Searchinger T, Heimlich R (2015) Avoiding bioenergy competition for food crops and land. Working Paper, Installment 9 of Creating a Sustainable Food Future, World Resources Institute, Washington, D.C. Available at: http://www.worldresourcesreport.org (accessed 2 October 2015).

Smith LC, Haddad LJ (2000) Explaining child malnutrition in developing countries: a cross-country analysis. IFPRI research report. Available at: http://www.ifpri.org/publication/explaining-child-malnutrition-developing-countries-0 (accessed 13 May 2016).

Souza GM, Victoria R, Joly C, Verdade L (2015) (eds) *Bioenergy & Sustainability: Bridging the Gaps*, Vol. 72. SCOPE, Paris, France. ISBN 978-2-9545557-0-6. Available at: http://www.bioenfapesp.org/scopebioenergy (accessed 2 October 2015).

Sulle E, Nelson F (2009) *Biofuels, Land Access and Rural Livelihoods in Tanzania*. IIED, London. ISBN: 978-1-84369-749-7.

Sumner DA (2009) Recent commodity price movements in historical perspective. *American Journal of Agricultural Economics*, **91**, 1250–1256.

Sunderland TCH, Powell B, Ickowitz A et al. (2013) Food Security and Nutrition: the Role of Forests. CIFOR Discussion Paper. Center for International Forestry, Bogor, Indonesia.

Tarka-Sanchez S, Woods J, Akhurst M et al. (2012) Accounting for indirect land use change in the life cycle assessment of biofuel supply chains. *Journal of the Royal Society, Interface*, **9**, 1105–1119.

Tay K (2013) Guatemala biofuels annual update on ethanol and biodiesel issues. Global Agricultural Information Network Report GT13006, USDA Foreign Agricultural Service. Available at: http://gain.fas.usda.gov/ (accessed 7 October 2015).

Tenenbaum DJ (2008) Food vs. fuel: diversion of crops could cause more hunger. *Environmental Health Perspectives*, **116**, A254–A257.

The Economist (2015) *Climate Change: Clear Thinking Needed*, Print edition, Nov 28th 2015. The Economist Newspaper Limited. London, UK.

The Economist Intelligence Unit (2015) The Global Food Security Index. Available at: http://foodsecurityindex.eiu.com/ (accessed 2 October 2015).

Thurow R, Kilman S (2009) *Enough: Why the World's Poor Starve in an Age of Plenty*. BBS Public Affairs, New York, NY.

Trostle R, Marti D, Rosin S, Wescott P (2011) Why have food commodity prices risen again? Report from the Economic Research Service, USDA, WRS-1103.

UNCTAD (2014) The global biofuels market: energy security, trade and development. Policy Brief No. 30 United Nations Publication. Available at: http://unctad.org/en/PublicationsLibrary/presspb2014d3_en.pdf (accessed 2 October 2015).

United Nations (2015) Transforming our world: the 2030 Agenda for Sustainable Development. United Nations General Assembly Resolution, 2015 September 18. Available at: https://sustainabledevelopment.un.org/post2015/transformingourworld (accessed 13 May 2016).

United Nations Conference on Trade and Development (UNCTAD) (2013) Wake up before it is too late: make agriculture truly sustainable now for food security in a changing climate. Trade and Environmental Review 2013. UN Symbol: UNCTAD/DITC/TED/2012/3.

USDA Economic Research Service (2013) U.S. Drought 2012: Farm and Food Impacts. Available at: http://www.ers.usda.gov/topics/in-the-news/us-drought-2012-farm-and-food-impacts.aspx (accessed 13 May 2016).

USDA Economic Research Service (2015) Definitions of Food Security: Ranges of Food Security and Food Insecurity. U.S. Department of Agriculture. Available at: http://www.ers.usda.gov/topics/food-nutrition-assistance/food-security-in-the-us/definitions-of-food-security.aspx (accessed 5 May 2016).

USDA-GATS (2015) USDA Foreign Agricultural Service's Global Agricultural Trade System data base. Available at: http://apps.fas.usda.gov/GATS/ (7 October 2015).

U.S. Energy Information Administration, International Energy Statistics, Biofuels Production, https://www.eia.gov/cfapps/ipdbproject/IEDIndex3.cfm?tid=79&pid=79&aid=1 (accessed 22 December 2015).

USG (2015) Feed the Future: 2015 Achieving Impact, U.S.Government's Global Hunger and Food Security Initiative. Available at: http://www.feedthefuture.gov/ (accessed 13 May 2016).

Wenzlau S (2013) Global food prices continue to rise. WRI. Available at: http://www.worldwatch.org/global-food-prices-continue-rise-0 (accessed 2 October 2015).

Woods J, Kalas N (2014) Can energy policy drive sustainable land use? lessons from biofuels policy development over the last decade. In: *Plants and BioEnergy* (ed. McCann M, Buckeridge M, Carpita N), pp. 13–33. Springer, New York, NY.

Woods J, Williams A, Hughes JK, Black MJ, Murphy RJ (2010) Energy and the food system. *Philosophical Transactions of the Royal Society of London. Series B, Biological sciences*, **365**, 2991–3006.

Woods J, Lynd LR, Laser M et al. (2015) Land and bioenergy. In: *Bioenergy & Sustainability: Bridging the Gaps* (eds Souza GM, Victoria R, Joly C, Verdade L), Chapter 9. Vol. 72, pp. 258–301.SCOPE, Paris.

World Bank Group (2015) Global Monitoring Report 2014/2015: Ending Poverty and Sharing Prosperity. World Bank, Washington, DC. Available at: http://www.worldbank.org/content/dam/Worldbank/gmr/gmr2014/GMR_2014_Full_Report.pdf (accessed 2 October 2015).

World Food Programme (WFP) (2015) Hunger glossary. Rome, Italy. Available at: https://www.wfp.org/hunger/glossary (accessed 2 October 2015).

World Health Organization (WHO) (2015) World Health Statistics 2015. Global Health Observatory (GHO) data and report. Available at: http://apps.who.int/iris/bitstream/10665/170250/1/9789240694439_eng.pdf?ua=1&ua=1 (accessed 8 October 2015).

World Hunger Education Service (2015) Available at: http://www.worldhunger.org/articles/Learn/world%20hunger%20facts%202002.htm (accessed 2 October 2015).

Wright B (2011) Biofuels and food security: time to consider safety valves? IPC Policy Focus, International Food and Agricultural Trade Policy Council. Available at: http://www.agritrade.org/Publications/BiofuelsandFoodSecurity.html (accessed 8 October 2015).

Zhang Z, Lohr L, Escalante C, Wetzstein M (2010) Food versus fuel: what do prices tell us? *Energy Policy*, **38**, 445–451.

Zilberman D, Hochman G, Rajagopal D, Sexton S, Timilsina G (2013) The impact of biofuels on commodity and food prices: assessment of findings. *American Journal of Agricultural Economics*, **95**, 275–281.

Zumkehr A, Campbell JE (2015) The potential for local croplands to meet US food demand. *Frontiers in Ecology and the Environment*, **13**, 244–248.

14

Competing uses for China's straw: the economic and carbon abatement potential of biochar

ABBIE CLARE[1,2], SIMON SHACKLEY[1], STEPHEN JOSEPH[3,4,5], JAMES HAMMOND[6,7], GENXING PAN[4] and ANTHONY BLOOM[1]

[1]University of Edinburgh, Crew Building, Kings Buildings, Edinburgh EH9 3JN, UK, [2]Scotland's Rural College, Kings Buildings, Edinburgh EH9 3JG, UK, [3]Discipline of Chemistry, University of Newcastle, Callaghan, NSW 2308, Australia, [4]Nanjing Agricultural University, Nanjing 210095, China, [5]School of Materials Science and Engineering, University of New South Wales, Sydney, NSW 2052, Australia, [6]Key Laboratory of Biodiversity and Biogeography of East Asia, Kunming Institute of Botany, Chinese Academy of Sciences, Kunming 650201, China, [7]World Agroforestry Centre, East, and Central Asia, Kunming 650201, China

Abstract

China is under pressure to improve its agricultural productivity to keep up with the demands of a growing population with increasingly resource-intensive diets. This productivity improvement must occur against a backdrop of carbon intensity reduction targets, and a highly fragmented, nutrient-inefficient farming system. Moreover, the Chinese government increasingly recognizes the need to rationalize the management of the 800 million tonnes of agricultural crop straw that China produces each year, up to 40% of which is burned in-field as a waste. Biochar produced from these residues and applied to land could contribute to China's agricultural productivity, resource use efficiency and carbon reduction goals. However competing uses for China's straw residues are rapidly emerging, particularly from bioenergy generation. Therefore it is important to understand the relative economic viability and carbon abatement potential of directing agricultural residues to biochar rather than bioenergy. Using cost-benefit analysis (CBA) and life-cycle analysis (LCA), this paper therefore compares the economic viability and carbon abatement potential of biochar production via pyrolysis, with that of bioenergy production via briquetting and gasification. Straw reincorporation and in-field straw burning are used as baseline scenarios. We find that briquetting straw for heat energy is the most cost-effective carbon abatement technology, requiring a subsidy of $7 $MgCO_2e^{-1}$ abated. However China's current bioelectricity subsidy scheme makes gasification (NPV $12.6 million) more financially attractive for investors than both briquetting (NPV $7.34 million), and pyrolysis ($-1.84 million). The direct carbon abatement potential of pyrolysis (1.06 $MgCO_2e$ per odt straw) is also lower than that of briquetting (1.35 $MgCO_2e$ per odt straw) and gasification (1.16 $MgCO_2e$ per odt straw). However indirect carbon abatement processes arising from biochar application could significantly improve the carbon abatement potential of the pyrolysis scenario. Likewise, increasing the agronomic value of biochar is essential for the pyrolysis scenario to compete as an economically viable, cost-effective mitigation technology.

Keywords: biochar, bioenergy, biomass, briquetting, China, gasification, pyrolysis

Introduction

In the next two decades, China must increase gross agricultural productivity by an estimated 30–50% to keep pace with a growing population and their progressively resource intensive diets (Zhang *et al.*, 2011). Moreover, it must achieve this on arable land that is diminishing in size and fertility due to industrial-contamination of soils (Chen, 2007) and which suffers from low soil organic matter levels (Pan, 2008; Fan *et al.*, 2012).

Correspondence: Abbie Clare
e-mail: Abbie.Clare@ed.ac.uk

Additionally China needs to tackle the current widespread overuse of chemical fertilizers and pesticides, which is leading to significant eutrophication of water bodies (Zhang *et al.*, 2013a), alongside substantial air pollution and associated climate change from anthropogenic emissions of reactive nitrogen (Liu *et al.*, 2013a).

In principal, biochar is a technology that may be able to address many of these challenges. Biochar is the charred by-product of biomass pyrolysis, which is the heating of plant-derived material in the absence of oxygen (Sohi *et al.*, 2009). The pyrolysis process also produces combustible gases (predominantly H_2, CO, CH_4) that can be captured and used for energy (Brown, 2009).

The biochar product has a porous latticed structure, formed from stable aromatic rings of carbon that are more resistant to decomposition than the biomass from which they were initially created. Evidence suggests that fractions of this initial biochar product may stay stable for hundreds (Haberstroh et al., 2006) or even thousands (Masiello, 1998; Lehmann et al., 2008) of years, inferring potential for biochar as a carbon sequestration and climate mitigation tool. Indeed, some studies even suggest that the conversion of available biomass to biochar could reduce annual net global emissions of carbon dioxide, methane and nitrous oxide by 12%, without endangering food security, habitat or soil conservation (Woolf et al., 2010).

In addition to this global warming mitigation potential, biochar also has positive agronomic impacts when applied to agricultural soils, specifically by increasing soil organic carbon (SOC) levels (Kimetu et al., 2008; Zimmerman et al., 2011); stimulating higher crop productivity or maintaining yields with lower input costs (Biederman & Harpole, 2013; Crane-Droesch et al., 2013; Liu et al., 2013b); improving fertilizer-use efficiency (Steiner et al., 2008; Chan & Xu, 2009; Van Zwieten et al., 2010); and/or remediating contaminated soils (Beesley et al., 2011; Bian et al., 2013; Houben et al., 2013). Moreover, China appears to have soils upon which biochar's impact on crop yields may be most significant, as demonstrated in a recent global meta-analysis of biochar studies (Crane-Droesch et al., 2013); research on the decline of SOC in China's soils, particularly on non-paddy land (Lal, 2002; Tang et al., 2006); and many China-based agronomic trials (Zhang et al., 2010; Bian et al., 2013; Lashari et al., 2013).

Additionally, existing biochar systems analyses report strong economic and environmental preferences for the use of waste biomass materials as biochar feedstocks, rather than using wood or other virgin biomass (Roberts et al., 2010a; Shackley et al., 2011). China demonstrates significant potential in this regard, producing an annual 800 million tonnes of agricultural straw residues, of which an estimated 505 million tonnes are available after retaining sufficient straw to maintain soil quality (Jiang et al., 2012). Moreover, many studies report that high proportions of straw are burned in field. For example, Wu et al. (2001) report that 33% of crop straw was burned in Jiangsu province, compared to 32.4% for Guangdong province (Lin & Song, 2002), 40% for Fuzhou city (Yu, 2003), and 39.6% for Shanghai (Yao et al., 2001). This is a consequence of low mechanization rates (Tang et al., 2006) and farmer demographic characteristics, (Cao et al., 2006) with farmers of greater income tending to burn more straw because of reduced demand for straw as a household fuel, and a scarcity of on-farm labour for straw collection. This

in-field straw burning emits high levels of particulate matter (PM), hydrocarbons and other pollutant gases to the atmosphere, resulting in significant local and regional air quality deterioration (Duan et al., 2004; Yan et al., 2006).

However, despite currently being plentiful, these straw residues are increasingly in demand as a result of China's bioelectricity subsidies. Recognizing the adverse environmental and health consequences of in-field straw burning, the Chinese government is providing financial incentives to promote the mechanized collection and conversion of straw to electrical energy that is fed into the national grid. The financial incentives offered are structured as a feed-in-tariff ($0.12 kWh^{-1} produced from agricultural and waste forestry biomass), subsidized loans, tax breaks and/or grants (Zhang et al., 2014). The feed-in-tariff rate is comparable to western bioenergy policies, (for example, UK energy companies can typically sell renewably-generated electricity for between $0.08 and 0.25 kWh^{-1}), however opinion is divided on whether these incentives are sufficient to create economically viable bioenergy projects (Lu & Zhang, 2010a; Zhang et al., 2013b, 2014).

In addition the extent to which these bioenergy subsidies might affect the economic viability of biochar projects is unknown. This therefore raises questions about how the agronomic results of biochar field trials translate into the development of biochar as a commercial product, and additionally whether commercial biochar projects can contribute to GHG emission reductions in China.

We therefore investigate and contrast the economics and carbon abatement potential of using China's straw resources for biochar production via pyrolysis with two bioenergy technologies: straw briquetting and straw gasification. These scenarios are compared against two reference cases (straw reincorporation and in-field straw burning) and are analyzed in terms of their relative profitability from a business perspective, and in terms of their environmental benefits from a global GHG balance perspective.

Materials and methods

Cost-benefit analysis (CBA) is used to compare the economic viability [net present value (NPV) per oven dry tonne (odt) straw], and life cycle analysis (LCA) is used to compare the environmental (MgCO$_2$e per odt straw) outcomes associated with three straw utilization scenarios: straw briquetting and subsequent combustion for heat energy (S$_{Briq}$); straw gasification for electrical energy (S$_{Gas}$); and straw pyrolysis for biochar and electrical energy (S$_{Pyr}$). These are compared to two baselines of straw reincorporation (S$_{Rein}$) and straw burning (S$_{Burn}$). S$_{Rein}$ assumes that all straw is incorporated into the field whereas S$_{Burn}$ assumes that straw is burned in-field.

Technology scenario selection

Straw briquetting (S_{Briq}) was chosen as a comparison scenario based on observations of straw briquettes on sale in Chinese town markets and online. Briquetting has much lower capital and technological expertise requirements than gasification and pyrolysis, and is therefore likely to be perceived as lower risk by investors and as an accessible option for small businesses. However it does not qualify for government bioelectricity subsidies, as briquettes tend to be bought for local heat and cooking applications rather than burned for commercial electricity generation. In contrast, straw gasification (S_{Gas}) was chosen on the basis that gasification is identified as a priority bioenergy technology in Chinese national policy documents (Han *et al.*, 2008; Zhang *et al.*, 2014), has been implemented in many technological development projects across China (Kirkels & Verbong, 2011), and is reportedly a viable economic proposition for Chinese businesses (Lu & Zhang, 2010a). Although co-firing with coal has also been found to be an economic use of straw residues, (Lu & Zhang, 2010a), it was not included as an option because the Chinese government does not currently provide financial incentives for bioelectricity produced through co-firing. This is due to concerns over the accurate verification of biomass co-firing rates at existing coal-fired power stations (Gan & Yu, 2008; Dong, 2012).

The pyrolysis (S_{Pyr}) scenario investigates the use of slow pyrolysis technology to produce biochar and a relatively small amount of electricity. Slow pyrolysis always delivers less electricity than other bioenergy options, because a proportion of the feedstock is converted to biochar and not into heat or electrical energy (Brown, 2009).

Each of the S_{Briq}, S_{Gas} and S_{Pyr} technology scenarios is guided by interviews conducted in summer 2012 at the Sanli New Energy bioenergy-plant in Henan Province, China. Sanli New Energy has capitalized upon the combination of a local straw-burning ban, related straw-burning avoidance subsidies ($28 Mg^{-1}$ straw paid to businesses that use straw for livestock rearing, paper production or bioenergy generation) and national bioelectricity subsidies (Zhang *et al.*, 2014), to build a 4 MW pyrolysis unit and straw briquetting plant. Data on Sanli's economics, straw collection system and size guided the choice of parameters used to structure and assess the S_{Briq}, S_{Gas} and S_{Pyr} scenarios. Table 1 provides an overview of these parameters. More detailed information on technology configuration is available in the Data S1 (S9–S17, and Figures S1 and S2).

The Technology Readiness Levels (TRL) for each technology (straw briquetting, gasification and pyrolysis) are also estimated, based on expert opinion and observations of the deployment of these technologies in rural Chinese settings. A TRL is a scale from one to nine that indicates the maturity of a given technology (Mankins, 1995; UK Ministry of Defence, 2014). Table S1 provides a description for each TRL. Briquetting scores the highest (9), as a mature 'off the peg' technology, followed by gasification at stages 7–8, and then pyrolysis at stages 5–6.

Cost benefit analysis

Published literature, industry reports, policy documents, interviews and online market estimates were used to develop appropriate pricing structures for S_{Briq}, S_{Gas} and S_{Pyr}, adjusted to 2014 prices. The CBA combines these values to generate an estimate of scenario profitability from the perspective of a business or potential investor, taking account of government bioelectricity and avoided straw burning subsidies.

The agronomic value for biochar is estimated by combining data on the microeconomics of farms in Henan (Clare *et al.*, 2014) with data from the latest published meta-analyses on biochar's yield impacts (Jeffery *et al.*, 2011; Crane-Droesch *et al.*, 2013), the findings from which are also consistent with results from China-based biochar experiments (Wang *et al.*, 2012; Zhang *et al.*, 2012a). Biochar's agronomic value is calculated as the value of the yield improvement seen in one growing year, per unit of biochar applied, assuming that biochar is applied once and that its effects last across two growing seasons. It should be noted that this estimate does not take spreading and transportation costs into account, and that therefore the commercial sale price of biochar to farmers will need to be less

Table 1 Overview of technical parameters for briquetting, gasification and pyrolysis

	Briquetting	Gasification	Pyrolysis
Technology Readiness Level (TRL)	9	7–8	5–6
Lifetime of operation (yrs)	20	20	20
Straw processed (odt yr^{-1})	28 000	28 000	28 000
Annual output	28 000 Mg briquettes	26 680 MWh bioelectricity	8400 MWh bioelectricity; 8300 Mg biochar
Energy offset	Equivalent MJ heat energy from coal briquettes	Equivalent MWh electrical energy from central China's grid	Equivalent MWh electrical energy from central China's grid
National bioelectricity subsidies	None	Feed-in-tariff for bioelectricity ($0.12 kWh^{-1}); subsidized capital loans; tax breaks (Zhang *et al.*, 2014);	Feed-in-tariff for bioelectricity ($0.12 kWh^{-1}); subsidized capital loans; tax breaks (Zhang *et al.*, 2014);
Local straw-burning subsidies	Avoided straw burning ($28 Mg^{-1})	Avoided straw burning ($28 Mg^{-1})	Avoided straw burning ($28 Mg^{-1})

than this figure. The baseline agronomic value for biochar of $110 Mg^{-1} is calculated according to the latest meta-analysis by Crane-Droesch *et al.* (2013), who report a 10% yield increase for a 3 Mg ha^{-1} application rate. However, the more conservative estimate of Jeffery *et al.* (2011), assuming that a 10 Mg ha^{-1} application stimulates 10% yield increases, gives biochar an agronomic value of just $33 Mg^{-1}. This is a significant price difference, and therefore the retail price of biochar is varied in the sensitivity analysis, reflecting this uncertainty and investigating the extent to which it impacts the overall profitability of S_{Pyr}.

Briquette market value is calculated based on the typical energy density of straw briquettes (McKendry, 2002; Roberts *et al.*, 2010b) and the value of this energy based on the spot price of coal in China at the time of writing ($95 Mg^{-1}; Zhao & Che, 2012; Yang, 2014). Finally, the market value of bioelectricity is set in line with the current Chinese bioelectricity subsidy of $0.12 kWh^{-1} (Zhang *et al.*, 2014).

The NPV of each scenario is calculated at the project level, over a 20 year lifetime, taking subsidized loans and tax breaks into account where relevant (Zhang *et al.*, 2014). The discount rate is set at 3.5% (Federal Reserve Bank of St Louis, 2014).

Life-cycle analysis

A GHG-oriented attributional LCA was performed, based on the ISO 14040 (2006) guidelines, and using a 100 year global warming potential. The three main GHGs were accounted for [carbon dioxide (CO_2), methane (CH_4), and nitrous oxide (N_2O)], and these are henceforth displayed in terms of their carbon dioxide equivalent global warming potential (CO_2e), calculated according to IPCC guidelines of CO_2e equivalence as 25 for CH_4 and 298 for N_2O (IPCC, 2007). The GHG abatement potentials of S_{Briq}, S_{Gas} and S_{Pyr} were calculated using S_{Rein} as the baseline scenario, however the S_{Burn} scenario is also displayed for reference. The analysis initially focuses on directly-attributable CO_2e emissions from each phase of the life cycle (raw material acquisition, production, distribution, energy offset and dismantling processes) before moving on to consider the indirect CO_2e abatement potential of reduced soil N_2O emissions and avoided fertilizer use as a result of biochar application.

Soil N_2O reductions following biochar application have been widely debated for some years, however a recent meta-analysis (Cayuela *et al.*, 2014) provides greater clarity on the extent of this effect. Cayuela *et al.*, report that biochars derived from woody and herbaceous feedstocks, including agricultural straws, demonstrate the highest emission reduction potential, with a 27% reduction in N_2O emissions for a 1–2% (by soil weight) biochar application rate. Data from this study is then combined with a China-specific field trial demonstrating a similar effect (Zhang *et al.*, 2012b) to calculate the additional contribution that N_2O emission reduction may have on the S_{Pyr} LCA result.

A similar approach is taken to calculating additional GHG abatement as a result of avoided fertilizer application. Recent trials in China suggest that the application of a combined biochar-NPK-clay compound [a biochar-mineral-chemical-composite (BMCC)] may be an economic option for farmers, where

~25% of NPK is replaced by biochar, on a weight basis (Joseph *et al.*, 2013). This data is combined with data on the carbon intensity of China's domestic fertilizer production industry, which emits 13.5 $MgCO_2e$ MgN^{-1} fertilizer as compared to an average of 9.7 $MgCO_2e$ MgN^{-1} in Europe (Zhang *et al.*, 2013c). The nitrogen (N) fertilizer is assumed to contribute to a standard NPK (16:16:16) mix. Emissions from potassium (K) and phosphorus (P) production in synthetic fertilizers are excluded, as they are an order of magnitude lower (West & Marland, 2002). Figure 1 displays the processes included in the direct and indirect abatement potential calculations.

The CO_2e offsets from avoided fossil fuel energy are calculated according to the carbon emission factor (CEF) of the fuel that straw-derived bioenergy is expected to replace. Straw briquettes are assumed to replace coal briquettes that are typically burned for heat and/or cooking purposes in local applications such as homes, schools and hospitals. In S_{Gas} and S_{Pyr}, each MWh of bioelectricity produced is assumed to replace one MWh of electricity in the central grid, which services Henan province and has an estimated carbon intensity of 1.133 $MgCO_2e$ MWh^{-1} (World Resources Institute, 2014).

The details of GHG emissions associated with different phases of the lifecycle are given in the supplementary material (S9–S17). Many of the parameters used to estimate these emissions are considered uncertain, therefore published literature and expert opinion were also used to estimate the uncertainty range and probability distribution of each parameter. An uncertainty analysis was then undertaken using a Monte Carlo method. 10 000 simulations were performed to derive median points and 95% confidence intervals for $MgCO_2e$ emitted per odt feedstock. The impact of each parameter's value on the final result was investigated using sensitivity analysis.

Results

Economic viability of briquetting, gasification and pyrolysis

Removing both national bioelectricity and local avoided straw-burning subsidies renders S_{Briq}, S_{Gas} and S_{Pyr} unprofitable, with project NPVs of $-2.88 million (m), $-19.0 m, and $-20.3 m, respectively (see black bars in Fig. 2). When including local avoided straw burning subsidies (see grey bars in Fig. 2), S_{Briq} becomes profitable (NPV $7.34 m), whereas S_{Gas} and S_{Pyr} still generate significant losses (NPV $-8.14 m and $-9.36 m respectively). However, the inclusion of income from China's national bioelectricity subsidy programme (see white bars in Fig. 2) has a significant impact on S_{Gas} profitability (NPV $12.60 m), increasing it above the unchanged S_{Briq} NPV ($7.34 m). Meanwhile, S_{Pyr} remains unprofitable (NPV $-1.84 m), due to the relatively lower electricity volume yielded per odt straw by pyrolysis as compared with gasification.

However the NPV of S_{Pyr} is strongly influenced by the agronomic value of biochar, which is one of the

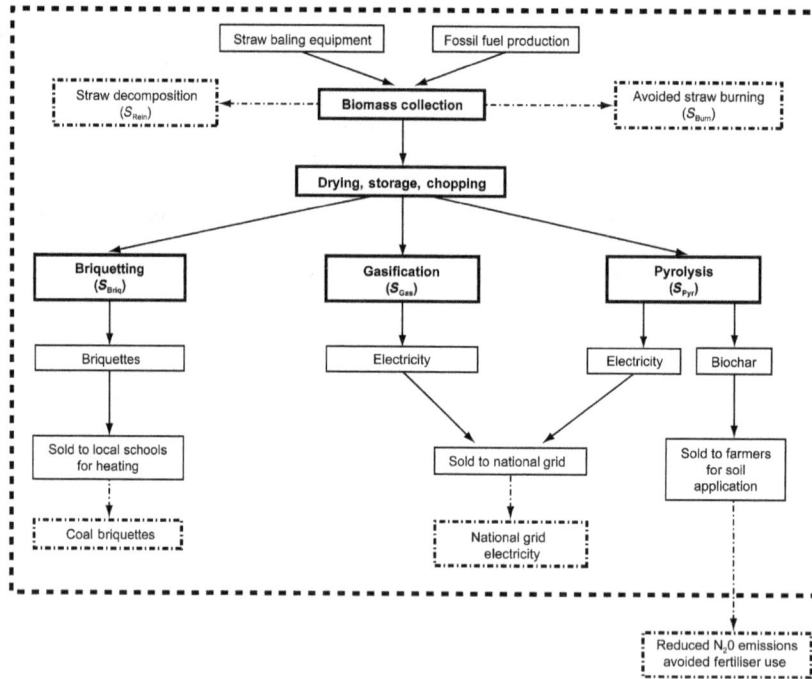

Fig. 1 Diagram of LCA boundaries: bold boxes indicate processes that emit CO_2e, dashed boxes indicate CO_2e offset or abatement processes. Processes within the bold dashed line are considered direct impacts of each scenario, and processes outside the bold dashed line are considered indirect impacts.

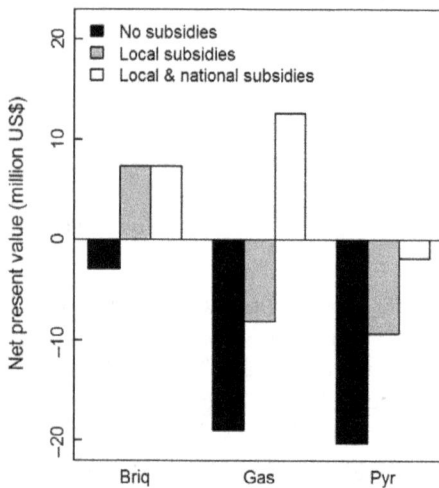

Fig. 2 Net present value (million US$), with and without Chinese government subsidies, for S_{Briq}, S_{Gas} and S_{Pyr}.

most uncertain parameters modelled in this CBA. At the baseline agronomic value of $110 Mg^{-1}$, (based on the results of Crane-Droesch et al. (2013)) the S_{Pyr} NPV (including all available subsidies) is $-1.84 m. However, assuming the more conservative estimate of $33 Mg^{-1}$, (based on the results of Jeffery et al. (2011)) the S_{Pyr} NPV drops even further to $-10.1 m. For S_{Pyr} to break even, biochar must sell for $128 Mg^{-1}$ if all

other factors remain equal, or for $206 Mg^{-1}$, if bioenergy subsidies are excluded. For the NPV of S_{Pyr} to equal that of S_{Gas}, biochar must sell for $238 Mg^{-1}$. Interestingly, in 2014 Sanli New Energy Company reported their biochar retail price as $259 Mg^{-1}$, which exceeds the break-even prices that we report as being necessary for pyrolysis profitability. However, this high sale price is at odds with current understanding of biochar's agronomic value in soil (as outlined above) and studies on agricultural economics and farmer-perspectives of biochar in the area (Clare et al., 2014).

Direct CO_2e abatement potential of briquetting, gasification and pyrolysis

Figure 3 outlines the CO_2e abatement potential of S_{Burn}, S_{Briq}, S_{Gas} and S_{Pyr}, including only direct processes in the analysis, all implicitly compared against S_{Rein} as the baseline scenario. The results suggest that, when including offsets from avoided fossil-fuel energy emissions (see black bars in Fig. 3), S_{Briq} offers the greatest carbon abatement (1.35 $MgCO_2e$ per odt straw) followed by S_{Gas} (1.16 $MgCO_2e$ per odt straw) and S_{Pyr} (1.06 $MgCO_2e$ per odt straw). This carbon abatement potential increases by 0.04 $MgCO_2e$ per odt straw for each scenario, if referenced to the S_{Burn} baseline rather than S_{Rein}. Interestingly this means that, despite only

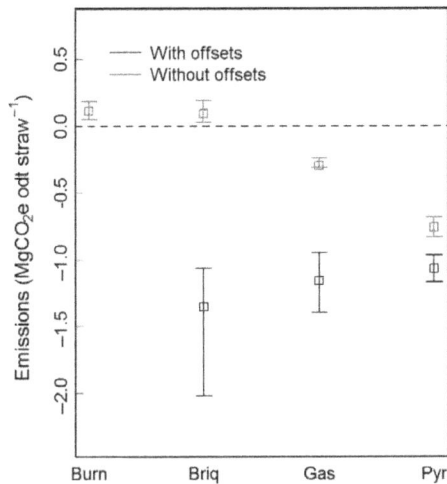

Fig. 3 Median and confidence interval estimates of $MgCO_2e$ abated per odt straw processed in S_{Briq}, S_{Gas} and S_{Pyr}, including and excluding offsets from avoided fossil-fuel energy (black bars and grey bars, respectively). Uses S_{Rein} as the baseline, and displays S_{Burn} for reference.

receiving local and not national subsidies, S_{Briq} appears to offer the greatest CO_2e abatement potential. However, S_{Briq} also displays the most variance in its carbon abatement, as a result of the wide variability in data available for comparing emissions from straw and coal briquettes in small stoves (Zhang et al., 2000; Wang et al., 2013).

If emissions offsets from avoided fossil fuel use are not included (see grey bars in Fig. 3), both S_{Gas} and S_{Pyr} still provide some carbon abatement. In the case of S_{Gas} this is because approximately 20% of feedstock carbon is initially stabilized in the ashy char produced during the gasification process (Lu & Zhang, 2010b) with 90% remaining stable over the 100 year time-scale of this analysis (Cross & Sohi, 2013). In the case of S_{Pyr}, 50% of feedstock carbon is initially stabilized in the biochar, with 80% of that amount (39% of the initial feedstock carbon) still remaining in the soil after 100 years (Singh et al., 2012; Crombie et al., 2013). This persistence is a pertinent point, as it can be argued that offset fossil fuel emissions are not avoided for long, because the fossil fuel still remains to be consumed. From these perspectives, it can therefore be argued that S_{Pyr} offers a more permanent GHG reduction than the other options.

Indirect CO2e abatement potential of pyrolysis

The application of biochar to agricultural land may contribute to the abatement potential of S_{Pyr} via indirect processes, which generally have a higher level of uncertainty and variability than the direct factors already

discussed. This can result from reduced certainty regarding biochar's impact on a given outcome (i.e., in the case of biochar's effect on N_2O emissions) and/or because the process relies on human behaviour change (i.e., the reduction in fertilizer application, or the application of biochar to land). Indirect environmental consequences of biochar application have been variously reported in past LCA studies (Roberts et al., 2010a; Hammond et al., 2011; Sparrevik et al., 2013), but recent evidence has improved the evidence base for the effect magnitude that might be expected for a given biochar application rate. Specifically, two indirect effects that have received increased attention are reduced N_2O emissions from soil and improved fertilizer use efficiency.

Reduced N_2O emissions from soil. Table 2 combines data from a recent meta-analysis of biochar's impact on soil N_2O emissions (Cayuela et al., 2014) with the baseline and reduced N_2O emission reductions reported in a China-based biochar field trial (Zhang et al., 2012a). According to these data, and assuming a one-year effect of biochar on N_2O emissions, the abatement potential of S_{Pyr} could be increased by 0.004–0.012 $MgCO_2e$ yr^{-1}. This represents a 1% increase in S_{Pyr}'s abatement potential, and we therefore suggest that the absolute contribution of biochar-induced soil N_2O emission reductions is relatively small.

Improved fertilizer use efficiency. If biochar were to aid the reduction of fertilizer application in China, the resulting GHG mitigation potential is large. Using data from Joseph et al. (2013) and Zhang et al. (2013c) we calculate that each Mg of biochar that replaces chemical fertilizer could abate an additional 1.33 $MgCO_2e$, and thus that each odt of straw feedstock being used to produce biochar could abate an additional 0.39 $MgCO_2e$.

Including these indirect effects of biochar application on avoided emissions from soil N_2O and fertilizer use reduction, the total abatement potential of S_{Pyr} increases to 1.46 $MgCO_2e$ per odt straw, which puts it ahead of

Table 2 Calculations of avoided N_2O emissions per tonne feedstock pyrolysed

Biochar application rate (%)	0.5*	2*	1–2†
% N_2O reduction from baseline	−40	−51	−27
N_2O avoided (kg per odt)	0.021	0.007	0.007
Abatement potential ($MgCO_2e$ per odt)	0.012	0.004	0.004

*Data taken from Zhang et al. (2012a).
†Data taken from Cayuela et al. (2014).

both S_{Gas} (1.16 MgCO$_2$e per odt straw) and S_{Briq} (1.35 MgCO$_2$e per odt straw) in terms of carbon abatement.

Sensitivity analysis

Figures 4 and 5 graphically display the results of sensitivity analysis undertaken on key parameters influencing the NPV and carbon abatement potential, respectively, of the S_{Briq}, S_{Gas} and S_{Pyr} scenarios. Both figures present the baseline NPV/carbon abatement value and a surrounding range, calculated by varying key economic/carbon abatement parameters by ±20%, whilst keeping all other parameter values constant. The parameter values used in these sensitivity analyses are available in S19 and S20 of Data S1.

Figure 4 displays the influence of the following economic parameters on the overall NPV for each scenario:

straw price, local straw burning subsidies, capital cost, labour cost, and the sale price of outputs (briquettes; electricity; electricity and biochar, for S_{Briq}, S_{Gas} and S_{Pyr}, respectively). All NPVs displayed include the financial support currently available from both local and national subsidy programmes.

The results in Fig. 4 suggest that sales prices for output products are very influential on the overall economic viability of briquetting, gasification and pyrolysis projects. Likewise, varying the capital cost of pyrolysis and gasification units has a significant impact on the economic viability of S_{Gas} and S_{Pyr}, even tipping S_{Pyr} into profitability where capital costs alone decrease by 20%. This is particularly relevant when considering the early stage of technological readiness of pyrolysis and the subsequent drop in capital cost that might be expected as this technology reaches higher stages of

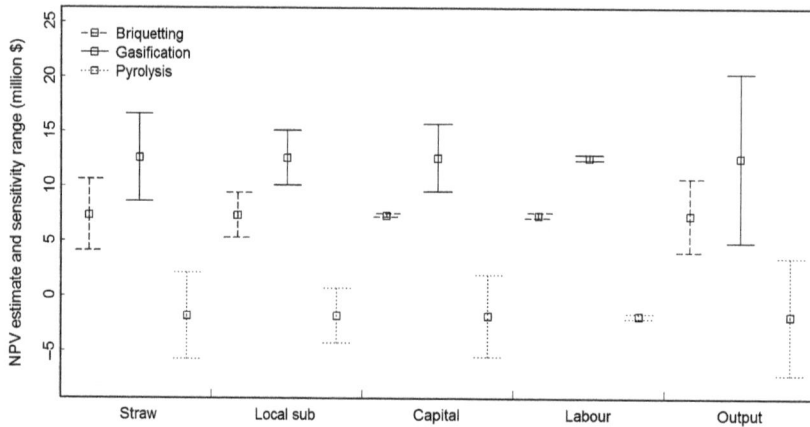

Fig. 4 Baseline NPV estimates (million US$) and sensitivity analyses for key parameters determining the economic viability of briquetting, gasification and pyrolysis. Ranges are produced by independently varying key parameters (*x*-axis) by ±20% and recording the impact on the overall NPV value.

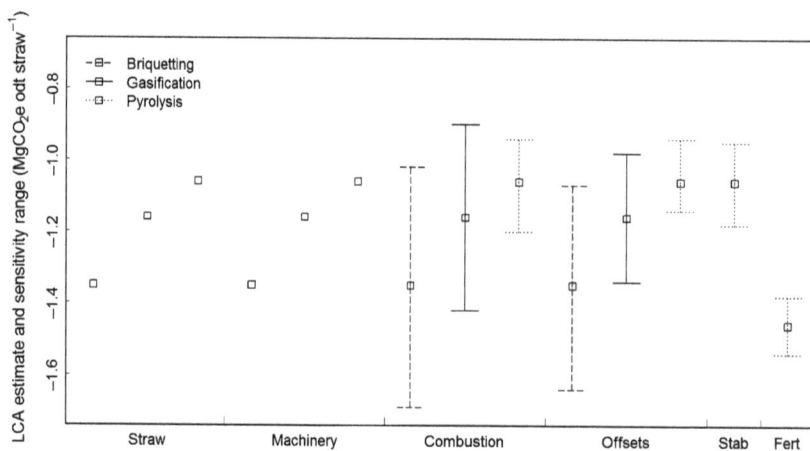

Fig. 5 Baseline carbon abatement estimates (MgCO$_2$e abated odt per straw) and sensitivity analyse for key parameters determining the carbon abatement potential of briquetting, gasification and pyrolysis. Ranges are produced by independently varying key parameters (*x*-axis) by ±20% and recording the impact on overall carbon abatement potential.

maturity (Utterback, 1996; Shackley *et al.*, in press). However, it must also be noted that the top range of S_{Pyr}'s NPVs do not overlap with the bottom range of the NPVs of S_{Briq} or S_{Gas}, suggesting that pyrolysis will require significant improvements in multiple economic parameters before it can compete with briquetting or gasification.

Figure 5 displays the results of a ±20% sensitivity analysis conducted on the following key parameters influencing the carbon abatement potential of S_{Briq}, S_{Gas} and S_{Pyr}: straw collection emissions; embedded emissions within machinery; direct emissions from the combustion of straw briquettes/gasification of straw/ pyrolysis of straw; offset emissions from avoided fossil fuel energy; the stability of carbon sequestered within biochar; and offset emissions from avoided fertilizer use. The results suggest that direct emissions from combustion of straw briquettes/gasification of straw/pyrolysis of straw, and offset emissions from avoided fossil fuel use, have the greatest impact on the carbon abatement potential of each scenario. This suggests that gasification and pyrolysis units must be well designed, maintained and managed by staff with appropriate expertise, and that improvements to the efficiency of boilers that combust straw briquettes could also improve their carbon abatement potential. Variation in emissions from straw collection and machinery/building construction has a negligible impact on overall carbon abatement balance. However, variability in fertilizer use and the stability of carbon sequestered within biochar have modest effects on the overall carbon abatement potential of S_{Pyr}.

Carbon abatement cost-effectiveness

In light of the Chinese government's carbon intensity reduction targets, it is important to consider the cost-effectiveness of S_{Briq}, S_{Gas}, and S_{Pyr} in terms of CO_2e abatement. Our results show that all three technologies require assistance from carbon pricing to break-even, although S_{Briq} requires a significantly lower price than S_{Gas} and S_{Pyr} (see Table 3, where S_{Pyr} (D) includes only direct effects and S_{Pyr} (I) includes both direct and indirect processes discussed in this paper). Requiring a car-

bon price of $7 $MgCO_2e$ per abated, S_{Briq} is the only technology studied here that can produce carbon abatement for less than $25 $MgCO_2e^{-1}$, as outlined in the Stern Report (Stern, 2006). Moreover, early price indications from China's nascent emissions trading scheme (which currently covers five municipal areas and two provinces; (Lo, 2012) suggest that domestic carbon prices (currently ranging between $5 and $20 $MgCO_2$ per abated) would only provide sufficient support to make S_{Briq} profitable (Song & Lei, 2014).

Discussion

We find that the briquetting of straw for sale as a local fuel in heating and cooking appliances to be the most efficient use of China's straw residue resources. S_{Briq} has the greatest carbon abatement potential (1.35 $MgCO_2e$ per odt straw as compared to 1.16 and 1.06 $MgCO_2e$ per odt for S_{Gas} and S_{Pyr}, respectively), and the highest economic abatement efficiency (requiring a relatively small carbon price of $7 $MgCO_2e^{-1}$ abated, compared to $61 $MgCO_2e^{-1}$ or $51–71 $MgCO_2e^{-1}$ abated, for S_{Gas} and S_{Pyr}, respectively.) Straw briquetting also has the highest technology readiness level (TRL), making it attractive for small businesses and village level industry. This technology also leads to the direct use of biomass energy for heat in boilers and heating systems of local communities, thus negating the need for expensive equipment and avoiding the inevitable energy wastage when converting heat energy into electricity.

However, the apparent success of straw briquetting is subject to two important caveats. Firstly, this scenario relies on the sale of straw briquettes to local households, schools and hospitals for combustion in relatively inefficient, small-scale boilers and stoves. However, as China's energy system modernizes, there may be a move towards more efficient district heating and power systems, which will reduce market demand for straw briquettes to be processed and sold in this way. Secondly, the heat energy produced from locally sold briquettes is not as fungible as electricity, which is socially a highly valued commodity.

This may explain why current Chinese bioenergy subsidies focus on bioelectricity generation, and supports our finding that national bioelectricity subsidies increase the NPV of gasification (NPV $12.60 m) above that of briquetting (NPV $7.34 m). However, pyrolysis remains unprofitable even when receiving local and national subsidy support (NPV $−1.84 m). For pyrolysis and associated biochar production to be able to compete with alternative uses of feedstock such as briquetting and gasification, the agronomic value of biochar will need to increase considerably. The current evidence suggests that biochar has an agronomic value of

Table 3 Comparing CO_2e abatement cost effectiveness for briquetting, gasification and pyrolysis

	S_{Briq}	S_{Gas}	S_{Pyr} (D)	S_{Pyr} (I)
Subsidy required ($ tonne feedstock per processed)	5	34	36	36
Subsidy required ($ $MgCO_2e^{-1}$ abated)	7	61	71	51

approximately $110 Mg^{-1} in central, grain-growing Chinese provinces such as Henan (Crane-Droesch *et al.*, 2013; Clare *et al.*, 2014). However, we find that biochar must sell for at least $238 Mg^{-1} in the presence of subsidies for the NPV of S_{Pyr} to equal that of S_{Gas}, which is far above what current research suggests is its agronomic value in the first year after application. Moreover, our LCA analysis suggests that pyrolysis is unlikely to attract financial support from the Chinese government on carbon abatement grounds alone, unless the abatement potential of indirect processes such as avoided fertilizer use are included and can be increased.

There are three considerations that may affect these findings. The first relates to the indirect mitigation potential of avoided fertilizer use. In fact, fertilizer application rates in China are so high that fertilizer application can be reduced by up to 27% with no impact on yields, and without requiring biochar application (Huang *et al.*, 2008). This calls into question the necessity of biochar to stimulate this particular indirect carbon abatement mechanism because, although replacement of NPK with biochar to produce a biochar-mineral-chemical-composite (BMCC) could theoretically reduce fertilizer application rates (Joseph *et al.*, 2013; Clare *et al.*, 2014), biochar is not essential to achieving this goal.

Secondly, there are anecdotal reports of two factories in central China producing 60 000Mg yr^{-1} of BMCC products for local agricultural markets. Field trials in China have recently suggested that BMCCs (which premix low application rates of biochar with inorganic fertilizer and clay) can produce yield increases of up to 40% (Joseph *et al.*, 2013). Applying this data to agricultural market conditions in Henan province, biochar's value as a soil amendment would be $5740 Mg^{-1}, increasing the S_{Pyr} NPV to over 50 times that of S_{Gas}. If these results are reproducible, this is a significant game-changer for the field of biochar research and application, however extensive field trials are necessary to ensure that such impacts can be replicated consistently.

Thirdly, the technological advancement, appropriate management and successful deployment of pyrolysis and gasification technologies will have an important impact both on their carbon abatement and economic potential. Improved technological maturity and deployment should improve the conversion efficiency from straw energy to electrical energy and/or biochar. This is a significant determinant of the overall economic viability and emissions balance of S_{Gas} and S_{Pyr}, both by increasing the units of economic output produced per unit of feedstock, and by avoiding emissions of strong climate forcing GHGs resulting from incomplete combustion. Also, the 'technological readiness' of pyrolysis currently lags behind gasification, making it potentially

more risky and less attractive for investors. As such, innovative technological advancements are needed for pyrolysis technology to compete with gasification and briquetting, both in terms of economic viability and carbon abatement potential.

Whether Chinese policy makers provide financial support to the advancement of pyrolysis technology is likely to depend on the outcomes of biochar trials related to land remediation [currently a significant issue in China (Chen, 2007; Khan *et al.*, 2008; Bian *et al.*, 2013)] and to BMCCs for food production. If early BMCC trial results of high yield impacts for low biochar application and reduced fertilizer application rates (Joseph *et al.*, 2013) are further substantiated, the Chinese government may consider pyrolysis/biochar technology as something that policy should support, even if this increases competition for straw feedstocks in China.

Acknowledgements

This work was supported by the Policy Innovation Systems for Clean Energy Security (PISCES) programme and the Sustainable Agriculture Innovation Network (SAIN), both of which are funded by the UK Department for International Development (DFID). It was also funded by the EPSRC (Science and Innovation Award) and Scotland's Rural College (SRUC). The authors would also like to thank Meng Qingnan for her assistance with data collection; Saran Sohi, Andrew Cross, and colleagues at the UK Biochar Research Centre for their constructive feedback; and the very helpful comments from anonymous reviewers of this article.

References

Beesley L, Moreno-Jiménez E, Gomez-Eyles JL, Harris E, Robinson B, Sizmur T (2011) A review of biochars' potential role in the remediation, revegetation and restoration of contaminated soils. *Environmental Pollution*, 159, 3269–3282.

Bian R, Chen D, Liu X *et al.* (2013) Biochar soil amendment as a solution to prevent Cd-tainted rice from China: results from a cross-site field experiment. *Ecological Engineering*, 58, 378–383.

Biederman L, Harpole WS (2013) Biochar and its effects on plant productivity and nutrient cycling: a meta-analysis. *Global Change Biology Bioenergy*, 5, 202–214.

Brown R (2009) Biochar production technology. In: *Biochar for Environmental Management - Science and Technology* (eds Lehmann J, Joseph S), pp. 127–146. Earthscan, London, UK.

Cao G, Zhang X, Zheng F (2006) Inventory of black carbon and organic carbon emissions from China. *Atmospheric Environment*, 40, 6516–6527.

Cayuela ML, van Zwieten L, Singh BP, Jeffery S, Roig A, Sánchez-Monedero M (2014) Biochar's role in mitigating soil nitrous oxide emissions: a review and meta-analysis. *Agriculture, Ecosystems & Environment*, 191, 5–16.

Chan K, Xu Z (2009) Biochar–nutrient properties and their enhancement. In: *Biochar for Environmental Management: Science and Technology* (eds Lehmann J, Joseph S), pp. 67–84. Earthscan, London, UK.

Chen J (2007) Rapid urbanization in China: a real challenge to soil protection and food security. *Catena*, 69, 1–15.

Clare A, Barnes A, McDonagh J, Shackley S (2014) From rhetoric to reality: farmer perspectives on the economic potential of biochar in China. *International Journal of Agricultural Sustainability*, 12, 440–458.

Crane-Droesch A, Abiven S, Jeffery S, Torn MS (2013) Heterogeneous global crop yield response to biochar: a meta-regression analysis. *Environmental Research Letters*, 8, 44–49.

Crombie K, Mašek O, Sohi SP, Brownsort P, Cross A (2013) The effect of pyrolysis conditions on biochar stability as determined by three methods. *Global Change Biology Bioenergy*, 5, 122–131.

Cross A, Sohi SP (2013) A method for screening the relative long-term stability of biochar. *Global Change Biology Bioenergy*, 5, 215–220.

Dong N (2012) *Support Mechanisms for Cofiring Secondary Fuels*. IEA Clean Coal Centre, London, UK.

Duan F, Liu X, Yu T, Cachier H (2004) Identification and estimate of biomass burning contribution to the urban aerosol organic carbon concentrations in Beijing. *Atmospheric Environment*, 38, 1275–1282.

Fan M, Shen J, Yuan L, Jiang R, Chen X, Davies WJ, Zhang F (2012) Improving crop productivity and resource use efficiency to ensure food security and environmental quality in China. *Journal of Experimental Botany*, 63, 13–24.

Federal Reserve Bank of St Louis (2014) Discount rate for China. Available at: http://research.stlouisfed.org/fred2/series/INTDSRCNM193N (accessed 6 March 2014).

Gan L, Yu J (2008) Bioenergy transition in rural China: policy options and co-benefits. *Energy Policy*, 36, 531–540.

Haberstroh PR, Brandes JA, Gélinas Y, Dickens AF, Wirick S, Cody G (2006) Chemical composition of the graphitic black carbon fraction in riverine and marine sediments at sub-micron scales using carbon X-ray spectromicroscopy. *Geochimica et Cosmochimica Acta*, 70, 1483–1494.

Hammond J, Shackley S, Sohi S, Brownsort P (2011) Prospective life cycle carbon abatement for pyrolysis biochar systems in the UK. *Energy Policy*, 39, 2646–2655.

Han J, Mol APJ, Lu Y, Zhang L (2008) Small-scale bioenergy projects in rural China: lessons to be learnt. *Energy Policy*, 36, 2154–2162.

Houben D, Evrard L, Sonnet P (2013) Mobility, bioavailability and pH-dependent leaching of cadmium, zinc and lead in a contaminated soil amended with biochar. *Chemosphere*, 92, 1450–1457.

Huang J, Hu R, Cao J, Rozelle S, Korea S (2008) Training programs and in-the-field guidance to reduce China's overuse of fertilizer without hurting profitability. *Journal of Soil and Water Conservation*, 63, 163–165.

IPCC (2007) Climate change 2007: working group I: the physical science basis. IPCC Fourth Assessment Report: Climate Change.

ISO (2006) *ISO 14040: Environmental Management—Life Cycle Assessment— Principles and Framework*. International Standards Organisation, Geneva.

Jeffery S, Verheijen FG, van der Velde M, Bastos AC (2011) A quantitative review of the effects of biochar application to soils on crop productivity using meta-analysis. *Agriculture, Ecosystems & Environment*, 144, 175–187.

Jiang D, Zhuang D, Fu J, Huang Y, Wen K (2012) Bioenergy potential from crop residues in China: availability and distribution. *Renewable and Sustainable Energy Reviews*, 16, 1377–1382.

Joseph S, Graber E, Chia C et al. (2013) Shifting paradigms on biochar: micro/nanostructures and soluble components are responsible for its plant growth promoting ability. *Carbon Management*, 4, 323–343.

Khan S, Cao Q, Zheng YM, Huang YZ, Zhu YG (2008) Health risks of heavy metals in contaminated soils and food crops irrigated with wastewater in Beijing, China. *Environmental Pollution (Barking, Essex: 1987)*, 152, 686–692.

Kimetu JM, Lehmann J, Ngoze SO et al. (2008) Reversibility of soil productivity decline with organic matter of differing quality along a degradation gradient. *Ecosystems*, 11, 726–739.

Kirkels AF, Verbong GPJ (2011) Biomass gasification: still promising? A 30-year global overview. *Renewable and Sustainable Energy Reviews*, 15, 471–481.

Lal R (2002) Soil carbon sequestration in China through agricultural intensification, and restoration of degraded and desertified ecosystems. *Land Degradation & Development*, 13, 469–478.

Lashari MS, Liu Y, Li L et al. (2013) Effects of amendment of biochar-manure compost in conjunction with pyroligneous solution on soil quality and wheat yield of a salt-stressed cropland from Central China Great Plain. *Field Crops Research*, 144, 113–118.

Lehmann J, Skjemstad J, Sohi S et al. (2008) Australian climate–carbon cycle feedback reduced by soil black carbon. *Nature Geoscience*, 1, 832–835.

Lin RQ, Song DL (2002) Utilizing status and problems of crop straw on Guangdong province (in Chinese). *Soil and Environmental Sciences*, 11, 110.

Liu X, Zhang Y, Han W et al. (2013a) Enhanced nitrogen deposition over China. *Nature*, 494, 459–462.

Liu X, Zhang A, Ji C et al. (2013b) Biochar's effect on crop productivity and the dependence on experimental conditions - a meta-analysis of literature data. *Plant and Soil*, 373, 583–594.

Lo AY (2012) Carbon emissions trading in China. *Nature Climate Change*, 2, 765–766.

Lu W, Zhang T (2010a) Life-cycle implications of using crop residues for various energy demands in China. *Environmental Science & Technology*, 44, 4026–4032.

Lu W, Zhang T (2010b) Supporting Information: life-cycle implications of using crop residues for various energy demands in China. *Environmental Science and Technology*, 1–62.

Mankins JC (1995) *Technology Readiness Levels: A White Paper*. NASA, USA.

Masiello C (1998) Black carbon in deep-sea sediments. *Science*, 280, 1911–1913.

McKendry P (2002) Energy production from biomass (Part 2): conversion technologies. *Bioresource Technology*, 83, 47–54.

Pan G (2008) Soil organic carbon stock, dynamics and climate change mitigation of China (in Chinese). *Advances in Climate Change Research*, 5, 1–7.

Roberts KG, Gloy B, Joseph S, Scott NR, Lehmann J (2010a) Life cycle assessment of biochar systems: estimating the energetic, economic, and climate change potential. *Environmental Science & Technology*, 44, 827–833.

Roberts KG, Gloy BA, Joseph S, Scott NR, Lehmann J (2010b) Life cycle assessment of biochar systems: estimating the energetic, economic, and climate change potential: supplementary information. *Environmental Science & Technology*, 44, S1–S32.

Shackley S, Hammond J, Gaunt J, Ibarrola R (2011) The feasibility and costs of biochar deployment in the UK. *Carbon Management*, 2, 335–356.

Shackley SJ, Clare AJ, Joseph S, McCarl BA, Schmidt H-P (in press) Chapter 31: economic evaluation of biochar systems: current evidence and challenges. In: *Biochar for Environmental Management* (eds Lehmann J, Joseph S). Earthscan, London, UK.

Singh N, Abiven S, Torn MS, Schmidt MWI (2012) Fire-derived organic carbon in soil turns over on a centennial scale. *Biogeosciences*, 9, 2847–2857.

Sohi S, Lopez-capel E, Krull E, Bol R (2009) Biochar, climate change and soil: a review to guide future research. *CSIRO Land and Water Science Report*, 5, 17–31.

Song R, Lei H (2014) Emissions trading in China: first reports from the field. World Resources Institute website. Available at: http://www.wri.org/blog/2014/01/emissions-trading-china-first-reports-field (accessed 1 August 2014).

Sparrevik M, Field JL, Martinsen V, Breedveld GD, Cornelissen G (2013) Life cycle assessment to evaluate the environmental impact of biochar implementation in conservation agriculture in Zambia. *Environmental Science & Technology*, 47, 1206–1215.

Steiner C, Glaser B, Geraldes WT, Lehmann J, Blum WEH, Zech W (2008) Nitrogen retention and plant uptake on a highly weathered central Amazonian Ferralsol amended with compost and charcoal. *Journal of Plant Nutrition and Soil Science*, 171, 893–899.

Stern N (2006) *The Economics of Climate Change*. HM Treasury, London, UK.

Tang H, Qiu J, Van Ranst E, Li C (2006) Estimations of soil organic carbon storage in cropland of China based on DNDC model. *Geoderma*, 134, 200–206.

UK Ministry of Defence (2014) Technology readiness levels (TRLs) in the project lifecycle. Available at: http://www.publications.parliament.uk/pa/cm201011/cmselect/cmsctech/619/61913.htm#note221 (accessed 23 April 2014).

Utterback JM (1996) *Mastering the Dynamics of Innovation*. Harvard Business Press, Boston, USA.

Van Zwieten L, Kimber S, Downie A, Morris S, Petty S, Rust J, Chan KY (2010) A glasshouse study on the interaction of low mineral ash biochar with nitrogen in a sandy soil. *Australian Journal of Soil Research*, 48, 569–576.

Wang J, Pan X, Liu Y, Zhang X, Xiong Z (2012) Effects of biochar amendment in two soils on greenhouse gas emissions and crop production. *Plant and Soil*, 360, 287–298.

Wang Q, Geng C, Lu S, Chen W, Shao M (2013) Emission factors of gaseous carbonaceous species from residential combustion of coal and crop residue briquettes. *Frontiers of Environmental Science & Engineering*, 7, 66–76.

West TO, Marland G (2002) A synthesis of carbon sequestration, carbon emissions, and net carbon flux in agriculture : comparing tillage practices in the United States. *Agriculture, Ecosystems & Environment*, 91, 217–232.

Woolf D, Amonette JE, Street-Perrott FA, Lehmann J, Joseph S (2010) Sustainable biochar to mitigate global climate change. *Nature Communications*, 1, 56.

World Resources Institute (2014) Getting every ton of emissions right. Available at: http://www.wri.org/sites/default/files/calculation_spreadsheet_of_china_regional_grid_emission_factors.xlsx (accessed 3 May 2014).

Wu L, Chen J, Zhu XD, Xu YP, Feng B, Yang L (2001) Straw-burning in rural areas of China: caused and controlling strategy (in Chinese). *China Population, Resources and Environment*, 11, 110–112.

Yan X, Ohara T, Akimoto H (2006) Bottom-up estimate of biomass burning in mainland China. *Atmospheric Environment*, 40, 5262–5273.

Yang J (2014) China benchmark spot coal price drops first time in three months. Available at: http://www.bloomberg.com/news/2014-01-06/china-benchmark-spot-coal-price-drops-first-time-in-three-months.html (accessed 6 January 2014).

Yao Z, Wang SH, Jiang XH (2001) The current situation and approach of return straw to field in suburb of Shanghai (in Chinese). *Agro-Environment and Development*, **3**, 40–41.

Yu Z (2003) The developing trend of resources treatment of crop stalk in Fuzhou city (in Chinese). *Fujian Environment*, **20**, 31–32.

Zhang J, Smith K, Ma Y *et al.* (2000) Greenhouse gases and other airborne pollutants from household stoves in China: a database for emission factors. *Atmospheric Environment*, **34**, 4537–4549.

Zhang A, Cui L, Pan G *et al.* (2010) Effect of biochar amendment on yield and methane and nitrous oxide emissions from a rice paddy from Tai Lake plain, China. *Agriculture, Ecosystems & Environment*, **139**, 469–475.

Zhang F, Cui Z, Fan M, Zhang W, Chen X, Jiang R (2011) Integrated soil-crop system management: reducing environmental risk while increasing crop productivity and improving nutrient use efficiency in China. *Journal of Environmental Quality*, **40**, 1051–1057.

Zhang A, Liu Y, Pan G, Hussain Q, Li L, Zheng J, Zhang X (2012a) Effect of biochar amendment on maize yield and greenhouse gas emissions from a soil organic carbon poor calcareous loamy soil from Central China Plain. *Plant and Soil*, **351**, 263–275.

Zhang A, Bian R, Pan G *et al.* (2012b) Effects of biochar amendment on soil quality, crop yield and greenhouse gas emission in a Chinese rice paddy: a field study of 2 consecutive rice growing cycles. *Field Crops Research*, **127**, 153–160.

Zhang F, Chen X, Vitousek P (2013a) An Experiment for the World. *Nature*, **497**, 33–35.

Zhang Q, Zhou D, Zhou P, Ding H (2013b) Cost analysis of straw-based power generation in Jiangsu Province, China. *Applied Energy*, **102**, 785–793.

Zhang W-F, Dou Z-X, He P *et al.* (2013c) New technologies reduce greenhouse gas emissions from nitrogenous fertilizer in China. *Proceedings of the National Academy of Sciences of the United States of America*, **10**, 8375–8380.

Zhang Q, Zhou D, Fang X (2014) Analysis on the policies of biomass power generation in China. *Renewable and Sustainable Energy Reviews*, **32**, 926–935.

Zhao G, Che K (2012) A study on predicting coal market price in china based on time sequence models. *International Journal of Business and Social Science*, **3**, 31–36.

Zimmerman AR, Gao B, Ahn M-Y (2011) Positive and negative carbon mineralization priming effects among a variety of biochar-amended soils. *Soil Biology and Biochemistry*, **43**, 1169–1179.

Permissions

List of Contributors

Jessica E. Abernathy, Dustin R. J. Graham and Mark E. Sherrard
Department of Biology, University of Northern Iowa, 144 McCollum Science Hall, Cedar Falls, IA 50614, USA

Daryl D. Smith
Tallgrass Prairie Center, 2412 West 27th Street, Cedar Falls, IA 50614-0294, USA

Jon P. Mccalmont, Paul Robson, Iain S. Donnison and John Clifton-Brown
Institute of Biological, Environmental and Rural Sciences (IBERS), Aberystwyth University, Gogerddan, Aberystwyth, Wales SY23 3EQ, UK

Astley Hastings
Institute of Biological and Environmental Science, University of Aberdeen, 24 St Machar Drive, Aberdeen AB24 3UU, UK

Niall P. Mcnamara
Centre for Ecology and Hydrology, Lancaster Environment Centre, Library Avenue, Bailrigg, Lancaster LA1 4AP, UK

Goetz M. Richter
Rothamsted Research, West Common, Harpenden, Hertfordshire AL5 2JQ, UK

Johannes Rahlf and Rasmus Astrup
Norwegian Institute of Bioeconomy Research (NIBIO), Ås 1431, Norway

Blas Mola-Yudego
Norwegian Institute of Bioeconomy Research (NIBIO), Ås 1431, Norway
School of Forest Sciences, University of Eastern Finland (UEF), Joensuu FI-80 101, Finland

Ioannis Dimitriou
Department of Crop Production Ecology, Swedish University of Agricultural Sciences (SLU), Uppsala S-750 07, Sweden

Mark Richards, Marta Dondini, Edward O. Jones, Astley Hastings, Dagmar N. Henner, Jo U. Smith and Pete Smith
Institute of Biological and Environmental Sciences, University of Aberdeen, 23 St Machar Drive, Aberdeen, AB24 3UU, UK

Mark Pogson
Institute of Biological and Environmental Sciences, University of Aberdeen, 23 St Machar Drive, Aberdeen, AB24 3UU, UK
Academic Group of Engineering, Sports and Sciences, University of Bolton, Deane Road, Bolton, BL3 5AB, UK

Suzanne Milner and Gail Taylor
Centre for Biological Sciences, University of Southampton, Life Sciences Building, Southampton, SO17 1BJ, UK

Matthew J. Talli S
Centre for Biological Sciences, University of Southampton, Life Sciences Building, Southampton, SO17 1BJ, UK
School of Biological Sciences, University of Portsmouth, King Henry Building, King Henry I Street, Portsmouth, PO1 2DY, UK

Eric Casella, Robert W. Matthews and Paul A. Henshall
Centre for Sustainable Forestry and Climate Change, Forest Research, Farnham, Surrey, GU10 4LH, UK

Niall P. Mcnamara
Centre for Ecology and Hydrology, Lancaster Environment Centre, Library Avenue, Bailrigg, Lancaster, LA1 4AP, UK

Jeanette Whitaker and Niall P. Mcnamara
Centre for Ecology and Hydrology, Lancaster Environment Centre, Library Avenue, Bailrigg, Lancaster LA1 4AP, UK

Christian A. Davies
Shell International Exploration and Production, Shell Technology Center Houston, 3333 Highway 6 South, Houston, TX 77082- 3101, USA

Pete Smith
Institute of Biological and Environmental Sciences, University of Aberdeen, 23 St Machar Drive, Aberdeen AB24 3UU, UK

Andy D. Robertson
Centre for Ecology and Hydrology, Lancaster Environment Centre, Library Avenue, Bailrigg, Lancaster LA1 4AP, UK

Shell International Exploration and Production, Shell Technology Center Houston, 3333 Highway 6 South, Houston, TX 77082- 3101, USA
Institute of Biological and Environmental Sciences, University of Aberdeen, 23 St Machar Drive, Aberdeen AB24 3UU, UK
Department of Soil and Crop Sciences, Colorado State University, Fort Collins, CO 80523, USA

Ross Morrison
Centre for Ecology and Hydrology, Maclean Building, Wallingford OX10 8BB, UK

Carmenza Robledo-Abad
Department of Environmental Systems Science, USYS TdLab, ETH Zürich, Universitätstrasse 22, 8092 Zurich, Switzerland
Helvetas Swiss Intercooperation, Maulbeerstr. 10, CH-3001 Bern, Switzerland

Hans-Jörg Althaus
Foundation for Global Sustainability (ffgs), Reitergasse 11, 8004 Zürich, Switzerland
Lifecycle Consulting Althaus, Bruechstr. 132, 8706 Meilen, Switzerland

Göran Berndes and Maria Nordborg
Department of Energy and Environment, Chalmers University of Technology, SE 41296 Gothenburg, Sweden

Simon Bolwig and Jay S. Gregg
DTU Management Engineering, Technical University of Denmark, 4000 Roskilde, Denmark

Esteve Corbera
Institute of Environmental Science and Technology, and Department of Economics and Economic History, Universitat Autònoma de Barcelona, 08193 Barcelona, Spain

Felix Creutzig
Mercator Research Institute on Global Commons and Climate Change and Technical University Berlin, 10829 Berlin, Germany

John Garcia-Ulloa
Institute of Terrestrial Ecosystems, ETH Zürich, Universitätstrasse 22, 8092 Zurich, Switzerland

Anna Geddes and Johan Lilliestam
Institute for Environmental Decisions, ETH Zürich, Climate Policy Group, Universitätstrasse 22, 8092 Zurich, Switzerland

Helmut Haberl, Christian Lauk and Stefan Leitner
Institute of Social Ecology Vienna (SEC), Alpen-Adria Universitaet (AAU), Schottenfeldgasse 29, 1070 Vienna, Austria

Susanne Hanger
Institute for Environmental Decisions, ETH Zürich, Climate Policy Group, Universitätstrasse 22, 8092 Zurich, Switzerland International Institute for Applied Systems Analysis, Schlossplatz 1, Laxenburg, Austria

Richard J. Harper
School of Veterinary and Life Sciences, Murdoch University, South Street, Murdoch, WA 6150, Australia

Carol Hunsberger
Department of Geography, University of Western Ontario, London, ON N6A 5C2, Canada

Rasmus K. Larsen
Stockholm Environment Institute (SEI), Linnégatan 87D, 115 23 Stockholm, 104 51 Stockholm, Sweden

Alexander Popp and Lena Scheiffle
Potsdam Institute for Climate Impact Research (PIK), 14412 Potsdam, Germany

Hermann Lotze-Campen
Potsdam Institute for Climate Impact Research (PIK), 14412 Potsdam, Germany
Humboldt-University zu Berlin, Unter den Linden 6, 10099 Berlin, Germany

Bart Muys
Division of Forest, Nature and Landscape, University of Leuven (KU Leuven), Celestijnenlaan 200E BE- 3001 Leuven, Belgium

Maria Ölund
Centre for Environment and Sustainability – GMV, University of Gothenburg, Aschebergsgatan 44, Göteborg, Sweden

Boris Orlowsky
Climate-Babel

Joana Portugal-Pereira
Energy Planning Program, COPPE, Federal University of Rio de Janeiro, Centro de Tecnologia, Sala C-211, C.P. 68565, Cidade Universitária, Ilha do Fundão, 21941-972 Rio de Janeiro, RJ, Brazil

Jürgen Reinhard
Informatics and Sustainability Research Group, Swiss Federal Institute for Material Testing and Research, Empa, Ueberlandstrasse 129, 8600 Duebendorf, Switzerland

Pete Smith
Institute of Biological and Environmental Sciences, ClimateXChange and Scottish Food Security Alliance-Crops, University of Aberdeen, 23 St Machar Drive, Aberdeen AB24 3UU, UK

Aidan M. Keith, Dafydd Elias, Jonathan Oxley and Niall P. Mcnamara
Centre for Ecology and Hydrology, Lancaster Environment Centre, Library Avenue, Bailrigg, Lancaster, LA1 4AP, UK

Rebecca L. Rowe
Centre for Ecology and Hydrology, Lancaster Environment Centre, Library Avenue, Bailrigg, Lancaster, LA1 4AP, UK
School of GeoSciences, University of Edinburgh, The King's Buildings, Alexander Crum Brown Road, Edinburgh, EH9 3FF, UK

Marta Dondini and Pete Smith
Institute of Biological and Environmental Sciences, University of Aberdeen, 23 St Machar Drive, Aberdeen, AB24 3UU, UK

Theodoros Skevas
Gulf Coast Research and Education Center, University of Florida, 14625 County Road 672, Wimauma, FL 33598, USA

Scott M. Swinton and Sophia Tanner
Department of Agricultural, Food, and Resource Economics, Michigan State University, Justin S. Morrill Hall of Agriculture 446 West Circle Dr., East Lansing, MI 48824-1039, USA

Gregg Sanford
Department of Agronomy, University of Wisconsin-Madison, 1575 Linden Dr, Madison, WI 53706, USA, 4Department of Plant, Soil and Microbial Sciences, Michigan State University, 1066 Bogue St A286, East Lansing, MI 48824, USA

Toby J. Townsend and Paul Wilson
Division of Agricultural and Environmental Sciences, School of Biosciences, University of Nottingham, Sutton Bonington Campus, Loughborough LE12 5RD, UK

Debbie L. Sparke
Division of Plant and Crop Sciences, School of Biosciences, University of Nottingham, Sutton Bonington Campus, Loughborough LE12 5RD, UK

Maud Viger and Gail Taylor
Centre for Biological Sciences, University of Southampton, Southampton SO17 1BJ, UK

Robert D. Hancock
Cell and Molecular Sciences, The James Hutton Institute, Invergowrie, Dundee DD2 5DA, UK

Franco Migl Ietta
Institute of Biometeorology (IBIMET), National Research Council (CNR), Via Caproni 8, Firenze 50145, Italy
FoxLab, Forest and Wood Science, San Michele All'Adige 38010, Italy

Søren Larsen and Niclas S. Bentsen
Department of Geosciences and Natural Resource Management, University of Copenhagen, Rolighedsvej 23, 1958 Frederiksberg C, Denmark

Deepak Jaiswal
Energy Bioscience Institute, University of Illinois, Urbana, IL 61801, USA

Stephen P. Long
Energy Bioscience Institute, University of Illinois, Urbana, IL 61801, USA
Departments of Crop Sciences and of Plant Biology, University of Illinois, Urbana, IL 61801, USA

Dan Wang
International Center for Ecology, Meteorology and Environment, School of Applied Meteorology, Nanjing University of Information Science and Technology, Nanjing 210044, China

Alison J. Haughton, Mark D. Mallott, Suzanne J. Clark and Angela karp
Rothamsted Research, West Common, Harpenden, Hertfordshire AL5 2JQ, UK

David A. Bohan
INRA, UMR 1347 Agroecologie, Pôle ECOLDUR, 17 rue Sully, Dijon CEDEX 21065, France

Victoria Mallott
168 Putteridge Road, Luton LU2 8HJ, UK

Rufus Sage
Game and Wildlife Conservation Trust, Burgate Manor, Fordingbridge, Hampshire SP6 1EF, UK

Keith L. Kline
Environmental Science Division, Climate Change Science Institute, Oak Ridge National Laboratory, TN 37831, USA

Siwa Msangi
International Food Policy Research Institute, 2033 K St NW, Washington, DC 20006, USA

Virginia H. Dale
Center for Bioenergy Sustainability, Environmental Science Division, ORNL, Oak Ridge, TN 37831, USA

Jeremy Woods
Centre for Environmental Policy, Imperial College London, Exhibition Road, London SW7 1NA, UK

Glaucia M. Souza
Instituto de Quimica, Universidade de São Paulo, Av. Prof. Lineu Prestes 748, São Paulo, Brazil

Patricia Osseweijer
Department of Biotechnology, Delft University of Technology, 2628 BC Delft, The Netherlands

Joy S. Clancy
CSTM, University of Twente, 7500AE Enschede, The Netherlands

Jorge A. Hilbert
Instituto de Ingeniería Rural INTA, cc 25 1712 Castelar, Buenos Aires, Argentina

Francis X. Johnson
Stockholm Environment Institute Africa Centre, World Agroforestry Centre (ICRAF), United Nations Avenue, Gigiri Nairobi, Kenya

Patrick C. Mcdonnell
BEE Energy, 2000 Nicasio Valley Rd., Nicasio, CA 94946, USA

Harriet K. Mugera
World Bank, 1818 H Street NW, Washington, DC 20433, USA

Simon Shackley and Anthony Bloom
University of Edinburgh, Crew Building, Kings Buildings, Edinburgh EH9 3JN, UK

Abbie Clare
University of Edinburgh, Crew Building, Kings Buildings, Edinburgh EH9 3JN, UK
Scotland's Rural College, Kings Buildings, Edinburgh EH9 3JG, UK

Stephen Joseph
Discipline of Chemistry, University of Newcastle, Callaghan, NSW 2308, Australia
Nanjing Agricultural University, Nanjing 210095, China
School of Materials Science and Engineering, University of New South Wales, Sydney, NSW 2052, Australia

Genxing Pan
Nanjing Agricultural University, Nanjing 210095, China

James Hammond
Key Laboratory of Biodiversity and Biogeography of East Asia, Kunming Institute of Botany, Chinese Academy of Sciences, Kunming 650201, China
World Agroforestry Centre, East, and Central Asia, Kunming 650201, China

Index

www.ingramcontent.com/pod-product-compliance
Lightning Source LLC
Chambersburg PA
CBHW082033190326
41458CB00010B/3357